T0192137

Lecture Notes in Computer Science

Lecture Notes in Artificial Intelligence   **14471**

Founding Editor

Jörg Siekmann

Series Editors

Randy Goebel, *University of Alberta, Edmonton, Canada*
Wolfgang Wahlster, *DFKI, Berlin, Germany*
Zhi-Hua Zhou, *Nanjing University, Nanjing, China*

The series Lecture Notes in Artificial Intelligence (LNAI) was established in 1988 as a topical subseries of LNCS devoted to artificial intelligence.

The series publishes state-of-the-art research results at a high level. As with the LNCS mother series, the mission of the series is to serve the international R & D community by providing an invaluable service, mainly focused on the publication of conference and workshop proceedings and postproceedings.

Tongliang Liu · Geoff Webb · Lin Yue ·
Dadong Wang
Editors

AI 2023: Advances in Artificial Intelligence

36th Australasian Joint Conference on Artificial Intelligence, AI 2023
Brisbane, QLD, Australia, November 28 – December 1, 2023
Proceedings, Part I

Springer

Editors
Tongliang Liu (iD)
The University of Sydney
Darlington, NSW, Australia

Geoff Webb (iD)
Monash University
Clayton, VIC, Australia

Lin Yue (iD)
The University of Newcastle
Callaghan, NSW, Australia

Dadong Wang (iD)
CSIRO Data61
Sydney, NSW, Australia

ISSN 0302-9743 ISSN 1611-3349 (electronic)
Lecture Notes in Artificial Intelligence
ISBN 978-981-99-8387-2 ISBN 978-981-99-8388-9 (eBook)
https://doi.org/10.1007/978-981-99-8388-9

LNCS Sublibrary: SL7 – Artificial Intelligence

This Springer imprint is published by the registered company Springer Nature Singapore Pte Ltd.
The registered company address is: 152 Beach Road, #21-01/04 Gateway East, Singapore 189721, Singapore

Paper in this product is recyclable.

Preface

This volume contains the papers presented at the 36th Australasian Joint Conference on Artificial Intelligence, AJCAI 2023. The conference was held during November 28 – December 1, 2023, and was hosted by the University of Queensland in Brisbane, Australia. This annual conference is one of the longest running conferences in artificial intelligence, with the first conference held in Sydney in 1987. The conference remains the premier event for artificial intelligence in Australasia, offering a forum for researchers and practitioners across all subfields of artificial intelligence to meet and discuss recent advances.

AJCAI 2023 received 213 submissions and each submission was reviewed by at least two Program Committee (PC) members or external reviewers in a double-blind process (over 90% of the submissions had three reviews). After a thorough discussion and rigorous scrutiny by the reviewers, 24 papers were accepted for long oral presentation and 58 papers were accepted for oral presentation at the conference. In total, 82 submissions were accepted for publication as full papers in these proceedings with an acceptance rate of 38% (the acceptance rate of the long oral presentations was 11%). AJCAI 2023 had six keynote talks by the following distinguished scientists: Ling Chen from the University of Technology Sydney, Australia; Manik Varma from Microsoft Research India, India; Peter Soyer from the University of Queensland, Australia; Maria Garcia De La Banda from Monash University, Australia; Mengjie Zhang from Victoria University of Wellington, New Zealand; and Dadong Wang from Data61, Australia.

The following are notable aspects of the AJCAI 2023 conference:

- AJCAI 2023 was jointly held with the Defence Artificial Intelligence 2023 Symposium (November 27, 2023). The Defence Artificial Intelligence Symposium is an exciting opportunity for Defence and AI researchers to come together and explore priorities, opportunities, and commonalities.
- AJCAI 2023 included a day with a special industry focus. Panel discussions allowed industry and academia to share challenges and research directions.
- AJCAI 2023 included four workshops, held on November 28: Foundations for Robust AI: Self-Supervised Learning, organised by Saimunur Rahman, David Hall, Stephen Hausler, and Peyman Moghadam; Federated Learning in Australasia: When FL Meets Foundation Models, organised by Guodong Long, Han Yu, and Tao Shen; Artificial Intelligence Enabled Trustworthy Recommendations, organised by Shoujin Wang, Rocky Tong Chen, Hongzhi Yin, Lina Yao, and Fang Chen; and Machine Learning for Data-Driven Optimization, organised by Xilu Wang, Xiangyu Wang, Shiqing Liu, and Yaochu Jin.
- AJCAI 2023 included three tutorials, held on November 28: Reinforcement Learning for Automated Negotiation Supply Chain Management League as an Example, presented by Yasser Mohammad; Towards Communication-Efficient and Heterogeneity-Robust Federated Learning, presented by Guodong Long and Yue Tan; and Decoding

the Grammar of DNA Using Natural Language Processing, presented by Tyrone Chen and Sonika Tyagi.

- AJCAI 2023 included a PhD Forum, held on November 28, to mentor and assist post-graduate students developing their research, with mentorship provided by research leaders. Limited travel support was provided.

We especially appreciate the work of the members of the Program Committee and the external reviewers for their expertise and tireless effort in assessing the papers within a strict timeline. We are also very grateful to the members of the Organising Committee for their efforts in the preparation, promotion, and organisation of the conference, especially the General Chairs, Dacheng Tao, Sally Cripps, and Janet Wiles, for coordinating the whole event.

Lastly, we thank the National Committee for Artificial Intelligence of the Australian Computer Society; Springer, for the professional service provided by the Lecture Notes in Artificial Intelligence editorial and publishing teams; and our conference sponsors: the Australian Computer Society; the Defence Artificial Intelligence Research Network; Pioneer Computers; the School of Computer Science at the University of Sydney; the School of Electrical Engineering and Computer Science at the University of Queensland; the Human Technology Institute at the University of Technology Sydney; the Adelaide University; and the UNSW AI Institute.

October 2023

Tongliang Liu
Miao Xu
Geoff Webb

Organization

General Chairs

Dacheng Tao	The University of Sydney, Australia
Sally Cripps	University of Technology Sydney, Australia
Janet Wiles	The University of Queensland, Australia

Program Chairs

Tongliang Liu	The University of Sydney, Australia
Miao Xu	The University of Queensland, Australia
Geoff Webb	Monash University, Australia

Proceedings Chairs

Weitong Chen	The University of Adelaide, Australia
Lin Yue	The University of Newcastle, Australia
Dadong Wang	Data61, Australia

Senior Program Committee

Jing Jiang	University of Technology Sydney, Australia
Mingyu Guo	The University of Adelaide, Australia
Jonathan Kummerfeld	University of Sydney, Australia
Hua Zuo	University of Technology Sydney, Australia
Shuo Chen	RIKEN, Japan
Zhongyi Han	Mohamed Bin Zayed University of Artificial Intelligence, United Arab Emirates
Runnan Chen	The University of Hong Kong, China
Jingfeng Zhang	University of Auckland, New Zealand
Feng Liu	The University of Melbourne, Australia
Huong Ha	RMIT University, Australia
Soyeon Han	University of Western Australia, Australia
Zhanna Sarsenbayeva	The University of Sydney, Australia
Mingming Gong	The University of Melbourne, Australia
Yu Yao	Usyd
Yuxuan Du	The University of Sydney, Australia

Clément Canonne	University of Sydney, Australia
Miaomiao Liu	Australian National University, Australia
Dawei Zhou	Xidian University, China
Yadan Luo	University of Science and Technology of China, China
Xiaobo Xia	The University of Sydney, Australia
Guanfeng Liu	Macquarie University, Australia
Zhen Fang	University of Technology Sydney, Australia
Hien Nguyen	University of Queensland, Australia

Program Committee

Ravneet Singh Arora	Block Inc, USA
Adnan Mahmood	Macquarie University, Australia
Yue Yuan	Shandong University, China
Seyedamin Pouriyeh	Kennesaw State University, USA
Yexiong Lin	The University of Sydney, Australia
Jiahui Gao	The University of Hong Kong, China
Xianzhi Wang	University of Technology Sydney, Australia
Zhuo Huang	Nanjing University of Science and Technology, China
Alex Chu	Beihang University, China
Ruihong Qiu	The University of Queensland, Australia
Qingzheng Xu	National University
Qiang Qu	The University of Sydney, Australia
Lynn Miller	Monash University, Australia
Zhuonan Liang	The University of Sydney, Australia
Kun Han	The University of Queensland, Australia
Tim Miller	The University of Queensland, Australia
Zhuoxiao Chen	The University of Queensland, Australia
Kun Wang	University of Technology Sydney, Australia
Changqin Huang	South China Normal University, China
Peng Yuwei	Wuhan University, China
Brendon J. Woodford	University of Otago, New Zealand
Weihua Li	Auckland University of Technology, New Zealand
Mingzhe Zhang	The University of Queensland, Australia
Peter Baumgartner	CSIRO, Australia
Manolis Gergatsoulis	Ionian University, Greece
Dianhui Wang	La Trobe University, Australia
Jianan Fan	University of Sydney, Australia
Xueping Peng	University of Technology Sydney, Australia
Kairui Guo	University of Technology Sydney, Australia

Zehong Cao   University of South Australia, Australia
Wenhao Yang   Nanjing University, China
Yi Gao   Southeast University, China
Yi Mei   Victoria University of Wellington, New Zealand
Chenhao Zhang   University of Queensland, Australia
Youquan Liu   Hochschule Bremerhaven, Germany
Wenhua Zhang   Shanghai University, China
Yu Yao   MBZUAI, UAE & CMU, USA
Hao Hou   Nanjing University of Science and Technology,
                                China
Yuan Liu   The University of Hong Kong, China
Jianlong Zhou   University of Technology Sydney, Australia
Ran Wang   University of Technology Sydney, Australia
Jun Wang   The University of Sydney, Australia
Weijia Zhang   The University of Newcastle, Australia
Zhuoyun Ao   Defence Science and Technology Organisation
Xinheng Wu   University of Technology Sydney, Australia
Abdul Sattar   Griffith University, Australia
Daokun Zhang   Monash University, Australia
Ge-Peng Ji   Wuhan University, China
Dongting Hu   The University of Melbourne, Australia
Chengbin Du   The University of Sydney, Australia
Ying Bi   Victoria University of Wellington, New Zealand
Rafal Rzepka   Hokkaido University, Japan
Cong Lei   The University of Sydney, Australia
Yue Tan   University of Technology Sydney, Australia
Hongwei Sheng   The University of Queensland, Australia
M. A. Hakim Newton   University of Newcastle, Australia
Shaokun Zhang   Penn State University, USA
Pengqian Lu   The University of Sydney, Australia
Peng Yan   Nanjing University of Post and
                                Telecommunication, China
Weidong Cai   The University of Sydney, Australia
Huan Huo   University of Technology Sydney, Australia
Yuhao Wu   The University of Sydney, Australia
Rui Dai   University of Science and Technology of China,
                                China
Fangfang Zhang   Victoria University of Wellington, New Zealand
Xiaobo Xia   The University of Sydney, Australia
Giorgio Gnecco   IMT - School for Advanced Studies, Lucca, Italy
Yu Zheng   The Chinese University of Hong Kong, China
Ickjai Lee   James Cook University, Australia

Jiepeng Wang	The University of Hong Kong, China
Qizhou Wang	Hong Kong Baptist University, China
Chen Liu	University of Technology Sydney, Australia
Yuanyuan Wang	The University of Melbourne, Australia
Wei Duan	The Australian Artificial Intelligence Institute (AAII), and University of Technology Sydney, Australia
Aoqi Zuo	The University of Melbourne, Australia
Yiming Ren	ShanghaiTech University, China
Stephen Chen	York University, Canada
Wenjie Wang	The University of Melbourne, Australia
Zhiyuan Li	University of Sydney, Australia
Tao Shen	Microsoft, China
Guangzhi Ma	University of Technology Sydney, Australia
Haodong Chen	The University of Sydney, Australia
Yu Lu	University of Technology Sydney, Australia
Angus Dempster	Monash University, Australia
Jing Teng	North China Electric Power University, China
Yawen Zhao	The University of Queensland, Australia
Harith Al-Sahaf	Victoria University of Wellington, New Zealand
Pengxin Zeng	Sichuan University, China
Hangyu Li	Xidian University, China
Huaxi Huang	CSIRO, Australia
Bernhard Pfahringer	University of Waikato, New Zealand
Huiqiang Chen	University of Technology Sydney, Australia
Xin Yu	University of Technology Sydney, Australia
Yanjun Zhang	University of Technology Sydney, Australia
Bach Nguyen	Victoria University of Wellington, New Zealand
Peng Mi	Xiamen University, China
Jiyang Zheng	University of Sydney, Australia
Rundong He	Shandong University, China
Shikun Li	Chinese Academy of Sciences, China
Kevin Wong	Murdoch University, Australia
Xiu-Chuan Li	Chinese Academy of Science, China
Jianglin Qiao	Western Sydney University, Australia
Maurice Pagnucco	The University of New South Wales, Australia
Bing Wang	The University of New South Wales, Australia
Zhaoqing Wang	The University of Sydney, Australia
Mark Reynolds	The University of Western Australia, Australia
Xuyun Zhang	Macquarie University, Australia
Zige Wang	Peking University, China
Chang Wei Tan	Monash University, Australia

Muyang Li	The University of Sydney, Australia
Guangyan Huang	Deakin University, Australia
Liangchen Liu	Xidian University, China
Nayyar Zaidi	Deakin University, Australia
Erdun Gao	The University of Melbourne, Australia
Chuyang Zhou	The University of Sydney, Australia
Shaofei Shen	The University of Queensland, Australia
Yixuan Qiu	The University of Queensland, Australia
Jianhua Yang	UWS, Australia
Keqiuyin Li	University of Technology Sydney, Australia
Yanjun Shu	Harbin Institute of Technology, China
Lingdong Kong	National University of Singapore, Singapore
Jingyu Zhang	City University of Hong Kong, China
Sung-Bae Cho	Yonsei University, South Korea
Shuxiang Xu	University of Tasmania, Australia
Wan Su	Shandong University, China
Markus Wagner	The University of Adelaide, Australia
Xiaoying Gao	Victoria University of Wellington, New Zealand
William Bingley	The University of Queensland, Australia
Sishuo Chen	Peking University, China
Hao Sun	Shandong University, China
Ming Zhou	Hefei University of Technology, China

Sponsors

Contents – Part I

Deep Learning

Machine Learning and Data Mining

Optimization

Medical AI

Contents – Part II

Explainable AI

Reinforcement Learning

Genetic Algorithm

Computer Vision

Multi-graph Laplacian Feature Mapping Incorporating Tag Information for Image Annotation

Yan Liu[1], Qianqian Shao[1], Rui Cheng[2], Weifeng Liu[1], and Baodi Liu[1(✉)]

[1] China University of Petroleum (East China), Qingdao, China
liuwf@upc.edu.cn, thu.liubaodi@gmail.com
[2] Guangzhou Maritime University, Guangzhou, China

Abstract. Automatic image annotation has exerted a tremendous fascination of many researchers with the development of multimedia and computer vision. However, most methods employ graph learning omit to combine the label information with the manifold structure between samples or utilize the single graph when computing the sample graph. Furthermore, the visual content is ignored in the process of computing the manifold structure between labels. These drawbacks lead to incomplete and inaccurate manifold information. To this end, we propose a Multi-graph Laplacian Feature Mapping Incorporating Tag Information method for image annotation. Our method firstly combines the label information with the Laplacian eigenmaps, and multi-graphs are utilized to maintain the local geometric structure of samples. Then the visual content is taken into account to obtain tag correlations. Afterward, a sea of empirical evaluations is conducted on three benchmark datasets to prove the effectiveness of the proposed method.

Keywords: Automatic image annotation · Laplacian feature mapping · multi-graphs · label information · visual content

1 Introduction

There are growing images on the Internet with the rapid development of Internet technology and digital imaging technology. How to quickly and accurately retrieve user-required images has became a pressing problem. There are currently two main image retrieval categories [2]: Testbased image retrieval(TBIR) and Content-based image retrieval(CBIR). However, TBIR consumes massive time to manually label each image with some keywords and CBIR triggers the inaccurate retrieval results due to the semantic gap. To solve these problems, an automatic image annotation (AIA) is proposed. AIA refers to this process that the computer automatically gives keywords that reflect the content of images and are called tags or labels. Therefore, AIA is regarded as the bridge between visual content and their semantic meanings.

T. Liu et al. (Eds.): AI 2023, LNAI 14471, pp. 3–14, 2024.
https://doi.org/10.1007/978-981-99-8388-9_1

The algorithms [4,13,15,20,31] of AIA are divided into two categories: models without graph structure and models with graph structure.It concerns AIA as the traditional machine learning for the model without graph structure [11,12, 23,27]. Models without graph structure can be divided into three categories [2]. The generative models [4,8,22,27] compute numerous conditional probabilities to obtain multiple tags for an image with multi-label. Then the independent classifiers are trained for the discriminative model [3,5,9,12]. nearest-neighbor based model [1,10,19,23] predicts the labels of the test image according to the selected similar neighbors. More recent studies by [17,26] and [11] manifest that the algorithms depending on deep learning, such as GAN and CNN, are applied to AIA.

The models with graph structures [6,7,14,16,21] consider manifold structure information of samples or labels. The manifold structures which are computed by Locally Linear Embedding (LLE) [15,16], Laplacian [14,28–30], Hypergraph [21,25] and Hessian [18] are embedded as a regularization term into the objective function to solve image annotation problems. However, these methods with graph structure still suffer from some drawbacks:

(1) The manifold structure is one-sided or incomplete for some existing methods which only obtain a single and global graph [16,21,25,28–30] to contain manifold structure information between all samples.
(2) The label information is ignored when computing the graph of samples for a host of methods [16,18,24]. However, it is a benefit for constructing the precise manifold structure between samples.
(3) Some methods do not consider the manifold structure between labels [14]. Other methods [6,14–16] compute tag correlation via the features of tags. And the feature vector of each tag is composed of the presence or absence of this label in all samples.

Considering the above shortcomings, we propose a new method, multi-graph Laplacian feature mapping incorporating tag information for image annotation in this paper. The main contributions are shown as follows:

(1) Multiple Laplacian graphs are computed to obtain the complete manifold information of samples. Then all graphs are embedded in the objective function by graph Laplacian regularization.
(2) Each Laplacian graph of samples is computed by the label information. More precisely, a dataset consists of two labels "sky" and "clouds". The first Laplacian graph is constructed by the samples labeled by "sky" and then the second graph is composed of the samples "clouds". Therefore, the neighbors of a sample are fixed and are not affected by sample features for each Laplacian graph via our method combining with tag information. Furthermore, our method possesses rosy performance in supervised learning due to depending on tag information.
(3) The manifold structure between labels is computed by the visual content. More exactly, each label is expressed by a cluster center which is the average feature of samples labeled by the label. Then the Laplacian graph of labels

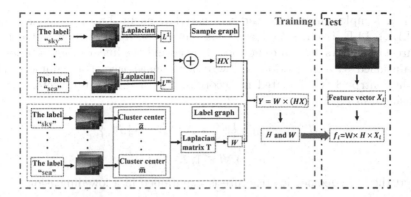

Fig. 1. The framework of proposed method. The training process is that manifold structures of samples and labels are embedded to obtain the pleasurable parameters W and H. Then W and H are employed to predict the tags of test sample.

is obtained via cluster centers. The tag graph Laplacian with visual content contains richer information comparing with the tag graph which is computed by the features of tags (Fig. 1).

It presents the model which is relevant to our method in Sect. 2. Then Sect. 3 describes the implementation details of the proposed algorithm. Moreover, the experimental results and contrast algorithms are illustrated in Sect. 5. Finally, Sect. 6 draws a conclusion for this paper.

2 Related Work

The model mentioned by the paper [16] introduces the local training set, low-rank matrix factorization, and the pairwise ranking loss. Meanwhile, it applies graph structure for preserving sample similarities and preserving tag correlation in the framework of semi-supervised learning.

X^T shows the transpose of matrix X and the trace of the matrix is indicated by Tr. The local feature matrix of samples is expressed by $X \in R^{d \times n}$, where d is the dimension of image representation and the number of local training images is n. $Y \in R^{m \times n}$ is the ground-truth label matrix and m is the number of labels.

The paper [16] shows a model that learns the coefficient matrix $Z \in R^{m \times d}$ to predict the multiple tags of a sample via $F = Z \times X$. The local training set is formed with n images that possess the smallest Euclidean distance from a query image. Corresponding to the local training set, the coefficient matrix Z is factorized as $Z = W \times H$ for avoiding overfitting due to $d >> n$. $W \in R^{m \times r}$ is the new coefficient and $H \in R^{r \times d}$ is the projection matrix that projects a high-dimensional matrix X to a low dimension. For the intensive model [16], they are provided for the sample similarities and tag correlations which severally restrain variables W and H.

Then, sample similarities and tag correlations by Locally Linear Embedding(LLE). LLE assumes that each sample can be linearly represented by its K nearest neighbors in high dimensional space and the linear representation can be maintained as a projection to low dimensional space. For Sample similarities, the matrix S is calculated in accordance with the feature vectors in high dimensional in Eq. 1.

$$S^* = \arg\min_{S} \left\{ \|X - XS\|_F^2 + \alpha\|S\|_1 \right\}$$

$$\text{s.t.} \quad S_{ll} = 0, \forall l \in 1, 2, \cdots, n \tag{1}$$

And the local geometric structure is maintained in a low dimension.

$$R(H) = \|HX - HXS\|_F^2 \tag{2}$$

H is constrained by the regularization function R (H). Similar to it, the regularization about W is described as follows. And T calculated by locally linear embedding regulates coefficient matrix W.

$$T = \arg\min_{T} \sum_{i=1}^{m} \left\{ \|Y_i \bullet -T_i Y\|_2^2 + \varphi\|T_i\|_1 \right\}$$

$$\text{s.t.} \quad T_{jj} = 0, \forall j = 1, 2, \cdots, m \tag{3}$$

where $\|T\|_1$ is l_1 norm of matrix and diagonal elements are zero so as to select the most relevant tags. Then the dependency of tags is shifted from Y to W.

$$Q \mid (W) = \|W - TW\|_F^2 \tag{4}$$

Finally, the pairwise ranking loss is proposed by the paper [16]. The following formulas denote the detail of pairwise ranking loss.

$$L(WHX, Y) = \sum_{i=1}^{n} \alpha_i \sum_{j,k=1}^{m} \varepsilon_{j,k}(WHX_i, Y_i) \tag{5}$$

$$\varepsilon_{j,k}(f, y) = l(f_j - f_k, y_j - y_k) \tag{6}$$

where $f_i - f_j$ denotes the difference between each two predicted labels for the same sample. Relatively $y_i - y_j$ represents the real difference. The pairwise ranking loss sorts labels on the basis of ground-truth tags. Hence, the predicted values of labels also maintain this ordering relation. More precisely, there are two tags y_i and y_j for a sample, $y_i = 1$ and $y_j = 0$. The predicted results of these tags tend to $f_i > f_j$ via pairwise ranking loss function.

3   Propoesd Method

In this section, it is revealed for the detail of multi-graph Laplacian incorporating tag information in the subsection IV-A. Then the subsection IV-B manifests the tag graph Laplacian with visual content. Afterward, loss function and the global objective function are given in subsection IV-C. Eventually, the subsection III-D demonstrates the parameter optimization.

3.1   Multi-graph Laplacian Incorporating Tag Information

The multiple Laplacian graphs are obtained to preserve complete and accurate sample structure information. And a total number of training images is described by N. Exactly, $X^a \in X$ is the subset of the training set where each sample is annotated by label a. Then for label a, the Laplacian graph can be computed to maintain the local manifold structure information during mapping high-dimensional feature space to a low dimension. However, the number of annotated samples is unequal for various tags. Hence, we extend the adjacency matrix to ensure dimensional consistency. The adjacency matrix is calculated by the following equation for label a.

$$S_{ij}^a = \begin{cases} \exp\left(-\frac{\|x_i - x_j\|_2^2}{\sigma^2}\right) & x_i \in X^a \text{ and } x_j \in X^a \\ 0 & x_i \notin X^a \text{ or } x_j \notin X^a \end{cases} \tag{7}$$

Therefore, the dimension of the adjacency matrices $W \in \mathbb{Z}^{n \times n}$ is identical for diverse tags. And N expresses the quantity of the whole training sample. The local manifold structure information of samples which are annotated by label a is maintained via Eq. 8.

$$\min \sum_{i,j=1}^{N} S_{ij}^a \left\| \frac{Hx_i}{\sqrt{D_{ii}^a}} - \frac{Hx_j}{\sqrt{D_{jj}^a}} \right\|_2^2$$
$$\Leftrightarrow \min \text{Tr}\left\{ HXL^a(HX)^T \right\} \tag{8}$$

where $D^a = \sum_{j=1}^{N} S_{ij}^a$ is a metric matrix. $L^a = I - (D^a)^{-\frac{1}{2}} S^a (D^a)^{-\frac{1}{2}}$ is the normalized Laplacian matrix of label a where I is the unit matrix. The paper shows a model that learns the coefficient matrix $Z \in R^{m \times d}$ to predict the multiple tags of a sample via $F = Z \times X$. The local training set is formed with n images that possess the smallest Euclidean distance from a query image.Corresponding to the local training set, the coefficient matrix Z is factorized as $Z = W \times H$ for avoiding overfitting due to d >> n. $W \in R^{m \times r}$ is the new coefficient and $H \in R^{r \times d}$ is the projection matrix that projects a high-dimensional matrix X to a low dimension.

For other tags, diverse Laplacian matrices, $L^b, L^c, \cdots L^m$, is computed according to the uniform approach. The subgraphs are extracted from each L^* according to the local training set samples number sequence corresponding to the local training set. More precisely, the subgraph l^a about label a is composed by ith,jth rows and ith, jth columns of L a when the local training set includes the sample i and sample j. And l^* is a matrix. The regularization of H is described by Eq. 9.

$$
\begin{aligned}
P(H) &= \mathrm{Tr}\left\{HXl^a(HX)^T\right\} \\
&+ \mathrm{Tr}\left\{HXl^b(HX)^T\right\} + \cdots + \mathrm{Tr}\left\{HXl^m(HX)^T\right\} \\
&= \mathrm{Tr}\left\{HX\left(l^a + l^b + \cdots + l^m\right)(HX)^T\right\}
\end{aligned}
\tag{9}
$$

Multiple graphs that are computed with regard to various tags present the more abundant structural information than only a graph. Therefore, our method is a more complete graph structure comparing with traditional single graph structure.

Our method respectively calculates a Laplacian matrix for each label and it is fixed for neighbors of a sample because an amount of samples annotated by a specific tag is invariable on a certain data set. Meanwhile, we consider the label information to obtain the Laplacian matrix. Therefore, our method is more accurate graph structural comparing with traditional Laplacian eigenmaps.

3.2   Tag Graph Laplacian with Visual Content

As mentioned in the previous section, H is constrained by the regularization function P (H). Similar to it, the regularization of W is described in this section.

Some methods propose TY substitutes for Y as the truth label set where T is a mapping matrix to express and maintain the correlations of tags. However, those approaches introduce noise and cause inaccurate prediction labels. In addition, the product of a coefficient matrix and feature matrix, W × HX, is predicted tag values. Thus the correlations of label Y can be transferred to coefficient matrix W when H is fixed. We adapt tag graph Laplacian with visual content to avoid label features that are composed of the presence or absence of this label in all samples.

The clustering center of tag a is the average sample of samples labeled by tag a in the first place. The clustering center of tag a is solved in (10)

$$
\overline{x^a} = \frac{1}{N^a}\sum_{j=1}^{N^a} x_j^a
\tag{10}
$$

where N^a is the number of samples labeled by tag a. Then, the same algorithm is applied to obtain m cluster centers for m labels. Afterward, the clustering center $\overline{x^*}$ replaces the features of tags.

There are one more point those tag correlations approximate clustering centers correlations which are computed by the Laplacian matrix. Then, the adjacency matrix is defined in (11).

$$
A_{ij}^a = \begin{cases} \exp\left(-\frac{\|\overline{x_i}-\overline{x_j}\|_2^2}{\sigma^2}\right) & i \neq j \\ 0 & i = j \end{cases}
\tag{11}
$$

And the dependency of tags is shifted from Y to W. Therefore, we have the following regularization:

$$\min \sum_{i,j=1}^{m} A_{ij} \left\| \frac{w_i}{\sqrt{E_{ii}}} - \frac{w_j}{\sqrt{E_{jj}}} \right\|_2^2 \tag{12}$$
$$\Leftrightarrow \min \mathrm{Tr}\left(W^T T W\right)$$

where $E = \sum_{j=1}^{m} A_{ij}$ is a metric matrix. $T = I - E^{-\frac{1}{2}} A E^{-\frac{1}{2}}$ is the normalized Laplacian matrix where I is the unit matrix.

The last but not the least, the regularization term for W is defined:

$$Q(W) = \mathrm{Tr}\left(W^T T W\right) \tag{13}$$

3.3   Loss Function and Objective Function

The pairwise ranking loss proposed by the paper [21] is applied in this paper.

$$\alpha_i = \exp[-\omega \cdot \mathrm{dist}(i)] \tag{14}$$

where dist(i) is Euclidean distance between sample i and the query sample and ω is the adjustment parameter.

In summary, the overview of our methods is reached.

$$\min_{W,H} L(WHX, Y) + \lambda P(H) + \gamma Q(W)$$
$$+ \eta \left(\|W\|_F^2 + \|H\|_F^2 \right) \tag{15}$$

where L(WHX, Y) is the pairwise loss function which is showed at the fore and the predicted label sets the product W, H and X. P (H) and Q (W) which respectively norm W and H are as regularizers. Meanwhile, λ, γ and η are the trade-off parameters.

4   Optimization

We apply the gradient descent algorithm to optimize this method. The respective derivatives of W and H are computed to optimize variables. One variable is fixed during processing another one. The paper [21] denotes derivative of loss function about W and H. And the derivatives of regularization terms P (H) and Q (W) are described in the following formulas.

$$\frac{dP}{dH} = 2\,HXLX^T, \frac{dQ}{dW} = 2\,W^T T W \tag{16}$$

5   Experimental Results

The experiment setting and the tagged results are given in this section. Subsection 5.1 describes the practical datasets, feature extraction method and evaluation criteria about experiment results. Subsection 5.2 provides the concrete results about each dataset. And then the analysis of parameters is showed in Sect. 5.3.

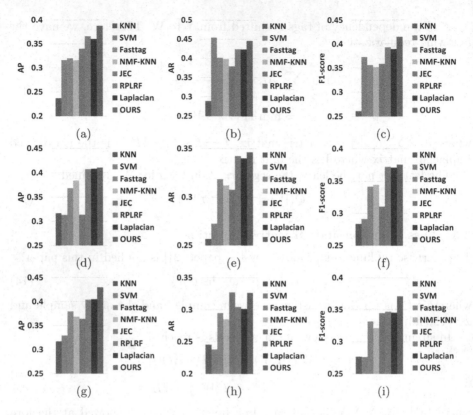

Fig. 2. Annotation results. (a)–(c): Corel5k; (d)–(f): ESPGame; (g)–(i): IAPRTC12.

5.1   Experiment Settings

In this subsection, the datasets are introduced firstly. Secondly, it is described for feature extraction method and evaluation standards.

We employ three image datasets, including Corel5K, ESPGame, and IAPRTC12 to demonstrate the effectiveness of the proposed method.

Corel5K dataset includes 4493 images in the training set and 499 images in the test set. The samples whose number of labels is zero are removed in the training set.

19, 187 images are included in this dataset provided by TagProp for the ESP Game dataset [8]. 500 images are randomly selected from the test set of ESP Game as the test set and the entire training samples except labeled by zero tags are involved in our experiment.

IAPRTC12 [8] consists of 17, 663 images as training set. The test set includes 500 images randomly selected from the test set of IAPRTC12.

In this experiment, the pre-trained Resnet is employed in the feature extraction process. Each feature vector is normalized ensuring that its ℓ_2 norm is 1 before the features are entered into the model. The conventional AP(average

Table 1. Experimental results on three datasets.

	Corel5k			ESP Game			IAPRTC12		
	AP	AR	F1 Score	AP	AR	F1 Score	AP	AR	F1 Score
lambda = 0 and $\gamma = 0$	38.1%	42.45%	40.18%	39.41%	32.96%	35.90%	42.36%	31.84%	36.35%
lambda = 0	38.50%	42.33%	40.33%	40.99%	33.74%	37.01%	42.58%	32.01%	36.55%
$\gamma = 0$	38.70%	**44.58%**	41.43%	40.90%	34.52%	37.44%	42.53%	32.55%	36.88%
OURS	**38.77%**	44.57%	**41.43%**	**41.50%**	**34.63%**	**37.57%**	**42.98%**	**32.64%**	**37.11%**

precision), AR(average recall) and F1-score are the evaluation indicators comparing each approach. The top 5 labels for each sample are considered to constitute the final predicted labels.

5.2 Experimental Performance

In this section, some experimental results are provided with respect to different datasets and methods. We choose some excellent and novel methods as methods of comparison, "RPLRF" [16], the method that replaces the LLE with laplacian in the paper, called "Laplacian", NMF-KNN [10], Fasttag [3], JEC [19] SVM and KNN. In addition, we test "$\lambda = 0$ and $\gamma = 0$", "$\lambda = 0$" and "$\gamma = 0$" in (15) to prove validity of multi-graphs Laplacian and tag graph Laplacian. In Fig. 2, the AP, AR and F1-score of diverse methods are revealed in different data sets.

As shown in Fig. 2, results indicate the superiority of our graph structures comparing with "RPLRF" and "Laplacian". More exactly, our method improves the AP by an average of 2.17%, 0.28%, and 2.59% on Corel5K, ESP Game, and IAPRTC12 comparing with "RPLRF". Similarly, the AP is increased by 2.77%, 0.24% and 2.51% in comparison with "Laplacian". And manifold structure information is rewarding to improve the performance of the model by comparing "OURS" with NMF-KNN [10], Fasttag [3], JEC [19], SVM and KNN. Specifically, our method improves 4.78%, 2.64% and 5.05% on Corel5k, ESPGame and IAPRTC12 compared with the second-best method among NMF-KNN, Fasttag, JEC, SVM, and KNN.

Results imply that the effectiveness of proposed graph structures is proved to compare "OURS" with "$\lambda = 0$ and $\gamma = 0$", "$\lambda = 0$" and "$\gamma = 0$" in Table 1.

5.3 The Analysis Parameters

The effect of many parameters, λ, γ, η and ω in (15), are involved in this session. Meanwhile, this session also describes the processes of adjusting the number of local training set n and the dimension of new feature matrix r. We adjust different parameters separately by fixing other parameters. Figure 3, it can be shown to ensure the values of parameters based on the AP and AR on corel5k dataset.

As Fig. 3 shown, AP and AR increase with the addition of λ. This phenomenon proves the effectiveness of multigraph Laplacian incorporating tag information. Meanwhile, ω enjoys a similar trend. There is the same tendency of η

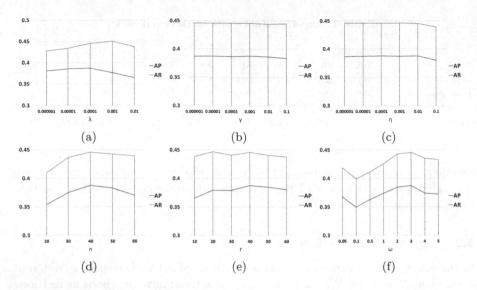

Fig. 3. The influence of parameters in Corel5k. (a): λ; (b): γ; (c): η; (d): n; (e): r; (f): w.

and γ. Specifically, η and γ have a thin effect on AP and AR when both values are small. However, the values of AP and AR decrease rapidly with η and γ increasing from 0.01 to 0.1. We select the suitable value of λ, ω, and γ for the best values of AP and AR. More precisely, $\lambda = 0.0001, \gamma = 0.001, \eta = 0.0001$ and $\omega = 3$ are assigned in our experiment. For n which is the number of the local training sets, we elect n = 40 and r = 40 to guarantee the best AP and AR.

6   Conclusion

In this paper, we propose a multi-graph Laplacian feature mapping method for automatic image annotation, which simultaneously incorporates tag information and tag graph Laplacian with visual content into the feature mapping. The label information is combined with the Laplacian eigenmaps during preserving similarities of samples. It solves the trouble of adjusting the number of neighbors as a parameter. Then, the local geometric structure employed via multi-graphs is more comprehensive in comparison with the single graph. Afterward, the correlations of visual content substitute for tag correlations. Our method achieves significant improvements in three benchmark datasets over the competing methods.

Acknowledgements. The paper was supported by the Natural Science Foundation of Shandong Province, China (Grant No. ZR2019MF073), the Fundamental Research Funds for the Central Universities, China University of Petroleum (East China) (Grant No. 20CX05001A), the Major Scientific and Technological Projects of CNPC (Grant No. ZD2019-183-008), and the Creative Research Team of Young Scholars at Universities in Shandong Province (Grant No. 2019KJN019).

References

1. Bakliwal, P., Jawahar, C.V.: Active learning based image annotation. In: 2015 Fifth National Conference on Computer Vision, Pattern Recognition, Image Processing and Graphics (NCVPRIPG), pp. 1–4. IEEE (2015)
2. Bhagat, P.K., Choudhary, P.: Image annotation: then and now. Image Vis. Comput. **80**, 1–23 (2018)
3. Chen, M., Zheng, A., Weinberger, K.: Fast image tagging. In: International Conference on Machine Learning, pp. 1274–1282. PMLR (2013)
4. Chen, Y., et al.: The image annotation algorithm using convolutional features from intermediate layer of deep learning. Multimed. Tools Appl. **80**, 4237–4261 (2021)
5. Fan, J., Gao, Y., Luo, H.: Integrating concept ontology and multitask learning to achieve more effective classifier training for multilevel image annotation. IEEE Trans. Image Process. **17**(3), 407–426 (2008)
6. Feng, S., Lang, C.: Graph regularized low-rank feature mapping for multi-label learning with application to image annotation. Multidimension. Syst. Signal Process. **29**, 1351–1372 (2018)
7. Ge, H., Yan, Z., Dou, J., Wang, Z., Wang, Z.: A semisupervised framework for automatic image annotation based on graph embedding and multiview nonnegative matrix factorization. Math. Probl. Eng. **2018** (2018)
8. Guillaumin, M., Mensink, T., Verbeek, J., Schmid, C.: TagProp: discriminative metric learning in nearest neighbor models for image auto-annotation. In: 2009 IEEE 12th International Conference on Computer Vision, pp. 309–316. IEEE (2009)
9. Huang, S.-J., Chen, J.-L., Mu, X., Zhou, Z.-H.: Cost-effective active learning from diverse labelers. In: IJCAI, pp. 1879–1885 (2017)
10. Kalayeh, M.M., Idrees, H., Shah, M.: NMF-KNN: image annotation using weighted multi-view non-negative matrix factorization. In: Proceedings of the IEEE Conference on Computer Vision and Pattern Recognition, pp. 184–191 (2014)
11. Ke, X., Zou, J., Niu, Y.: End-to-end automatic image annotation based on deep CNN and multi-label data augmentation. IEEE Trans. Multimed. **21**(8), 2093–2106 (2019)
12. Kong, D., Ding, C., Huang, H., Zhao, H.: Multi-label ReliefF and F-statistic feature selections for image annotation. In: 2012 IEEE Conference on Computer Vision and Pattern Recognition, pp. 2352–2359. IEEE (2012)
13. Koohbanani, N.A., Unnikrishnan, B., Khurram, S.A., Krishnaswamy, P., Rajpoot, N.: Self-path: self-supervision for classification of pathology images with limited annotations. IEEE Trans. Med. Imaging **40**(10), 2845–2856 (2021)
14. Li, J., Feng, S., Lang, C.: Graph regularized low-rank feature learning for robust multi-label image annotation. In: 2016 IEEE 13th International Conference on Signal Processing (ICSP), pp. 102–106. IEEE (2016)
15. Li, X., Shen, B., Liu, B.-D., Zhang, Y.-J.: A locality sensitive low-rank model for image tag completion. IEEE Trans. Multimed. **18**(3), 474–483 (2016)
16. Li, X., Shen, B., Liu, B.-D., Zhang, Y.-J.: Ranking-preserving low-rank factorization for image annotation with missing labels. IEEE Trans. Multimed. **20**(5), 1169–1178 (2017)
17. Li, Y., Song, Y., Luo, J.: Improving pairwise ranking for multi-label image classification. In: Proceedings of the IEEE Conference on Computer Vision and Pattern Recognition, pp. 3617–3625 (2017)
18. Liu, W., Tao, D.: Multiview hessian regularization for image annotation. IEEE Trans. Image Process. **22**(7), 2676–2687 (2013)

19. Makadia, A., Pavlovic, V., Kumar, S.: A new baseline for image annotation. In: Forsyth, D., Torr, P., Zisserman, A. (eds.) ECCV 2008, Part III. LNCS, vol. 5304, pp. 316–329. Springer, Heidelberg (2008). https://doi.org/10.1007/978-3-540-88690-7_24

20. Sie, E., Vasisht, D.: RF-annotate: automatic RF-supervised image annotation of common objects in context. In: 2022 International Conference on Robotics and Automation (ICRA), pp. 2590–2596. IEEE (2022)

21. Tang, C., Liu, X., Wang, P., Zhang, C., Li, M., Wang, L.: Adaptive hypergraph embedded semi-supervised multi-label image annotation. IEEE Trans. Multimed. **21**(11), 2837–2849 (2019)

22. Verma, Y.: Diverse image annotation with missing labels. Pattern Recogn. **93**, 470–484 (2019)

23. Verma, Y., Jawahar, C.V.: Image annotation by propagating labels from semantic neighbourhoods. Int. J. Comput. Vision **121**, 126–148 (2017)

24. Wang, F., Liu, J., Zhang, S., Zhang, G., Li, Y., Yuan, F.: Inductive zero-shot image annotation via embedding graph. IEEE Access **7**, 107816–107830 (2019)

25. Wang, L., Ding, Z., Fu, Y.: Adaptive graph guided embedding for multi-label annotation. In: IJCAI (2018)

26. Wu, B., Chen, W., Sun, P., Liu, W., Ghanem, B., Lyu, S.: Tagging like humans: diverse and distinct image annotation. In: Proceedings of the IEEE Conference on Computer Vision and Pattern Recognition, pp. 7967–7975 (2018)

27. Xiang, Y., Zhou, X., Chua, T.-S., Ngo, C.-W.:. A revisit of generative model for automatic image annotation using Markov random fields. In: 2009 IEEE Conference on Computer Vision and Pattern Recognition, pp. 1153–1160. IEEE (2009)

28. Xue, Z., Junping, D., Zuo, M., Li, G., Huang, Q.: Label correlation guided deep multi-view image annotation. IEEE Access **7**, 134707–134717 (2019)

29. Zhang, J., He, Z., Zhang, J., Dai, T.: Cograph regularized collective nonnegative matrix factorization for multilabel image annotation. IEEE Access **7**, 88338–88356 (2019)

30. Zhang, J., Rao, Y., Zhang, J., Zhao, Y.: Trigraph regularized collective matrix tri-factorization framework on multiview features for multilabel image annotation. IEEE Access **7**, 161805–161821 (2019)

31. Zhang, Y., Wu, J., Cai, Z., Philip, S.Y.: Multi-view multi-label learning with sparse feature selection for image annotation. IEEE Trans. Multimed. **22**(11), 2844–2857 (2020)

Short-Term Solar Irradiance Forecasting from Future Sky Images Generation

Hoang Chuong Nguyen[✉] and Miaomiao Liu

Australian National University, Acton, Australia
{hoangchuong.nguyen,miaomiao.liu}@anu.edu.au

Abstract. Solar irradiance prediction is critical for the integration of the solar power to the existing power system. A recent trend in the literature is to adopt deep learning-based methods to predict future solar irradiance from sky images. While these models have achieved significant improvements, they are only capable of making predictions for a fixed forecasting horizon due to their non-recurrent nature. To this end, we propose a deep learning network that is capable of predicting solar irradiance in an autoregressive manner, which allows predictions across a long time horizon. Particularly, we reduce the problem to first generating future sky images which are then used to predict future solar irradiance. We evaluate our models on TSI880 and ASI16 datasets, and show that our model achieves superior performance compared to previous works for 4-h ahead-of-time predictions. Furthermore, we also demonstrate that the solar irradiance forecast of our model is not limited to only 4 h, but can be extended for even longer horizon.

Keywords: Deep learning · Computer Vision · Solar irradiance forecasting

1 Introduction

Fossil fuel burning is one of the main cause for a huge amount of carbon dioxide (CO_2) emitted into the environment, leading to an increasing severity of global warming and climate changes [1]. In lieu of burning fossil for energy generation, utilizing environmentally friendly energy source, such as solar power, could help to alleviate the (CO_2) emission, which further helps to protect the environment. A wide adoption of solar energy requires an integration of solar power to the existing power grid system while maintaining its stability [2]. However, solar power generation is unstable due to the uncontrollable nature of the environment. This renders predicting future solar power, namely solar power forecasting, a critical task. Existing methods for this task focus on forecasting solar irradiance because of its strong correlation with solar power.

Supplementary Information The online version contains supplementary material available at https://doi.org/10.1007/978-981-99-8388-9_2.

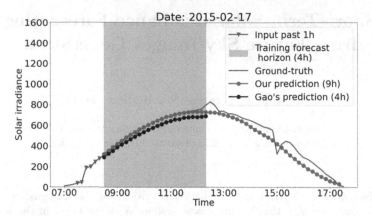

Fig. 1. Advantage of our model over previous work [4]. Although both models are trained to use data in the past 1 h to predict solar irradiance for the next 4 h, [4] is only able to produce predictions for a fixed horizon during inference, which is the future 4 h. On the other hand, our model is not limited to a certain forecast horizon and is capable of predicting solar irradiance for the rest of the day.

Existing methods for this task could be categorized into two families: physical methods [5–7] and data-driven methods [4,10,17]. In particular, physical approaches mainly rely on the physical process to predict future solar irradiance. For example, [6] proposed to adopt a clear-sky model, which is a function of zenith angle, to calculate future solar irradiance on clear-sky days. These physical models tend to make strong assumptions on the environment, which is not well-generalized as solar irradiance highly depends on real-time weather conditions.

With a rapid increase in the volume of historical weather data and sky images, deep learning-based approaches are proposed recently to extract useful information and predict solar irradiance from observed sky images. In particular, [17] uses a pretrained CNN model to encode each image into a features vector. A sequence of feature vectors representing a sequence of sky images are then fed into a LSTM model to encode temporal information and predict future solar irradiance. Instead of directly regressing solar irradiance, [4] proposed to predict residuals between the ground-truth solar irradiance and the one produced by the clear-sky model [6]. Due to the non-recurrent natures of their models, [4,17] are limited to a fixed forecast horizon defined at the beginning of the training process. In other words, given an input of historical sky images in the past 1 h, these models are trained to predict future solar irradiance for the next 4 h. At test time, these models are only able to produce predictions for a forecast horizon of 4 h and this forecast horizon cannot be extended further. This limits these models to predict the solar irradiance for a longer horizon.

As [4,17] have demonstrated that sky images provide useful information to estimate solar irradiance, we propose to reduce the problem to future sky images generation in order to achieve predictions across even long-term horizon. Specifically, given a sequence of historical sky images, we first predict future sky images in an autoregressive manner. Future auxiliary data (including zenith angle, azimuth angle, clear-sky irradiance, date and time) are also calculated using closed-form physical equations [11]. Then, the predicted future images and the calculated future auxiliary data are used to predict future solar irradiance. Thanks to our design, our model is not limited to a fixed forecasting horizon at test time even though the forecasting horizon is fixed during training. This advantage of our approach over the previous work [4] is visualized in Fig. 1. The contributions of this paper are summarized as follows:

- We propose a deep learning model that predicts future solar irradiance autore-gressively. Unlike previous works, our model is not limited to a fixed forecast horizon but can predict solar irradiance for an arbitrary number of timesteps in the future.
- Our model achieves superior performance compared to previous works, espe-cially on the ASI16 dataset in which there is no noise caused by the broken suntracker in the test set.

2   Related Works

Physical Methods. Solar irradiance forecasting has been studied for decades. In the early days when there was lack of sensors collecting weather data, existing methods relied on geometric information such as position, time and location to forecast solar irradiance [5]. From this, geometric information has been proven to have a high correlation with solar irradiance on clear-sky days. These methods has formed a foundation of the field and they are still used in practice. Neverthe-less, early methods tend to make strong assumptions on geometric information, making them not well-generalized and special cares must be taken while deploy-ing these methods in other locations.

Deep Learning-Based Methods. With a rapid growth in the volume of weather data, deep-learning approaches is taking a lead in this task. In particu-lar, [16] presented a convolutional LSTM network for precipitation nowcasting. This model take advantages of convolution neural network (CNN) to process spatial information in an image, and long short-term memory network (LSTM) to extract useful temporal information from a sequence of images. Additionally, [15] utilizes LSTM to capture the dependency between hourly solar irradiance and forecast hourly solar irradiance for the next day.

A recent trend in the field is to exploit sky images to forecast future solar irradinace [4,17]. Particularly, [4] uses Vision Transformer [3] to shows that sky images provide useful information (such as a position of the sun in an image) to predict future solar irradiance. Instead of directly regressing solar irradiance, [4] proposes to predict residuals between ground-truth solar irradiance and the one produced by a clear-sky model [6]. On the other hand, [17] first use CNN to encode a sky image into a features vector. Then a sequence of features vectors representing a sequence of sky images are fed into a LSTM model to forecast future solar irradiance. [20] stacks the red channel of all observed sky images together, which are then used by a CNN model to predict solar irradiance. This method shows that solar irradiance forecasting could be achieved without the need for a recurrent model. [14] leverages cloud motions in a form of optical flows to improve future solar irradiance predictions. A study on how different deep learning architectures (including CNN, LSTM, 3D CNN and convolutional LSTM) perform on this task was presented by [13]. Most recently, [9] proposes to use transformer encoder to extract features from observed images. The features are then fed into a transformer decoder to predict features associated with sky images in the future. These are, in turn, used to predict future solar irradiance.

Due to their non-recurrent natures of all aforementioned methods, they are only able to predict solar irradiance for a fixed forecast horizon defined at the beginning of the training process, making them not totally fit in the context of solar irradiance forecasting. In contrast, our forecasting model is capable of forecasting solar irradiance for an arbitrary forecast horizon thanks to its autoregressive prediction mechanism.

3   Method

Let us now formulate the solar irradiance forecasting problem and introduce our approach. We denote $X_{1:M} = \{x_1, x_2, \ldots, x_M\}$ as an observed sequence of sky images and $A_{1:M} = \{a_1, a_2, \ldots, a_M\}$ as an observed sequence of auxiliary data at the first M timesteps, with x_t and a_t denoting an image and auxiliary data at timestep t, respectively. The auxiliary data is a seven-dimensional vector containing measurements including zenith angle, azimuth angle, clear-sky irradiance, month, date, hour and minute.

Given $X_{1:M}$ and $A_{1:M}$, our goal is to predict future solar irradiance for the next T timesteps denoted as $\hat{S}_{M+1:M+T} = \{\hat{s}_{M+1}, \hat{s}_{M+2}, \ldots, \hat{s}_{M+T}\}$, where $\hat{s}_t$ is the predicted solar irradiance at timestep t. To achieve this, we propose a forecasting network consisting of the following modules:

– A future auxiliary calculator that takes date and time information to calculate future values for the auxiliary data in the next T timesteps, denoted as $A_{M+1:M+T}$. These deterministic data can be computed accurately by using closed-form physical equations [11].
– An image prediction model takes the observed sky images $X_{1:M}$ as input to predict future sky images $\hat{X}_{M+1:M+T}$.
– Lastly, the forecasting framework which is achieved by a nowcasting model taking the computed auxiliary data $A_{M+1:M+T}$ and the predicted images $\hat{X}_{M+1:M+T}$ to predict solar irradiance $\hat{S}_{M+1:M+T}$ for the future T timesteps.

3.1   Nowcasting Model

The nowcasting model takes an input image x_t and an input auxiliary data a_t at timestep t to estimate solar irradiance $\hat{s}_t$ at that timestep. We adapted the *Alexnet* [8] as a nowcasting model in our design. Additionally, the last layer of this network is modified to regress solar irradiance instead of classification. At the first stage of the training process, the nowcasting model is trained using L_1 loss between the predicted solar irradiance $\hat{s}_t$ and the ground-truth one s_t as shown in Eq. 1. More details about the training procedure are described in Sect. 3.3.

$$\mathcal{L}_s = |\hat{s}_t - s_t| \tag{1}$$

3.2   Image Prediction Model

We adapted the *PredRNN* [19] model as a module predicting future sky images. Thanks to its great capability in encoding temporal and spatial information, this model is capable of predicting future sky images in an autogressive manner. Specifically, this model takes a sequence of observed sky images $X_{1:M}$ at the first M steps to predict future sky images $\hat{X}_{M+1:M+T}$ for the next T timesteps. To train this model, we use L_1 loss between the ground-truth image and the predicted image as shown below.

$$\mathcal{L}_x = \frac{1}{(M+T-1)|\Omega|} \sum_{t=2}^{M+T} \sum_{i \in \Omega} |\hat{x}_t^i - x_t^i| \tag{2}$$

where Ω defines the set of pixel coordinate in each image and $|\Omega|$ is cardinality of the set.

3.3   The Forecasting Framework

The final forecasting framework is constructed from the three components: the image prediction model, the auxiliary calculator and the nowcasting model. The prediction process is presented as follows. Firstly, the auxiliary calculator takes the date time information of the future timesteps to compute the future auxiliary data $A_{M+1:M+T}$. Likewise, the image prediction model takes the observed sky images $X_{1:M}$ to predict future sky images $\hat{X}_{M+1:M+T}$. Ultimately, the nowcasting model takes each pair of predicted future image $\hat{x}_t$ and computed future auxiliary data a_t to estimate future solar irradiance $\hat{s}_t$ at the corresponding timestep. The full forecasting model and its autoregressive prediction process is shown in Fig. 2.

The training process of the forecasting model consists of two stages. At the first stage, each model is trained separately using their loss functions described in previous sections. Regarding the nowcasting model, at the first stage, it takes a real image x_t and an auxiliary data a_t to estimate solar irradiance $\hat{s}_t$. As for the second stage, we combine all modules together in order to fine-tune the forecasting model. Particularly, the nowcasting model are kept fixed at this stage and we only fine-tune the image prediction model. As the nowcasting model acts as a solar irradiance estimator in our design, it is not reasonable to continue training this model at the second stage. The loss function used to fine-tune the image prediction model at the second stage includes L_1 loss on predicted images and predicted solar irradiance as described below.

$$\mathcal{L}_{forecasting} = \mathcal{L}_x + \frac{\lambda}{M+T-1} \sum_{t=2}^{M+T} |\hat{s}_t - s_t|, \tag{3}$$

where λ is a hyper-parameter balancing the importance of the two losses.

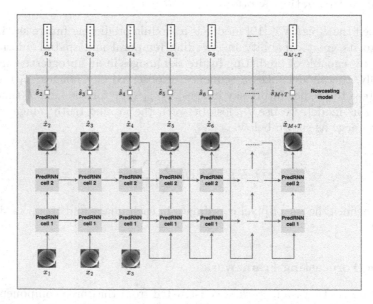

Fig. 2. An overview of the proposed forecasting model. The nowcasting model takes the predicted future sky images and the computed auxiliary data to predict future solar irradiance.

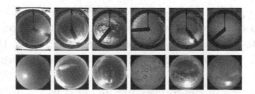

Fig. 3. Examples of sky images under different weather conditions in the TSI880 dataset (top) and the ASI16 dataset (bottom)

4   Experiments and Results

4.1   Datasets

The TSI880 and ASI16 datasets are collected by the Solar Radiation Research Laboratory (SRRL) [18] of the National Renewable Energy Laboratory [12] in

Golden, Colorado, the United States. The two datasets include sky images and twelve weather measurements including azimuth angle, tower dry bulb temperature, tower relative humidity, total cloud cover, average wind speed, station pressure, zenith angle and clear-sky irradiance, month, date, hour and minute. We show examples of sky images under different weather conditions in Fig. 3. The auxiliary data are collected using a Baseline Measurement System (BMS) which has more than 80 meteorological sensors.

TSI880. From 2004 to 2016, a total of 304,309 sky images are captured using a commercial camera (TSI). Each sky image is captured every 10 min. Moreover, a sun tracker is deployed to block sunlight, preventing sky images from being saturated. Following the setup in [4,17], data collected in 2015 and 2016 are used as a test set. In addition, we split data from 2004 to 2014 into a training set and a validation set. During training, our model takes 6 consecutive frames in the past 1 h ($M = 6$) to predict 24 values of solar irradiance for the future 4 h ($T = 24$).

ASI16. Images in this dataset are captured using EKO All Sky Imager (ASI-16) camera from September 26, 2017 to the end of 2020. Similarly to the TSI880 dataset, one image is taken every 10 min but without a sun tracker. We also use data in the past 1 h to predict solar irradiance for the future 4 h. Following [4], we use data collected from 2017 to 2019 as our training and validation set, whereas data collected in 2020 is used as a test set.

4.2 Evaluate Metrics, Data Processing and Hyper-parameters

Similar to [4,17], normalized mean absolute percentage ($nMAP$) error of predictions is used as an evalution metrics in our experiments.

$$nMAP = \frac{1}{N} \sum_{n=1}^{N} \frac{|s_n - \hat{s}_n|}{\frac{1}{N} \sum_{n=1}^{N} s_n} \times 100 \tag{4}$$

where s_n is the ground-truth solar irradiance, $\hat{s}_n$ is the predicted one of the nth sample in the dataset.

We only use seven out of twelve weather measurements available in the dataset including month, date, hour, minute, zenith angle, azimuth angle and clear-sky irradiance. These data are normalized to have zero mean and unit standard deviation. Additionally, we downsample sky images to a size of 128×128 in order to reduce the memory requirement of the image prediction model.

At the first training stage, we train the nowcasting model for 11 epochs on the TSI880 dataset and 24 epochs on the ASI16 using Adam optimizer with a learning rate of 0.002. Regarding the image prediction model, we train it for 3 epochs on the TSI880 dataset and 6 epochs on the ASI16 dataset. This model is further fine-tuned for 1 epoch for both datasets at the second stage. Adam optimizer with a learning rate of 0.003 and 0.0005 is used at the first stage and the second stage respectively. The hyper-parameter λ in Equa. 3 is set to 0.0007.

4.3   Nowcasting Results

The comparision between our nowcasting model and previous works [4,17] is shown in Table 1. Although this is not a main contribution of this paper, we present the performance of the nowcasting model to illustrate that with a simple *Alexnet* model, the performance of our nowcasting model is already on par with [4] in which Vision Transformer [3] is used.

Table 1. Nowcasting nMAP errors (%) (*-reproduced result reported in [4])

	TSI880 2015	TSI880 2016	ASI16
Siddiqui et al. [17]	14.6	15.7	13.1*
Gao et al. [4]	7.7	7.7	**6.9**
Ours	**7.3**	**6.8**	7.2

Table 2. Forecasting nMAP errors (%) (future 4 h) (*-reproduced result reported in [4]). Fine-tuning indicates the second training stage.

	TSI880 2015				TSI880 2016				ASI16			
	+1 h	+2 h	+3 h	+4 h	+1 h	+2 h	+3 h	+4 h	+1 h	+2 h	+3 h	+4 h
Siddiqui et al. [17]	**17.9**	**25.2**	31.6	39.1	**16.9**	25.0	31.9	39.5	21.6*	27.9*	33.0*	36.9*
Gao et al. [4]	22.6	26.3	30.1	33.7	19.9	**23.8**	**27.2**	**30.7**	17.4	20.9	25.1	29.2
Ours	22.6	26.6	30.6	34.7	19.8	24.3	28.9	32.9	17.5	21.6	26.0	30.6
Ours (fine-tuning)	22.1	26.0	**29.9**	**33.1**	19.4	24.0	28.0	31.7	**16.1**	**20.0**	**24.1**	**28.1**

4.4   Forecasting Results

Table 2 shows the $nMAP$ error for the forecasting task on the TSI880 and ASI16 dataset. After fine-tuning, our model produces more accurate predictions than others on the predictions for the last two hours of the TSI880 2015 test set. In particular, our model gains approximately 0.2% and 0.6% reduction in the $nMAP$ error on this test set for 3-h and 4-h predictions. As for the ASI16 dataset, our model achieves the best performance on all timesteps. Compared to the previous state-of-the-art [4], our approach reduces the $nMAP$ error by 1.3%, 0.9%, 1.0% and 1.1% for 1-h, 2-h, 3-h and 4-h predictions respectively.

Additionally, it can be seen that our model gains some improvements after being fine-tuned at the second stage of the training process. Before fine-tuning, the performance of our model is already on par with [4] on the ASI16 dataset, and outperforms [17] on almost all predictions.

Qualitative Results. The predicted solar irradiance and their corresponding predicted images are shown in Fig. 4 and Fig. 5, respectively. It can be observed that our image prediction model can predict deterministic movements of the sun and the sun tracker (i.e. black bars in the images) correctly. However, on cloudy days, our model fails to predict cloud motions. This is understandable since our image prediction model [19] is a deterministic model, thus it is unable to capture uncertainties in cloud movements, but has a tendency to predict blurry images in order to minimize the loss function. Nevertheless, the predicted images still provide useful information such as the sun's position in an image, helping the nowcasting model to produce more plausible predictions than [4] (see Fig. 4).

Predicting Solar Irradiance for the Future 8 h. Although both [4,17] and our forecasting model are trained to predict solar irradiance for the future 4 h, our model can predicts solar irradiance for further timesteps thanks to the autoregressive prediction mechanism of our model. However, this cannot be achieved by [4] and [17]. We illustrate this advantage of our model in Table 3, which shows the error in predicting solar irradiance for the future 8 h.

Why our Model Performs Worse than [4] on the TSI880 Dataset? Our nowcasting model incorporates additional information from the predicted future images, unlike [4] and [17] which only use observed images. However, the result on the TSI880 dataset is not as expected, especially on the TSI880 2016 test set. We then investigated the reason behind this and realized that the sun tracker worked correctly from 2004 up to April 2015 and then stopped working starting from May 2015 until the end of 2016. This noise results in a sun tracker always standing at the same location in the predicted image as shown in Fig. 6. A non-working sun tracker, which is a wrong indicator of the sun positions, could

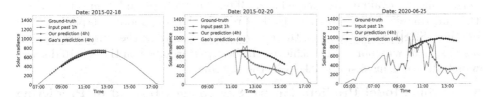

Fig. 4. Forecasting results produced our model and Gao's model [4].

Table 3. Forecasting nMAP errors (%) (future 8 h)

	TSI880 2015				TSI880 2016				ASI16			
	+5 h	+6 h	+7 h	+8 h	+5 h	+6 h	+7 h	+8 h	+5 h	+6 h	+7 h	+8 h
Ours (fine-tuning)	36.8	41.8	46.0	48.4	34.9	38.9	43.1	48.2	30.1	35.9	43.8	52.4

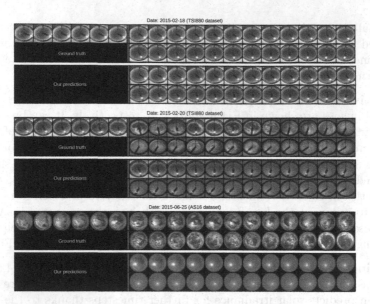

Fig. 5. Qualitative result of the predicted images produce by our image prediction model on three dates. The red separator separates 6 input timesteps in the past hour from 24 target timesteps in the future 4 h (Color figure online).

Table 4. Forecasting nMAP errors (%) on the TSI880 validation set

2-5	TSI880 (2011 and 2012)			
	+1 h	+2 h	+3 h	+4 h
Gao et al. [4]	18.3	22.0	25.1	28.6
Ours (after fine-tuning)	**17.4**	**21.3**	**24.5**	**28.2**

negatively affect the ability to predict accurate solar irradiance of our model. To verify this, we evaluate our model and [4] on the TSI880 validation set in which the sun tracker works properly. Table 4 reveals that our model outperforms [4] on the TSI880 dataset in case there is no noise caused by the sun tracker. This also explains why our model achieves the best performance on the ASI16 dataset which does not have sun trackers in the sky images.

Why Our Model Predicts Future Sky Images Instead of Predicting Low Dimensional Features of the Future Images? Low dimensional features are less computationally expensive but it contains less information of the sky and the amount of information loss increases as we predict further into the future. This renders low dimensional features not suitable for autoregressive predictions. Moreover, as the sun positions in sky images has been proven crucial for solar irradiance prediction [4], it is necessary to understand whether we can predict the sun's motions correctly using low dimensional features. To this end,

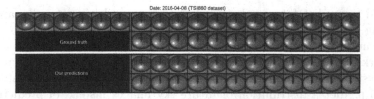

Fig. 6. Example of sky images having non-working sun trackers

Table 5. Predicting images versus predicting features (ASI16 dataset)

	+1 h	+2 h	+3 h	+4 h
Predict features	17.2	20.9	25.1	29.2
Predict images	**16.1**	**20.0**	**24.1**	**28.1**

Fig. 7. Sky images produced by different approaches.

we first use the last hidden layer of the nowcasting model to extract a 512-dimentional features vector f_t from each sky image x_t. Next, we train a decoder $D(\cdot)$ to reconstruct a sky image x_t from each extracted features f_t. We then keep $D(\cdot)$ fixed and train a $LSTM$ that takes a sequence of features $F_{1:M}$ in the past 1 h to predict a sequence of features vectors $F_{M+1:M+T}$ for the future 4 h using the loss function below.

$$\mathcal{L}_f = \frac{1}{(M+T-1)} \sum_{t=2}^{M+T} 0.04||\hat{f}_t - f_t||_1 + ||D(\hat{f}_t) - x_t||_1 \tag{5}$$

For a fair comparison, this $LSTM$ model also undergoes the fine-tuning process similar to the one applied to the forecasting model that predicts future sky images. Table 5 shows that predicting future sky images results in a lower error, while the model predicting image features achieves a similar performance as [4] which also uses the features extracted from the nowcasting model to predict future solar irradiance. Figure 7 contains ground-truth and predicted images obtained from different approaches. The last row in the figure indicates that the decoder works correctly as it can reconstruct the original images from features f_t extracted from real images. Nevertheless, the images decoded from the predicted features $\hat{f}_t$ have incorrect sun positions. This shows that using low dimensional features fails to capture correct motions of the sun, which negatively affects predictions of the nowcasting model. Conversely, predicting future images could capture the sun motions correctly, as can be observed in the second row of the figure.

5   Conclusion

In this work, we propose a framework that can autoregressively predict future solar irradiance. To achieve this, we first predict future sky images which are then used to predict future solar irradiance. Our model is not only capable of producing predictions for an arbitrary forecast horizon, but also able to produce more accurate predictions compared to previous works. One limitation of our framework is its inability to predict cloud motions, which are highly uncertain. An interesting research direction to be explored in future is to use a stochastic model to predict a distribution of future sky images instead of predicting the most-likely one.

Acknowledgment. This research was supported by the (ARC) fellowship (DE180100628).

References

1. The cause of climate change. https://climate.nasa.gov/causes/
2. Antonanzas, J., Osorio, N., Escobar, R., Urraca, R., Martinez-de Pison, F.J., Antonanzas-Torres, F.: Review of photovoltaic power forecasting. Sol. Energy **136**, 78–111 (2016)
3. Dosovitskiy, A., et al.: An image is worth 16×16 words: transformers for image recognition at scale. arXiv preprint arXiv:2010.11929 (2020)
4. Gao, H., Liu, M.: Short-term solar irradiance prediction from sky images with a clear sky model. In: Proceedings of the IEEE/CVF Winter Conference on Applications of Computer Vision, pp. 2475–2483 (2022)
5. Haurwitz, B.: Isolation in relation to cloud type. J. Atmos. Sci. **5**(3), 110–113 (1948)
6. Haurwitz, B.: Insolation in relation to cloudiness and cloud density. J. Atmos. Sci. **2**(3), 154–166 (1945)
7. Inman, R.H., Pedro, H.T., Coimbra, C.F.: Solar forecasting methods for renewable energy integration. Prog. Energy Combust. Sci. **39**(6), 535–576 (2013)
8. Krizhevsky, A., Sutskever, I., Hinton, G.E.: ImageNet classification with deep convolutional neural networks. Commun. ACM **60**(6), 84–90 (2017)
9. Mercier, T.M., Rahman, T., Sabet, A.: Solar irradiance anticipative transformer. In: Proceedings of the IEEE/CVF Conference on Computer Vision and Pattern Recognition, pp. 2064–2073 (2023)
10. Munkhammar, J., van der Meer, D., Widén, J.: Probabilistic forecasting of high-resolution clear-sky index time-series using a Markov-chain mixture distribution model. Sol. Energy **184**, 688–695 (2019)
11. NOAA: Global forecast system. national centers for environmental prediction (2019). www.ncdc.noaa.gov
12. NREL: SRRL BMS NREL transforming energy. National Renewable Energy Laboratory (NREL). https://www.nrel.gov/
13. Paletta, Q., Arbod, G., Lasenby, J.: Benchmarking of deep learning irradiance forecasting models from sky images-an in-depth analysis. Sol. Energy **224**, 855–867 (2021)
14. Paletta, Q., Arbod, G., Lasenby, J.: Cloud flow centring in sky and satellite images for deep solar forecasting. In: WCPEC, pp. 1325–1330 (2022)
15. Qing, X., Niu, Y.: Hourly day-ahead solar irradiance prediction using weather forecasts by LSTM. Energy **148**, 461–468 (2018)

16. Shi, X., Chen, Z., Wang, H., Yeung, D.Y., Wong, W.K., Woo, W.C.: Convolutional lstm network: A machine learning approach for precipitation nowcasting. In: Advances in Neural Information Processing Systems, vol. 28 (2015)
17. Siddiqui, T.A., Bharadwaj, S., Kalyanaraman, S.: A deep learning approach to solar-irradiance forecasting in sky-videos. In: 2019 IEEE Winter Conference on Applications of Computer Vision (WACV), pp. 2166–2174. IEEE (2019)
18. SRRL: SRRL BMS daily plots and raw data files. NREL Solar Radiation Research Laboratory (SRRL). https://midcdmz.nrel.gov/apps/sitehome.pl?site=BMS
19. Wang, Y., Long, M., Wang, J., Gao, Z., Yu, P.S.: PredRNN: recurrent neural networks for predictive learning using spatiotemporal LSTMs. In: Proceedings of the 31st International Conference on Neural Information Processing Systems, pp. 879–888 (2017)
20. Wen, H., et al.: Deep learning based multistep solar forecasting for PV ramp-rate control using sky images. IEEE Trans. Industr. Inf. 17(2), 1397–1406 (2020)

No Token Left Behind: Efficient Vision Transformer via Dynamic Token Idling

Xuwei Xu[1], Changlin Li[2], Yudong Chen[1], Xiaojun Chang[2], Jiajun Liu[3], and Sen Wang[1(✉)]

[1] School of Electrical Engineering and Computer Science, The University of Queensland, St Lucia, QLD 4066, Australia
{xuwei.xu,yudong.chen,sen.wang}@uq.edu.au
[2] University of Technology Sydney, Sydney, NSW 2007, Australia
[3] DATA61, CSIRO, Pullenvale, QLD 4069, Australia
ryan.liu@data61.csiro.au

Abstract. Vision Transformers (ViTs) have demonstrated outstanding performance in computer vision tasks, yet their high computational complexity prevents their deployment in computing resource-constrained environments. Various token pruning techniques have been introduced to alleviate the high computational burden of ViTs by dynamically dropping image tokens. However, some undesirable pruning at early stages may result in permanent loss of image information in subsequent layers, consequently hindering model performance. To address this problem, we propose IdleViT, a dynamic token-idle-based method that achieves an excellent trade-off between performance and efficiency. Specifically, in each layer, IdleViT selects a subset of the image tokens to participate in computations while keeping the rest of the tokens idle and directly passing them to this layer's output. By allowing the idle tokens to be re-selected in the following layers, IdleViT mitigates the negative impact of improper pruning in the early stages. Furthermore, inspired by the normalized graph cut, we devise a token cut loss on the attention map as regularization to improve IdleViT's token selection ability. Our method is simple yet effective and can be extended to pyramid ViTs since no token is completely dropped. Extensive experimental results on various ViT architectures have shown that IdleViT can diminish the complexity of pretrained ViTs by up to 33% with no more than 0.2% accuracy decrease on ImageNet, after finetuning for only 30 epochs. Notably, when the keep ratio is 0.5, IdleViT outperforms the state-of-the-art EViT on DeiT-S by 0.5% higher accuracy and even faster inference speed. The source code is available in the supplementary material.

Keywords: Efficient Vision Transformer · Token Idle

Supplementary Information The online version contains supplementary material available at https://doi.org/10.1007/978-981-99-8388-9_3.

1 Introduction

Vision Transformers (ViTs) have demonstrated remarkable performance in various vision tasks, including classification [7,26], object detection [13,14] and segmentation [3,8]. Despite ViTs' achievements, the high computational complexity of ViTs hinders their deployments in real-world scenarios where computing resources are usually limited. As a result, there is a growing demand for efficient methods that strike a balance between performance and computational efficiency, enabling ViTs in resource-constrained environments.

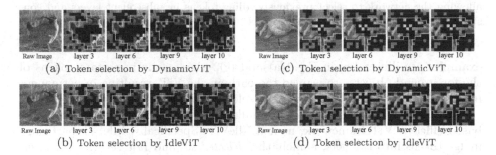

(a) Token selection by DynamicViT   (c) Token selection by DynamicViT

(b) Token selection by IdleViT   (d) Token selection by IdleViT

Fig. 1. Visualized examples of self-correcting ability for IdleViT. We take DeiT-S [19] as the backbone and compare the token selection results between a token pruning method, DynamicViT [17], and our IdleViT. Tokens containing the foreground object have been manually labelled with red borders for comparison. IdleViT can re-select the tokens of the foreground object which are unselected in the early layers.

Various approaches have been proposed to address the problem, such as constructing lightweight self-attention architectures [1,2,15] and integrating efficient convolutions with the self-attention mechanism [4,5,23]. However, these methods often necessitate dedicated architecture design and training from scratch, which impose constraints on resource-constrained devices. Alternatively, some studies concentrate on reducing the computational complexity for pretrained ViTs while maintaining high performance. They identify the token redundancy issue in ViTs [17,25] and point out that not all the image tokens contribute equally to the final prediction [12,17]. Consequently, dynamic token pruning techniques [9,11,12,16,17,24] have been introduced to progressively eliminate those less informative tokens in a pretrained ViT without significantly compromising its performance.

However, existing token pruning methods encounter an essential challenge. Empirical observations on the pruning results indicate that some tokens pruned in the early layers could be critical for accurate prediction. Unfortunately, these pruned tokens can never be re-selected in token-pruning-based methods. The information within these tokens is too early to abandon completely, yet there is no way to reintroduce them into subsequent computations. Imperfect pruning

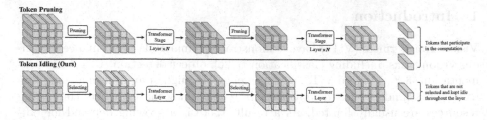

Fig. 2. Comparison between token pruning and token idling methods. Both the token-pruning-based and token-idle-based methods reduce the number of tokens involved in computations. However, our token-idle-based method retains all the tokens, allowing the network to select previously idling tokens in subsequent layers and construct skip connections of the idling tokens.

examples are illustrated in Fig. 1(a) and (c), where some important tokens of the foreground objects are pruned too early in the ViT, resulting in permanent information loss and even worse token selections in deeper layers.

To mitigate the aforementioned challenges, we present a novel token-idle-based efficient ViT framework, named IdleViT. Specifically, IdleViT partitions image tokens into two sets, namely the *Selected* set and *Idle* set, in each layer. Only the *Selected* tokens participate in the self-attention computation, thereby reducing the computational complexity. The *Idle* tokens remain unchanged until the end of each layer, where they are concatenated back to the *Selected* tokens. Unlike previous token pruning methods, IdleViT is capable of selecting tokens from those virtually pruned in earlier layers. In Fig. 1(b) and (d), given the same keep ratio in each layer, our proposed method preserves more informative foreground patches. IdleViT also differs from the previous methods in the network's topological structure and receptive field. As Fig. 2 depicts, token pruning methods progressively decrease the regions visible to the ViT as the tokens are removed and eventually construct a pyramid-shaped ViT structure. On the contrary, IdleViT can completely maintain the receptive field and establish skip connections of the image tokens in the backbone. Moreover, we demonstrate that the skip connections in IdleViT can alleviate the over-smoothing problem in ViTs and contribute to the effectiveness of the token idle strategy.

Additionally, inspired by the normalized graph cut theory [18], we introduce a novel token cut loss as a regularization term on the attention map. The token cut loss aims to maximize pairwise attention scores within the *Selected* set while minimizing attention scores between tokens from different sets. This fosters stronger intra-relationships among the *Selected* tokens and restricted inter-relationships between the two sets, resulting in more distinguishable token sets. It is worth noting that the token cut loss is only applied during finetuning and does not affect the inference speed.

We have conducted extensive experiments on representative ViT models, including DeiT [19] and LV-ViT [10], on the ImageNet-1K [6] dataset. The experimental results demonstrate that IdleViT can reduce ViTs' computational complexity while maintaining high accuracy. For instance, DeiT-S with IdleViT achieves 79.6% in top-1 accuracy with a 36% inference speed boost and a 33% reduction in computational complexity.

The main contributions of this paper are as follows:

- We propose a novel token-idle-based efficient ViT framework called IdleViT.
- We devise a token cut loss as a regularization term to enhance the token partition results and improve the performance.
- We prove that the token idle strategy can mitigate the over-smoothing problem in existing token pruning methods.

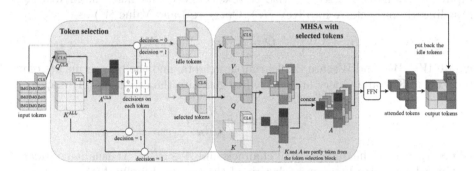

Fig. 3. IdleViT framework. At the beginning of each layer, IdleViT selects tokens with respect to the class attention where tokens with higher attention scores are chosen to participate in computations. The idling tokens are directly passed to the output of each layer, generating the input for the next layer.

2   Related Work

Vision Transformer (ViT) has attracted significant attention in the computer vision area since the success of [7]. However, the heavy computational cost of the self-attention mechanism hinders ViT's deployments in computing resource-constrained environments. As a result, dynamic token pruning methods [9,11,12, 16,17,22,24] have been introduced to expedite ViTs by progressively reducing the number of tokens involved in the self-attention calculation. DVT [22] achieves dynamic token numbers by early exiting from a cascade of ViTs with different token numbers. DynamicViT [17] proposes a learnable predictor to dynamically prune unimportant tokens. EViT [12] proposes a token reorganization method based on class attention without introducing extra network parameters. ATS [9] adaptively determines the number of tokens to prune in each stage. AdaViT [16] introduces a lightweight decision network in each block to predict whether the image patches, heads and blocks should be pruned. Evo-ViT [24] proposes a slow-fast update module that can update the tokens not involved in the computation. Different from existing dynamic token pruning methods, our IdleViT reduces the computational cost by minimizing the participation of unimportant tokens, without actually dropping them. Our method can be regarded as an extension to the previous methods.

3   Methods

3.1   Preliminaries

Vision Transformer (ViT). ViT first divides and projects the input image into a number of image tokens. Analogous to the vanilla Transformer [20], ViT also adds a special class token [CLS], which wraps the image information and is used for classification. Each ViT block has two layers: the multi-head self-attention (MHSA) layer and the feed-forward network (FFN) layer. Given an input feature map $X \in \mathbb{R}^{N \times C}$ with N tokens and C channels, the MHSA layer first linearly projects X into Key (K), Query (Q) and Value (V) as

$$K = XW_K, \quad Q = XW_Q, \quad V = XW_V, \tag{1}$$

where W_K, W_Q and W_V are the learnable weights, and the bias terms are omitted. Then, Key and Query are utilized to generate the attention (A) map by

$$A = \text{softmax}(\frac{QK^\top}{\sqrt{d_K}}), \tag{2}$$

where $d_K = C$ is the dimension of channels and $A \in \mathbb{R}^{N \times N}$ is usually considered as the relationships between each pair of the tokens. Finally, MHSA computes the self-attended feature map by the attention-weighted sum of V and employs a linear projection to activate each token.

Normalized Graph Cut. In graph theory, a cut divides the vertices in a graph into two disjoint subgraphs. Specifically, if a directed weighted graph $\mathcal{G} = (\mathcal{V}, \mathcal{E})$, consisting of a set of vertices $\mathcal{V}$ and a set of edges $\mathcal{E}$, is partitioned into two subsets $\mathcal{S}_1$ and $\mathcal{S}_2$, where $\mathcal{S}_1 \cup \mathcal{S}_2 = \mathcal{V}$, the graph cut can be measured by

$$\text{Cut}(\mathcal{S}_1, \mathcal{S}_2) = \sum_{i \in \mathcal{S}_1, j \in \mathcal{S}_2} \mathcal{E}_{i,j}. \tag{3}$$

The minimum cut in a graph is defined as the smallest cut among all possible cuts in the given graph. Identifying the minimum cut allows us to find meaningful partitions or boundaries in the graph, which is valuable for tasks like image segmentation, network flow analysis, and clustering. To prevent potential trivial solutions where one subset barely contains vertices, the normalized graph cut [18] is introduced as a constrained version of graph cut, formulated as

$$\text{NCut}(\mathcal{S}_1, \mathcal{S}_2) = \frac{\text{Cut}(\mathcal{S}_1, \mathcal{S}_2)}{\text{Assoc}(\mathcal{S}_1)} + \frac{\text{Cut}(\mathcal{S}_2, \mathcal{S}_1)}{\text{Assoc}(\mathcal{S}_2)}, \tag{4}$$

where $\text{Cut}(\mathcal{S}_1, \mathcal{S}_2)$ represents the graph cut between the two subsets and $\text{Assoc}(\mathcal{S}_i)$ is the association of $\mathcal{S}_i$ that ensures the scale of each subset is non-trivial. The association of subset $\mathcal{S}_i$ is defined as the sum of weights of all the edges touching vertices in $\mathcal{S}_i$, which is formulated as

$$\text{Assoc}(\mathcal{S}_i) = \sum_{j \in \mathcal{S}_i, k \in \mathcal{V}} \mathcal{E}_{j,k}. \tag{5}$$

The normalized graph cut was widely used as an optimization method for the image segmentation task. In image segmentation, methods based on the normalized graph cut usually aim to identify semantically consistent components by minimizing the normalized cut of image pixels.

3.2 Token Selection and Idling

As illustrated in Fig. 3, the IdleViT framework reduces the computational cost for ViTs by dynamic token selection at the beginning of each layer and preserves the token information by token idling throughout the layer.

Dynamic Token Selection. To determine the most informative tokens that participate in the calculation, we utilize the attention scores between the [CLS] token and image tokens by default. At the beginning of a ViT block, image tokens with top-$\mathcal{K}$ attention scores towards the [CLS] token are selected and straightforwardly named the *Selected* tokens. We relax the *Selected* token set to include the [CLS] token itself, which always involves in the MHSA calculation. Meanwhile, the unselected image tokens remain the same throughout the layer and are called *Idle* tokens. Specifically, we denote the set of *Selected* token indices and *Idle* token indices as S and I, respectively. Consequently, the *Selected* tokens and *Idle* tokens are denoted by X_S and X_I, respectively.

It is worth noting that the token idle strategy is independent of any particular token selection algorithm. Instead, it can be incorporated into different token selection methods, such as the token predictor in DynamicViT [17]. By default, we apply a parameter-free method in accordance with recent research [12], which achieves better performance and does not expand the model size.

Additionally, we use $\mathcal{K}$ to represent the number of *Selected* tokens and lowercase k as the base keep ratio so that $\mathcal{K} = \lfloor kN \rfloor$. Following the previous work, Idle-ViT also selects tokens in a hierarchical manner where the keep ratio decreases geometrically when the layer increases. We divide a ViT network into four stages and set the real keep ratio for layers in the ith stage as k^{i-1}.

Token Idling. At the end of each layer, the *Idle* tokens are concatenated back to the feature map based on the corresponding indices I. Notably, the input size and output size of a feature map in one layer remain constant.

3.3 Token Cut Loss

Minimizing the normalized graph cut has been recognized to enhance semantic consistency in various computer vision tasks. IdleViT separates tokens into two sets (i.e., *Selected* and *Idle* sets), which is analogous to a binary graph cut. Therefore, we can approximate the most semantically consistent separation of tokens by achieving the minimum normalized cut of tokens, with a slightly constrained setting where the sizes of the two sets are fixed. Based on the normalized graph cut described in Eq. 4, we propose a token cut loss on the attention map to enhance the semantic consistency of the *Selected* tokens. In the scenario of ViT, the attention map A can be regarded as the edge set $\mathcal{E}$ in Eq. 3 since they are naturally similar to each other with all non-negative values reflecting the relationships between data points.

We apply the normalized cut on the attention map with two adjustments to accommodate its peculiarities compared to the graph edges. One primary concern is the speciality of the [CLS] token. Enforcing the attention scores from the [CLS] token to *Idle* tokens to be 0 can cause training collapse since the [CLS] token encodes global information. Therefore, we relax this constraint to allow the [CLS] token to have non-zero attention towards the *Idle* tokens. After substituting $\mathcal{E}$ with A, the normalized cut on the attention map for the *Selected* set S and the *Idle* set I is formulated as

$$\text{NCut}(S, I) = \frac{\text{Cut}(S, I)}{\text{Assoc}(S)} + \frac{\text{Cut}(I, S)}{\text{Assoc}(I)} = \frac{\sum_{i \in S} \sum_{j \in I} A_{i,j}}{\sum_{g \in S, h \in U} A_{g,h}} + \frac{\sum_{i \in I} \sum_{j \in S \setminus \{0\}} A_{i,j}}{\sum_{g \in I, h \in U} A_{g,h}},$$
(6)

where $U = S \cup I$ is the universal set of tokens.

Besides, we analyze the association denominator in Eq. 6. Due to the softmax function in Eq. 2, the sum of attention scores towards a single token is always 1. Therefore, the two associations can be simplified as

$$\text{Assoc}(S) = \sum_{g \in S, h \in U} A_{g,h} = \sum_{g \in S} \left(\sum_{h \in U} A_{g,h} \right) = \sum_{g \in S} 1 = |S| = \mathcal{K}, \qquad (7)$$

$$\text{Assoc}(I) = \sum_{g \in I, h \in U} A_{g,h} = \sum_{g \in I} \left(\sum_{h \in U} A_{g,h} \right) = \sum_{g \in I} 1 = |I| = N - \mathcal{K}, \qquad (8)$$

where $\mathcal{K}$ and N are the pre-defined number of selected tokens and the total number of tokens, respectively. Eventually, by taking Eq. 7 and 8 into 6, we propose an inter loss to minimize the modified normalized cut as

$$L_{\text{inter}} = \frac{1}{\mathcal{K}} \sum_{i \in S} \left(\sum_{j \in I} A_{i,j} \right)^2 + \frac{1}{N - \mathcal{K}} \sum_{i \in I} \left(\sum_{j \in S \setminus \{0\}} A_{i,j} \right)^2. \qquad (9)$$

Moreover, since our approach replaces the association term in the normalized cut with a fixed scalar, the original constraint for maximizing connections within a set no longer applies. Therefore, we introduce an additional intra loss to reinforce the intra-relationship of the *Selected* set during MHSA computation. The intra loss minimizes the distance between 1 and the sum of attentions for each selected token with other selected tokens as

$$L_{\text{intra}} = \frac{1}{\mathcal{K}} \sum_{i \in S} \left(1 - \sum_{j \in S} A_{i,j} \right)^2. \qquad (10)$$

Finally, the token cut loss for training IdleViT is the sum of intra loss and inter loss in each layer as

$$L_{\text{cut}} = \sum (L_{\text{intra}} + L_{\text{inter}}). \qquad (11)$$

And we would like to emphasize that the token cut loss is only applied during finetuning as a regularization term and does NOT influence the inference speed.

3.4 Finetuning

As stated in the Introduction section, our study aims to expedite ViTs for computing resource-constrained environments, where the training cost is also a significant constraint. Due to this scenario, our method is designed to work on pretrained ViTs with a few finetuning epochs, which is remarkably more computing resource-friendly than training from scratch.

We follow the conventional finetuning pipeline as [17] to incorporate knowledge distillation as a training technique, where the full-size ViT model is utilized as the teacher to distil its pruned version. We employ knowledge distillation on both the logits and features, and apply the token cut loss as an additional optimization target. As a result, the total objective of finetuning IdleViT on a pretrained ViT is formulated as

$$L = L_{cls} + \alpha L_{logit} + \beta L_{feature} + \theta L_{cut}, \qquad (12)$$

where L_{cls}, L_{logit}, $L_{feature}$ and L_{cut} are the cross entropy loss between output logits and ground truths, the KL divergence on output logits between the teacher and student model, the mean squared error on token features between the teacher and student model, and our proposed token cut loss, respectively. α, β and θ are the coefficients for the these loss functions.

Table 1. IdleViT main results. We report the accuracy, computational complexity (measured in GMACs) and inference speed (measured in image/second) of IdleViT. The blue values reflect the differences compared to the full-size model.

Model	Keep ratio	Top-1 acc (%)	Top-5 acc (%)	GMACs	Speed (image/s)
DeiT-S [19]	1.0	79.8	94.9	4.6	2476.8
	0.9	79.9 (+0.1)	95.0 (+0.1)	4.0 (-13%)	2662.1 (+7%)
	0.8	79.9 (+0.1)	95.0 (+0.1)	3.5 (-24%)	3031.4 (+22%)
	0.7	79.6 (-0.2)	94.9 (+0.0)	3.1 (-33%)	3361.3 (+36%)
IdleViT-DeiT-S	0.6	79.3 (-0.5)	94.7 (-0.2)	2.7 (-41%)	3693.0 (+49%)
	0.5	79.0 (-0.8)	94.5 (-0.4)	2.4 (-48%)	4071.7 (+64%)
	0.4	78.4 (-1.4)	94.2 (-0.7)	2.1 (-54%)	4362.7 (+76%)
	0.3	77.3 (-2.6)	93.6 (-1.3)	1.9 (-59%)	4686.3 (+89%)
LV-ViT-S [10]	1.0	83.3	96.3	6.6	700.7
	0.9	83.3 (+0.0)	96.3 (+0.0)	5.8 (-13%)	752.5 (+7%)
	0.8	83.2 (-0.1)	96.3 (+0.0)	5.1 (-24%)	855.3 (+22%)
IdleViT-LV-ViT-S	0.7	83.1 (-0.2)	96.3 (+0.0)	4.5 (-32%)	937.5 (+34%)
	0.6	82.9 (-0.4)	96.2 (-0.1)	4.0 (-40%)	1040.3 (+48%)
	0.5	82.6 (-0.7)	96.1 (-0.2)	3.6 (-46%)	1131.4 (+61%)

Notably, IdleViT does not actually select a subset of tokens to participate in the calculation during finetuning. Instead, the MHSA layer calculates the full-size attention map of X for the token cut loss. After each MHSA layer, we filter the *Idle* tokens in the output feature map and retain them the same as the input ones. On the contrary, during testing, IdleViT only performs the MHSA and FFN on the *Selected* tokens, as Fig. 3 illustrates.

4 Experiments

4.1 Implementation Settings

Dataset. We choose ImageNet-1K [6] as the finetuning and testing dataset, which contains around 1.28 million images for training and 50 thousand images for validation. We compare the performance of IdleViT with other models that are also trained and finetuned on ImageNet-1K for fair comparisons.

Backbone Models. Two representative ViTs, the DeiT [19] and LV-ViT [10], are selected as the backbones for IdleViT. These two models are well-known in ViT families and are widely used as backbone models for token pruning methods [12,17,24]. Specifically, we only present the performance of DeiT-S (12 layers) and LV-ViT-S (16 layers) in this paper. Both the backbones are evenly divided into four stages with keep ratios $[1, k, k^2, k^3]$ at each stage, respectively.

Finetuning Configurations. We follow the same image augmentations and finetuning recipes in [19] for both DeiT-S and LV-ViT-S, but set the base learning rate to 2×10^{-5} and minimum learning rate to 2×10^{-6}. We set the finetuning batch size to 1024 for the base keep ratios between 0.9-0.5 and 2048 for the base keep ratios between 0.4-0.3. The coefficients α, β and θ for the total loss are set to 5, 500 and 20, respectively. All the models are finetuned for only 30 epochs.

Hardware. We finetune IdleViT on 2 NVIDIA Tesla V100 GPUs and measure the speed on a single NVIDIA Tesla V100 GPU with the batch size fixed to 128.

Table 2. Comparisons among token reduction methods on pretrained DeiT-S. We compare the top-1 accuracy (Acc), computational complexity (measured in GMACs) and inference speed (measured in image/second). For different methods, we adjust their corresponding token reduction ratios to achieve similar computational complexity.

Method	#Param	Acc	GMACs	Speed	Acc	GMACs	Speed	Acc	GMACs	Speed	Acc	GMACs	Speed
		$k = 0.8$			$k = 0.7$			$k = 0.6$			$k = 0.5$		
DyViT [17]	22.8M	79.6	**3.4**	**3405.0**	79.3	**3.0**	**3889.6**	78.5	**2.5**	**4474.3**	77.5	**2.2**	**5147.3**
EViT [12]	**22.1M**	79.8	3.5	2285.5	79.5	3.0	2621.8	78.9	2.6	3045.1	78.5	2.3	3383.3
Evo-ViT [24]	22.4M	78.4	3.5	2292.9	78.2	3.0	2605.8	78.0	2.6	2997.7	77.7	2.4	3172.6
ATS [9]	**22.1M**	79.6	**3.4**	2035.7	79.2	3.1	2161.3	78.9	2.7	2228.6	78.2	2.3	2351.7
IdleViT	**22.1M**	**79.9**	3.5	3031.4	**79.6**	3.1	3361.3	**79.3**	2.7	3693.0	**79.0**	2.4	4071.7

4.2 Results

Main Results. Table 1 presents the main results of our method, demonstrating IdleViT's ability to reduce computational complexity with minimal accuracy loss. For example, IdleViT achieves even higher accuracy with a 24% complexity reduction and 22% speed-up on DeiT-S when $k = 0.8$. Overall, our method is capable of cutting down a ViT's complexity by approximately 33% while incurring no more than 0.2% accuracy loss. Results on larger models are provided in the supplementary material.

Comparisons with Token Pruning Methods. As stated in the Introduction, we target expediting ViTs in computing resource-constrained scenarios where the training cost is also a significant burden. Therefore, we only compare with dynamic token pruning methods that can be finetuned on pretrained ViT backbones for 30 epochs and exclude those methods which necessitate training from scratch for 300 epochs. As a result, we compare IdleViT with DynamicViT [17], EViT [12], ATS [9] and Evo-ViT [24] on DeiT-S. Table 2 evinces that our approach outperforms existing token pruning methods at all keep ratios. More comparisons on other keep ratios and finetuning costs are provided in the supplementary material. We also present comparisons of IdleViT with other ViTs and convolutional neural networks in the supplementary material.

Results on Pyramid ViT. The token idle strategy, which is independent of the token selection method, can be regarded as an extension of the existing token-pruning models. To signify the superiority of the token idle strategy, we deploy IdleViT on a pyramid ViT, with DynamicViT [17] and Swin-Ti [14] as the token selection method and the backbone model, respectively. Table 3 indicates that the token idle strategy improves the top-1 accuracy on ImageNet under various keep ratios compared to vanilla DynamicViT. Notably, EViT [12] and other token pruning methods based on the [CLS] attention cannot be employed in this scenario due to the absence of the [CLS] token in the Swin Transformer.

Table 3. Token idle strategy on a pyramid ViT. We choose pretrained Swin-Tiny [14] as the backbone and use a predictor as DynamicViT [17] to select tokens. We compare the finetuned accuracy with and without token idle.

Idle	$k = 1$	$k = 0.9$	$k = 0.7$	$k = 0.5$
×	81.2%	78.9%	74.2%	65.1%
√		79.9%	79.6%	79.5%

Table 4. Effects of token cut loss. We provide finetuning results on DeiT-S with and without token cut loss on various keep ratios.

Loss type		Top-1 acc (%)						
inter	intra	$k = 0.9$	$k = 0.8$	$k = 0.7$	$k = 0.6$	$k = 0.5$	$k = 0.4$	$k = 0.3$
×	×	79.8	79.7	79.5	79.0	78.5	78.0	76.7
√	×	79.9	79.8	79.6	79.3	78.9	78.3	77.0
×	√	79.9	79.8	79.6	79.3	78.9	78.2	77.0
√	√	79.9	79.9	79.6	79.3	79.0	78.4	77.3

4.3   Analysis of Token Cut Loss

Ablation Study of the Token Cut Loss. Table 4 shows the effects of our proposed token cut loss. The experiments demonstrate that the combination of both intra and inter loss yields an average accuracy improvement of 0.3%, which represents a modest yet meaningful gain in this field. Moreover, the efficacy of the token idle strategy signifies as the keep ratio decreases. For instance, at the base keep ratio of 0.3, IdleViT achieves top-1 accuracy of 77.3% with token cut loss, surpassing the finetuning results without token cut loss at 76.7% by 0.6%, which is a significant improvement in this field. It is worth noting that token cut loss is only adopted during finetuning and does not affect the inference speed.

Effect on the Attention Map. Furthermore, we provide insights into the impact of token cut loss on the self-attention mechanism through Fig. 5, which illustrates the attention maps of the image tokens of Fig. 1(c). Figure 5(a) indicates that using both inter and intra loss regularization enables image tokens to concentrate on their respective sets during MHSA computation. Limited attentions between the *Selected* set and the *Idle* set indicate a clear separation of tokens and strong semantic consistency within each set. In contrast, Fig. 5(b) shows that training solely with inter loss causes tokens to primarily focus on themselves, hindering global interactions in MHSA. Figure 5(c) illustrates cross-set attentions, where the *Idle* set also interacts with the *Selected* set, suggesting inadequate separability of the two sets from a semantic consistency perspective.

4.4   Analysis of Token Idle Strategy

We investigate the reasons behind the superior performance of the token idle strategy compared to existing token pruning methods and find that IdleViT's network structure can alleviate the oversmoothing problem in ViTs. A prior study [21] observes that ViT's performance does not consistently improve with deeper layers and may even decline in very deep layers due to the oversmoothing problem. This oversmoothing problem, commonly observed in graph neural networks, results in similar image tokens as the layers deepen. In current token pruning methods, as the number of tokens progressively decreases, such oversmoothing problem becomes more severe. We compare the average cosine similarity among tokens in Table 6, where token pruning methods all lead to very similar tokens in the deep layers, which draws negative effects on the performance. However, IdleViT can reintroduce the tokens from previous layers to the deep layers and subsequently relieve the oversmoothing problem (Fig. 4).

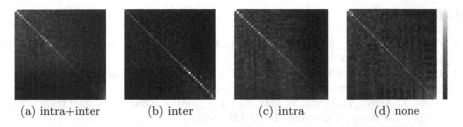

(a) intra+inter   (b) inter   (c) intra   (d) none

Fig. 4. Heat maps of the attention map finetuned with different token cut loss. The tokens are sorted in the order of class attention score. The left-top corner represents the tokens with the highest class attention while the right-bottom corner stands for the token with the lowest class attention.

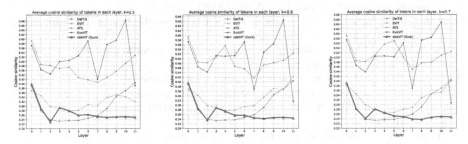

Fig. 5. Visualization of the oversmoothing problem. We calculate the average cosine similarity of image tokens in each layer for DeiT-S [19] (blue), EViT [12] (orange), Evo-ViT [24] (purple), ATS [9] (green) and our IdleViT (red). A smaller average cosine similarity represents a less severe oversmoothing problem. The results clearly demonstrate that IdleViT effectively avoids increasing similarity among image tokens when compared to existing token pruning methods and vanilla DeiT-S. (Color figure online)

5   Conclusion

In this paper, we present IdleViT, a token-idle-based approach that reduces the computational cost of Vision Transformer without significantly compromising its performance. In each layer, a subset of tokens is selected for participation in the multi-head self-attention and feed-forward network calculation, while the unselected tokens are idled and directly sent to the end of each layer. Unlike existing token-pruning-based methods, IdleViT avoids information loss by preserving all the tokens. Additionally, we propose a token cut loss to regularize the attention map in the multi-head self-attention module, contributing to a better division of the tokens. Extensive experiments have demonstrated that our model can accelerate various ViTs with minimal accuracy loss, resulting in an excellent balance between efficiency and performance.

References

1. Chen, M., Peng, H., Fu, J., Ling, H.: AutoFormer: searching transformers for visual recognition. In: ICCV (2021)
2. Chen, Y., Dai, X., Chen, D., Liu, M., Dong, X., Yuan, L., Liu, Z.: Mobile-former: bridging MobileNet and transformer. In: CVPR (2022)
3. Chen, Z., Duan, Y., Wang, W., He, J., Lu, T., Dai, J., Qiao, Y.: Vision transformer adapter for dense predictions. In: ICLR (2023)
4. Chen, Z., Xie, L., Niu, J., Liu, X., Wei, L., Tian, Q.: Visformer: the vision-friendly transformer. In: ICCV (2021)
5. Dai, Z., Liu, H., Le, Q.V., Tan, M.: CoAtNet: marrying convolution and attention for all data sizes. In: NeurIPS (2021)
6. Deng, J., Dong, W., Socher, R., Li, L.J., Li, K., Fei-Fei, L.: ImageNet: a large-scale hierarchical image database. In: CVPR (2009)
7. Dosovitskiy, A., et al.: An image is worth 16×16 words: transformers for image recognition at scale. In: ICLR (2021)
8. Fang, Y., et al.: EVA: exploring the limits of masked visual representation learning at scale. In: CVPR (2023)
9. Fayyaz, M., et al.: Adaptive token sampling for efficient vision transformers. In: Avidan, S., Brostow, G., Cissé, M., Farinella, G.M., Hassner, T. (eds.) ECCV 2022. LNCS, vol. 13671, pp. 396–414. Springer, Cham (2022). https://doi.org/10.1007/978-3-031-20083-0_24
10. Jiang, Z.H., et al.: All tokens matter: token labeling for training better vision transformers. In: NeurIPS (2021)
11. Kong, Z., et al.: SPViT: enabling faster vision transformers via latency-aware soft token pruning. In: Avidan, S., Brostow, G., Cissé, M., Farinella, G.M., Hassner, T. (eds.) ECCV 2022. LNCS, vol. 13671, pp. 620–640. Springer, Cham (2022). https://doi.org/10.1007/978-3-031-20083-0_37
12. Liang, Y., Chongjian, G., Tong, Z., Song, Y., Wang, J., Xie, P.: EViT: expediting vision transformers via token reorganizations. In: ICLR (2021)
13. Liu, Z., et al.: Swin transformer V2: scaling up capacity and resolution. In: CVPR (2022)
14. Liu, Z., et al.: Swin transformer: hierarchical vision transformer using shifted windows. In: ICCV (2021)
15. Mehta, S., Rastegari, M.: MobileViT: light-weight, general-purpose, and mobile-friendly vision transformer. In: ICLR (2022)
16. Meng, L., et al.: AdaViT: adaptive vision transformers for efficient image recognition. In: CVPR (2022)
17. Rao, Y., Zhao, W., Liu, B., Lu, J., Zhou, J., Hsieh, C.J.: DynamicViT: efficient vision transformers with dynamic token sparsification. In: NeurIPS (2021)
18. Shi, J., Malik, J.: Normalized cuts and image segmentation. TPAMI **22**, 888–905 (2000)
19. Touvron, H., Cord, M., Douze, M., Massa, F., Sablayrolles, A., Jégou, H.: Training data-efficient image transformers & distillation through attention. In: ICML (2021)
20. Vaswani, A., et al.: Attention is all you need. In: NeurIPS (2017)
21. Wang, P., Zheng, W., Chen, T., Wang, Z.: Anti-oversmoothing in deep vision transformers via the Fourier domain analysis: from theory to practice. In: ICLR (2022)
22. Wang, Y., Huang, R., Song, S., Huang, Z., Huang, G.: Not all images are worth 16×16 words: dynamic vision transformers with adaptive sequence length. In: NeurIPS (2021)

23. Wu, H., et al.: CvT: introducing convolutions to vision transformers. In: ICCV (2021)
24. Xu, Y., et al.: Evo-ViT: slow-fast token evolution for dynamic vision transformer. In: AAAI (2022)
25. Yuan, L., et al.: Tokens-to-token ViT: training vision transformers from scratch on ImageNet. In: ICCV (2021)
26. Zhai, X., Kolesnikov, A., Houlsby, N., Beyer, L.: Scaling vision transformers. In: CVPR (2022)

Story Sifting Using Object Detection Techniques

Wilkins Leong(✉)🆔, Julie Porteous🆔, and Jonathan Thangarajah🆔

RMIT University, Melbourne, Australia
leong.wilkins@gmail.com, {julie.porteous,john.thangarajah}@rmit.edu.au

Abstract. Our focus is on the problem of *story sifting* (or *story recognition*), which is the automated sifting of interesting stories that emerge from the interactions between virtual characters in virtual storyworld environments. To date, approaches to story sifting have been either: manual, with the burden of authoring sifting patterns; or automated, but of limited efficiency and scalability. In this paper, we address these shortcomings via a novel approach that recasts the problem as one of object detection. We demonstrate how an object detection model can be trained to detect story arcs of prominent story types emerging from an interactive virtual storyworld which can occur anywhere in the storyworld's timeline. We evaluate our approach using synthetic virtual story environments that show our approach is able to: detect story arcs anywhere in the storyworld's timeline with a high degree of accuracy and more efficiently than the state-of-the-art ARC SIFT, making it scalable in real-time.

Keywords: Story Sifting · Emergent Narratives · Interactive Narratives

1 Introduction

Story sifting (or *story recognition*) [26] describes the problem of how a user (human or computer) can filter out from a larger collection of events specific events that match some criteria of narrative interest. Events can be generated by the interactions between characters in virtual storyworld environments such as interactive emergent narrative systems and games. Examples of these systems include areas such as entertainment [24,25], healthcare [27] and games such as The Sims series, Dwarf Fortress, and RimWorld which provide very rich narrative environments. Given the potential scale of such systems, this offers great scope for the sifting of interesting stories for use in real-time allowing users to experience interesting stories as they occur in a simulated storyworld.

This research was supported by funding from the Commonwealth of Australia.

Supplementary Information The online version contains supplementary material available at https://doi.org/10.1007/978-981-99-8388-9_4.

Until recently, story sifting systems tended to rely on the use of sifting patterns that were manually authored using a range of query languages (examples include [12,25]). Not only is the authoring burdensome for non-technical users, it also limits the scope of these approaches to be used in interactive narrative systems. Recently, the ARC SIFT system introduced an approach to story sifting that removed the need for manual query authoring through the use of visual specification, or story arcs, as sifting patterns [15]. ARC SIFT allowed users to draw the "shape" of story arcs of interest. The notion of story arcs is something that has been widely used by narrative theorists and screenwriters [17,29] and drawing them has been shown to be intuitive for users. ARC SIFT tracked changing relationships between virtual characters in a virtual storyworld environment (represented using Dynamic Character Networks [14]), plotted these against time as character story arcs, and then attempted to match user drawn target story arcs against character story arcs to sift output stories using the time-series comparison algorithm Dynamic Time Warping (DTW) [18]. Although ARC SIFT removed the burden of manually authoring queries, the approach to target story arc matching led to a number of shortcomings that limit its applicability: (i) the use of DTW, which has a quadratic time and space complexity, limits its use to only small story sequences and environments with few characters. Hence, their approach is not scalable and efficient as the size of the virtual environment grows, nor is it viable to be used in real-time allowing for a user to sift stories as they occur; and (ii) ARC SIFT is only able to identify a target story arc if it matches the story arc that spanned across the full timeline of the virtual environment, in other words, it is not able to find story arcs that are segments of longer virtual storyworlds. This is a significant limitation of their work.

We have developed an approach that overcomes these shortcomings. It is based on the observation that the matching of user-desired story arcs, to character story arcs, is an object detection problem, and this allows us to recast the problem and exploit efficient object detection algorithms. To this end, we have used the object detection algorithm YOLOv5s [8], a single-stage, deep learning algorithm, to train an object detection model to detect story arcs of prominent story types [21] that emerge from pairs of characters in an simulated storyworld.

This approach is implemented in a system, referred to as ARC DETECTOR. We have evaluated it using synthetic data which allowed us to: (i) test our techniques in terms of model performance in sifting story arcs; and (ii) benchmark the computational efficiency against ARC SIFT with varied story arc lengths. We report results that show our approach is able to: (i) detect story arcs with a high degree of accuracy, anywhere in the virtual environment; and (ii) perform more efficiently than the state-of-the-art ARC SIFT, which makes it more scalable, and opens the possibility of sifting to occur in real-time instead of only after the virtual environment has ended. The results also demonstrate the ability of ARC DETECTOR to sift stories that match patterns of interest, that occur anywhere in the virtual environment and are not limited to starting and ending at the same points as the virtual environment. We also illustrate the output of the approach with examples of stories sifted from a virtual storyworld.

Table 1. Story Arc classes, their descriptions and the formulae used to generate each type of story arc used during Model Development: n is the length of the desired arc, and x is between 0 and n. This results in y being between 0 and 100. Story Arc names and shapes are from [21].

Story Arc Class	Description	Shape	Story Arc Formulae
0: Rags to Riches	Fortunes constantly rise	/	$y = \dfrac{100x}{n}$
1: Riches to Rags	Fortunes constantly fall	\	$y = -\dfrac{100x}{n}$
2: Man in a Hole	Fortunes fall then rise	\\/	$y = (\dfrac{20x}{n} - 10)^2$
3: Icarus	Fortunes rise then fall	/\	$y = -(\dfrac{20x}{n} - 10)^2 + 100$
4: Oedipus	Fortunes fall, rise then fall	\\/\	$y = 50(\sin(\dfrac{3\pi x}{n} + \dfrac{\pi}{2})) + 50$
5: Cinderella	Fortunes rise, fall then rise	/\\/	$y = -50(\sin(\dfrac{3\pi x}{n} + \dfrac{\pi}{2})) + 50$

2   Background and Related Work

The concept of a story arc has been widely employed in the writing industry, from early narrative theorists such as Aristotle and Freytag through to screenwriters such as Vonnegut [28], McKee [17] and Weiland [29]. This same idea has been referred to as a character arc, narrative arc or emotional arc, and describes the shape of a story characterised by the changing *fortunes* of a central character over time due to the character's interactions with the world it is in. These fortunes are story specific and can relate to aspects such as wealth, health, safety, physiological needs, esteem, self-actualization, transcendence, love, or belonging.

Story arcs have been used in narrative generation as a mechanism for user specification of story requirements, as this has been shown to be intuitive. For example, as input to a plan-based narrative generator [20], to specify the shape of stories of interest for story sifting [15], and to guide the co-creation of stories using text generation algorithms, such as GPT-3 [4]. Recent work has also sought to extract story arcs from a range of narrative media, including text and film. For example [3] trained deep convolutional neural networks to compute emotional arcs from movies. Reagan et al [21] used matrix decomposition, supervised learning, and unsupervised learning to classify the emotional arcs of 1,327 stories from Project Gutenberg's fiction collection. They showed that *six core story arc shapes* formed the essential building blocks of all the complex emotional trajectories in their dataset [21]. We have selected these arcs for use in our work with the development of Arc Detector. They are included for reference in Table 1.

Interactive emergent narrative systems and games allow for the emergence of stories, or narratives, through the interactions between simulated characters within them. Story sifting was identified as one of the key challenges that faced emergent interactive narrative systems [26]. Whilst these systems provide a rich environment for the emergence of stories, a key challenge is how to identify stories that correspond to desired story criteria, i.e. resonate with a user. This is something which has garnered recent interest in terms of story "curation" or "sifting" [25, 26]. Recent examples of story sifting systems addressing this challenge include [1,6,9–11,13,19,25]. However, they share a need for the authoring of sifting queries, specified using some form of technical language, such as the bespoke query language used in [9] or pre-defined chunks of procedural code [25].

The recent ARC SIFT approach overcomes the burden of manually authoring story sifting queries through the use of a visual specification which allowed users to draw the shape, the arc, of their story of interest [15]. Their approach is to use *Dynamic Character Networks*, [14], to track temporal changes in the changing relationships between pairs of characters in a virtual storyworld. Plotting these changing relationships over time allows them to generate collections of story arcs, which they call *simulation story arcs*. They then use a time series comparison algorithm Dynamic Time Warping (DTW) [18] to find the best match for the user desired story arc from the collection of simulation story arcs. However, a limitation of ARC SIFT is its use of DTW to match user arcs, as this was limited to matching over the entire virtual storyworld timeline and does not allow identification of interesting arcs embedded within it. Another limitation of DTW is its speed and scalability, being of quadratic time and space complexity. This makes it infeasible for real-time story sifting as stories occur. In this work, we address these limitations.

3 Approach

3.1 Recasting Story Sifting as Object Detection

Based on the observation that sifting stories which display a specific desired shape, or story arc, has similarities to the object detection problem in images, we recast the problem as such. From this perspective, the entire virtual environment with all of its events which we are aiming to sift from is the *image* in which objects are to be detected, and the specific desired sifted stories are *objects* which are detected in those images. If an entire virtual storyworld is expressed as a collection of story arcs for each character in that storyworld, (as in ARC SIFT using Dynamic Character Networks [15]) then the desired *objects* to be detected are a specific subset of story arcs within that virtual storyworld.

We assume as a starting point, the presence of a collection of story arcs from which we can attempt to detect individual story arcs that match a target shape, and where a matching arc can be an entire arc from the collection, or a segment of an arc. This is illustrated in Fig. 1, where on the left, the entire arc is of the "Man in a hole" story arc shape. The middle image shows a story arc where only the middle third of the story arc fits the "Man in a hole" story arc shape. This is highlighted with a bounding box in the right image.

In our work, we have used arcs built up from an emergent interactive narrative virtual environment, using the approach of [15] as outlined in the previous

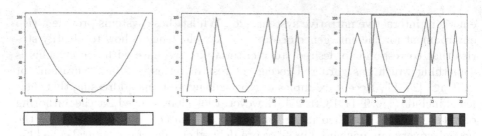

Fig. 1. A toy example of recasting story sifting as an object detection problem. Left top depicts 10 elements of a story arc plot that fits the "Man in a hole" story arc type. Left bottom is the encoding of this story arc as an image of pixels arranged sequentially. The Middle depicts the same story arc now embedded in background noise. Right most depicts a bounding box that identifies where a "Man in a hole" story exists inside this story arc.

section. However, we note that the collection of arcs from which to sift could come from other sources, such as event logs from other games and agent simulations, or from the datasets of thousands of emotional arcs automatically extracted from video in the work of Chu et al [3], and from works of fiction, as in the output of the work of Reagan et al [21].

3.2   Representing Story Arcs as Images

In order to apply object detection techniques to story sifting, a suitable representation for story arcs was required. Images prepared for object detection are typically stored as an array of "pixels", arranged in rows and columns with each pixel having 3 values to denote red, green and blue (RBG) intensity values within the range of 0 to 255. Thus, for application of object detection algorithms, the approach we adopted was to encode the information in story arcs by sequentially arranging pixels, with each pixel adopting the colour intensity same value as the nth element of a story arc. For example, a simple story arc where the fortunes of the central character gradually rise over 255 events can be encoded as an image array with 255 pixels as [0,0,0], [1,1,1], ... , [255,255,255]. Figure 1 shows the correspondence between the plot of a story arc and its representation using an array of pixels, for the classic story arc shape "Man in a Hole".

Our working hypothesis was that, given sufficient labelled training images representing story arcs, an object detection model could be trained to detect each of the six core story arc types identified by Reagan et al. [21] from large collections of images representing story arcs. These arc types, their shapes, along with a description of each arc are shown in Table 1.

3.3   Choice of YOLOv5 Model

The YOLO (an acronym for "You Only Look Once") family of neural network architectures perform object detection in a "single stage". This differs to traditional object detection neural network architectures such as RCNN [5] which feature two stages: the first stage predicts where possible bounding boxes are in an image, while the second stage classifies the predicted bounding box region into classes. This is performed as two models.

YOLO [22] was introduced as an architecture that predicts a bounding box region and class probabilities all in one end-to-end neural network. The main advantage of YOLO is speed, and can run as fast as 45 frames per second. It is therefore often used for object detection in real-time (for example frame by frame object detection from a video feed). This also makes it an ideal candidate for real-time story sifting.

Multiple iterations of YOLO have been developed and presented since the first version was published, including [2, 7, 23].

YOLOv5 is a collection of object detection models and was selected as a candidate for our problem. The primary reason was that this was the most popular open-source implementation of YOLO compatible with common Python libraries such as PyTorch. YOLOv5 is actively maintained by Uralytics, and has over 13.7k forks and 38.2k stars on GitHub.

Of all the available models in the YOLOv5 collection, we chose YOLOv5s (small). While larger YOLOv5 models claimed to be able to perform better on more complex and larger data sets, the trade-off was that they took longer to train, and took longer when performing predictions. This is because their architectures (i.e., total number of layers and tunable parameters) were larger. A further justification to use YOLOv5s was that the images representing the story arcs) were much less complex than a typical image used for object detection, and so the use of a larger model was unjustified.

4   Model Development

In order to train and validate our story sifting model ARC DETECTOR, 6000 synthetic training instances of story arcs were created, corresponding to roughly 1000 for each of the six core story arc types. 1500 synthetic test instances were created for performance testing of the model, corresponding to roughly 250 for each of the six core story arc types. For each instance, a target arc was warped, notified, and embedded in background events noise. Refer to Section 1 and Figure 1 in the supplementary material for the steps used to generate the training and validation story arc instances.

The training instances were used to train a YOLOv5s model over 100 epochs (numbered 0 to 99), using random initial weights and batch sizes of 4. This was the highest batch size our hardware limitations allowed for, as suggested by the YOLOv5 documentation. Random initial weights were used as our synthetic story arc dataset was unique to any other dataset a pre-trained model could have been trained for.

We used the default hyperparameters found in the YOLOv5 repository[1], as specified in the file `hyp.scratch-low.yaml`, except for `fliplr` which describes the "flip left right probability". This was changed from the default of 0.5 to 0.0

---

[1] https://github.com/ultralytics/yolov5.

in order to discriminate between mirror images of each other, such as "Rags to riches" and "Riches to rags", and "Cinderella" and "Oedipus". These hyperparameters did not evolve throughout training.

5 Evaluating Model Performance

The metrics measured to describe model performance included precision (P), recall (R), mean average precision (MaP). Further descriptions and definitions for these metrics can be found in Section 2 of the supplementary material.

5.1 Model Performance

Training ran for 100 epochs. Each epoch produced a model that was validated against the dataset of 1500 validation instances. The results of each epoch are summarised in Figure 2 of the supplementary material.

YOLOv5 by default selects the best-performing epoch based on a combined sum of the mAP@0.5 and mAP@[0.5:0.95] scores, with 0.9 weighting given to mAP[0.5:0.95]. It was observed this peak sum was reached at epoch 82. Thus the model produced after epoch 82 was selected as the top-performing model for use with ARC DETECTOR.

The results of the model performance for ARC DETECTOR are reported in Table 2. The table reports values for Precision, Recall, and mean average precision (mAP@0.5 and mAP@[0.5-0.95]) for each of the 6 core story arc classes, and for overall. The results show that when tested against the 1500 validation instances, ARC DETECTOR demonstrated an overall precision score of 0.818, recall score of 0.765, mAP@50 score of 0.852 and mAP@[0.5:0.95] score of 0.764. ARC DETECTOR performed mostly consistent across all 6 story arc classes. It was observed that ARC DETECTOR had slightly lower scores for Oedipus and Cinderella, which can be explained by the fact that these were the most complex shapes out of the 6 classes.

These results support our expectation that ARC DETECTOR is able to, with a high degree of accuracy and confidence, draw a bounding box to indicate the

Table 2. Performance of the trained YOLOv5s object model ARC DETECTOR for each of the core arc classes and overall for all classes: for Precision, Recall, mAP@0.5 and mAP@[0.5:0.95]. See Sect. 5 for discussion.

Arc Class	Precision	Recall	mAP@0.5	mAP@[0.5:0.95]
Rags To Riches	0.841	0.802	0.883	0.782
Riches To Rags	0.893	0.846	0.925	0.838
Man In Hole	0.878	0.774	0.883	0.793
Icarus	0.766	0.814	0.853	0.782
Oedipus	0.734	0.732	0.791	0.704
Cinderella	0.794	0.624	0.776	0.685
Overall	0.818	0.765	0.852	0.764

Table 3. Timing Efficiency comparison between ARC DETECTOR and ARC SIFT. Times in ms. for Small, Medium and Large Instances, averaged, μ, over 3 runs. Overall averages, μ, show (i) ARC DETECTOR is significantly faster; (ii) efficiency constant as instance increases, unlike ARC SIFT which grows quadratically.

		ARC SIFT				ARC DETECTOR			
		run1	run2	run3	μ	run1	run2	run3	μ
SMALL INSTANCES	0	973.5	930.3	933.8	945.8	10.8	10.1	10.8	10.6
	1	917.9	932.9	928.9	926.6	10.2	10.7	14.5	11.8
	2	954.9	924.3	926.2	935.1	11.0	10.4	10.5	10.6
	3	911.6	923.9	929.9	921.8	10.6	10.5	10.0	10.4
	4	912.1	930.2	923.2	921.8	10.0	10.7	11.2	10.6
	5	917.0	923.2	914.0	918.1	10.1	10.4	9.9	10.1
	6	919.7	920.8	935.1	925.2	10.4	10.5	10.0	10.3
	7	921.4	930.1	916.2	922.6	10.1	10.2	10.8	10.4
	8	924.7	935.1	918.6	926.1	10.0	10.0	10.3	10.1
	9	925.6	923.5	913.3	920.8	10.3	10.4	10.6	10.4
	μ				926.4				10.5
MEDIUM INSTANCES	0	23149.0	23816.7	23583.3	23516.3	17.2	16.3	12.9	15.5
	1	23759.1	23773.1	23219.6	23584.0	10.8	10.9	10.6	10.8
	2	23224.7	23845.8	23287.5	23452.7	13.2	12.1	10.6	12.0
	3	23165.7	23725.0	23603.5	23498.1	10.1	10.0	27.6	15.9
	4	23204.8	23773.0	23210.5	23396.1	10.5	11.0	11.0	10.8
	5	23194.8	23891.2	23576.5	23554.2	11.3	12.0	10.8	11.4
	6	23197.6	25312.5	23209.0	23906.4	11.4	10.9	13.1	11.8
	7	23185.3	24668.0	23180.6	23678.0	14.9	10.6	20.1	15.2
	8	23263.9	24821.9	23604.4	23896.7	17.0	11.3	10.1	12.8
	9	23354.7	23598.5	23550.0	23501.0	10.6	10.1	10.7	10.5
	μ				23598.3				12.7
LARGE INSTANCES	0	94267.1	94256.9	95043.7	94522.5	11.3	11.2	10.7	11.1
	1	94383.8	94242.4	94528.5	94384.9	13.5	10.7	10.2	11.5
	2	98780.0	95154.9	94399.6	96111.5	12.2	10.6	14.6	12.5
	3	97100.5	96555.6	94489.3	96048.4	9.8	10.9	15.6	12.1
	4	93124.1	94447.6	92859.2	93477.0	9.9	10.2	10.5	10.2
	5	92632.2	94350.7	92887.1	93290.0	10.3	10.0	10.3	10.2
	6	99034.3	94210.5	94391.7	95878.8	10.4	10.5	10.3	10.4
	7	98288.7	94312.6	93107.9	95236.4	10.4	10.5	10.8	10.6
	8	95343.8	94532.8	92783.9	94220.2	10.4	10.6	11.3	10.8
	9	94638.4	94924.6	94333.9	94632.3	11.0	11.5	11.8	11.4
	μ				94780.2				11.1

start and end points of a core story arc shape, and correctly label it when it is amongst a background of other noisy data. The examples in Table 4 illustrate the detection of story arc shapes which are segments of larger arcs.

6   Evaluating Time Efficiency

Synthetic story arcs were created by generating n random numbers between integers 0 and 100 inclusive, where n is the length of the arc. This represents the number of events in the story arc. We synthetically produced 10 small story arcs ($n = 1000$), medium ($n = 5000$) and large ($n = 10000$).

For time efficiency we compared: (i) an implementation using Dynamic Time Warping as in [15], referred to as ARC SIFT; with (ii) our trained model, ARC DETECTOR. For each story arc, we timed in milliseconds for ARC SIFT and ARC DETECTOR. Experiments were run on an Apple MacBook Pro (13-inch, M1, 2020). Each run was performed three times, to calculate an overall average.

The results are summarised in Table 3. The average time for ARC DETECTOR to process a small story arc was 10.5 ms, a medium story arc, 12.7 ms and a large story arc, 11.1 ms. This was compared to ARC SIFT which processed a small story arc in an average of 926.4 ms, 23598.3 ms for medium arcs and 94780.2 ms for large arcs. The time taken by ARC DETECTOR is significantly faster than ARC SIFT and doesn't grow with instance size in the same way as ARC SIFT which grows quadratically. This shows that ARC DETECTOR remains a scalable approach compared to ARC SIFT.

7   Detection from a Virtual Storyworld Environment

To demonstrate ARC DETECTOR on real stories, we used it to sift desired user story arcs from the outputs of a virtual storyworld, similar to [15], based on Prom Week [16]. This storyworld consisted of 200 virtual characters modelled as students on a school campus in the week before the prom. They were randomly assigned 1 of 50 possible dorm rooms to sleep at night, and 3 of 25 possible classes to attend throughout the day. At set intervals, they move throughout the day. The character's goals are to find a prom date. They are free to interact with each other if they are in the same vicinity. If the characters become relationally close enough, characters may choose to ask each other to the prom. In this storyworld, positive interaction events raise the relationship scores between two characters causing the resulting story arc to trend upwards, while negative interaction events decrease them causing the story arc to trend downward. The virtual story world ran for 1500 ticks, resulting in a total of 17061 interaction events.

Using Dynamic Character Networks [14], each pair of characters' relationships were plotted as story arcs, where different event types scored different points (positive or negative) for the relationships between each character. These were converted to arrays of RGB pixels (an image) as described in Sect. 3.2. This virtual storyworld assumed symmetrical relationships between each pair of characters, and thus there were 19900 possible relationships resulting in 19900 images. Each image was fed to the YOLOv5s model ARC DETECTOR, resulting in detections across all 6 story arc types. A selection of these detections along with their corresponding stories are shown in Table 4.

Table 4. Story arcs sifted using ARC DETECTOR from the event logs of a virtual Prom Week storyworld. For each story arc type: **Image** shows the sifted pixels; **Arc** shows the corresponding sifted story arc representing the changes in relationship between two characters in the Prom Week virtual storyworld; and corresponding snippets of stories sifted from the virtual storyworld.

Class	Image	Arc	Sifted Story Snippets
Cinderella			$\oplus$ "Nice to meet you", $\oplus$ "I'd love to go ..", $\oplus$ "Having a great time", $\oplus$ "I like you", $\odot$ "... not fun", $\oplus$ "... great time"
Icarus			$\odot$ "Sorry I'm not", $\oplus$ "I'd love to go ..", $\ominus$ "What classes do you have" $\odot$ "I'm not sure ...", $\odot$ "Sorry I'm not going ..."
ManInHole			$\odot$ "Sorry I'm not going ...", $\odot$ "This is not fun", $\ominus$ "What classes do you have", ..., $\oplus$ "I like you", ...
Oedipus			$\odot$ "I'm not sure ...", $\oplus$ "I like you", ..., $\ominus$ "What classes do you have", ..., $\oplus$ "I like you", ... $\odot$ "I'm not sure ...",
RagsToRiches			$\oplus$ "Nice to meet you", $\oplus$ "Having a great time" $\oplus$ "... great time", $\ominus$ "What classes do you have", $\oplus$ "... a great time"
RichesToRags			$\odot$ "This is not fun", $\odot$ "This is not fun", ...

8   Discussion

Time Efficiency and Scalability: In Sect. 6, it was shown that ARC DETECTOR scaled significantly better than ARC SIFT. There are two ways a virtual environment can scale which impacts the time efficiency of a story sifter. Firstly, a virtual environment can scale up by simply extending how long it runs for. The longer a virtual environment runs, the more events are captured, and therefore the *longer* the resulting story arcs. As demonstrated in Sect. 6, ARC DETECTOR scaled significantly better than ARC SIFT. Secondly, a virtual environment can scale up by increasing the number of virtual characters in the environment. As the number of virtual characters of scales up, the resulting amount of story arcs produced grows quadratically.

These two factors combined affect the time performance of a story sifter. Overall, our results demonstrate that ARC DETECTOR outperforms the state of the art on both factors. The results for ARC DETECTOR allow for the potential of real-time story sifting, where events can be sifted in real-time as they occur in a virtual environment, rather than sifting after the virtual environment has ended. This is something that cannot be done with ARC SIFT.

Regarding the Ability to Detect Segments: A key limitation to ARC SIFT was that it was only able to identify a target story arc if it matches the full length of a story arc, in other words, it is not able to find story arcs that pertain to segments of the virtual environment's entire timeline. This means that ARC SIFT became less and less useful the longer a virtual environment continues for. This limitation of ARC SIFT was a key motivation for the use of object detection techniques, as they are able to identify meaningful objects within a bigger picture. Performance of ARC DETECTOR clearly demonstrates its ability to detect story arcs which are segments of larger arcs. Examples of instances where ARC DETECTOR has detected a segment of a story arc matching a target arc are shown in Table 4: for arc types "Cinderella", "Icarus", "ManInHole", "Oedipus", "RagsToRiches", and "RichesToRags".

9 Conclusion

In this paper we have recast the problem of story sifting as an object detection problem. We have proposed a method of converting a story arc that tracks the changing relationships between virtual characters in a virtual storyworld environment into pixels arranged sequentially in an image. We trained a YOLOv5s object detection model called ARC DETECTOR that can detect 6 core story arc types in these images. Being an object detection model, ARC DETECTOR was able to detect for meaningful segments within a larger story arc, rather than classifying the entire story arc, which was a key improvement over ARC SIFT. ARC DETECTOR also demonstrated high recall, precision and mean average precision scores across all 6 classes of core story arc types. Furthermore, ARC DETECTOR has shown to be significantly more time efficient than ARC SIFT, and shows the potential to scale well as the size of a virtual storyworld grows, as well as the potential for automated real time story sifting.

The work presented in this paper provides a strong base for future research, for example, we plan to apply ARC DETECTOR to an events stream in order to perform real-time story sifting, and using the 6 core story arc types as building blocks to construct more complex target story arc shapes.

References

1. Behrooz, M., Swanson, R., Jhala, A.: Remember that time? Telling interesting stories from past interactions. In: Schoenau-Fog, H., Bruni, L.E., Louchart, S., Baceviciute, S. (eds.) ICIDS 2015. LNCS, vol. 9445, pp. 93–104. Springer, Cham (2015). https://doi.org/10.1007/978-3-319-27036-4_9
2. Bochkovskiy, A., Wang, C.Y., Liao, H.Y.M.: YOLOv4: optimal speed and accuracy of object detection (2020). arXiv:2004.10934 [cs, eess]
3. Chu, E., Roy, D.: Audio-visual sentiment analysis for learning emotional arcs in movies. In: 17th IEEE International Conference on Data Mining (ICDM) (2017)
4. Chung, J.J.Y., Kim, W., Yoo, K.M., Lee, H., Adar, E., Chang, M.: TaleBrush: sketching stories with generative pretrained language models. In: Proceedings of the 2022 CHI Conference on Human Factors in Computing Systems (2022)
5. Girshick, R., Donahue, J., Darrell, T., Malik, J.: Rich feature hierarchies for accurate object detection and semantic segmentation. In: 2014 IEEE Conference on Computer Vision and Pattern Recognition (2014)
6. Grinblat, J., Bucklew, C.B.: Subverting historical cause and effect: generation of mythic biographies in Caves of Qud. In: Foundations of Digital Games (FDG) (2017)
7. Huang, X., et al.: PP-YOLOv2: a practical object detector (2021). arXiv:2104.10419 [cs]

8. Jocher, G., et al.: ultralytics/YOLOv5: v6.2 - YOLOv5 Classification Models, Apple M1, Reproducibility, ClearML and Deci.ai integrations (2022)
9. Kreminski, M., Dickinson, M., Wardrip-Fruin, N.: Felt: a simple story sifter. In: Cardona-Rivera, R.E., Sullivan, A., Young, R.M. (eds.) ICIDS 2019. LNCS, vol. 11869, pp. 267–281. Springer, Cham (2019). https://doi.org/10.1007/978-3-030-33894-7_27
10. Kreminski, M., Dickinson, M., Mateas, M.: Winnow: a domain-specific language for incremental story sifting. In: Proceedings of the AAAI Conference on Artificial Intelligence and Interactive Digital Entertainment (AIIDE) (2021)
11. Kreminski, M., Dickinson, M., Wardrip-Fruin, N., Mateas, M.: Select the unexpected: a statistical heuristic for story sifting. In: Vosmeer, M., Holloway-Attaway, L. (eds.) ICIDS 2022. LNCS, vol. 13762, pp. 292–308. Springer, Cham (2022). https://doi.org/10.1007/978-3-031-22298-6_18
12. Kreminski, M., Samuel, B., Melcer, E.F., Wardrip-Fruin, N.: Evaluating AI-based games through retellings. In: Proceedings of the Fifteenth AAAI Conference on Artificial Intelligence and Interactive Digital Entertainment (AIIDE) (2019)
13. Kreminski, M., Wardrip-Fruin, N., Mateas, M.: Toward example-driven program synthesis of story sifting patterns. In: Proceedings of the AIIDE 2020 Workshops co-located with 16th AAAI Conference on Artificial Intelligence and Interactive Digital Entertainment (AIIDE) (2020)
14. Lee, O.J., Jung, J.J.: Story embedding: learning distributed representations of stories based on character networks. Artif. Intell. (AIJ) **281** (2020)
15. Leong, W., Porteous, J., Thangarajah, J.: Automated sifting of stories from simulated storyworlds. In: Proceedings of the Thirty-First International Joint Conference on Artificial Intelligence, (IJCAI) (2022)
16. McCoy, J., Treanor, M., Samuel, B., Reed, A.A., Wardrip-Fruin, N., Mateas, M.: Prom week. In: Proceedings of the International Conference on the Foundations of Digital Games (FDG) (2012)
17. McKee, R.: Story: Substance, Structure, Style, and the Principles of Screenwriting. ReganBooks, Los Angeles (1997)
18. Müller, M.: Dynamic Time Warping. Springer, Heidelberg (2007)
19. Osborn, J.C., Samuel, B., Mateas, M., Wardrip-Fruin, N.: Playspecs: regular expressions for game play traces. In: Eleventh Artificial Intelligence and Interactive Digital Entertainment Conference (AIIDE) (2015)
20. Porteous, J., Teutenberg, J., Pizzi, D., Cavazza, M.: Visual programming of plan dynamics using constraints and landmarks. In: Proceedings of the Twenty-First International Conference on International Conference on Automated Planning and Scheduling (ICAPS) (2011)
21. Reagan, A.J., Mitchell, L., Kiley, D., Danforth, C.M., Dodds, P.S.: The emotional arcs of stories are dominated by six basic shapes. EPJ Data Sci. **5**(1), 1–12 (2016). https://doi.org/10.1140/epjds/s13688-016-0093-1
22. Redmon, J., Divvala, S., Girshick, R., Farhadi, A.: You only look once: unified, real-time object detection. In: Proceedings of the IEEE Conference on Computer Vision and Pattern Recognition (CVPR) (2016)
23. Redmon, J., Farhadi, A.: YOLOv3: an incremental improvement (2018). arXiv:1804.02767 [cs]
24. Riedl, M.O., Bulitko, V.: Interactive narrative: an intelligent systems approach. AI Mag. **34**(1), 67–77 (2013)
25. Ryan, J.O.: Curating simulated storyworlds. Ph.D. thesis, UC Santa Cruz (2018)

26. Ryan, J.O., Mateas, M., Wardrip-Fruin, N.: Open design challenges for interactive emergent narrative. In: Schoenau-Fog, H., Bruni, L.E., Louchart, S., Baceviciute, S. (eds.) ICIDS 2015. LNCS, vol. 9445, pp. 14–26. Springer, Cham (2015). https://doi.org/10.1007/978-3-319-27036-4_2
27. Siddle, J., et al.: Visualization of patient behavior from natural language recommendations. In: Proceedings of the Knowledge Capture Conference (K-CAP) (2017)
28. Vonnegut, K.: Kurt Vonnegut on the Shapes of Stories (1995). Accessed 31 July 2023
29. Weiland, K.: Creating Character Arcs: The Masterful Author's Guide to Uniting Story Structure, Plot, and Character Development. PenForASword Publishing (2016)

SimMining-3D: Altitude-Aware 3D Object Detection in Complex Mining Environments: A Novel Dataset and ROS-Based Automatic Annotation Pipeline

Mehala Balamurali[✉][iD] and Ehsan Mihankhah

Australian Centre for Robotics, University of Sydney, Camperdown, Australia
{mehala.balamurali,ehsan.mihankhah}@sydney.edu.au

Abstract. Accurate and efficient object detection is crucial for safe and efficient operation of earth-moving equipment in mining. Traditional 2D image-based methods face limitations in dynamic and complex mine environments. To overcome these challenges, 3D object detection using point cloud data has emerged as a comprehensive approach. However, training models for mining scenarios is challenging due to sensor height variations, viewpoint changes, and the need for diverse annotated datasets.

This paper presents novel contributions to address these challenges. We introduce a synthetic dataset SimMining-3D [1] specifically designed for 3D object detection in mining environments. The dataset captures objects and sensors positioned at various heights within mine benches, accurately reflecting authentic mining scenarios. An automatic annotation pipeline through ROS interface reduces manual labor and accelerates dataset creation.

We propose evaluation metrics accounting for sensor-to-object height variations and point cloud density, enabling accurate model assessment in mining scenarios. Real data tests validate our model's effectiveness in object prediction. Our ablation study emphasizes the importance of altitude and height variation augmentations in improving accuracy and reliability.

The publicly accessible synthetic dataset [1] serves as a benchmark for supervised learning and advances object detection techniques in mining with complimentary pointwise annotations for each scene. In conclusion, our work bridges the gap between synthetic and real data, addressing the domain shift challenge in 3D object detection for mining. We envision robust object detection systems enhancing safety and efficiency in mining and related domains.

Keywords: Simulation to Real · Viewpoint Diversity · Mining Automation

---

Supported by Rio Tinto Centre for Mine Automation, Australian Centre for Robotics.

T. Liu et al. (Eds.): AI 2023, LNAI 14471, pp. 55–66, 2024.
https://doi.org/10.1007/978-981-99-8388-9_5

1   Introduction

Effective object detection is vital for ensuring the safe and efficient operation of earth-moving equipment in the mining industry. However, traditional 2D image-based methods often encounter limitations in dynamic and complex mine environments where objects can be occluded or obscured. To overcome these challenges, 3D object detection utilizing point cloud data provides a more comprehensive representation of objects and the environment, leading to improved accuracy and efficiency.

Training models for 3D object detection poses unique challenges, including adapting pretrained models to new datasets and effectively handling viewpoint variations. Additionally, collecting real-world data for training in mining scenarios can be particularly challenging due to complex terrains, cluttered surroundings, safety concerns, and logistical difficulties in active mining environments. However, simulation offers a valuable solution by generating large and diverse datasets without incurring the risks and costs associated with real-world data collection.

Simulated environments provide researchers with precise control over data variability and complexity, facilitating the training and evaluation of algorithms specifically tailored for the detection of earth-moving equipment in mining contexts. Furthermore, simulations offer the opportunity for automatic annotation, significantly reducing the manual labor and costs involved in accurately labeling real-world data (Fig. 1).

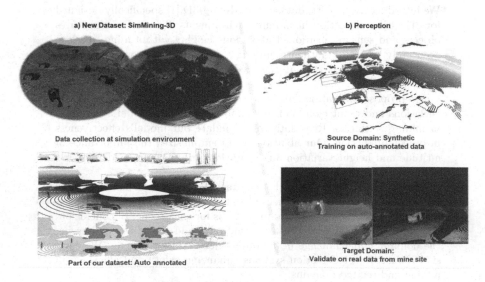

Fig. 1. Proposed workflow illustrates the scope of this study

In this work, we propose a novel approach that addresses the domain shift between synthetic and real data in 3D object detection for complex mining environments. Our methodology involves training a model on a synthetic dataset generated from representative simulated mine environments. This approach effectively overcomes the challenges inherent to object detection in complex mining scenarios, accounting for variations in sensor height and other critical factors.

One of our significant contributions is the development of a comprehensive synthetic dataset explicitly designed for 3D object detection in mining environments. This dataset includes objects and sensors placed at different heights within pit benches, accurately capturing the complexities of real mining operations. Furthermore, we introduce novel evaluation metrics that consider sensor-to-object height and point cloud density variations for more accurate model performance assessment in mining scenarios.

Moreover, through extensive experiments, we successfully predict the presence of objects in real data captured from actual mining environments and evaluate the accuracy of our approach. The results demonstrate the effectiveness and practicality of our 3D object detection model in real-world mining scenarios.

Additionally, we have incorporated an automatic annotation pipeline leveraging the ROS interface. This pipeline includes new algorithmic solution to automate the annotation process, reducing manual labor and accelerating the dataset creation for 3D object detection in complex mining environments.

To assess the impact of altitude and height variation, we conduct an extensive ablation study. This study showcases the significance of an altitude shift augmentations in improving the overall accuracy and reliability of 3D object detection models specifically tailored for complex mining environments.

Our key contributions are as follows:

1) The development of a comprehensive synthetic dataset capturing the complexities of mining environments, including objects and sensors placed at different heights within pit benches.
2) Introduction of novel evaluation metrics that consider sensor-to-object height and point cloud density variations for more accurate model performance assessment in mining scenarios.
3) Successful prediction of objects in real data captured from actual mining environments, demonstrating the effectiveness of our approach.
4) The implementation of an automatic annotation pipeline using the ROS interface, significantly reducing manual labor and expediting the dataset creation process.
5) An extensive ablation study showcasing the importance of altitude and height variation augmentations in enhancing the accuracy and reliability of 3D object detection models for mining environments.

To support our research and foster collaboration, we have made our comprehensive synthetic dataset publicly accessible. Researchers can utilize this dataset as a benchmark for supervised learning, enabling the evaluation and advancement of object detection techniques in mining environments. Additionally, we provide a video summarizing our experimental trials, and the dataset is available at [1], ensuring accessibility and encouraging further exploration in this field.

By effectively bridging the gap between synthetic and real data, our work demonstrates the potential of synthetic data and simulation-based methodologies in overcoming domain shift challenges. We envision the development of robust and reliable object detection systems that can be practically deployed in mining and related domains, enhancing safety and operational efficiency.

2   Related Study

Automatic annotation and data generation have been the focus of extensive research in recent years [2–5], with several state-of-the-art methods available for generating labeled data. One of the most popular techniques for data generation is through simulation environments. Simulations are widely used to collect data for machine learning applications, especially for perception systems.

Gazebo [6] and CARLA [7] are two popular simulation environments used for autonomous vehicle research. These environments offer realistic virtual environments with a large number of sensors, including cameras and LiDARs, for data collection. Publicly available datasets from simulation environments have become a crucial source for training machine learning algorithms. Several datasets have been made publicly available for research purposes, including the CARLA [7], LGSVL [8], and Udacity datasets [9], which include labeled data for autonomous vehicle perception systems. These datasets cover a wide range of scenarios, including urban and highway driving, and offer various sensor modalities, such as LiDAR, camera, and radar. In addition, these datasets offer accurate ground-truth labels for different perception tasks such as object detection, semantic segmentation, and lane detection.

In [10,11], synthetic multimodal 3D raw data and automated semantic labeled data have been generated from Gazebo simulations of a ground vehicle operating in diverse natural environments and off-road terrains. The aim is to expedite software development and improve the generalization of models to new scenes.

However, the proposed approach for automatic 3D annotation aims to tackle the challenges of generating labeled data to increase automation in earth moving operations at mine sites. By leveraging the advantages of simulation environments, a large volume of labeled data can be generated. This approach has the potential to significantly enhance the performance of machine learning models for various applications in dynamic and degraded environments, such as mining.

3   New Dataset: SimMining3D

3.1   Data Collection at Simulated Environment

The data collection process took place within a representative simulation environment based on the Yandicoogina mine site, as described in [12]. This simulated environment accurately replicates the real-world conditions and characteristics of the mine site, that contained within a rectangular area, spanning 583 m in longitude and 379 m in latitude. Additionally, the environment has an elevation of 63.5 m, ensuring a comprehensive representation of the mine site's terrain and topography. The CAD models of earthmoving equipment and sensors were imported into Gazebo (Fig. 2). The point clouds in this dataset were acquired with a simulated MST system [12]. The system consists of a simulated 128-line OS2 LiDAR sensor and a RGB camera. LiDAR sensor can capture point clouds at up to 700,000 points per second at a vertical field of view of 22.5° with an accuracy of ±2.5–8 cm.

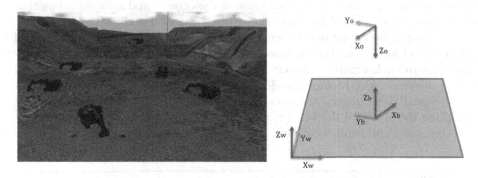

Fig. 2. Left: Representative Yandicoogina mine simulation environment in Gazebo, Right: Coordinate system used: x, y, z denote the coordinate system, with subscripts o, b, and w referring to Ouster sensor, mobile study trailer base footprint, and world coordinate, respectively.

Six excavators were placed at random locations in the simulation environment, and their movements and rotations were automatically controlled and repeated for each sensor position. The data collection of an object in the sensor coordinate frame is demonstrated in Fig. 2.

The data collection process involved capturing information from distinct sensor heights in different scenarios. In one scenario, both the sensor and excavators were placed within the bench. In the second scenario, the sensor was positioned inside the pit, observing excavators within the pits and benches taller than 10 m. In the third scenario, the sensor was placed on the benches, observing excavators within the pit. Excavators were continuously rotated at 0.2 rad per second and moved from their initial location by approximately 75 m with a speed of 0.5 m/s. In total, 818, 617, and 690 frames were captured for each scenario, respectively, resulting in a total of 2125 frames. This approach allowed us to gather diverse data across different sensor heights and excavator locations within the complex mining environment.

3.2 Automatic Annotation

Once the data is collected from the simulation environment, the next step in the automatic annotation pipeline is to determine the position, orientation, and dimension of the objects in the simulation environment. As detailed in the Algorithm 1 this information was obtained by using ROS tools and packages such as "tf" or "odometry" that provide the pose information of objects in the environment. Unlike other autonomous vehicle or robots with sensors, the proposed data collection platform in this study will remain stable during data collection at multiple locations. Hence the base-footprint coordinate frame of the MST will not change constantly. Then ROS package was used to publish the object pose information via object recognition messages. Using the published object pose information, the next step is to generate 3D bounding boxes around the objects. This was done by inferring the dimension of the objects and creating a box that encompasses the entire object from the ros messages as described in Algorithm 1. The generated bounding box information were used to crop the point clouds of interested objects from the environment and stored in separates folders corresponding to each different objects as annotations in a format suitable for 3d object detection. Furthermore, as described in the original custom object format in the OpenPCDet library, the bounding box information in LiDAR coordinates, including the center of the box (x, y, and z), the dimensions of the box (dx, dy, and dz), yaw value, and the object class name, was saved for all objects in each

---

Algorithm 1. Automatic Annotation

1: **Input:** Objects and sensor pose information, LiDAR scans, objects' dimensions (dx, dy, dz), Number of object classes=3
2: **Output:** kitty format.txt, Semantic Labels .csv, gt_database
3: **for all** f_i in frames **do**
4:     bbox = []
5:     pcd ← read_points(f_i, field_names=['x', 'y', 'z', 'rgb'])
6:     **for all** o_i in objects **do**
7:         label ← name(o_i)
8:         center ← center(o_i, field_names=['x', 'y', 'z'])
9:         size ← Size(o_i, field_names=['dx', 'dy', 'dz'])
10:         rotation ← RotationMatrix(o_i, field_names=['w', 'qx', 'qy', 'qz'])
11:         center[z] ← center[z] + size[dz]/2
12:         roll_X, pitch_Y, yaw_Z ← quaternion_to_euler_angle(o_i, field_names= ['w', 'qx', 'qy', 'qz'])
13:         Write to text file ('x', 'y', 'z', 'dx', 'dy', 'dz', yaw_Z, label)
14:         Bbox ← OrientedBoundingBox(center, rot, size)
15:         Crop_3d ← crop(pcd, Bbox)
16:         Save_point_cloud(gt_database/Crop_3d_f_i_o_i, Crop_3d)
17:         **for all** c_i in [c_1, c_2, c_3] **do**
18:             **if** $o_i = c_i$ **then**
19:                 color ← [r_i, g_i, b_i]
20:                 write to csv (frame_f_i.csv, Crop_3d (x, y, z, color, c_i))
21:             **end if**
22:         **end for**
23:     **end for**
24: **end for**

---

scene. In addition, we introduced a new column in the above text files to indicate the objects' difficulties, labeled as 0 for easy, 1 for moderate, and 2 for hard difficulties. The corresponding changes were made in the OpenPCDet files to read and utilize this additional information from the text files. Similarly, the semantic values correspond to each 3D LiDAR points (coordinates) were saved in the csv files (Fig. 3).

The annotation pipeline generated a large dataset of annotated objects in the simulation environment. To avoid close similarities between frames, we strategically selected total of 933 frames from continuous recordings, ensuring diversity in the dataset with corresponding labels provided with each difficulty level. The dataset includes ground truth information for 5353 excavators observed from multiple perspectives, ensuring comprehensive coverage and robust analysis.

In addition to the 3D bounding box annotations used for object detection evaluation in this paper, complementary semantic point-wise labels were provided for 933 frames.

This dataset was used to train state of the art machine learning model for 3D object detection and semantic segmentation tasks. The dataset is available at [1].

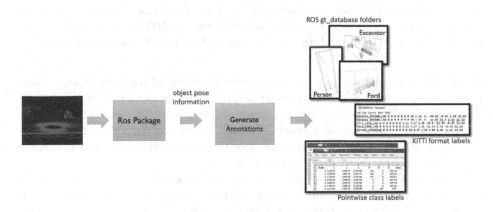

Fig. 3. Automatic annotation pipeline

4 Perception: Baseline Experiment

4.1 Experimental Setup

Baseline results are provided using OpenPCDet framework [13], Poinpillar 3D object detection model trained on only the source domain of simulation environment, ideally representing the worst-case and best-case respectively for performance on the target domain-real data. PointPillars is an end-to-end model that uses a set of pillar-based representations for 3D object detection in point clouds. It processes each pillar individually and applies a set of convolutions and max-pooling operations to generate features for object detection [14].

This study focuses on the exacavator static model. Unlike in other 3d object detection models studied in other literature the excavators are dimensionally huge and the small change in the 3d box orientation can result noticeably different in the prediction. Given the large discrepancy in the size of the ground truth anchor boxes and the points from the objects, we filter out the ground truths which consists of less than 100 points from the objects.

As our datasets contain only one object class that needs to be detected, we adapt the output layers of the algorithms to predict a single class only. The anchor size is set to the ground truth dimensions of the Hitachi excavator, with l = 8.65 m, w = 23.9 m, and h = 10.02 m for length l, width w, and height h, respectively. We empirically test different network parameters, such as the voxel size for PointPillars, the number of filters in network. For all other parameters, we choose the voxel size of 2.19, 2.19, 14 and the maximum point per voxels of 32. For data augmentation, we use random horizontal flip, scale augmentations. Although the LiDAR sensors capture reflections at over 200 m distance, we limit the detection range of our network to a horizontal range of 175.2 m in the dimensions x and y and −12 and 4 m in z direction in order to capture the large variation of mine pit walls and the tall objects. We remove the intensity channel from network and only use x, y, and z as input features. The network was trained on 1235 object samples chosen from all senarios with batch size of 2 and learning rate for 120 epochs and validated on 509 object samples using modified-Kitty format validation as discussed at section Evaluation metric.

Data Augmentation. In this study, we propose the 'altitude shift' augmentation technique tailored for point cloud data. Unlike the previous Random World Translation (RWT) method, our approach utilizes a uniform distribution to generate random values within a specified range. By simulating altitude variations, the altitude shift augmentation enhances the adaptability and generalization capabilities of 3D object detection models.

Unlike RWT technique, which introduces noise based on standard deviations for each axis, our Altitude Shift augmentation guarantees an equal probability for all values within the specified range. This approach allows the models to adapt to various object heights commonly found in real-world scenarios. Additionally, the Altitude Shift is implemented on-the-fly during each run, providing the flexibility to either apply a fixed shift (Constant_Altitude_Shift) or randomly sample values (Random_Altitude_Shift) within the range.

Implemented along the z-axis, the altitude shift function modifies the vertical position of points and their corresponding ground truth bounding boxes. It takes three input parameters: ground truth boxes (gt_box), points, and an offset range determining the permissible shift values.

Algorithm 2. Altitude Shift

1: **Input:** LiDAR point cloud $L \in \mathbb{R}^{N \times 3}$ with N points, ground truth box gt_box, Offset range [min, max]
2: **Output:**Shifted LiDAR point cloud L
3: **for** $l \in L$ **do**
4:     offset $\leftarrow$ RandomUniform(offset_range[0], offset_range[1])
5:     $l.z \leftarrow l.z +$ offset
6:     $gt\_box.z \leftarrow gt\_box.z +$ offset
7: **end for**

We further evaluate other augmentation techniques based on OpenPCDet [13] to compare the impact of altitude variation on the overall accuracy in complex object detection environments. These environments are characterized by varying heights, presenting unique challenges for accurate object detection. By conducting a comprehensive analysis, we aim to understand how different augmentation strategies, including altitude variation, specifically address the complexities associated with height variation.

Evaluation Metric. We assess excavator detection difficulty using specific criteria based on height variation between the sensor and object, as well as point cloud density. The difficulty levels are as follows:

Level 0 - Easy: Height variation < 10 m, point cloud density > 750 points.

Level 1 - Moderate: Height variation < 10 units, point cloud density 100 to 750 points, or height variation > 10 m, point cloud density > 750 points.

Level 2 - Hard: Height variation > 10 m, point cloud density 100 to 750 points.

By categorizing excavator instances into these levels, we evaluate detection algorithm performance in complex mining environments. Our modifications in OpenPCDet enable accurate evaluation aligned with mining-specific challenges.

4.2   Results and Discussion

Evaluations on Simulated Data. The evaluation was conducted on a synthetic dataset captured in a complex mining environment using a simulation environment. The 3D object detection performance of Pointpillar for the Hitachi excavator is presented in Table 1, reporting Mean Average Precision (mAP) results for 3D and Bird's Eye View (BEV) detection at IoU 0.7. The evaluation includes 40 recall positions, providing comprehensive assessment across difficulty levels: Easy, Moderate, and Hard proposed in this paper. The results demonstrate the impact of different augmentation techniques compared to no augmentation on object detection performance in complex environments. The Random_Altitude_Shift (RAS) augmentation is highly effective for improving object detection accuracy. It consistently outperforms the baseline across difficulty levels and evaluation metrics, providing significant improvements in both

BEV and 3D detection. Comparing RAS with other techniques like RWT, CAS, and standard augmentation, RAS performs better, especially in the Hard level. It captures altitude variations in the dataset, resulting in more accurate object detection. CAS and standard augmentation show improvements but are not as consistent as RAS. RWT, focusing on z-axis offset, lags behind RAS in accuracy. Considering altitude-specific variations is crucial for improved detection, as shown by RAS outperforming RWT. The controlled altitude shifts introduced by RAS allow better adaptation to real-world height variations. RAS has practical implications for robust object detection in complex environments, enhancing accuracy and reliability in critical applications like autonomous driving and robotics.

Table 1. Object Detection Performance Evaluation with Augmentation (BEV and 3D at 0.7 IoU)

Method	BEV at 0.7 IoU			3D at 0.7 IoU		
	Easy	Mod.	Hard	Easy	Mod.	Hard
None	94.96	90.44	61.05	91.57	87.38	58.23
RAS ($[-2, 2]$)	99.00	94.16	64.35	94.31	90.10	60.86
RAS ($[-0.5, 0.5]$)	95.42	91.00	61.40	94.45	87.93	60.91
CAS (0.5)	93.23	86.92	60.25	56.38	35.19	37.73
RWT_z_only (0.5)	95.47	91.00	61.47	88.64	84.65	57.93
Standard	98.11	91.20	61.59	97.85	91.00	61.43
Standard + RAS	98.47	91.33	61.65	94.52	90.12	60.82
Standard + CAS	97.94	90.97	61.45	85.66	79.82	53.88
Standard + RWT_z_only	97.79	90.93	61.39	93.88	87.37	60.56

Validation on Real Data. Predictions on both simulated and real data were presented in Fig. 4. Figure 4(b) and (c) showcase the video and point cloud captured at the real minesite, respectively. Figure 4(c) demonstrates the successful transfer of the trained model on synthetic data to a real-world scenario. Considering the disparity in lidar positions between simulation and reality in terms of height, the input point cloud was transformed to accommodate the sensor's height variation during model prediction. Videos depicting the predictions from simulation to simulation and simulation to real for various scenarios can be found at [1].

Fig. 4. Model predictions on both simulated data (a), (c) and real data (d)

5   Conclusion

In conclusion, our study emphasizes the significant contributions of synthetic data generation, automatic annotation, altitude shift augmentation, and sim-to-real transformation in enhancing object detection models for mining environments.

Furthermore, excavators' complex shapes during operations should be accommodated in future studies. Addressing the impact of random values on augmentations through systematic work will provide valuable insights.

The integration of these approaches improves the accuracy and reliability of object detection models for mining applications. These advancements have the potential to enhance safety, efficiency, and productivity in mining operations, addressing challenges related to domain gap and limited real data availability.

Acknowledgements. This work has been supported by the Australian Centre for Robotics and the Rio Tinto Centre for Mine Automation, the University of Sydney.

References

1. SimMining-3D. https://github.com/MehalaBala/SimMining_3D
2. Nikolenko, S.I.: Synthetic simulated environments. In: Nikolenko, S.I. (ed.) Synthetic Data for Deep Learning. SOIA, vol. 174, pp. 195–215. Springer, Cham (2021). https://doi.org/10.1007/978-3-030-75178-4_7
3. Yue, X., Wu, B., Seshia, S.A., Keutzer, K., Sangiovanni-Vincentelli, A.L.: A LiDAR point cloud generator: from a virtual world to autonomous driving. In: Proceedings of the ACM International Conference on Multimedia Retrieval, Yokohama, Japan, 11–14 June 2018, pp. 458–464 (2018)

4. Smith, A., et al.: A deep learning framework for semantic segmentation of underwater environments. In: OCEANS 2022, Hampton Roads, pp. 1–7 (2022)
5. Saputra, R.P., Rakicevic, N., Kormushev, P.: Sim-to-real learning for casualty detection from ground projected point cloud data. In: 2019 IEEE/RSJ International Conference on Intelligent Robots and Systems (IROS), Macau, China, pp. 3918–3925 (2019). https://doi.org/10.1109/IROS40897.2019.8967642
6. Koenig, K., Howard, A.: Design and use paradigms for Gazebo an open-source multi-robot simulator. In: Proceedings of the IEEE-RSJ International Conference on Intelligent Robots and Systems, pp. 2149–2154 (2004)
7. Dworak, D., Ciepiela, F., Derbisz, J., Izzat, I., Komorkiewicz, M., Wójcik, M.: Performance of LiDAR object detection deep learning architectures based on artificially generated point cloud data from CARLA simulator. In: 2019 24th International Conference on Methods and Models in Automation and Robotics (MMAR), Miedzyzdroje, Poland, pp. 600–605 (2019). https://doi.org/10.1109/MMAR.2019.8864642
8. Rong, G., et al.: LGSVL simulator: a high fidelity simulator for autonomous driving. In: 2020 IEEE 23rd International Conference on Intelligent Transportation Systems (ITSC), Rhodes, Greece, pp. 1–6 (2020). https://doi.org/10.1109/ITSC45102.2020.9294422
9. Udacity Dataset (2018). https://github.com/udacity/self-driving-car/tree/master/datasets
10. Sánchez, M., Morales, J., Martínez, J.L., Fernández-Lozano, J.J., García-Cerezo, A.: Automatically annotated dataset of a ground mobile robot in natural environments via gazebo simulations. Sensors 22, 5599 (2022). https://doi.org/10.3390/s22155599
11. Tallavajhula, A., Meriçli, Ç., Kelly, A.: Off-road lidar simulation with data-driven terrain primitives. In: 2018 IEEE International Conference on Robotics and Automation (ICRA), Brisbane, QLD, Australia, pp. 7470–7477 (2018). https://doi.org/10.1109/ICRA.2018.8461198
12. Balamurali, M., et al.: A framework to address the challenges of surface mining through appropriate sensing and perception. In: 17th International Conference on Control, Automation, Robotics and Vision (ICARCV), pp. 261–267 (2022). https://doi.org/10.1109/ICARCV57592.2022.10004309
13. OpenPCDet Development Team. OpenPCDet: An opensource toolbox for 3D object detection from point clouds (2020). https://github.com/open-mmlab/OpenPCDet
14. Lang, A.H., Vora, S., Caesar, H., Zhou, L., Yang, J., Beijbom, O.: PointPillars: fast encoders for object detection from point clouds. CoRR, abs/1812.05784 (2018)

Oyster Mushroom Growth Stage Identification: An Exploration of Computer Vision Technologies

Lipin Guo[1](✉)[iD], Wei Emma Zhang[1][iD], Weitong Chen[1][iD], Ni Yang[2][iD],
Queen Nguyen[3][iD], and Trung Duc Vo[3]

[1] The University of Adelaide, Adelaide, SA 5005, Australia
{lipin.guo,wei.e.zhang,weitong.chen}@adelaide.edu.au
[2] Division of Food, Nutrition and Dietetics, University of Nottingham,
Sutton Bonington Campus, Loughborough LE12 5RD, UK
ni.yang@nottingham.ac.uk
[3] Clever Mushroom, Gunning, NSW 2581, Australia
{queen.nguyen,trung.vo}@cleveragriculture.com.au

Abstract. Mushrooms play a pivotal role in bolstering Australia's economy, impacting key sectors like agriculture, food production, and medicinal advancements. To meet the escalating need for sustainable food options and enhance mushroom harvesting efficiency, this research: i) introduces an innovative dataset featuring three growth stages of oyster mushrooms; ii) designs a monitoring system which consists of image acquisition, cloud storage, label map and applications to achieve effective monitoring; and iii) proposes a label map method to monitor different stages within panoramic images captured from the real mushroom cultivation environment. Our preliminary studies show that the label map with state-of-art VGG-16 model emerges as the optimal choice, achieving an impressive accuracy of 82.22%. Our dataset can be obtained upon request.

Keywords: Oyster Mushroom · Growth · Computer Vision

1 Introduction

Edible mushrooms are highly valued for their nutritional benefits, which encompass reducing the risk of cancer, enhancing antioxidant levels, bolstering immunity, improving neurocognition, and augmenting Vitamin D intake [1]. Mushrooms hold a significant position as a vital food source in Australia, contributing substantially to the economy. As of June 2022, the country had produced an impressive 66,236 tons of mushrooms, with a substantial production value of $434.2 million [5]. Small mushroom companies aspire to monitor the growth of mushrooms to improve growth conditions, enable timely and efficient harvesting, reduce waste, and optimize labor allocation. However, the expense associated with labeling is not affordable. Driven by industrial demand and a scarcity of relevant research in Australia, this paper undertakes an exploration of computer

T. Liu et al. (Eds.): AI 2023, LNAI 14471, pp. 67–78, 2024.
https://doi.org/10.1007/978-981-99-8388-9_6

(a)   (b)

Fig. 1. (a) Mushroom cultivation in a shipping container at a mushroom company; (b) Three growth stages of oyster mushroom: Stage one is the early stage (top); Stage two is the intermediate stage (middle); Stage three is the mature stage (bottom).

vision technologies for monitoring the growth of mushrooms. Due to oyster mushroom's low environmental control prerequisites and limited vulnerability to fruiting body-affecting pests and diseases [6], it can be cultivated in a straightforward and cost-effective manner. Accordingly, this study concentrates on monitoring the growth of oyster mushrooms by using image classification algorithms.

In the realm of mushroom classification, prevailing studies predominantly focus on binary classification to differentiate between poisonous and edible mushrooms [8], or encompass multi-class classification for mushroom species categorization [12]. Notably, within the oyster mushroom domain, research endeavours have concentrated on aspects like freshness assessment [11,17] and automated harvesting [13–15]. Exploring the growth stages of oyster mushrooms remains a scarcely explored avenue. Even when venturing into the broader realm of plant growth, analogous endeavours are infrequent. Hence, the primary objective of this paper is to delve into potential solutions for effectively monitoring the growth of oyster mushrooms under the real-world setting.

Our work is a preliminary study to explore the automatic solution for a small company which grows exotic mushrooms in containers. There are two main challenges for this work: (1) The challenge lies in obtaining data. Due to the lack of appropriate existing data for monitoring of mushroom growth, the acquisition of data from a local small-scale mushroom company necessitates careful considerations: the optimal choice of an image collection device, the intricacies of image labeling, and the allocation of resources within a defined budget. (2) Navigating the challenge of image feature identification of mushrooms in different stages proves distinctive. Prior studies have predominantly concentrated on individual mushrooms positioned at the centre of the images, whereas this research extends the scope to capture the holistic representation of the growth stages in a complex setting. Our panoramic images (shown in Fig. 1(a)) encapsulate diverse mushroom stages dispersed across varying positions within each image.

To address the challenges, we design a oyster mushroom monitoring system, consisting of image acquisition, cloud storage, label map, and applications. We collect a small dataset of encompassing various growth stages of oyster mushrooms and collaborate with the staff at the mushroom company to label the dataset. The dataset contains oyster mushroom images in three growth stages as shown in Fig. 1(b). To accomplish the objective of monitoring oyster mushroom growth, we introduce label map to show the stage information from the patches of a panoramic image. The essence of the label map lies in equitably partitioning a panoramic image into N patches, subjecting these patches to an optimal feature extraction methodology and classifier, resulting in distinct labels. Later, these labels are subsequently organized into a vector, which is reshaped into a label map. Finally, our experimental findings highlight VGG-16 as the optimal architecture for feature extractor and classifier within our label map method by comparing the machine learning and deep learning image classification algorithms, achieving an accuracy rate of 82.22%.

Our main contributions are:

- We design a solution for classifying multiple oyster mushroom growth stages within a panoramic image in a real-world complex setting.
- We perform preliminary studies on recognizing oyster mushroom growth stages by exploring both traditional machine learning and deep learning models.
- We address the data gap by curating and meticulously labeling a dataset encompassing various growth stages of oyster mushrooms.

Next, the subsequent sections encompass an exploration of related work (Sect. 2), a comprehensive elucidation of the research design (Sect. 3), and an insightful interpretation of empirical studies (Sect. 4).

2   Related Works

Data. In recent years, research endeavors in the oyster mushroom domain have spanned a diverse array of subjects, encompassing valorization and waste management [20], automated harvesting [13–15], freshness evaluation [11,17], grading assessment [21], growth enhancement [7], as well as IoT-based monitoring systems [19]. This research focuses on monitoring the growth stage of oyster mushrooms using image classification algorithms. However, the work for the oyster mushroom stage image classification is rare. The most relevant work is Surige et al. [19] in which, the authors proposed to classify five different stages of the oyster mushroom life cycle, consisting of stage one (ten hours to harvest), stage two (five hours to harvest), stage three (harvest now), stage four (one day past - suitable for consumption) and stage five (2 days past - not suitable for consumption). Our work condensed five stages proposed in Surige et al. [19] into three distinct stages with revised descriptions, specifically highlighting the key phases that contribute to its successful cultivation, shown in Fig. 1(b). In the growth stages of oyster mushrooms, stage one is characterised by the readiness of mushroom grow kits for pinning or their presence in the pinning stage, where small pin-like structures emerge on the substrate as an early sign of mushroom development. Stage two is characterised by the emergence of small pin-like structures that reach a cap scanning of 3–4 cm, while stage three represents the maturation phase of the mushroom with a cap size ranging between 5–7 cm. Moreover, this paper aims to categorize the panoramic view of the entire oyster mushroom growing environment into three stages, which presents a more complex and challenging task compared to the classification of individual mushroom images.

Algorithm. In the past decade, computer vision has predominantly embraced deep learning algorithms, especially convolutional neural networks (CNNs). Regardless of architectural variations, CNNs fundamentally include convolutional layers (with or without ReLU activation and pooling) for feature extraction and fully connected layers for classification. The famous architectures consist of Visual Geometry Group (VGG) [18], MobileNet [16] and residual network (ResNet) [4]. Some researchers designed system to monitor or measure the growth of mushroom and used CNNs to recognize and localize mushroom. Lu et al. proposed a mushroom growth measurement system for common mushrooms in greenhouse encompassing image capture, mushroom recognition (using CNNs), position correction, size measurement, growth rate estimation, quantity assessment, harvest time calculation, data recording, and harvest notifications throughout the mushroom fruiting phase [10]. Surige et al. developed an IoT-based monitoring system for oyster mushroom featuring four functions: Environmental Monitoring utilizing long short term memory (LSTM), Harvest Time Detection using CNNs with the MobileNet V2 model, Disease Detection and Control Recommendation based on CNNs with MobileNet V2, and Yield Prediction employing LSTM [19]. This paper also proposes a oyster mushroom monitoring system with different components, including image acquisition, cloud storage, label map, and applications (shown in Fig. 2(a)). Zarifie et al. used pretrained VGG-16 to extract features and classify different grades based on quality of grey oyster mushroom [21]. However, in oyster mushroom domain, algorithms not only use deep learning methodologies, but also use machine learning methodologies (shown in Table 1).

Table 1. Oyster Mushroom Works

Ref.	Classification Tasks	Dataset Size	Model	Performance (Acc.)
[11]	Freshness	120	ANN, SVM	ANN (95%), SVM (98%)
[17]	Freshness	240	ANN	94.4%
[21]	Grading	600	VGG-16	90%
[19]	Growing Stages	1,887	MobileNetV2	92%

Some researchers realized colour, texture and morphology are important mushroom features, and extracted the important features manually by colour maps, then used ANN or the combination of ANN and SVM to classify the freshness of oyster mushrooms [11,17]. Additionally, Vision Transformer (ViT)

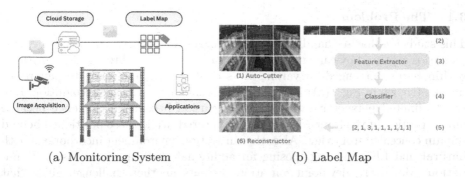

(a) Monitoring System   (b) Label Map

Fig. 2. (a) Oyster mushroom monitoring system; (b) Label map procedure: (1) Auto-Cutter: automatically divide a panoramic image into $P \times P$ patches; (2) Arrange patches in order from left to right, top to bottom and labelled by mushroom company staff; (3) Feature Extractor: extract features from the patches respectively; (4) Classifier: classify features into one of three distinct stages and generate corresponding labels, where the labels are represented as 0, 1, or 2; (5) Create a concatenated 1D vector of labels and increment the label values by 1, aligning them with the corresponding stage indices (1, 2, 3); (6) Reconstructor: Reshape the 1D vector into $P \times P$ label map.

[3], a deep learning based algorithm which utilises self-attention mechanisms to extract feature inspired by Transformer models used in natural language processing, have gained popularity in the last three years. There is no paper explore ViT for oyster mushroom image classification. This paper delves into two distinct types of image classification algorithms. On one hand, given the small dataset and the significance of morphology as a feature, machine learning-based image classification algorithms exhibit promising potential. On the other hand, recognizing the subtle differences within and between classes, pretrained deep learning-based image classification algorithms may demonstrate exceptional discriminatory power in distinguishing various stages of oyster mushrooms.

3   The Monitoring System

To facilitate oyster mushroom growth monitoring, we introduce a comprehensive system illustrated in Fig. 2(a), comprising four key components: image acquisition, cloud storage, label map, and applications. Initially, panoramic oyster mushroom images are captured via a camera and transmitted to cloud storage via WIFI. Subsequently, these cloud-stored images are processed through a supervised image classification method, referred to as the "label map", which selects an optimal model. Finally, this model is leveraged for applications. Given the unique challenges posed by panoramic images, as discussed in Sect. 3.1, we employ label map for panoramic monitoring, with detailed insights provided in Sects. 3.2.

3.1 The Problem

This study emphasises handling complex images that closely resemble real-world scenarios captured by cameras. In our scenario, oyster mushrooms are cultivated within bottles arranged on various tiers of shelves. These intricate panoramic images (shown in Fig. 1(a)) introduce some challenges in image recognition. Due to the limited perspective of the lens, occlusion becomes a challenge because only a portion of the front-row bottles is captured, while those positioned behind remain concealed from view. Also, the cultivation environment incorporates both natural and LED lighting, posing an additional challenge in terms of illumination. Moreover, viewpoint variation presents another challenge, given that the three shelves are arranged in a left, middle, and right configuration. The toughest challenge is oyster mushrooms can span across three distinct growth stages, further complicating the task of accurately identifying each stage within a panoramic image. Due to the large amounts of bottles and the closely spaced arrangement of them, identifying a bottle of mushroom become difficult. Thus we propose label map as a solution to automatically identify the mushroom stages. We first split the panoramic images into patches, and then based on the split images, we classify them into different stages and reconstruct the classification results to automatically monitor the panoramic images.

3.2 The Label Map

Instead of obtaining the stage information of individual mushroom bottles, using label map method can achieve global modelling by integrating of the stage information from patches. As depicted by Fig. 2(b), a panoramic RGB image ($H \times W \times 3$) can be divided into $P \times P$ patches ($\frac{H}{P} \times \frac{W}{P} \times 3$) by auto-cutter. These patches are arranged from left to right, and then top to bottom. Because the patches include positional information of the original images, the sequential order of the patches is important. Within these $P \times P$ patches, each patch exclusively corresponds to a single growth stage rather than encompassing all three stages. Then, these patches are passed through feature extractor and classifier sequentially. Each patch will generate a corresponding label based on the probability outcome. After concatenating these labels, the output ($1 \times P^2$) will be a label representative of the original image. To match the original stage indices, add 1 to the output. The addition will not increase the model complexity. Reconstructor reshapes the output into a $P \times P$ grid which yields a label map that effectively delineates the growth stages present within the panoramic image. A related issue with this method is that within a single patch, it's possible to encounter a combination of two stages, typically a mixture of stage one and stage two. To address this challenge, during the ground-truth labeling process, the mushroom company staff assigns a label to only one stage based on either the majority of mushroom stages within the patch or by considering the misclassification cost and assigning it to the stage with the lowest cost.

To measure the distance between the true growth stage probabilities y and predicted growth stage probabilities $\hat{y}$, this paper uses multi-class cross entropy loss function shown in the Eq. (1) where M is the number of panoramic samples, C is the number of classes and $\hat{y}_i$ is the predicted probability of a specify class j to the sample x_i.

$$CE(\boldsymbol{y}, \hat{\boldsymbol{y}}) = - \sum_{i=1}^{M} \boldsymbol{y}_i \log(\hat{\boldsymbol{y}}_i) = - \sum_{i=1}^{M} \sum_{j=1}^{C} y_{ij} \log(p_\theta(y_{ij}|\boldsymbol{x}_i)) \tag{1}$$

To minimize the loss, our primary objective shifts towards identifying an optimal feature extractor and classifier within the label map method. This optimal model needs to be robust against challenges such as occlusion, varying illumination, and changes in viewpoint. In this paper, we explore existing machine learning and deep learning-based image classification algorithms to find the optimal feature extractor and classifier for our label map method.

Machine Learning Based Image Classification Algorithm. We first explore the classic machine learning solution, in which we use scale invariant feature transform (SIFT) technique [9] to extract features and then we apply support vector machine (SVM) [2] to classify the patches. For M panoramic RGB images, a total of $M \times P \times P$ RGB patches can be obtained. These patches constitute a dataset, which is subsequently divided into training and testing data. To obtain SIFT features, we transform the RGB patches from three dimensions to grayscale images with two dimensions, as The SIFT technique [9] handles grayscale images. For each patch, SIFT [9] employs various levels of the Gaussian pyramid, in which multi-scale patches apply Gaussian smoothing and downsampling, to detect the key points. Subsequently, SIFT [9] computes gradients within a 16×16 window centred on an identified keypoint, generating an orientation histogram in vector form to construct a keypoint descriptor, thus creating SIFT features. Next, we apply k-means clustering to establish the visual vocabulary (Bag of Features) using the training SIFT features. Then, we associate each SIFT descriptor of a patch with the closest visual word in the BoF vocabulary and create a visual word histogram for the patch. Later, we combine these BoF histograms into a unified feature matrix for both training and testing purposes. Finally, we use the SVM classifier [2] to find the hyperplane that maximally separates three classes while aiming to minimize classification errors.

Deep Learning Based Image Classification Algorithm. In terms of deep learning based image classification algorithms, this paper used pretrained VGG-16 [18], ResNet18 [4], ResNet34 [4], ResNet50 [4], MobileNetV2 [16] and ViT-B-16 [3]. All the models are pretrained on ImageNet-1k dataset, and then fine-tuned on Oyster Mushroom dataset. $P \times P$ patches from a panoramic image can be set as a batch of the whole $M \times P \times P$ dataset to the model, which eliminates the need for any further image processing and does not increase the computation complexity. VGG [18] employs sequential 3×3 convolution/ReLU blocks and 2×2 max pooling, progressively increasing the channel count from 64 to 512, and culminating in three fully connected layers, yielding networks with 16–19 layers. ResNet [4] works by first passing a patch image through a convolutional layer that detects the basic features like edges and corners, then feed forward through several residual blocks. These blocks consist of multiple convolutional layers with shortcut connection, which performs identity mapping and add the result to the output of stacked layers. Finally, ResNet [4] ends with a global average pooling and fully connected layer to classify the input. In contrast to VGG [18], ResNet [4] has a similar structure but is significantly deeper, ranging from 18 to 152 layers, due to its utilization of direct connections across convolutional layers, addressing the accuracy degradation problem when layers increase. MobileNetV2 [16], begins with an initial convolutional layer for extracting low-level features from a patch image, followed by seven bottleneck residual blocks with varying strides. Each block involves a combination of pointwise and depthwise convolutional layers to capture spatial information while managing computational efficiency. The architecture concludes with a 1×1 convolutional layer, a global average pooling layer, and a fully connected linear classifier with dropout to classify the input. ViT-B-16 [3] transforms an oyster mushroom patch image by converting it into a sequence of 2D 16×16 patches, then flatten the 2D patches and process through a linear projection layer. Position embeddings and an additional class token are incorporated, then forward to multiple Transformer Encoder which has a self-attention mechanism for global context information. Lastly, the extra class token is fed through an MLP Head (two-layer classification network) to predict the stage.

4   Empirical Studies

4.1   Settings

For image acquisition in the monitoring system, images for this research were collected from two different shipping containers at a small mushroom company, where oyster mushrooms were cultivated under controlled environmental conditions (temperature: 18–22 °C, humidity: 70–90%, CO_2 levels: 800–1500 ppm). The shipping containers were illuminated with RGB LED strip lights. The panoramic images were captured using a Tapo C310 IP camera connected via WIFI. Due to the unstable WIFI signal, the camera often went offline, resulting in the inability to capture images. Also, variations in natural and LED lighting conditions can lead to image blurring. Moreover, considering the approximately 14-day life cycle of oyster mushrooms, detecting growth changes occurring within 1-hour intervals proves challenging for human observations. Due to previous concerns, our current dataset is approximate to the patches divided from panoramic images. Images from this dataset were captured using an iPhone 11 in the high-efficiency HEIC format, featuring a resolution of $4,032 \times 3,024$ pixels and utilizing the RGB colour space. The images were captured within different distances between the lens and the samples. The dataset size is a balanced dataset with 150 images so far (we are continuing the image capturing). We used 70% for training and 30% for testing.

4.2   Performances

After training and testing the model, we used Accuracy, Macro Precision, Macro Recall, Macro F1 and Macro Area Under the Receiver Operating Characteristic Curve (AUC-ROC) for evaluation.

Table 2. Model Performances. Acc. denotes accuracy, Pre. means precision and Rec. represents recall.

Model	Pretrained	Acc.(%)	Pre.(%)	Rec.(%)	F1(%)	AUC-ROC(%)
SIFT-SVM	No	48.89	51.43	48.89	48.52	–
VGG-16	Yes	**82.22**	82.24	82.22	**82.00**	**92.07**
ResNet18	Yes	66.67	67.86	66.67	66.05	86.89
ResNet34	Yes	80.00	82.78	80.00	80.41	87.85
ResNet50	Yes	**82.22**	82.73	82.22	81.79	91.33
MobileNetV2	Yes	68.89	69.34	68.89	68.43	84.00
ViT-B-16	Yes	62.22	66.67	62.22	59.98	79.56

As Table 2 shows, the deep learning-based feature extractors and classifiers surpass the traditional methodology SIFT-SVM. Even the least performing deep learning-based model exhibits superior results compared to the SIFT-SVM model. One factor is that images have RGB channels, which have three dimensions. However, SIFT-SVM converts colourful images into grey-scale images, resulting in the loss of valuable information by dimension reduction. Another factor lies in transfer learning because the pretrained models have learnt rich features from other huge datasets, they already have useful information, and fine-tuning helped inject domain knowledge.

Among the various deep learning models, VGG-16, ResNet18, ResNet34, ResNet50 and MobileNetV2, the variants of CNNs demonstrate higher accuracy compared to transformer-based ViT-B-16. Essentially, the inherent inductive biases of CNNs, such as translation equivariance and locality, outperform the self-attention mechanism of ViT-B-16 on this small dataset. When features are extracted at an earlier layer, the translation equivariance principle makes sure the neural network's response remains consistent for the same image patch, regardless of its position. And the locality principle makes sure the network focuses on local regions, without paying attention to the distant regions. As channel numbers and layer depth increase, the features capturing local information are aggregated to make predictions. Hence, CNNs have the ability to capture fine-grained image details, which are crucial due to the subtle differences in both intra-class and inter-class variations. ViT-B-16 acquires these inherent biases by training on a large dataset.

Among the CNN variants, VGG-16 stands out as the top performer with an accuracy of 82.22%, slightly surpassing ResNet50 in terms of macro F1 and

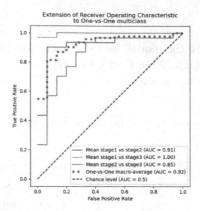

Extension of Receiver Operating Characteristic
to One-vs-One multiclass

Mean stage1 vs stage2 (AUC = 0.91)
Mean stage1 vs stage3 (AUC = 1.00)
Mean stage2 vs stage3 (AUC = 0.85)
One-vs-One macro-average (AUC = 0.92)
Chance level (AUC = 0.5)

Fig. 3. ROC curve at One-vs-One Multiclass Classification

macro ROC scores, 82% and 92.07% respectively. In our case, diverse impacts of misclassifications incur varying costs. For example, misclassifying stage two as stage three bears fewer adverse consequences due to their shared importance, warranting harvest. Conversely, misclassifying stage three as stage two carries greater repercussions, given the diminished quality, loss of nutrients, and reduced selling potential, underscoring the significance of accurate classification. Furthermore, misidentifying stage one as stage two not only squanders labour time but also poses challenges in labour scheduling and rearrangement. Due to the diverse costs associated with different types of misclassifications and the discriminative ability between minute intra-class and inter-class variances, both macro F1 and marco AUC-ROC are crucial metrics. The macro-F1 score represents the harmonic mean of precision and recall across all stages of mushroom growth. It takes into account both false positives and false negatives for each individual stage, providing a balanced assessment of the model's performance across all classes. Thus, a higher macro-F1 score is important as it indicates a better balance between accurately identifying different stages of mushroom growth while minimizing the overall misclassification rate. Additionally, the Receiver Operating Characteristic (ROC) curve (depicted in Fig. 3) is a graphical representation that illustrates the trade-off between the true positive rate (sensitivity) and the false positive rate (1-specificity) as the classification threshold for each stage is varied. The ROC curve helps assess the model's ability to discriminate between different stages of mushroom growth by showing how well it can distinguish between positive and negative samples for each class. The area under the ROC curve (AUC-ROC) is often used as a quantitative measure of the model's overall performance, where a higher AUC-ROC value indicates better discriminatory power and accuracy in distinguishing between different growth stages of oyster mushrooms. As Fig. 3 shows, the VGG-16 classifier effectively discriminates between stage one and three with an AUC score of 1.00, while distinguishing between stage two and three proves to be the most challenging with an AUC score of 0.85. This observation is logical, given that the delicate pin-like structures of the mushroom during stage one are distinctly different from the matured mushroom in stage three. In contrast, stage two is close to the maturation process of stage three. Therefore, we select VGG-16 as our optimal feature extractor and classifier for our dataset.

5 Conclusion and Future Works

In conclusion, this study introduces the Oyster Mushroom dataset, encompassing three distinct growth stages of the oyster mushroom, and addresses the challenge of classifying panoramic images through the innovative use of a monitoring system. Our experimental results indicate that the label map method within the monitoring system exhibits remarkable performance, achieving an accuracy of 82.22%. In future endeavours, we aim to delve into alternative approaches for predicting harvest timing, including treating growth stage images as time series data and leveraging techniques such as regression or recurrent neural networks.

Acknowledgments. This research work is supported by Australian Research Council Early Career Industry Fellowship (IE230100119).

References

1. Australian Mushroom Growers Association: Mushroom research. https:// australianmushroomgrowers.com.au/learn-about-australian-mushrooms/ mushroom-research/. Accessed 22 Sept 2023
2. Cortes, C., Vapnik, V.: Support-vector networks. Mach. Learn. **20**, 273–297 (1995). https://doi.org/10.1007/BF00994018
3. Dosovitskiy, A., et al.: An image is worth 16 × 16 words: transformers for image recognition at scale. In: 9th International Conference on Learning Representations (2021). https://doi.org/10.48550/arXiv.2010.11929
4. He, K., Zhang, X., Ren, S., Sun, J.: Deep residual learning for image recognition. In: 2016 IEEE Conference on Computer Vision and Pattern Recognition (CVPR), pp. 770–778 (2016). https://doi.org/10.1109/CVPR.2016.90
5. Hort Innovation: Australian horticulture statistics handbook 2021/22. https:// www.horticulture.com.au/growers/help-your-business-grow/research-reports-publications-fact-sheets-and-more/australian-horticulture-statistics-handbook/. Accessed 22 Sept 2023
6. Kües, U., Liu, Y.: Fruiting body production in basidiomycetes. Appl. Microbiol. Biotechnol. **54**(2), 141–152 (2000). https://doi.org/10.1007/s002530000396
7. Kumari, S., Naraian, R.: Enhanced growth and yield of oyster mushroom by growth-promoting bacteria Glutamicibacter arilaitensis MRC119. J. Basic Microbiol. **61**(1), 45–54 (2021). https://doi.org/10.1002/jobm.202000379
8. Li, H., et al.: Reviewing the world's edible mushroom species: a new evidence-based classification system. Compr. Rev. Food Sci. Food Saf. **20**(2), 1982–2014 (2021)
9. Lowe, D.G.: Distinctive image features from scale-invariant keypoints. Int. J. Comput. Vis. **60**, 91–110 (2004). https://doi.org/10.1023/B:VISI.0000029664.99615.94
10. Lu, C.P., Liaw, J.J., Wu, T.C., Hung, T.F.: Development of a mushroom growth measurement system applying deep learning for image recognition. Agronomy **9**(1) (2019). https://doi.org/10.3390/agronomy9010032

11. Mukherjee, A., et al.: Development of artificial vision system for quality assessment of oyster mushrooms. Food Anal. Methods **15**(6), 1663–1676 (2022)
12. Picek, L., et al.: Danish fungi 2020 - not just another image recognition dataset. In: Proceedings of the IEEE/CVF Winter Conference on Applications of Computer Vision (WACV), pp. 1525–1535 (2022)
13. Qian, Y., Jiacheng, R., Pengbo, W., Zhan, Y., Changxing, G.: Real-time detection and localization using SSD method for oyster mushroom picking robot. In: Proceedings of the 2020 IEEE International Conference on Real-time Computing and Robotics (RCAR 2020), pp. 158–163 (2020). https://doi.org/10.1109/RCAR49640.2020.9303258
14. Rahmawati, D., Ibadillah, A., Ulum, M., Setiawan, H.: Design of automatic harvest system monitoring for oyster mushroom using image processing. Atlantis Highlights Eng. **1**, 143–147 (2018). https://doi.org/10.2991/icst-18.2018.31
15. Rong, J., Wang, P., Yang, Q., Huang, F.: A field-tested harvesting robot for oyster mushroom in greenhouse. Agronomy **11**(6) (2021). https://doi.org/10.3390/agronomy11061210
16. Sandler, M., Howard, A., Zhu, M., Zhmoginov, A., Chen, L.C.: MobileNetv 2: inverted residuals and linear bottlenecks. In: 2018 IEEE/CVF Conference on Computer Vision and Pattern Recognition, pp. 4510–4520 (2018)
17. Sarkar, T., et al.: Comparative analysis of statistical and supervised learning models for freshness assessment of oyster mushrooms. Food Anal. Methods **15**(4), 917–939 (2022)
18. Simonyan, K., Zisserman, A.: Very deep convolutional networks for large-scale image recognition. In: 3rd International Conference on Learning Representations (2015). https://doi.org/10.48550/arXiv.1409.1556
19. Surige, Y.D., Perera, W.S., Gunarathna, P.K., Ariyarathna, K.P., Gamage, N., Nawinna, D.: IoT-based monitoring system for oyster mushroom farming. In: Proceedings of the 3rd International Conference on Advancements in Computing (ICAC 2021), pp. 79–84 (2021). https://doi.org/10.1109/ICAC54203.2021.9671112
20. Wan Mahari, W.A., et al.: A review on valorization of oyster mushroom and waste generated in the mushroom cultivation industry. J. Hazard. Mater. **400**, 1–15 (2020)
21. Zarifie Hashim, N.M., et al.: Grey oyster mushroom classification toward a smart mushroom grading system for agricultural factory. In: Proceedings of the 2nd International Conference on Intelligent Technologies (CONIT), pp. 1–6 (2022). https://doi.org/10.1109/CONIT55038.2022.9847864

Handling Heavy Occlusion in Dense Crowd Tracking by Focusing on the Heads

Yu Zhang, Huaming Chen, Zhongzheng Lai, Zao Zhang, and Dong Yuan[✉]

The University of Sydney, Camperdown, Australia
{yu.zhang1,huaming.chen,zhongzheng.lai,zao.zhang,
dong.yuan}@sydney.edu.au

Abstract. With the rapid development of deep learning, object detection and tracking play a vital role in today's society. Being able to identify and track all the pedestrians in the dense crowd scene with computer vision approaches is a typical challenge in this field, also known as the Multiple Object Tracking (MOT) challenge. Modern trackers are required to operate on more and more complicated scenes. According to the MOT20 challenge result, the pedestrian is 4 times denser than the MOT17 challenge. Hence, improving the ability to detect and track in extremely crowded scenes is the aim of this work. In light of the occlusion issue with the human body, the heads are usually easier to identify. In this work, we have designed a joint head and body detector in an anchor-free style to boost the detection recall and precision performance of pedestrians in both small and medium sizes. Innovatively, our model does not require information on the statistical head-body ratio for common pedestrians detection for training. Instead, the proposed model learns the ratio dynamically. To verify the effectiveness of the proposed model, we evaluate the model with extensive experiments on different datasets, including MOT20, Crowdhuman, and HT21 datasets. As a result, our proposed method significantly improves both the recall and precision rate on small and medium sized pedestrians, and achieves state-of-the-art results in these challenging datasets.

Keywords: Crowd · Detection · Tracking

1 Introduction

Tracking by detection is one of the most critical framework for the Multiple Object Tracking (MOT) challenge. Since the modern MOT challenge is now prone to happen in more and more crowded scenes, the tracker performance is limited by the design of the object detectors. With a latest detector, it is possible to achieve a near state-of-the-art performance in terms of tracking accuracy with a traditional tracking algorithm. In the meantime, many approaches with novel architectures have been proposed for the pedestrian detection task. Earlier works like Faster-RCNN [14], Mask-RCNN [7], SSD [11] utilize anchor-based model

© The Author(s), under exclusive license to Springer Nature Singapore Pte Ltd. 2024
T. Liu et al. (Eds.): AI 2023, LNAI 14471, pp. 79–90, 2024.
https://doi.org/10.1007/978-981-99-8388-9_7

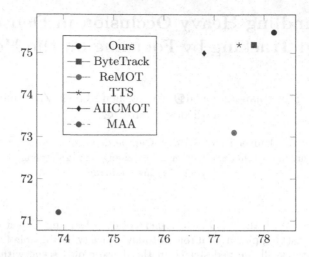

Fig. 1. MOTA-IDF1 comparisons of different trackers on the test set of MOT20. With the assist of our supplementary body detector, the proposed tracking framework achieves 78.2 MOTA and 75.5 IDF1 score, outperforming all previous trackers.

framework while recently researches about anchor-free detection have started to attract more and more attention [21] (Fig. 1).

The key challenge of the pedestrian detection is occlusion in the scene. In crowded scenes, people overlap with each other most of the times. The challenge of occlusion makes it hard for the detection model to extract useful features to identify the person effectively. It also causes problem with the Re-ID functions with some models because the visual features required to identify the person constantly changes due to occlusion. Multiple previous researches have stated that exploiting multiple part of a pedestrian helps to improve the performance of pedestrian detection [2,3,26].

According to previous research [3], human body detection are often missing due to the large overlaps in extremely crowded scenes. The model either failed to identify enough features for the occluded person or eliminate them during the NonMaximum Suppression (NMS) post-process procedure. Compared to body detection, the occlusion between heads are less likely to occur. With this consideration, the feature map can become more distinguishable and consistent for training. By comparing the head detection results and body detection results on the same MOT challenge sequence, we find out that the head detector generally identifies more people than the body detector (except for those people not showing full bodies in the scene). Recently, [17] introduced a new head tracking dataset to further assist tracking humans efficiently in densely crowded environments. Hence, by combining head tracking with body tracking detectors, we should be able to improve the performance of the MOT challenge.

However, adopting head tracking to body tracking is not a trivial task. First of all, many pedestrians do not show their heads in the camera sight. Head detector alone cannot fully replace the body detector in the context of MOT. Also, in a traditional detector, human heads are detected separately. Manually generating body bounding boxes from head detection introduce lots of false positive as people have different poses. Hence, detecting both head and body simultaneously then linking them together is challenging. Bi-Box [26] proposed a model to estimate both visible part and the full body at the same time. JointDet [3] and HBAN [13] are two anchor-based solution which rely on a fixed head-body ratio to generate the head proposal from the regular body proposal. The problem with these methods is that, they still rely on anchors or external method to generate head and body proposals, leading to low recall rate and inaccurate body Region of Interest (RoI). Although most standing human follows a static head-body ratio, it still varies a lot due to different human poses and camera positions. Using a fixed head-body proposal to predict pedestrians in complex scenes leads to a less satisfying result.

Motivated by the findings, we propose JointTrack as a novel solution. The contributions of this work are three folds: 1). We adopt the joint head and body detector in an anchor-free style to further boost its detection performance in extremely crowded scenes, and a supplementary body detector is leveraged to overcome its limitation in the MOT challenge; 2). Instead of relying on a fixed head-body ratio, a new module to include both the head and body prediction based on SimOTA is proposed, which learns the relationship dynamically during training; 3). the proposed model achieves state-of-the-art head detection performance on CrowdHuman dataset, and the joint head and body tracking framework achieves state-of-the-art performance in the MOT20, HT21 challenges.

2   Related Work

Head detection is difficult due to the small size of head compared to pedestrian body, hence it is firstly used for crowd counting [20,24] by estimating the crowd density. The development of Feature Pyramid Network (FPN) [10] enables a more accurate detection of objects in different scales. Since head detection scenes are generally more challenging than body detection scenes in terms of crowd density, many researchers have discovered that combining head detection with body detection usually help to improve the detector performance.

Lu et al. [13] propose a head-body alignment net to jointly detect human head and body. They use two parallel RPN branches to propose the head body RoIs and use an additional Alignment Loss to enforces body boxes to locate compactly around the head region. PedHunter [2] train the model to predict a head mask while predicting the pedestrian bodies. It serves as an attention mechanism to assist the feature learning in CNN. It helps to reduce the false positive by learning more distinguishable pedestrian features. However, the method does not eliminate the false negatives as the RPN in the model is trained just like traditional body detector. Chi et al. propose JointDet [3] to use pedestrian heads to assist in the body detection which shares a similar idea to this work. However, they use a static body-head ratio to generate body proposals from head proposals

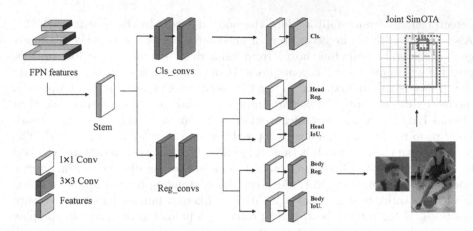

Fig. 2. Overview of the proposed model. The image is first processed by a FPN structure to extract features of different scales. Then, a 1 × 1 stem conv layer followed by two 3 × 3 branches extract the classification and regression features of the image. For the regression branch, we add two additional prediction heads to predict the head and body simultaneously. The predictions are evaluated by a Joint SimOTA module to filter out the grids that have either bad head prediction or bad body prediction. We calculate losses for the rest of the grids.

to reduce the computational workload in tiling the anchors for body proposals. In our work, we abandon the previous anchor-based design and propose a much simpler and more efficient solution for the joint head-body detection problem.

3   Methodology

3.1   Framework Overview

The overall framework is shown in Fig. 2. For the object detection part, we train a head detector adopting the YOLOX [6] structure. We add an additional regression branch to the decoupled head and the classification head remains the same. The detector aims to classify and regress human heads in the scene. It also generates a body prediction along with each head detection. An overview graph is included in appendix. To combine the main detector with the supplementary body detector in the process of multiple object tracking, we first do a bipartite matching to pair our head detection with the external body detection, then the unmatched head detection can contribute to the tracking process.

A heat map comparison is shown in Fig. 3. For the input image Fig. 3a, we demonstrate two heat maps: one from the baseline body detector and the other is from our joint head-body detector. The two detectors are built with identical backbone network and are trained with the same training dataset & equal epochs for fair comparison. From the figure, we can see that the head detection model can extract more distinguishable features from the image than the body detection

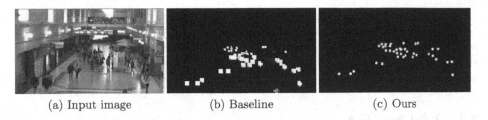

(a) Input image   (b) Baseline   (c) Ours

Fig. 3. Heatmap comparison between head detection and body detection model under the same model structure. (a) is the input image from MOT20 dataset. (b) is the heatmap from the baseline. (c) is the heatmap from our model. Comparing to the baseline, our heatmap is more distinguishable and can detect more pedestrians in the scene.

model. The head heat map also contains features that are missed by the baseline body detector (e.g. the man sitting on the right side and the man standing in the middle). Hence, it should perform better on extremely crowded scenes with a proper architecture design.

The advantages of the proposed approach are: 1). To address the head and body proposal relationship problem existed in previous work. Since a person can have various posture, generating the body proposal according to the common standing posture ratio is not always reliable. A bad proposal often leads to a bad prediction. Hence, we solve this issue by training an anchor-free detector and predict the head and body from the same feature maps simultaneously. 2). In crowded scenes, human heads are usually easier to detect. In some heavily occluded scenes, the head of a person can be easily observed while the rest of the body is hidden. However, only detecting heads is not sufficient for MOT problems since sometimes heads are not visible. Thus, we use an additional body detector to serve as a complementary model.

3.2   Anchor-Free Head-Body Detection

In the previous design of JointDet [3], pedestrian head and body are predicted from two separate RoIs following a specific head-body ratio. This causes two problems: 1). The ratio is statistically obtained based on all human head-body pairs in the CrowdHuman dataset, which are dominantly from a horizontal camera angle, resulting in a standing posture head-body ratio. The body proposal generated from this ratio is not suitable for the top-down camera angle sequences in MOT. 2). The extra hyperparameters introduced by the anchor boxes heavily influences the performance, impacting the overall recall and precision score.

Therefore, we adopt the design of the recent YOLOX [6] and propose an anchor-free style joint head-body detector. For each location, the regression value (top left x and y, width and height) of the head and body are directly predicted by two parallel CNN branches. An illustration is shown in Fig. 4.

With the anchor-free design, our model can directly predict the pedestrian body box without using a predefined body anchor box while greatly reducing the requested efforts to heuristically tune the parameters for training. Supported by the training advantage, we also eliminate the performance impact by the body anchors.

3.3 Joint SimOTA

We use both the head prediction and body prediction to perform label assignment during training. Specifically, the pair-wise matching degree between a ground truth g_i and prediction p_j is calculated as:

$$C_{ij} = L_{ij}^{cls} + \lambda_1 L_{ij}^{hreg} + \lambda_2 L_{ij}^{breg} \tag{1}$$

where λ_1 and λ_2 are the balancing coefficient. L_{ij}^{cls} is the classification loss between the ground truth g_i and prediction p_j. L_{ij}^{hreg} and L_{ij}^{breg} are the regression losses for head and body prediction respectively. For a ground truth g_i, we select the top k predictions with the least cost within a fixed center region of the head ground truth box as its positive samples. The value k is determined dynamically according to the IoUs of the predicted boxes [5]. The grids that contains these positive samples are considered as positive grids and all other grids are consider negative. Only positive grids are saved up for loss computing to reduce the computing resource cost.

The Joint SimOTA makes sure that during the grid sample assignment process, only those with both a good head prediction and a good body prediction are selected as positive samples. It serves as a function similar to the combination of the head-body relationship discriminating module (RDM) and the statistical head-body ratio proposal generation method introduced in [3]. Instead of generating the body proposals from heads based on a statistical ratio, all head-body prediction pairs are generated based on the features extracted by the stem network and dynamically evaluated by the joint simOTA process. This helps to simplify the prediction process by reducing the extra hyper-parameters introduced by RDM module and boost the average detection precision.

3.4 Tracking Framework

We use a simple yet efficient two-step tracking framework inspired by ByteTrack [22]. First, we need to combine the detection results from both detectors to eliminate the duplicated detection and split the detection set into two parts: first class detection and second class detection. We do a bipartite matching between the body prediction from the traditional body detector and our joint head-body detector.

We use Hungarian Algorithm for the bipartite matching step. The matched head and body are all classified as first class detection. For the unmatched head

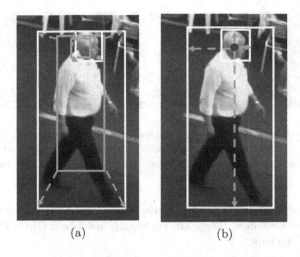

(a)   (b)

Fig. 4. Illustration of our anchor-free approach vs. traditional static head-body ratio anchor-based approach. a). Anchor-based approach, where head and body boxes are regressed from separate RoIs. b). Our anchor-free approach, where head and body boxes are predicted simultaneously from the same grid box, achieving a faster inference speed and less parameters for training.

and unmatched body, we further classify them according to their confidence score. Once we have the two sets of detection, we can perform tracking. The detailed tracking algorithm is available in appendix.

3.5 Loss Function

The loss function of the purposed method consists of four parts and is defined as follows:

$$\mathcal{L} = \mathcal{L}_{cls} + \mathcal{L}_{obj} + \alpha_1 \mathcal{L}_{head} + \alpha_2 \mathcal{L}_{body} \tag{2}$$

where $\mathcal{L}_{cls}$ and $\mathcal{L}_{obj}$ is the binary cross entropy loss for head classification and confidence score prediction. $\mathcal{L}_{head}$ and $\mathcal{L}_{body}$ are the regression loss for the predicted head and body bounding boxes. α_1 and α_2 are the balancing coefficients and we set them to 5 during our training and experiments. We use an additional L1 loss for the head and body detections for the last 10 epochs of training.

3.6 Training Details

Dataset Construction and Data Augmentation. Since there is few dataset containing head and body labels at the same time, we use existing datasets to build our own training data. We combine the ground truth for MOT20 and HT21 since they have two identical training sequences. We also use the Crowdhuman dataset during training since they explicitly label the pedestrian heads and

Fig. 5. Qualitative results on MOT20 dataset. Pedestrians detected by the baseline are shown in grey boxes. Pedestrians detected by our method which are not detected by the baseline are shown in red boxes. Best viewed in colour. (Color figure online)

bodies. Following the work from [6], we use Mosaic and MixUp strategies to boost the detecting performance. They are commonly used in YOLOv4 [1], YOLOv5 [8], and other detectors.

Training Parameters Setting. We use a YOLOX-x model pretrained on COCO dataset to initialize the training. The model was trained on two RTX3090 graphic cards with a batch size of 8 and epoch number of 80. Following the design of ByteTrack [22], the optimizer is SGD with a momentum ratio of 0.9. The learning rate is set to 0.0001 with 1 epoch warmup and use the cosine annealing strategy introduced in [12]. The input of the training images are set to 896×1600. The entire training takes about 42 h. For post processing, we use a NMS of threshold of 0.45 for the head prediction and 0.7 for the body prediction to eliminate overlapped detection.

4   Experiments

Since our approach focus on extremely crowded scenes, the experiments are mainly conducted on MOT20 dataset. Experiments are also conducted on Crowdhuman to demonstrate the robustness of the proposed approach. In the supplementary material, we also provide results on the HT21 dataset to demonstrate our model's performance on head tracking benchmarks.

4.1   MOT Challenge

MOTChallenge is a human tracking benchmark that provides carefully annotated datasets and clear metrics to evaluate the performance of tracking algorithms and pedestrian detectors. Earlier benchmark MOT17 provides over $29k$ annotations in a total of 14 sequences. MOT20 is a more recent benchmark that consist of 8 different sequences depicting very crowded challenging scenes.

Table 1. AP and MR^{-2} performance of different detection methods on MOT20

Method	AP↑	MR^{-2} ↓
Baseline	85.67	14.0
Ours w/o Joint SimOTA	91.16	9.0
Ours	92.99	7.0

MOT20. For the MOT20 challenge, we conduct two parts of experiments. The first one is the detection performance and the second one is the tracking performance. Similar to the "Pedestrian Detection in Complex and Crowded Events" in HiEve, we evaluate the detection performance on MOT20 dataset. Since the ground truth label for the MOT20 test data is not available, we conduct the experiment on the training set and show qualitative for the test set. The training dataset for our experiment is the Crowdhuman dataset and the mixed dataset we created for our model. We choose two sequences from the HT21 dataset and the MOT20 dataset and combine their ground truth label. We add offset to the label to match the image size differences.

For the detection performance metric, we choose the log-average miss rate over 9 points ranging from 10^{-2} to 10^{0} FPPI (the MR^{-2}) to evaluate the performance of our detector. The performance comparison is shown in Table 1. We use current top-performance detector used in ByteTrack [22] as our baseline. As shown in the table, our method achieves 92.99% AP which is 7.32% higher than the baseline. For the log average missing rate, we also achieves 7% MR^{-2} which is 7% better than the base line. The result has shown that our method reaches the supreme pedestrian detection performance by letting the model focus on the heads instead of the bodies.

For the tracking performance, the results are uploaded to the official website for evaluation. The main evaluation metrics are Multiple Object Tracking Accuracy (MOTA), IDF1 and Higher Order Tracking Accuracy (HOTA). The result is compared with FairMOT [23], TransCenter [18], TransTrack [16], CSTrack [9], SOTMOT [25], MAA [15], ReMOT [19] and ByteTrack [22]. Despite many detections are ignored in the process of evaluation [4], we still achieve state-of-the-art performance of 78.2% MOTA and 75.5% IDF1.

4.2   Qualitative Result on MOT20

We demonstrate the effectiveness of our model in this section. The results are shown in Fig. 5. We have randomly selected 3 tracking sequences from the MOT20 test set and perform detection with both the baseline approach and our proposed approach. It is clear that from Fig. 5, our approach has a stronger ability to detect pedestrians in crowded scenes. In Fig. 5, the red boxes are the pedestrians only detected by our method not from the baseline. It usually happens to people that are heavily overlapped with others or only showing part

Table 2. Performance comparison on the test set with state-of-the-art on MOT20

	MOTA↑	IDF1↑	HOTA↑	FP↓	FN↓	IDs↓
FairMOT	61.8	67.3	54.6	103440	88901	5243
TransCenter	61.9	50.4	43.5	45895	146347	4653
TransTrack	65.0	59.4	48.5	27197	150197	3608
CSTrack	66.6	68.6	54.0	25404	144358	3196
SOTMOT	68.6	71.4	57.4	57064	101154	4209
MAA	73.9	71.2	57.3	**24942**	108744	1331
ReMOT	77.4	73.1	61.2	28351	86659	2121
ByteTrack	77.8	75.2	61.3	26249	87594	**1223**
Ours	**78.2**	**75.5**	**61.9**	30187	**81119**	1325

Table 3. head and body AP performance of different detection methods on Crowdhuman

Method	head	body
Baseline-Head	54.2	–
Baseline-Body	-	57.8
Ours	55.0	57.9

of their body due to different postures. Since our model focuses on detecting the heads, it can extract the corresponding head features from the image and accurately predict the body boxes (Table 2).

4.3   Ablation Study on Joint SimOTA

Our Joint SimOTA module helps the model to learn head-body relationship dynamically which can improve the detection performance. We conduct ablation study by training our proposed model with the same settings except the Joint SimOTA module. According to Table 1, adding the Joint SimOTA module can boost the AP by 1.83% and reduce the log average missing rate by 2%. The performance improvement indicates that body regression contributing to the training process helps the model to predict more accurate body bounding boxes. Otherwise, the model learns features for the head and body separately leading to more false positives.

4.4   Crowdhuman

Crowdhuman is a public pedestrian detection dataset that contains various crowded scenes. It provides over $470k$ pedestrian labels in a total of 15000 training set and 4370 validation set. For each pedestrian, the label contains the head, visible body part and full body annotation. To demonstrate the robustness of our method, we conduct experiments on the validation set with both the head and body detection. We retrain the YOLOX model for the two tasks separately as our baseline. According to Table 3, our approach improves the head and body AP by 0.8% and 0.1% respectively. The result shows that combining the head and body detection helps to boost the performance of both tasks.

5   Conclusion

In this paper, we propose an anchor-free style joint head-body detection model to detect pedestrians' head and body simultaneously. By focusing on the heads, the model can detect pedestrians more effectively than the baseline body detectors due to high occlusion. With our proposed model, we can significantly reduce the detected false negative in extremely crowded scenes. We have conducted extensive experiments on MOT20, HT21 and Crowdhuman datasets. Our approach achieves state-of-the-art detection and tracking performance and is robust to various scenarios.

References

1. Bochkovskiy, A., Wang, C.Y., Liao, H.Y.M.: YOLOv4: optimal speed and accuracy of object detection. arXiv preprint arXiv:2004.10934 (2020)
2. Chi, C., Zhang, S., Xing, J., Lei, Z., Li, S.Z., Zou, X.: PedHunter: occlusion robust pedestrian detector in crowded scenes. In: Proceedings of the AAAI Conference on Artificial Intelligence, vol. 34, pp. 10639–10646 (2020)
3. Chi, C., Zhang, S., Xing, J., Lei, Z., Li, S.Z., Zou, X.: Relational learning for joint head and human detection. In: Proceedings of the AAAI Conference on Artificial Intelligence, vol. 34, pp. 10647–10654 (2020)
4. Dendorfer, P., et al.: MOT20: a benchmark for multi object tracking in crowded scenes. arXiv preprint arXiv:2003.09003 (2020)
5. Ge, Z., Liu, S., Li, Z., Yoshie, O., Sun, J.: OTA: optimal transport assignment for object detection. In: Proceedings of the IEEE/CVF Conference on Computer Vision and Pattern Recognition, pp. 303–312 (2021)
6. Ge, Z., Liu, S., Wang, F., Li, Z., Sun, J.: YOLOX: exceeding YOLO series in 2021. arXiv preprint arXiv:2107.08430 (2021)
7. He, K., Gkioxari, G., Dollár, P., Girshick, R.: Mask R-CNN. In: Proceedings of the IEEE International Conference on Computer Vision, pp. 2961–2969 (2017)
8. Jocher, G.: ultralytics/YOLOv5: v3.1 - Bug Fixes and Performance Improvements (2020). https://doi.org/10.5281/zenodo.4154370. https://github.com/ultralytics/yolov5
9. Liang, C., Zhang, Z., Zhou, X., Li, B., Zhu, S., Hu, W.: Rethinking the competition between detection and ReID in multiobject tracking. IEEE Trans. Image Process. **31**, 3182–3196 (2022)
10. Lin, T.Y., Dollár, P., Girshick, R., He, K., Hariharan, B., Belongie, S.: Feature pyramid networks for object detection. In: Proceedings of the IEEE Conference on Computer Vision and Pattern Recognition, pp. 2117–2125 (2017)
11. Liu, W., et al.: SSD: single shot multibox detector. In: Leibe, B., Matas, J., Sebe, N., Welling, M. (eds.) ECCV 2016. LNCS, vol. 9905, pp. 21–37. Springer, Cham (2016). https://doi.org/10.1007/978-3-319-46448-0_2

12. Loshchilov, I., Hutter, F.: SGDR: stochastic gradient descent with warm restarts. arXiv preprint arXiv:1608.03983 (2016)
13. Lu, R., Ma, H., Wang, Y.: Semantic head enhanced pedestrian detection in a crowd. Neurocomputing **400**, 343–351 (2020)
14. Ren, S., He, K., Girshick, R., Sun, J.: Faster R-CNN: towards real-time object detection with region proposal networks. In: Advances in Neural Information Processing Systems, vol. 28 (2015)
15. Stadler, D., Beyerer, J.: Modelling ambiguous assignments for multi-person tracking in crowds. In: Proceedings of the IEEE/CVF Winter Conference on Applications of Computer Vision, pp. 133–142 (2022)
16. Sun, P., et al.: TransTrack: multiple object tracking with transformer. arXiv preprint arXiv:2012.15460 (2020)
17. Sundararaman, R., De Almeida Braga, C., Marchand, E., Pettre, J.: Tracking pedestrian heads in dense crowd. In: Proceedings of the IEEE/CVF Conference on Computer Vision and Pattern Recognition, pp. 3865–3875 (2021)
18. Xu, Y., Ban, Y., Delorme, G., Gan, C., Rus, D., Alameda-Pineda, X.: TransCenter: transformers with dense queries for multiple-object tracking. arXiv preprint arXiv:2103.15145 (2021)
19. Yang, F., Chang, X., Sakti, S., Wu, Y., Nakamura, S.: ReMOT: a model-agnostic refinement for multiple object tracking. Image Vis. Comput. **106**, 104091 (2021)
20. Zhang, C., Li, H., Wang, X., Yang, X.: Cross-scene crowd counting via deep convolutional neural networks. In: Proceedings of the IEEE Conference on Computer Vision and Pattern Recognition, pp. 833–841 (2015)
21. Zhang, S., Chi, C., Yao, Y., Lei, Z., Li, S.Z.: Bridging the gap between anchor-based and anchor-free detection via adaptive training sample selection. In: Proceedings of the IEEE/CVF Conference on Computer Vision and Pattern Recognition, pp. 9759–9768 (2020)
22. Zhang, Y., et al.: ByteTrack: multi-object tracking by associating every detection box. arXiv preprint arXiv:2110.06864 (2021)
23. Zhang, Y., Wang, C., Wang, X., Zeng, W., Liu, W.: FairMOT: on the fairness of detection and re-identification in multiple object tracking. Int. J. Comput. Vis. **129**(11), 3069–3087 (2021)
24. Zhang, Y., Zhou, D., Chen, S., Gao, S., Ma, Y.: Single-image crowd counting via multi-column convolutional neural network. In: Proceedings of the IEEE Conference on Computer Vision and Pattern Recognition, pp. 589–597 (2016)
25. Zheng, L., Tang, M., Chen, Y., Zhu, G., Wang, J., Lu, H.: Improving multiple object tracking with single object tracking. In: Proceedings of the IEEE/CVF Conference on Computer Vision and Pattern Recognition, pp. 2453–2462 (2021)
26. Zhou, C., Yuan, J.: Bi-box regression for pedestrian detection and occlusion estimation. In: Proceedings of the European Conference on Computer Vision (ECCV), pp. 135–151 (2018)

SAR2EO: A High-Resolution Image Translation Framework with Denoising Enhancement

Shenshen Du, Jun Yu$^{(\boxtimes)}$, Guochen Xie, Renjie Lu, Pengwei Li, Zhongpeng Cai, and Keda Lu

University of Science and Technology of China, Hefei 230026, China
{dushens,gcxie,renjielu,lipw,zpcai,lukeda}@mail.ustc.edu.cn,
harryjun@ustc.edu.cn

Abstract. Synthetic Aperture Radar (SAR) to electro-optical (EO) image translation is a fundamental task in remote sensing that can enrich the dataset by fusing information from different sources. Recently, many methods have been proposed to tackle this task, but they are still difficult to complete the conversion from low-resolution images to high-resolution images. Thus, we propose a framework, SAR2EO, aiming at addressing this challenge. Firstly, to generate high-quality EO images, we adopt the coarse-to-fine generator, multi-scale discriminators, and improved adversarial loss in the pix2pixHD model to increase the synthesis quality. Secondly, we introduce a denoising module to remove the noise in SAR images, which helps to suppress the noise while preserving the structural information of the images. To validate the effectiveness of the proposed framework, we conduct experiments on the dataset of the Multi-modal Aerial View Imagery Challenge (MAVIC), which consists of large-scale SAR and EO image pairs. The experimental results demonstrate the superiority of our proposed framework, and we win the first place in the MAVIC held in CVPR PBVS 2023.

Keywords: Image Translation · High Resolution · Denoising

1 Introduction

Image translation aims at transferring images between the source domain and the target domain by mixing the content and style in an end-to-end manner. Typically, given an image, the image translation task is required to preserve its content while introducing various styles from images in different domains, which makes the task extremely useful in diverse applications, such as face attribute editing and scene style transferring. In the long run, the image translation task has attracted much research attention and many methods are proposed to tackle this problem. Early methods typically adopt style matrix as a mediator in style transferring and show good performance. But there is still much room for improvement.

Recent years, generative adversarial network (GAN) [5] based models emerge and gradually dominate the image translation tasks. Many effective methods are

© The Author(s), under exclusive license to Springer Nature Singapore Pte Ltd. 2024
T. Liu et al. (Eds.): AI 2023, LNAI 14471, pp. 91–102, 2024.
https://doi.org/10.1007/978-981-99-8388-9_8

proposed in this field and gain much popularity, include pix2pix [8], cycleGAN [22] and pix2pixHD [18]. These models adopt the encoder-decoder structure, where the encoder encodes the input image into a low-dimensional feature vector and the decoder decodes this feature vector into the target image with desired style. During training, the models use paired data and attempt to learn how to convert the style of input image into the target image style via minimizing the difference between the generated image and the real target image. As the GAN based image translation model is capable of generating photorealistic images in image style transferring, it has been widely used in many daily and entertainment scenes.

However, different from the daily tasks, the remote sensing image translation task, especially the Synthetic Aperture Radar(SAR) and electro-optical (EO) translation task, is still under exploration. Mechanically, the EO images are collected using electro-optical (EO) sensors by capturing images in the visible spectrum (such as RGB and grayscale images). And the SAR images are reproduced through radar signals and act as a complementation of the EO images in the severe conditions such as heavy fog or lack of visible light. The goal of the SAR2EO translation task is to produce high-quality and high-fidelity EO images with SAR images. However, due to the large gap between the SAR and EO images and the heavy noise of images in remote sensing senario, the translation results are often suboptimal.

In this work, we propose a simple but effective framework based on the pix2pixHD with key improvements. Based on the characteristics of SAR and EO images, we propose a denoising enhancement to suppress noise in SAR images. Compared with pix2pix and some other models, the quality of the generated images has been greatly improved. Finally, our solution shows excellent performance on three evaluation metrics: LPIPS [21], FVD [17], and L2 Norm, and ranks the first on the leaderboard of MAVIC held in the CVPR23 PBVS Workshop with a final score of 0.09. Our main contributions are as follows:

1. On the basis of pix2pixHD, we proposes an effective translation framework named SAR2EO, which demonstrates the capability of converting SAR images into EO images with high quality.
2. We propose a denoising enhancement module, which effectively suppresses noise in SAR images while retaining the structural information of the images.
3. Our proposed method exhibits outstanding performance, leading to the first-place in the MAVIC held in CVPR PBVS 2023.

2   Related Work

2.1   GAN

The GAN [5] model was first proposed in 2014, and generally includes two types of networks, G and D. G stands for generator, which generates images by taking a random code as input and outputting a fake image generated by a neural network. It provides an example of mapping random values to real data. The other network, D, stands for discriminator, which is trained to distinguish between real and fake data examples generated by the generator. Its input is the image outputted by G, and then it discriminates whether the image is real or fake. The G and D compete with each other to better achieve the task of generating more output that conforms to the mapping relationship for a given input.

Although GAN [5] has provided a direction for image translation tasks, traditional GAN [5] has some "obvious" flaws: (1) it lacks user control, meaning that in a cGANs [10,15], inputting random noise results in a random image. Random images can deceive the discriminator network and have the same features as real images, but they are often not what we want. (2) It has low resolution and quality issues. The generated images may look good, but when you zoom in, you will find that the details are quite blurry. However, over time, researchers have proposed a series of improved GAN network models, such as cGANs [10,15], Wasserstein GAN (WGAN) [2], and CycleGAN [22], which greatly improve the performance and stability of GAN networks in image translation tasks.

2.2 Image-to-Image Translation

The task goal of image translation is to learn the mapping relationship between the source domain images and target domain images. Depending on the input dataset, image translation can typically be classified into two types: paired and unpaired.

The purpose of supervised learning methods is to learn the mapping relationship between input and output images by training a set of paired image pairs [7,8,12,18,23], which is usually achieved through filename pairing. For example, the training dataset for a facial translation task needs to include pairs of photographs of the same person in different languages. Paired image translation tasks are usually easier to train than unpaired tasks because there is a definite correspondence between images in the dataset. Unpaired image-to-image translation often maps images between two or more domains [1,3,9,13,16,19,20], where image instances do not match. However, those models can be affected by unwanted images and cannot concentrate on the most usefull part of image. Therefore, the "explorers" of deep learning strive to find GAN [5] to solve this problem.

3   Proposed Method

This section presents a description of the proposed translation framework. The framework comprises coarse-to-fine generator, multi-scale discriminators, and a denoising enhancement, as illustrated in Fig. 1. The coarse-to-fine generator is designed to extract both global and local features by combining them. Meanwhile, the multi-scale discriminators are not limited to the input size but are designed to synthesize both the overall image and image details by using smaller and larger discriminators, respectively. The improved adversarial loss further improves the quality of the synthesized images by enhancing their realism. Additionally, the denoising enhancement module is capable of reducing noise in SAR images by applying a non-linear approach to replace the noise.

3.1   Preliminary: Pix2pixHD

The pix2pix model is unable to generate high resolution images, and the generated images lack details and realistic texture. Thus, pix2pixHD proposes the following solutions: coarse-to-fine generator, multi-scale discriminators and improved adversarial loss to improve the above problems [18].

Coarse-to-Fine Generator: The generator is split into two sub-networks, G_1 and G_2, with G_1 as the global generator and G_2 as the local enhancer, The generator is denoted as $G = G_1, G_2$, where G_1 operates at 1,024 × 512 resolution and G_2 outputs an image with 4x the previous output size for synthesizing. G_2 can be used to synthesize higher resolution images. G_1 is comprised three components: a convolutional front-end G_1^F, a set of residual blocks G_1^R, and a transposed convolutional back-end G_1^B. A semantic label map of resolution 1,024 × 512 is fed through these three components to generate an image of the same resolution. G_2 also has three components: a convolutional front-end G_2^F, a set of residual blocks [6] G_2^R, and a transposed convolutional back-end G_2^B. The input label map to G2 has a resolution of 2,048 × 1,024. The input to G_2^R is the element-wise sum of two feature maps: the output feature map of G_2^F and the last feature map of G_1^B. This integrates global information from G_1 to G_2. Training involves first training G_1, followed by training G_2 in order of resolution. Finally, all networks are fine-tuned together. This generator design is effective in aggregating global and local information for image synthesis.

Multi-scale Discriminators: For a generative network, designing a discriminator is a rather difficult task. Compared to low-resolution images, for high-resolution images, the discriminator requires a large receptive field, which requires a large convolutional kernel or a deeper network structure. Adding a large convolutional kernel or deepening the network is easy to cause overfitting and will increase the computational burden. To solve the above problems, multi-scale discriminators have been proposed. It consists of three different discriminators, $D = D_1, D_2, D_3$, which are the same network structure but operate on different image scales. Then, the generated images are downsampled with a factor of 2 and 4, resulting in three images with different resolutions, which are then inputted into the three identical discriminators. This way, the D corresponding to the image with the smallest resolution will have a larger receptive field, providing a stronger global sense for image generation, while the D corresponding to the image with the largest resolution will capture finer and more detailed features.

Improved Adversarial Loss: Since the discriminators have three different sizes and they all are multi-layer convolutional networks, the loss extracts convolutional features from different levels of the synthesized image and matches them with the features extracted from the real image. Then, the feature matching loss is obtained, and its equation is shown as follow:

$$\mathcal{L}_{\text{FM}}(G, D_k) = \mathbb{E}_{(\mathbf{s},\mathbf{x})} \sum_{i=1}^{T} \frac{1}{N_i} \left[\left\| D_k^{(i)}(\mathbf{s}, \mathbf{x}) - D_k^{(i)}(\mathbf{s}, G(\mathbf{s})) \right\|_1 \right] \tag{1}$$

$D_k^{(i)}$ denotes discriminator, where k refers to the k_{th} discriminator and i refers to the number of layers in each discriminator. N refers to the number of elements in each layer. T refers to the number of layers. This feature loss is combined with the GAN's real or fake loss to form the final loss:

$$\mathcal{L}_{total} = \min_{G} \left(\left(\max_{D_1, D_2, D_3} \sum_{k=1,2,3} \mathcal{L}_{\text{GAN}}(G, D_k) \right) + \lambda \sum_{k=1,2,3} \mathcal{L}_{\text{FM}}(G, D_k) \right) \tag{2}$$

where λ controls the importance of the two terms.

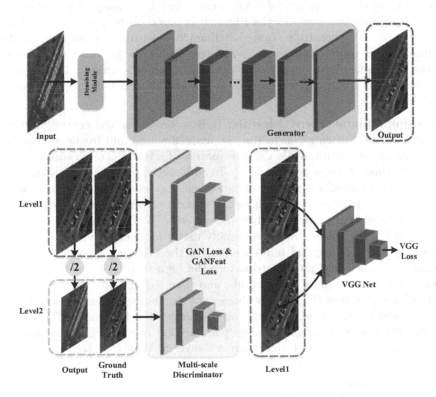

Fig. 1. The overview of our proposed framework.

3.2   SAR and EO Images

SAR is a type of high-resolution imaging radar that can obtain high-resolution radar images similar to optical photographs under extremely low visibility meteorological conditions. SAR image interpretation is very difficult for several reasons. The unique geometric characteristics of SAR images increase the difficulty interpretation [11]; the inherent coherent speckle noise of SAR images causes target edges to be blurred, and the clarity to decrease, requiring completely different methods for SAR image interpretation. SAR images also have multiple reflection effects, false phenomena, doppler frequency shifts, etc. In contrast, EO images can clearly display imaging edge features, and their resolution is higher than that of SAR images. The difficulty in converting SAR images to EO images lies in eliminating noise in SAR images and improving the resolution of the generated images.

3.3   Denoising Enhanced SAR2EO Framework

Based on the characteristics of SAR images and EO images, we propose the SAR2EO solution as shown in the Fig. 1. It is based on pix2pixHD and utilizes the coarse-to-fine generator to fuse global and local features, making the model's feature learning more comprehensive and specific. With the multi-scale discriminators, this method can enhance the details of generated images. The improved adversarial loss improves the quality of the synthesized images by enhancing their realism.

Denoising Enhancement Module: It is common for the energy of signals or images to be concentrated in the low-frequency and mid-frequency bands of the amplitude spectrum, while the information of interest is often submerged by noise in the higher frequency bands [4]. To address the problem of high frequency noise in SAR images, a denoising enhancement module is applied. This module is particularly effective in smoothing noise and can protect sharp edges of the image by choosing appropriate values to replace polluted points. It performs well for salt-and-pepper noise, and is especially useful for speckle noise.

During the training and inference stages, the SAR images will go through denoising enhancement module to reduce the high-frequency noise of the SAR images and preserve more edge details, and then fed into the generator. As for the EO images used as the label, we hope that the generated images of the model will be closer to this label, so we do no changes to the EO images. This is also the advantage of the image-to-image pair method over the unpaired method. The denoising enhancement algorithm is shown in Algorithm 1.

4   Experiments

In this section, we first discuss the dataset and evaluation metrics briefly, and then introduce the implementation details. We then quantitatively evaluate the performance of our approach on the dataset. Finally, some ablation experiments are conducted to demonstrate the effectiveness of each component.

Algorithm 1: The denoising enhancement algorithm.

Data: An input image I and a window size of $n \times m$.
Result: An output image I'.
I' is initialized to I;
for $i = 1$ *to* $height(I) - n$ **do**
 for $j = 1$ *to* $width(I) - m$ **do**
  M is a sub-image of size $n \times m$ centered at pixel (i, j);
  $I'_{i,j}$ is set to the median pixel value of M;
 end
end

4.1 Dataset

The data for this challenge consists of two types of small window regions (chips) taken from large images captured by several EO and SAR sensors mounted on an airplane. The EO chips are 256×256 pix images and belong to targets taken from an airplane. The SAR chips contain roughly the same field of view as the corresponding EO images and are of matching resolution to the EO images. The dataset is divided into:

* Train data: This set resembles the data which is non-uniform and imbalanced.
* Validation: This set is a uniformly distributed among all classes with about <100 of samples per class.
* Test data: This split resembles the validation test.

The purpose is to use the provided (SAR+EO) training image pairs to design and implement a method for translating SAR images to EO images.

4.2 Metrics

It is open and difficult to evaluate the quality of synthesized images [14]. L2 Norm, FVD [17] and LPIPS [21] are used as the metrics.

The $L2$ Norm measures the length of the vector, which is the distance from the origin to the point represented by the vector. In image processing, the $L2$ Norm is often used to calculate the pixel-wise difference between two images.

Learned Perceptual Image Patch Similarity (LPIPS) is a perceptual image quality metric that measures the similarity between two images based on the response of deep neural networks. LPIPS has been shown to correlate well with human perception of image quality. The LPIPS metric is calculated by passing two images through a pre-trained deep neural network and computing the distance between the feature representations of the two images. The distance metric used is typically the $L2$ Norm. The final LPIPS score is obtained by averaging the distances over multiple image patches. Where $I1$ and $I2$ are the two images being compared, fi is the feature extractor for the i-th image patch, and n is the total number of patches.

Table 1. PBVS 2023 Multi-modal Aerial View Imagery Challenges - Translation test set leaderboard.

Team	Final Score	LPIPS	FVD	L2
Ours	**0.09**	**0.25**	**0.02**	**0.01**
pokemon	0.14	0.35	0.04	0.01
wangzhiyu918	0.14	0.38	0.02	0.01
ngthien	0.18	0.43	0.10	0.01
Wizard001	0.26	0.50	0.27	0.02
hanhai	0.30	0.30	0.59	0.01
u7355608	0.33	0.54	0.43	0.02
jsyoon	0.33	0.46	0.53	0.01

Fréchet Video Distance (FVD) is a metric that measures the similarity between two sets of images. FVD is based on the distance between the feature representations of the images calculated by a pre-trained deep neural network. The FVD metric is calculated by first computing the mean and covariance of the feature representations of the real images and the generated images. The distance between the mean and covariance is then computed using the Fréchet distance, which is a measure of similarity between two multivariate Gaussian distributions. A lower FVD score indicates a higher similarity between the two sets of images. The formula for FVD can be represented as:

$$FVD(real, fake) = ||\mu_{real} - \mu_{fake}||_2^2 + Tr(\sigma_{real} + \sigma_{fake} - 2 * (\sigma_{real} * \sigma_{fake})^{1/2}) \tag{3}$$

where μ_{real} and σ_{real} are the mean and covariance of the feature representations of the real images, μ_{fake} and σ_{fake} are the mean and covariance of the feature representations of the generated images, and $Tr(\cdot)$ denotes the trace of a matrix.

The evaluation indicator on the final ranking is the average of the above three indicators.

4.3   Implementation Details

Development phase/learning: Since we have the validation set and the corresponding labels accessible at this phase, we perform the validation offline, train on the training set, and then test on the validation set. Final Evaluation: In this phase, the training and validation sets were jointly used as the training set to expand the data. The experiments were performed on NVIDIA GPU 3090.

4.4   Main Results

Challenge Overview: Sensor translation algorithms allow for dataset augmentation and allows for the fusion of information from multiple sensors. EO and SAR sensors provide a unique environment for translation. The motivation for this challenge is to understand how if and how data from one modality can be translated to another modality. This competition challenges participants to design methods to translate aligned images from the SAR modality to the EO modality.

Table 2. The effect of using different schemes on the experimental results, the LPIPS↓, FVD↓ and L2↓ metric on the local validation dataset.

Method	LPIPS	FVD	L2
pix2pix	0.484638	0.08	0.02
pix2pixhd	0.259611	0.02	0.01
ours	**0.253924**	**0.02**	**0.01**

In the context of the Multi-modal Aerial View Imagery Challenge (MAVIC), our team attained the first rank in the ultimate test set ranking. This feat was accomplished through the attainment of noteworthy performance metrics, specifically a LPIPS score of 0.25, a FVD score of 0.02, and a L2 score of 0.01. The composite score of the team was computed as 0.09, representing the average of the aforementioned performance indicators. The obtained results provide compelling evidence of the efficacy of the team's competition strategy. The final test set leaderboard is shown in Table 1.

Additionally, we compared our proposed method with other image translation models, namely pix2pix and pix2pixHD. The training data provided by the competition was utilized for training, and the validation set was used for evaluation, with LPIPS, FVD and L2 being the chosen metric. The performance of the translators is presented in a Table 2.

Table 3. Conducting the experiments on the pix2pixHD and pix2pixHD with denoising enhancement, the metric is LPIPS, FVD and L2 on the local validation dataset.

Method	LPIPS	FVD	L2
pix2pixHD	0.259611	0.02	0.01
+denoising enhancement	**0.253924**	**0.02**	**0.01**

The images generated by different methods are showcased in Fig. 2. Regarding the quality of the results, SAR images generally exhibit a high degree of noise and blurry object boundaries, leading to a patchy visual appearance. The second column showcases the results generated by the pix2pix model, which suffers from severe distortions, with object shapes being deformed and inaccurate color generation. However, the generated images still exhibit basic shape contours, albeit with blurry details. In contrast, the third column depicts the results generated by pix2pixHD, which exhibit a higher level of image clarity and closely resemble the labels in terms of shape, with improved color accuracy. Nevertheless, some details still require refinement when compared to the label images. Finally, the results produced by our framework exhibit even better quality compared to pix2pixHD, with shapes and colors that are more similar to the label images, and sharper details in the generated images.

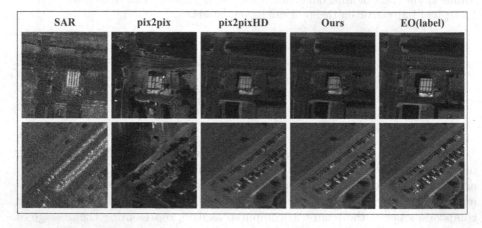

| SAR | pix2pix | pix2pixHD | Ours | EO(label) |

Fig. 2. The results generated by the different models and the corresponding EO images(labels). The first column of images depicts the original SAR images, while the second and third columns showcase the results generated by pix2pix and pix2pixHD, respectively. The fourth column presents the results generated by our framework, and the fifth column displays the corresponding EO images, which serve as the labels for the SAR images. Better viewed in color with zoom-in.

4.5   Ablation Studies

In this section, we conducted ablation experiments to demonstrate the effectiveness of our framework. The validation dataset was utilized for the experimentation. Furthermore, visualizations of the images in this section were presented to illustrate the efficacy of our denoising enhancement.

Effectiveness of Denoising Enhancement Module of Our Framework: In the Table 3, the metric is slightly improved, it shows the effectiveness of denoising enhancement.

5   Conclusion

This paper presents a novel framework for converting SAR images to EO images. Our framework is based on pix2pixHD architecture, which is known for its ability to generate high-quality images. The coarse-to-fine generator, multi-scale discriminators, and improved adversarial loss in pix2pixHD are utilized to enhance the resolution and quality of the generated EO images. To further improve the conversion quality, we analyzed the specific characteristics of SAR and EO images and proposed a denoising enhancement module to reduce the noise in SAR images and enhance the contours of objects in SAR images. This denoising enhancement module was incorporated into the overall framework and resulted in a significant improvement in image quality. The proposed framework has several advantages over existing methods. First, our framework is highly effective in generating high-quality EO images from SAR images, which is a challenging task due to the significant differences between the two modalities. Second, our denoising enhancement module effectively reduces the noise in SAR images, which is a common issue in SAR images. Third, our framework is based on the widely-used pix2pixHD architecture, making it easy to implement and deploy in various applications. We evaluated the performance of our framework in the MAVIC competition, where the goal was to generate high-quality EO images from SAR images. Our framework achieved a final score of 0.09, which was the best among all participating teams and secured the first place in the competition.

Acknowledgements. This work was supported by the Natural Science Foundation of China (62276242), National Aviation Science Foundation (2022Z071-078001), CAAI-Huawei MindSpore Open Fund (CAAIXSJLJJ-2021-016B, CAAIXSJLJJ-2022-001A), Anhui Province Key Research and Development Program (202104a05020007), USTC-IAT Application Sci. & Tech. Achievement Cultivation Program (JL06521001Y), Sci. & Tech. Innovation Special Zone (20-163-14-LZ-001-004-01).

References

1. Anoosheh, A., Agustsson, E., Timofte, R., Van Gool, L.: ComboGAN: unrestrained scalability for image domain translation. In: Proceedings of the IEEE Conference on Computer Vision and Pattern Recognition Workshops, pp. 783–790 (2018)
2. Arjovsky, M., Chintala, S., Bottou, L.: Wasserstein generative adversarial networks. In: International Conference on Machine Learning, pp. 214–223. PMLR (2017)
3. Benaim, S., Wolf, L.: One-sided unsupervised domain mapping. In: Advances in Neural Information Processing Systems, vol. 30 (2017)
4. Castleman, K.R.: Digital Image Processing. Prentice Hall Press, Hoboken (1996)
5. Goodfellow, I., et al.: Generative adversarial networks. Commun. ACM **63**(11), 139–144 (2020)
6. He, K., Zhang, X., Ren, S., Sun, J.: Deep residual learning for image recognition. In: Proceedings of the IEEE Conference on Computer Vision and Pattern Recognition, pp. 770–778 (2016)
7. Hoffman, J., et al.: CyCADA: cycle-consistent adversarial domain adaptation. In: International Conference on Machine Learning, pp. 1989–1998. PMLR (2018)
8. Isola, P., Zhu, J.Y., Zhou, T., Efros, A.A.: Image-to-image translation with conditional adversarial networks. In: Proceedings of the IEEE Conference on Computer Vision and Pattern Recognition, pp. 1125–1134 (2017)
9. Kim, T., Cha, M., Kim, H., Lee, J.K., Kim, J.: Learning to discover cross-domain relations with generative adversarial networks. In: International Conference on Machine Learning, pp. 1857–1865. PMLR (2017)
10. Mirza, M., Osindero, S.: Conditional generative adversarial nets. arXiv preprint arXiv:1411.1784 (2014)

11. Oliver, C., Quegan, S.: Understanding Synthetic Aperture Radar Images. SciTech Publishing (2004)
12. Park, T., Liu, M.Y., Wang, T.C., Zhu, J.Y.: Semantic image synthesis with spatially-adaptive normalization. In: Proceedings of the IEEE/CVF Conference on Computer Vision and Pattern Recognition, pp. 2337–2346 (2019)
13. Rosales, R., Achan, K., Frey, B.J.: Unsupervised image translation. In: ICCV, pp. 472–478 (2003)
14. Salimans, T., Goodfellow, I., Zaremba, W., Cheung, V., Radford, A., Chen, X.: Improved techniques for training GANs. In: Advances in Neural Information Processing Systems, vol. 29 (2016)
15. Sohn, K., Lee, H., Yan, X.: Learning structured output representation using deep conditional generative models. In: Advances in Neural Information Processing Systems, vol. 28 (2015)
16. Tang, H., Xu, D., Wang, W., Yan, Y., Sebe, N.: Dual generator generative adversarial networks for multi-domain image-to-image translation. In: Jawahar, C.V., Li, H., Mori, G., Schindler, K. (eds.) ACCV 2018. LNCS, vol. 11361, pp. 3–21. Springer, Cham (2019). https://doi.org/10.1007/978-3-030-20887-5_1
17. Unterthiner, T., van Steenkiste, S., Kurach, K., Marinier, R., Michalski, M., Gelly, S.: FVD: a new metric for video generation (2019)
18. Wang, T.C., Liu, M.Y., Zhu, J.Y., Tao, A., Kautz, J., Catanzaro, B.: High-resolution image synthesis and semantic manipulation with conditional GANs. In: Proceedings of the IEEE Conference on Computer Vision and Pattern Recognition, pp. 8798–8807 (2018)
19. Wang, Y., van de Weijer, J., Herranz, L.: Mix and match networks: encoder-decoder alignment for zero-pair image translation. In: Proceedings of the IEEE Conference on Computer Vision and Pattern Recognition, pp. 5467–5476 (2018)
20. Yi, Z., Zhang, H., Tan, P., Gong, M.: DualGAN: unsupervised dual learning for image-to-image translation. In: Proceedings of the IEEE International Conference on Computer Vision, pp. 2849–2857 (2017)
21. Zhang, R., Isola, P., Efros, A.A., Shechtman, E., Wang, O.: The unreasonable effectiveness of deep features as a perceptual metric. In: Proceedings of the IEEE Conference on Computer Vision and Pattern Recognition, pp. 586–595 (2018)
22. Zhu, J.Y., Park, T., Isola, P., Efros, A.A.: Unpaired image-to-image translation using cycle-consistent adversarial networks. In: Proceedings of the IEEE International Conference on Computer Vision, pp. 2223–2232 (2017)
23. Zhu, J.Y., et al.: Toward multimodal image-to-image translation. In: Advances in Neural Information Processing Systems, vol. 30 (2017)

A New Perspective of Weakly Supervised 3D Instance Segmentation via Bounding Boxes

Qingtao Yu[1,2], Heming Du[1,2], and Xin Yu[1(✉)]

[1] University of Queensland, Brisbane, Australia
{heming.du,xin.yu}@uq.edu.au
[2] Australian National University, Canberra, Australia
terry.yu@anu.edu.au

Abstract. Existing fully supervised method 3D point cloud segmentation methods heavily rely on carefully annotated point labels. In this work, we look at weakly-supervised 3D instance segmentation using bounding boxes supervision. Bounding boxes are much easier to annotate than dense point-wise labels. Moreover, they demonstrated high potential in addressing instance-level segmentation compared to other types of weak annotations. However, existing bounding-box supervised techniques have struggled to keep pace with the development of fully-supervised methods. To tackle this issue, we propose a simple-yet-effective approach to directly leverage the network architecture of fully-supervised methods for such weak supervision scenarios. We found that accurate instance labels for each point can be generated with the given bounding boxes by leveraging 3D geometric prior. Such a process is efficient and does not require any additional training or fine-tuning. The generated point-wise labels can be fed to any advanced fully-supervised model without re-designing specific networks for bounding-box supervision. In this fashion, our designed approach achieves on par performance of fully supervised methods in terms of AP, AP50 and AP25. Remarkably, we outperformed the state-of-the-art bounding-box supervised method by 21%. Compared with existing methods, our method is extremely simple and only involves two small heuristics in the data preprocessing step. In addition, our method is proven to be robust against noisy bounding box scenario through experiments.

Keywords: Weakly supervised learning · 3D point cloud

1 Introduction

Instance segmentation is one of the fundamental tasks in 3D point cloud analysis. The objective is to predict instance masks and corresponding semantic labels for each point cloud scene. Current 3D indoor instance segmentation methods [2, 8, 12, 15, 24, 27–30] reply on fully-supervised annotations, which requires assigning

T. Liu et al. (Eds.): AI 2023, LNAI 14471, pp. 103–114, 2024.
https://doi.org/10.1007/978-981-99-8388-9_9

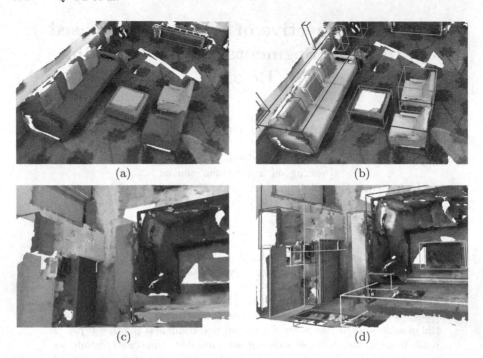

(a)   (b)

(c)   (d)

Fig. 1. Images (a), (c) represent point-wise annotation and images (b), (d) represent bounding-box annotation. The latter approach can significantly reduce the annotation effort and simplifies the labeling procedure.

semantic and instance labels to each 3D point However, such labelling process is often time-consuming and laborious due to the vast number of points in each scene. Unlike point-wise labelling, 3D instance bounding boxes offer a more efficient alternative, as it only involves drawing a single box around each object (Fig. 1).

Among different types of weak supervisions for point clouds [4, 7, 11, 13, 21, 22, 31–33, 40, 41], bounding boxes stand out as a more prospective direction. Each bounding box covers the full extent of the object, and it provides rough instance information such as object centre and size. It is naturally a rich instance representation, thus making it more capable of handling instance-level segmentation. Conversely, other forms of weak annotations, such as sparse points, are mostly restricted to semantic segmentation tasks. Hence, there has been a growing interest in investigating weakly-supervised point cloud instance segmentation methods with 3D bounding-box annotations.

While methods for fully-supervised instance segmentation methods developing rapidly, those supervised by bounding boxes haven't seen significant breakthroughs for a long duration. The performance gap between such two strategies tends to rise. The point-wise annotation provides a more detailed and informative supervision, thus offering researchers greater flexibility to implement diverse strategies, e.g. bottom-up and top-down strategies. In this fashion, if accurate point-wise annotation can be generated from given bounding boxes, we can directly utilise them on the advanced fully-supervised method. Existing work also attempts to address this issue following such strategy [7,13]. However, the designs of those methods are complex. Our method only involves two simple heuristics in the data preprocessing step but surpasses these methods in terms of performance. We expect that future work will utilise our method as a baseline and solve the bounding box supervised problem as a noisy labelling problem [23,36,37].

As a rough outer range of each instance, a bounding box usually over-include background points and have significant volume overlaps with other bounding boxes. Therefore, correctly identifying and re-allocating these ambiguous points becomes the main challenge of the task. We utilize given bounding boxes to initialize candidate points for each instance. Then we leverage 3D geometric consistency prior (superpoint) to refine the candidate points. Finally a smallest-box heuristic is implemented to uniquely assign the each superpoint to the instance. The whole process can be formulated as a flexible plug-in module and adaptive to any fully-supervised structure. Different from the previous works, our method is very simple and doesn't require re-designing the learning network structure from scratch. Moreover, we evaluated our approach on noisy bounding box scenarios. Even with a massive noise, the performance tends to be stable and doesn't exhit significant decrease.

2 Related Work

2.1 Fully Supervised Method

As one of the fundamental tasks of 3D [9,25,26], Point cloud instance segmentation methods aim to predict instance masks of 3D point clouds as well as their semantic categories. Early methods can be grouped into two classes: proposal-based and grouping-based paradigms. Proposal-based methods employ a top-down strategy to generates region proposals and then segments the instance within each proposal. For instance, [35] and [10] first regresses 3D bounding boxes for all instances and then leverage the point features to produce instance masks. In [20], a Gaussian instance center network has been proposed to estimate instance center heatmaps. Grouping-based methods implement a bottom-up pipeline that produces point-wise predictions and then cluster points into different instances. For example, MASC [17] leverage the mesh graph to cluster the instances after extracting the semantic features of points. To group instances more robustly, Jiang [12] propose to estimate point offsets with respect to object centers. Following such design, Liang [15] and Chen [2] improve segmentation performance by adopting hierarchical aggregation schemes. Furthermore, [8] introduces an additional graph convolutional network to refine the grouping outputs. Recently, SoftGroup [29,30] leverage the advantages of both proposal-based and grouping-based methods. They proposed a new 3D instance segmentation architecture with bottom-up soft grouping and a subsequent top-down refinement.

State-of-the-art methods further improve 3D instance segmentation performance by utilizing advancements of transformers. The latest works, SPFormer [28] and Mask3D [27], both implement a transformer decoder following the design of cutting-edge 2D segmentation methods [1,3]. Instead of clustering points, such methods learn a set of instance queries and compute instance masks directly based on the similarities between point features and query vectors. Such strategy can better model the relationship between objects and points, while accelerating the inference process at the same time. Moreover, an attention mask mechanism is incorporated to enhance training efficiency.

2.2 Weakly Supervised Method

Weakly supervised method doesn't reply on informative annotations [34,39]. Sparse-point weak supervision approaches only use a small portion of labeled point clouds to learn semantic segmentation. Xu et al. [33] propose to propagate gradients of labeled points to those of labeled points in optimizing their semantic segmentation network. Hou et al. [11] annotate 0.1% of points and train a 3D semantic segmentation network by contrastive learning. In addition, Liu et al. [22] generate semantic pseudo labels from one point per object by implementing a contrastive learning strategy. Note that the works [22,36] focus on semantic segmentation instead of instance segmentation.

Instance segmentation is a more challenging task [3,18,19]. Compared to sparse point supervision, 3D bounding box annotations are regarded as weak supervision and are suitable for the instance segmentation tasks since they naturally provide rough instance information. Liao et al. [16] propose a point cloud object detection and instance segmentation network learned from labeled bounding boxes. The network can also mine the categories of unlabeled bounding-boxes during training. Du et al. [7] and Kulharia et al. [13] both leverage 3D local geometric information to generate point-level labels from bounding-box annotations. Recently, [4] proposed a new approach that both generating point-level labels while using boxes as supervision. However, all these methods demonstrate a high performance gap between current fully-supervised methods.

3   Methodology

Our goal is to achieve precise point-wise instance labels from bounding boxes. The main challenge is how to identify these ambiguous points that are in multiple bounding boxes or background points included within the bounding boxes. Our proposed method is simple and effective.

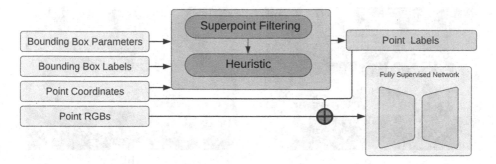

Fig. 2. Our pipeline is exhibited as above. It is straightforward but effective.

3.1 Problem Description

Instead of point-wise annotation, we assume only the bounding box annotation is given; that is, each foreground instance has one axis-aligned bounding box to cover all the points within the instance. Thus, each bounding box can be characterized with six parameters, XYZ coordinates of maximum corners and minimum corners. Each bounding box is paired with a unique instance ID along with a semantic label.

3.2 Cluster-Based Candidate Points Filtering

Each point contained within the instance bounding boxes might potentially belong to the corresponding instance. First, we consider all the contained points as raw candidate points of the instances. Secondly, we invoke 3D superpoints to filter out some unlikely points. Superpoints are clusters of points symbolising local geometric continuity, formed via a normal-based graph cluster technique [14], as shown in Fig. 3 All the points within a superpoint have similar geometric characters. Hence it can be presumed that these points belong to the same instance. Given that bounding boxes might only include additional points, we identify points as candidates only if the associated superpoints are entirely within the box. In other words, if any point in the superpoints lies outside the box, we can confidently exclude the entire superpoint. In this case, we can optimally minimise unlikely labelled points associated with each instance, especially the background points within the bounding boxes. Following this process, candidate points for each instance can be obtained.

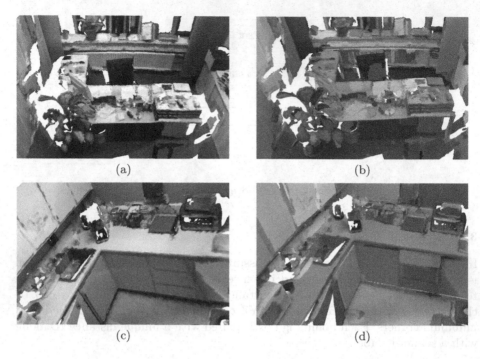

(a)   (b)

(c)   (d)

Fig. 3. Images (a), (c) show the original point cloud scene, images (b), (d) demonstrate the superpoints, where each color represents one superpoint. Superpoints can be considered as oversegmentations of the point clouds. All the points within one superpoint share similar geometric characters.

3.3   Smallest-Box Heuristic

Note that some points in the overlapping area of boxes might be candidate points for multiple instances, if the entire superpoint is included in the multiple bounding boxes. To overcome this issue, we designed an effective heuristic: Instance labels of such points are assigned to the bounding box with the smallest volume. Such heuristic is designed through our observation of the 3D scenes. In most cases, such as table and chair, sink and cabinet, the points in the overlapping area belong to the smaller object. The overview of our whole process is demonstrate as Fig. 2.

4   Experiment

To validate the effectiveness of the proposed method, we conduct experiments on the currently most popular 3D indoor point cloud dataset, ScanNet-V2. We achieve state-of-the-art 3D instance segmentation performance under bounding box annotations.

Table 1. Main results table

Supervision	Method	AP	$AP50$	$AP25$
Box	Box2Mask	0.391	0.597	0.718
	WISGP & SSTNet [7]	0.352	0.569	0.702
	Softgroup^{++} [29] & ours	0.393	0.623	0.765
	SPFormer [28] & ours	**0.473**	**0.689**	**0.787**
Mask	PointGroup [12]	0.348	0.517	0.713
	SSTNet [15]	0.494	0.643	0.740
	SoftGroup^{++} [29]	0.458	0.674	0.791
	ISBNet [24]	0.545	0.731	0.825
	Mask3D [27]	0.552	0.737	0.835
	SPFormer [28]	0.563	0.739	0.829

4.1 Implementation Details

The fully supervised method requires at least four inputs to the network, including point coordinates, point RGB colors, point-wise semantic labels and point-wise instance labels. The first two inputs are $N*3$ vectors, and the last two are $N*1$ vectors, where N is the number of points in the scene. In our weakly supervised bounding box settings, besides point coordinates and point RGB colors, we only have bounding box parameters, corresponding instances and semantic labels. Their sizes are $M*6$, $M*1$ and $M*1$, where M is the number of foreground instances. Point-wise semantic labels and instance labels can be achieved through the bounding box information through our method.

After that, we feed these inputs into two state-of-the-art representative fully supervised methods, superpoint transformer (SPFormer) and Softgroup^{++}. The first one is a transformer-based model. The second one generates final predictions through bottom-up grouping followed by a top-up refinement. To follow the weakly supervised setting, we removed the full-supervised pre-trained model and trained the network from scratch. In this case, we reduce the learning rate and enlarge the number of training epochs.

4.2 Dataset

ScanNet-V2 dataset [5] contains 1,613 scans with 3D semantic and instance annotations. The dataset is split into training, validation, and testing sets, with 1,201, 312, and 100 scans. Each scan provides point cloud reconstructed from 2D RGBD images [6,38]. It contains 18 object semantic categories. We train on the training set and report results on the validation set for comparison with other methods. An efficient normal-based graph cut image segmentation method [14] is utilized for superpoint generation (Fig. 4).

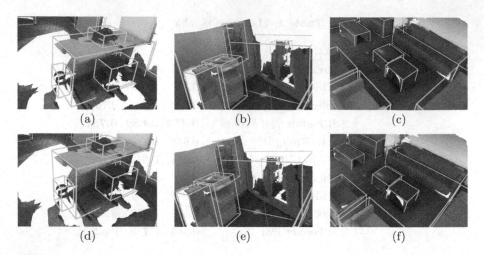

(a)   (b)   (c)

(d)   (e)   (f)

Fig. 4. These figures are our results visualisation, the upper figures show the generated point-wise labeled from bounding boxes using our method. The lower figures show the ground truth point-wise labels. There is only a minor difference between them. That's the reason we can achieve on-par performance with fully supervised methods.

4.3   Evaluation Metrics and Experiment Results

Following previous works, we adopt mean average precision (mAP) for primary performance evaluation. It calculates the average scores among different Intersection over Union (IoU) thresholds, ranging from 50% to 95%, with increments of 5%. And the final score is the mean value of all the semantic classes. In addition, AP50 and AP25 represent the scores corresponding to IoU thresholds of 50% and 25%, respectively. We present the mAP, AP50, and AP25 results for the Scan-Netv2 dataset on Table 1. The results demonstrate with our data-preprocessing and the advance of full-supervised network. Our model significantly outperformers the weak supervised network design.

4.4   Ablation Study

In this section, we validate the effectiveness of our geometic prior guidance and smallest-box heuristic. First, we aim to prove the effectiveness of geometric guidance. We only implement the smallest-box heuristic and each superpoint can be partially included in multiple instances. Since SPFormer also requires a unique label for each superpoints, we follow the implementation of published code in SPFormer to do the majority vote on each superpoint instead of our approach. Second, we show the effectiveness of smallest heuristic by comparing it with random assignments. Since fully-supervised methods require a unique instance and semantic label on each point, after implementing the geometric consistency operation, we randomly assign the ambiguous superpoints that are entirely included in multiple instances to any one of them. The result is demonstrated in Table 2:

Table 2. Ablation

Random assignment	Majority Vote	Ours	AP	AP_{50}	AP_{25}
✓			0.454	0.617	0.744
	✓		0.460	0.632	0.771
		✓	**0.473**	**0.689**	**0.787**

4.5   Robustness

With the power of superpoints, our method also demonstrates robustness against noisy bounding box scenarios. We randomly enlarge the bounding boxes according to different rates. The noise follows a Gaussian distribution of mean value $\lambda \cdot S$, where λ is the noise level while S is the side length of bounding boxes. And we set the standard deviation as $0.05 \cdot \lambda \cdot S$. The results in the Table 3 demonstrates our method is robust to noise. The chosen of network is SPF [28]. The reason is the usage of superpoints. We focus on the assignment of geometric clusters instead of individual points.

Table 3. Results under noisy bounding boxes

Noise level	AP	AP_{50}	AP_{25}
$\lambda = 0.0$	0.473	0.689	0.787
$\lambda = 0.1$	0.437	0.672	0.780
$\lambda = 0.2$	0.396	0.625	0.761
$\lambda = 0.3$	0.364	0.620	0.771

5   Conclusion

We proposed a new approach that can achieve highly accurate instance mask labels from bounding-box weak supervision by leveraging the 3D geomeric prior knowledge. With the power of advanced fully supervised method and a efficient data preprocessing step, we achieved state-of-the-art results of weakly supervised 3D instance segmentation using bounding box annotation.

Acknowledgements. This research is funded in part by ARC-Discovery grant (DP220200800 to XY) and ARC-DECRA grant (DE230100477 to XY). We thank all anonymous reviewers and ACs for their constructive suggestions.

References

1. Carion, N., Massa, F., Synnaeve, G., Usunier, N., Kirillov, A., Zagoruyko, S.: End-to-end object detection with transformers. In: Vedaldi, A., Bischof, H., Brox, T., Frahm, J.-M. (eds.) ECCV 2020. LNCS, vol. 12346, pp. 213–229. Springer, Cham (2020). https://doi.org/10.1007/978-3-030-58452-8_13
2. Chen, S., Fang, J., Zhang, Q., Liu, W., Wang, X.: Hierarchical aggregation for 3D instance segmentation. In: Proceedings of the IEEE/CVF International Conference on Computer Vision (ICCV), pp. 15467–15476 (2021)
3. Cheng, B., Choudhuri, A., Misra, I., Kirillov, A., Girdhar, R., Schwing, A.G.: Mask2Former for video instance segmentation. arXiv preprint arXiv:2112.10764 (2021)
4. Chibane, J., Engelmann, F., Anh Tran, T., Pons-Moll, G.: Box2Mask: weakly supervised 3D semantic instance segmentation using bounding boxes. In: Avidan, S., Brostow, G., Cissé, M., Farinella, G.M., Hassner, T. (eds.) ECCV 2022. LNCS, vol. 13691, pp. 681–699. Springer, Cham (2022). https://doi.org/10.1007/978-3-031-19821-2_39
5. Dai, A., Chang, A.X., Savva, M., Halber, M., Funkhouser, T., Nießner, M.: ScanNet: richly-annotated 3D reconstructions of indoor scenes. In: Proceedings of the IEEE Conference on Computer Vision and Pattern Recognition, pp. 5828–5839 (2017)
6. Dai, A., Nießner, M., Zollhöfer, M., Izadi, S., Theobalt, C.: BundleFusion: real-time globally consistent 3D reconstruction using on-the-fly surface reintegration. ACM Trans. Graph. (ToG) **36**(4), 1 (2017)
7. Du, H., Yu, X., Hussain, F., Armin, M.A., Petersson, L., Li, W.: Weakly-supervised point cloud instance segmentation with geometric priors. In: Proceedings of the IEEE/CVF Winter Conference on Applications of Computer Vision, pp. 4271–4280 (2023)
8. Engelmann, F., Bokeloh, M., Fathi, A., Leibe, B., Nießner, M.: 3D-MPA: multi-proposal aggregation for 3D semantic instance segmentation. In: Proceedings of the IEEE/CVF Conference on Computer Vision and Pattern Recognition, pp. 9031–9040 (2020)
9. Han, C., Yu, X., Gao, C., Sang, N., Yang, Y.: Single image based 3D human pose estimation via uncertainty learning. Pattern Recogn. **132**, 108934 (2022)
10. Hou, J., Dai, A., Nießner, M.: 3D-SIS: 3D semantic instance segmentation of RGB-D scans. In: Proceedings of the IEEE/CVF Conference on Computer Vision and Pattern Recognition, pp. 4421–4430 (2019)
11. Hou, J., Graham, B., Nießner, M., Xie, S.: Exploring data-efficient 3D scene understanding with contrastive scene contexts. In: Proceedings of the IEEE/CVF Conference on Computer Vision and Pattern Recognition, pp. 15587–15597 (2021)
12. Jiang, L., Zhao, H., Shi, S., Liu, S., Fu, C.W., Jia, J.: PointGroup: dual-set point grouping for 3D instance segmentation. In: Proceedings of the IEEE Conference on Computer Vision and Pattern Recognition (CVPR) (2020)
13. Kulharia, V., Chandra, S., Agrawal, A., Torr, P., Tyagi, A.: Box2Seg: attention weighted loss and discriminative feature learning for weakly supervised segmentation. In: Vedaldi, A., Bischof, H., Brox, T., Frahm, J.-M. (eds.) ECCV 2020. LNCS, vol. 12372, pp. 290–308. Springer, Cham (2020). https://doi.org/10.1007/978-3-030-58583-9_18
14. Landrieu, L., Boussaha, M.: Point cloud over segmentation with graph-structured deep metric learning. In: Proceedings of the IEEE/CVF Conference on Computer Vision and Pattern Recognition, pp. 7440–7449 (2019)

15. Liang, Z., Li, Z., Xu, S., Tan, M., Jia, K.: Instance segmentation in 3D scenes using semantic superpoint tree networks. In: Proceedings of the IEEE/CVF International Conference on Computer Vision, pp. 2783–2792 (2021)
16. Liao, Y., Zhu, H., Zhang, Y., Ye, C., Chen, T., Fan, J.: Point cloud instance segmentation with semi-supervised bounding-box mining. IEEE Trans. Pattern Anal. Mach. Intell. **44**, 10159–10170 (2021)
17. Liu, C., Furukawa, Y.: MASC: multi-scale affinity with sparse convolution for 3D instance segmentation. arXiv preprint arXiv:1902.04478 (2019)
18. Liu, C., et al.: Audio-visual segmentation, sound localization, semantic-aware sounding objects localization. arXiv preprint arXiv:2307.16620 (2023)
19. Liu, C., et al.: BAVS: bootstrapping audio-visual segmentation by integrating foundation knowledge. arXiv preprint arXiv:2308.10175 (2023)
20. Liu, S.H., Yu, S.Y., Wu, S.C., Chen, H.T., Liu, T.L.: Learning Gaussian instance segmentation in point clouds. arXiv preprint arXiv:2007.09860 (2020)
21. Liu, Y., Hu, Q., Lei, Y., Xu, K., Li, J., Guo, Y.: Box2Seg: learning semantics of 3D point clouds with box-level supervision. arXiv preprint arXiv:2201.02963 (2022)
22. Liu, Z., Qi, X., Fu, C.W.: One thing one click: a self-training approach for weakly supervised 3D semantic segmentation. In: Proceedings of the IEEE/CVF Conference on Computer Vision and Pattern Recognition, pp. 1726–1736 (2021)
23. Ma, F., Wu, Y., Yu, X., Yang, Y.: Learning with noisy labels via self-reweighting from class centroids. IEEE Trans. Neural Netw. Learn. Syst. **33**(11), 6275–6285 (2021)
24. Ngo, T.D., Hua, B.S., Nguyen, K.: ISBNet: a 3D point cloud instance segmentation network with instance-aware sampling and box-aware dynamic convolution. In: Proceedings of the IEEE/CVF Conference on Computer Vision and Pattern Recognition, pp. 13550–13559 (2023)
25. Qi, X., Liu, C., Li, L., Hou, J., Xin, H., Yu, X.: EmotionGesture: audio-driven diverse emotional co-speech 3D gesture generation (2023)
26. Qi, X., Liu, C., Sun, M., Li, L., Fan, C., Yu, X.: Diverse 3D hand gesture prediction from body dynamics by bilateral hand disentanglement. In: Proceedings of the IEEE/CVF Conference on Computer Vision and Pattern Recognition (CVPR), pp. 4616–4626 (2023)
27. Schult, J., Engelmann, F., Hermans, A., Litany, O., Tang, S., Leibe, B.: Mask3D for 3D semantic instance segmentation. arXiv preprint arXiv:2210.03105 (2022)
28. Sun, J., Qing, C., Tan, J., Xu, X.: Superpoint transformer for 3D scene instance segmentation. arXiv preprint arXiv:2211.15766 (2022)
29. Vu, T., Kim, K., Luu, T.M., Nguyen, T., Kim, J., Yoo, C.D.: SoftGroup++: scalable 3D instance segmentation with octree pyramid grouping. arXiv preprint arXiv:2209.08263 (2022)
30. Vu, T., Kim, K., Luu, T.M., Nguyen, T., Yoo, C.D.: SoftGroup for 3D instance segmentation on point clouds. In: Proceedings of the IEEE/CVF Conference on Computer Vision and Pattern Recognition, pp. 2708–2717 (2022)
31. Wu, Y., et al.: PointMatch: a consistency training framework for weakly supervised semantic segmentation of 3D point clouds. arXiv preprint arXiv:2202.10705 (2022)
32. Wu, Z., Wu, Y., Lin, G., Cai, J., Qian, C.: Dual adaptive transformations for weakly supervised point cloud segmentation. In: Avidan, S., Brostow, G., Cissé, M., Farinella, G.M., Hassner, T. (eds.) ECCV 2022. LNCS, vol. 13691, pp. 78–96. Springer, Cham (2022). https://doi.org/10.1007/978-3-031-19821-2_5
33. Xu, X., Lee, G.H.: Weakly supervised semantic point cloud segmentation: towards 10x fewer labels. In: Proceedings of the IEEE/CVF Conference on Computer Vision and Pattern Recognition, pp. 13706–13715 (2020)

34. Xu, Y., Yu, X., Zhang, J., Zhu, L., Wang, D.: Weakly supervised RGB-D salient object detection with prediction consistency training and active scribble boosting. IEEE Trans. Image Process. **31**, 2148–2161 (2022)

35. Yang, B., et al.: Learning object bounding boxes for 3D instance segmentation on point clouds. arXiv preprint arXiv:1906.01140 (2019)

36. Ye, S., Chen, D., Han, S., Liao, J.: Learning with noisy labels for robust point cloud segmentation. In: Proceedings of the IEEE/CVF International Conference on Computer Vision, pp. 6443–6452 (2021)

37. Yu, Q., Du, H., Liu, C., Yu, X.: When 3D bounding-box meets SAM: point cloud instance segmentation with weak-and-noisy supervision (2023)

38. Zhan, H., Zheng, J., Xu, Y., Reid, I., Rezatofighi, H.: ActiveRMAP: radiance field for active mapping and planning. arXiv preprint arXiv:2211.12656 (2022)

39. Zhang, J., Yu, X., Li, A., Song, P., Liu, B., Dai, Y.: Weakly-supervised salient object detection via scribble annotations. In: Proceedings of the IEEE/CVF Conference on Computer Vision and Pattern Recognition, pp. 12546–12555 (2020)

40. Zhang, Y., Li, Z., Xie, Y., Qu, Y., Li, C., Mei, T.: Weakly supervised semantic segmentation for large-scale point cloud. In: Proceedings of the AAAI Conference on Artificial Intelligence, vol. 35, pp. 3421–3429 (2021)

41. Zhou, Y., Zhu, Y., Ye, Q., Qiu, Q., Jiao, J.: Weakly supervised instance segmentation using class peak response. In: Proceedings of the IEEE Conference on Computer Vision and Pattern Recognition, pp. 3791–3800 (2018)

Large-Kernel Attention Network with Distance Regression and Topological Self-correction for Airway Segmentation

Yan Hu, Erik Meijering, and Yang Song[(✉)]

School of Computer Science and Engineering, The University of New South Wales,
Sydney, NSW 2052, Australia
yang.song1@unsw.edu.au

Abstract. Airway segmentation is a prerequisite for diagnosing and screening pulmonary diseases. While computer aided algorithms have achieved great success in various medical image segmentation tasks, it remains a challenge in keeping the continuity of airway branches due to the special tubular shape. Some existing airway-specific segmentation models introduce topological representations such as neighbor connectivity and centerline overlapping into deep models and some other methods proposed customized network modules or training strategies based on the characteristics of airways. In this paper, we propose a large-kernel attention block to enlarge the receptive field as well as maintain the details of thin branches. We reformulate the segmentation problem into pixelwise segmentation and connectivity prediction with a differentiable connectivity modeling technique, and also propose a self-correction loss to minimize the difference between these two tasks. In addition, the binary ground truth is transformed into distances from the boundary, and distance regression is used as additional supervision. Our proposed model has been evaluated on two public datasets, and the results show that our model outperforms other benchmark methods.

Keywords: Airway segmentation · tubular structure · large-kernel attention · self-correction loss · distance regression

1 Introduction

Airway segmentation from computed tomography (CT) images is a crucial step for pulmonary diseases including asthma, bronchiectasis, and emphysema. Accurate segmentation enables the analysis of bronchial morphology details branch direction, which is useful for diagnosing abnormalities and assisting surgery. Manual segmentation is however time-consuming and error-prone, and relies on the expertise of doctors. Therefore, automatic computer-aided segmentation algorithms are beneficial.

Deep learning methods have been widely studied in segmentation of organs such as the liver [3] and prostate [4], which exhibit spherical or ellipsoidal shapes.

© The Author(s), under exclusive license to Springer Nature Singapore Pte Ltd. 2024
T. Liu et al. (Eds.): AI 2023, LNAI 14471, pp. 115–126, 2024.
https://doi.org/10.1007/978-981-99-8388-9_10

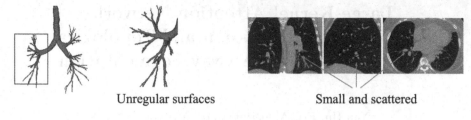

Unregular surfaces   Small and scattered

Fig. 1. The challenges of accurate airway segmentation.

On the other hand, airways are of tubular shapes and branches are small and scattered in cross-sections (see Fig. 1) so that universal models such as vanilla UNet and its variants [1,6,11,21] cannot obtain good performance.

To improve the performance of airway segmentation, various loss functions have been proposed for tubular structure segmentation [7,12,14,16]. For example, Shit et al. [12] proposed centerline Dice (clDice) loss for general tubular structure segmentation based on skeletonization overlapping error. They implemented the skeletonization through deep learning operations. For airway segmentation, Wang et al. [14] proposed a radial distance (RD) loss with the same definition as Dice loss but using radial distance map instead of binary ground truth. The radial distance map is obtained during preprocessing by computing the shortest Euclidean distance from each voxel of airway to the centerline.

There are also some studies proposing novel modules [8,14,17]. Wang et al. [14] incorporated 3D slice-by-slice convolutional layers within the UNet architecture to better capture the spatial relationships. Qin et al. [10] proposed a feature recalibration module to help discriminative feature learning and attention distillation module to focus on tubular structure feature learning.

Some other existing tubular structure segmentation methods reformulate the pixel-wise segmentation to another equivalent task focusing on the topological structures such as connectivity prediction [8,9], skeleton regression [14] and the combined approaches [13,15,18]. In particular, Qin et al. [8] proposed to model the binary ground truth into a 26-channel vector by modeling the connectivity in a 3×3 neighboring cube and performed connectivity prediction instead of original pixel-wise segmentation. Zhang et al. [18] combined key points detection, centerline regression and connectivity prediction using separate networks.

Despite these recent developments, a significant challenge for airway segmentation remains particularly for thin branches because they can be easily missing or broken due to pooling/upsampling operations. We propose to address the challenge by improving the network with dense skip connections to pass low-level features to high-level learning compared to the typically used backbone of UNet. Also, we design a large-kernel attention module on the skip connections to enlarge the receptive field as well as maintain the thin branches. In addition, since topological supervision has been proved effective for tubular segmentation [8,18], we design a new approach to embed the connectivity modeling and distance regression into the network and also design the corresponding topolog-

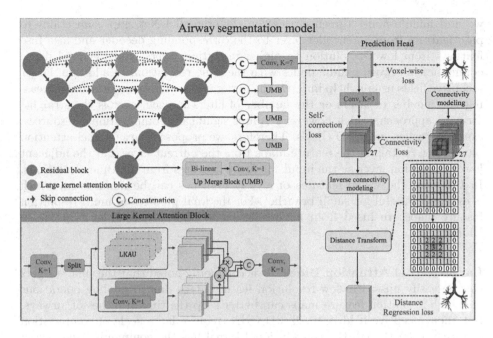

Fig. 2. Overview of the proposed model. (Color figure online)

ical self-correction loss and distance regression loss. We evaluated our proposed method on two public datasets and our method achieved the state-of-the-art performance. The visual results show our method can significantly alleviate the discontinuity in thin branches.

In summary, our contributions are as follows: (1) We propose a dense network with large-kernel attention block for airway segmentation. (2) In contrast to existing methods that incorporate connectivity modeling and distance regression in the pre-processing or post-processing steps [8,18], we incorporate pixel-wise segmentation and connectivity prediction into the network and propose the self-correction loss utilizing the consistency of these two tasks. (3) We propose a distance transformation approach on binary segmentation of airway based on morphology erosion and then propose the distance regression loss function.

2   Method

2.1   Network Architecture

Traditional U-shape networks [2,6,11] used in medical image segmentation consist of an encoder, a decoder and skip connections to pass low level features from the encoder to the decoder. Airway branches are scattered and small so that they are easily lost or merged with pooling/upsampling operations in traditional UNet. Therefore, we propose to use a dense-connected network (similar to UNet++ [21]) as the basic architecture. As shown in Fig. 2 (top left), we add middle nodes (green circles) on the original UNet skip connections to pass features from an encoder layer to the corresponding decoder and also fuse features from lower-resolution to higher-resolution layers, and we build dense connections between middle nodes with the same resolution. In addition, larger receptive fields usually help improve the performance of deep models and increasing the number of layers or the number of filters in each layer is the straightforward approach, which however requires significant computational resources, especially when using 3D CNNs. Therefore, we propose a large-kernel attention block (Fig. 2), taking the feature maps from the current layer and the adjacent lower-resolution layer as the input for fusing features from multiple resolutions. In this way, the visual features of airway branches can be maintained, which is especially useful for small branches. For the final prediction, multi-resolution feature maps are fused from low-resolution to high-resolution by "Up Merge Block".

Large-Kernel Attention Block. It is commonly noted that the small targets can be easily missed in low-resolution feature maps due to pooling operations and high resolution feature maps can better keep the tiny branches of airways, but their receptive field is small. 3D CNNs require large scale of computation resources even though the convolutional kernel has the commonly used size of 3. Our proposed a large-kernel attention block (LKAB) splits the features into multiple groups to perform large-kernel and small-kernel convolutions simultaneously, achieving their functions in representation learning as well as reducing the computation cost.

Each LKAB consists of two branches (Fig. 2). To reduce the computational cost, the input feature map X is first averagely split into three groups along the channel dimension as $\{X_1, X_2, X_3\}$. On the first branch, there are only three 3D convolution layers to process $\{X_1, X_2, X_3\}$ separately, obtaining $\{X_1', X_2', X_3'\}$. On the other branch, there are three large-kernel units (LKUs) adopted to process $\{X_1, X_2, X_3\}$ and obtain $\{Z_1, Z_2, Z_3\}$. Each LKU consists of three convolutional layers sequentially with the kernel size of $(1, 3, 5), (1, 5, 7)$ and $(1, 7, 9)$, respectively. The output is defined as $Y = Conv_{1 \times 1 \times 1}([X_1' \times Z_1, X_2' \times Z_2, X_3' \times Z_3])$.

2.2   Prediction Head

To learn the refined features with topology information, we propose a multi-task prediction head, containing three serial outputs: 1) voxel-wise segmentation, 2) connectivity prediction for topology structure learning, and 3) distance map regression for boundaries details refinement.

Voxel-Wise Segmentation. The breakage and missing segmentation of airway branches are due to false negatives. Union-based Tversky loss ($\mathbf{L_{tversky}}$) [13] controls the trade-off between false positives and false negatives by hyperparameters, α and β. The mathematic formulation is defined as:

$$\mathbf{L}_{\text{tversky}}(Y, \hat{Y}) = 1 - \frac{\sum_{i=1}^{N} Y_i \hat{Y}_i}{\sum_{i=1}^{N} Y_i \hat{Y}_i + \alpha \sum_{i=1}^{N} (1 - Y_i) \hat{Y}_i + \beta \sum_{i=1}^{N} Y_i (1 - \hat{Y}_i)} \quad (1)$$

where Y is the ground truth, $\hat{Y}$ is the output of network. In our experiments, α and β are set as 0.3 and 0.7, respectively, to emphasize the false negatives.

Voxel-based cross entropy loss takes each voxel in the image with equal weights, but voxels within and near airways are expected to have larger weights when computing loss in our study. Therefore, we introduce a weighted cross entropy loss with W to assign larger weight to voxels of and near airway, defined as:

$$\begin{cases} \mathbf{L}_{\text{wce}}(Y, \hat{Y}) = \frac{1}{N} \sum_{i=1}^{N} W(Y_i) \times (Y_i log(\hat{Y}_i) + (1 - Y_i)log(1 - \hat{Y}_i)) \\ \\ W(Y_i) = exp^{-P(Y_i)} \times M + 1 \end{cases} \quad (2)$$

where $P(\cdot)$ represents average pooling with the window size of 3 and stride of 1, and M is the dilated ground truth using soft dilation [12] to generate the dilated mask of airway. Then W makes the weights for boundaries, centerline and background decrease gradually.

The total voxel-wise loss is the combination of these two types of loss calculation, defined as:

$$L_{pw} = \mathbf{L}_{\text{tversky}}(Y, \hat{Y}) + \mathbf{L}_{\text{wce}}(Y, \hat{Y}) \quad (3)$$

Connectivity Prediction. The connectivity modeling [8] is capable of describing the connective relation between a center voxel and its nearest 26-neighbor voxels in a $3 \times 3 \times 3$ cube. Let p represent a given voxel labeled as airway, $Q_i \in Q$ represents the set of its 26 neighbors and p, and $V_p \in R^{27 \times 1}$ denotes the connectivity vector. If Q_i and p are both labeled as airway, they are connected and $V_p^i = 1$. Otherwise, they are unconnected and $V_p^i = 0$. Let Y_c and $\hat{Y}_c$ represent the connectivity ground truth and the output of connectivity prediction. The loss of connectivity prediction is defined as:

$$L_c = \mathbf{L}_{\text{Dice}}(Y_c, \hat{Y}_c) \quad (4)$$

Theoretically, connectivity maps and binary segmentation masks can be transformed to each other. For instance, only when the connectivity vector of both p and Q_i show they are connected, p and Q_i are airway. Given a 27-channel connectivity map V, we first organize it to $\tilde{V} \in R^{27 \times 1}$. Specifically, $\tilde{V}_p^i$ is the average of V_p^i and $V_{Q_i}^j$, where j represents the index of p when the center voxel is Q_i. Then, the binary segmentation label of p is defined as:

$$Y_p = \sum_{i=1}^{27} (\tilde{V}_p^i \times \mathbf{sign}(\tilde{V}_p^i - \delta)) / \sum_{i=1}^{27} \mathbf{sign}(\tilde{V}_p - \delta) \quad (5)$$

where δ is the thresholding to determine connected or unconnected. These two steps are referred to as inverse connectivity modeling C^{-1}.

Previous studies [8,9,18] have conducted the connectivity modeling in pre-processing, formulating the segmentation task as a connectivity prediction task, and used inverse connectivity modeling to obtain the binary segmentation in post-processing. Since connectivity modeling and inverse connectivity modeling are separate from the network training in these methods, the consistency between these two tasks cannot be exploited. Therefore, we implement them by using convolutional layers with pre-defined kernel, which are untrainable but support error back propagation because they are differentiable. More specifically, we define a $3 \times 3 \times 3$ kernel with the center voxel and one of its neighbors initialized as 1, then use this kernel to convolve a binary segmentation mask of airway. The corresponding two voxels are connected if the result is 2. Otherwise, they are unconnected. Similarly, we can also implement the fist step $(V \to \tilde{V})$ of inverse connectivity modeling C^{-1} by convolutional layers with the pre-defined kernel. Based on the consistency of pixel-wise segmentation and connectivity prediction, we propose a self-correction loss, i.e.,

$$L_{sc} = \mathbf{L}_{\text{dice}}(C^{-1}(\hat{Y}_c), \hat{Y}) \tag{6}$$

Distance Transformation. Previous studies [14,18] reformulated the binary segmentation task as a distance regression problem by converting the binary ground truth into the shortest distance from the centerline [14] or mask a Gaussian distribution on the centerline [18] as the new ground truth. Though the distance regression is effective for airway segmentation, it involves tuning of hyperparameters. In our approach, we propose a distance transformation approach based on erosion and it can be embedded into the network.

Specifically, erosion can eliminate the boundary voxels of a binary image and thin the object inward to the centerline until all foreground regions of the image have been eroded away. Therefore, we iteratively perform erosion ($\mathbf{Erode}_i, i = 1, ..., N$, N represents the maximum radius of airway) on the binary airway segmentation $\tilde{Y}$ and ground truth Y. A simple sample is illustrated in Fig. 2. Then i can represent the distance from the eroded voxels in $\mathbf{Erode}_i$ to the original boundary. The binary segmentation of airway from connectivity prediction can be formulated as $\tilde{Y} = \mathbf{sign}(C^{-1}(\hat{Y}_c))$. We conduct distance transformation on $\tilde{Y}$ and the ground truth Y, obtaining the distance maps $D(\tilde{Y})$ and $D(Y)$. The distance regression loss is then defined as:

$$L_{reg} = \mathbf{L}_2(exp^{-D(\tilde{Y})}, exp^{-D(Y)}) \tag{7}$$

Loss Function. The overall loss function is the combination of three predictions, defined as:

$$L = \alpha_1 L_{pw} + \alpha_2 L_c + \alpha_3 L_{sc} + \alpha_4 L_{reg} \tag{8}$$

where the weights are set to $\alpha_1 = 1, \alpha_2 = 1, \alpha_3 = 1, \alpha_4 = 5$.

2.3   Implementation Details

Datasets. In this study, we evaluated our segmentation model on two public datasets. *(1) The Binary Airway Segmentation Dataset (BAS)* [19]. It contains 90 CT cases, in which 20 cases are from the EXACT'09 Challenge [5] training set and 70 cases are from the Lung Image Database Consortium image collection (LIDC-IDRI). The intra-section resolution ranges from 0.5×0.5 mm^2 to 0.82×0.82 mm^2 and thickness ranges in 0.5–1 mm. We split these 90 CT cases into the training (50 cases), validation (20 cases) and testing (20 cases) sets following [20]. *(2) ATM'22 Challenge dataset.* This dataset contains 300 cases for training and 50 cases for online validation respectively. The intra-section resolution ranges from 0.5×0.5 mm^2 to 0.82×0.82 mm^2 and thickness ranges in 0.5–1 mm. However, only the annotation of training cases is available for our training and the online evaluation system is on the running for evaluation. In this study, we trained our model on the whole training set and then submitted the segmentation results to the online evaluation system to obtain quantitative evaluations for fair comparison with other methods.

Preprocessing. The intensity is first truncated to the range $[-1024, 600]$ (HU) and then linearly normalized to $[0, 1]$. Meanwhile, the coordinates are normalized to $[0, 1]$ and concatenated with CT images as the input of the segmentation network. Airway is first divided into segments at each bifurcation and these segments are classified into three classes based on the number of voxels, i.e., large (top rank 5), medium (top rank 40) and small (the rest). In each epoch, 48 patches are randomly cropped with their center voxel in large, medium and small at the rate of 10%, 40% and 50%, respectively. The training phase is divided into two stages with input size of $64 \times 64 \times 64$ for the first stage and $96 \times 96 \times 96$ for the second stage. In the second stage, data augmentation is employed by random rotation (angle ranges in $[-10, 10]$).

Training Details. Adam is adopted as the optimizer, with the learning rate of 10^{-4} and 10^{-6}, the number of epochs of 30 and 50, and the batch size of 4 and 1 for two stages, respectively. During the testing, the size of input patches is $96 \times 96 \times 96$ and the stride of sliding window is 48 in each direction. The morphology operations (erosion and dilation) are from [12]. Experiments are performed on 1 NVIDIA RTX 3090 with 24 GB memory using TensorFlow 2.4.0.

3   Experimental Results

3.1   Metrics

To evaluate our model and make a fair comparison with other methods, we follow the ATM'22 Challenge evaluation using the following metrics: (1) precision, (2) the Dice similarity coefficient (Dice), (3) the ratio of tree length detected (TD), and (4) the ratio of branches detected (BD).

3.2   Comparison with Other Methods

BAS Dataset. Results achieved by different methods are summarised in Table 1. We compared to four typical methods, i.e., AirwayNet [8] and RCL-RD [14] using topological learning, and Tubule-CNN [10] and Alleviate-Grad [20] using task-specific modules and learning strategies. AirwayNet and RCL-RD are reimplemented by ourselves. Tubule-CNN is trained using the officially released code. We directly quote the result of Alleviate-Grad.

Table 1. Results of BAS dataset.

Method	Dice	TD	BD	Precision
AirwayNet [8]	0.901	0.836	0.814	0.892
RCL-RD [14]	0.882	0.862	0.843	0.901
Tubule-CNN [10]	0.914	0.908	0.865	0.881
Alleviate-Grad [20]	–	0.914	0.887	–
Ours	**0.920**	**0.931**	**0.894**	**0.923**

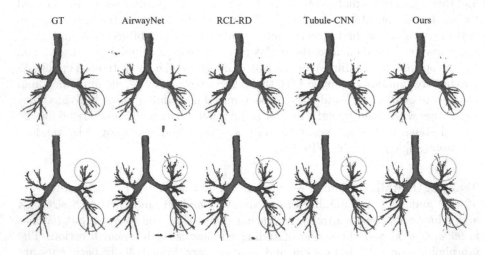

Fig. 3. Visual comparison with other methods.

The experimental results show that our method outperforms the other compared methods. Specifically, compared to AirwayNet using connectivity modeling prediction and RCL-RD using distance regression, the performance of Tubule-CNN and Alleviate-Grad are improved because they use task-specific loss functions and modules for the thin branches segmentation. Our method achieved substantial improvements on all the metrics with the advantages brought by topological learning and customized loss functions for thin branches. Figure 3 shows two examples of the segmentation results. It can be seen that our method

Table 2. Results of visual samples from BAS dataset.

Method		Dice	TD	BD	Precision
Sample 1	AirwayNet	0.934	0.869	0.721	0.938
	RCL_RD	0.944	0.893	0.737	0.925
	Tubule-CNN	0.911	0.844	0.730	0.856
	Ours	0.937	0.928	0.853	0.927
Sample 2	AirwayNet	0.918	0.930	0.893	0.898
	RCL_RD	0.874	0.962	0.951	0.803
	Tubule-CNN	0.901	0.954	0.926	0.846
	Ours	0.916	0.967	0.959	0.856

Table 3. Leaderboard of ATM'22 challenge validation dataset (Top-5).

Rank	Score	Dice	TD	BD	Precision
1st	5.5	0.960	**0.961**	**0.943**	0.953
2nd (Ours)	6.3	**0.961**	0.908	0.871	**0.977**
3rd	8.8	0.944	0.958	0.876	0.934
4th	9.3	0.916	0.958	0.876	0.948
5th	10	0.914	0.949	0.866	0.955

outperforms others for the thin branches with smoother surfaces and less breakages. That is because LKAB can pass more effective information with local details and our regression loss focuses on the accuracy of boundaries. In addition, there are abnormal points outside of the airway, especially for AirwayNet and RCL-RD due to using small training patches. Our method can alleviate this issue with the self-correction loss because it helps reduce the false positive rate (Table 2).

ATM Challenge 2022 Dataset. Also, we report the results of the official leaderboard of ATM'22 challenge in Table 3. Our results are ranked second among all the participants. Although TD and BD of our method are lower than highest results, we achieved the top-1 performance in Dice and precision. Such trade-offs have been observed in other studies as well [7].

Figure 4 shows the visual results we submitted to the online validation system and Table 4 shows the quantitative evaluation results returned. Both the visual and quantitative results indicate that the segmentation is relatively successful. However, we have observed that there are still some noise segments located outside the airway. It may be related to the distribution of training samples. It can be addressed through improvements in the sample strategy or post-processing techniques.

Fig. 4. Visual results of ATM Challenge validation dataset.

Table 4. Quantitative results corresponded to visual result of ATM Challenge

No.	DSC	TD	BD	Precision
Sample 1	0.976	0.977	0.967	0.959
Sample 2	0.959	0.913	0.878	0.970
Sample 3	0.956	0.973	0.961	0.948

Table 5. Results of the ablation study on the BAS dataset.

Setting	Dice	TD	BD	Precision
w/o LKAB	0.901	0.913	0.861	0.894
L_{pw}	0.885	0.913	0.846	0.868
$L_{pw} + L_c$	0.893	0.904	0.837	0.882
$L_{pw} + L_c + L_{sc}$	0.918	0.912	0.850	0.916
$L_{pw} + L_c + L_{reg}$	0.905	0.923	0.864	0.921
Ours	**0.920**	**0.931**	**0.894**	**0.923**

3.3   Ablation Study

Effectiveness of Large Kernel Attention. To prove the effectiveness of large kernel attention, we replaced all middle nodes (green circles in Fig. 2) with simplified attention blocks. Compared to LKAB, the feature map splitting was removed and the kernel size of the attention branch was set to 3 for all the three convolutional layers, and the kernel size of another branch remained 1. The results of our full model and "w/o LKAB" in Table 5 show LKAB brings large improvement in all the evaluation metrics because it helps maintain local details while providing a large receptive field, which is useful for thin branches segmentation.

Effectiveness of Connectivity Modelling and Self-correction Loss. The final segmentation is obtained from the output of pixel-wise prediction. Compared to only using pixel-wise supervision ("L_{pw}"), connectivity modelling

("$L_{pw} + L_c$") helps to improve the Dice because it considers neighbours' annotations. Consistency regularization helps models to make consistent predictions even when inputs undergo small perturbations. Self-correction loss utilizes consistency between voxel-wise segmentation and connectivity prediction to improve robustness against perturbations in inputs. Compared to the result of "$L_{pw}+L_c$", the ratios of detected branches and tree length remain roughly flat with a small improvement while Dice is improved by 0.013. This aligns with the expectation of improving the performance via reducing the false positives.

Effectiveness of Distance Regression Loss. Distance regression loss focuses more on boundaries because the gradient of it is more sensitive to the errors of boundaries. The ablation result of distance regression loss improves a bit in Dice with more improvement in BD and TD. This shows its effectiveness in helping the segmentation of thin branches.

4 Conclusion

In this paper, we propose an airway segmentation model based on the UNet++ architecture. Large-kernel attention blocks are used to merge features learned from different resolutions and maintain local details while providing a large receptive field. Moreover, we propose task-specific loss functions, i.e., self-correction utilizing the consistency between voxel-wise segmentation and connectivity prediction and distance regression loss focusing on the accuracy of boundaries. Our methods achieved superior performance on two public datasets, showing potential applicability in real-world clinical scenarios.

References

1. Çiçek, Ö., Abdulkadir, A., Lienkamp, S.S., Brox, T., Ronneberger, O.: 3D U-net: learning dense volumetric segmentation from sparse annotation. In: Ourselin, S., Joskowicz, L., Sabuncu, M.R., Unal, G., Wells, W. (eds.) MICCAI 2016. LNCS, vol. 9901, pp. 424–432. Springer, Cham (2016). https://doi.org/10.1007/978-3-319-46723-8_49
2. Diakogiannis, F.I., Waldner, F., Caccetta, P., Wu, C.: ResUNet-a: a deep learning framework for semantic segmentation of remotely sensed data. ISPRS J. Photogramm. Remote. Sens. **162**, 94–114 (2020)
3. Gul, S., Khan, M.S., Bibi, A., Khandakar, A., Ayari, M.A., Chowdhury, M.E.: Deep learning techniques for liver and liver tumor segmentation: a review. Comput. Biol. Med. 105620 (2022)
4. Khan, Z., Yahya, N., Alsaih, K., Al-Hiyali, M.I., Meriaudeau, F.: Recent automatic segmentation algorithms of MRI prostate regions: a review. IEEE Access **9**, 97878–97905 (2021)
5. Lo, P., et al.: Extraction of airways from CT (EXACT'09). IEEE Trans. Med. Imaging **31**(11), 2093–2107 (2012)

6. Milletari, F., Navab, N., Ahmadi, S.A.: V-Net: fully convolutional neural networks for volumetric medical image segmentation. In: 2016 Fourth International Conference on 3D Vision (3DV), pp. 565–571. IEEE (2016)
7. Nan, Y., et al.: Fuzzy attention neural network to tackle discontinuity in airway segmentation. arXiv preprint arXiv:2209.02048 (2022)
8. Qin, Y., et al.: AirwayNet: a voxel-connectivity aware approach for accurate airway segmentation using convolutional neural networks. In: Shen, D., et al. (eds.) MICCAI 2019. LNCS, vol. 11769, pp. 212–220. Springer, Cham (2019). https://doi.org/10.1007/978-3-030-32226-7_24
9. Qin, Y., Gu, Y., Zheng, H., Chen, M., Yang, J., Zhu, Y.M.: AirwayNet-SE: a simple-yet-effective approach to improve airway segmentation using context scale fusion. In: 2020 IEEE 17th International Symposium on Biomedical Imaging (ISBI), pp. 809–813. IEEE (2020)
10. Qin, Y., et al.: Learning tubule-sensitive CNNs for pulmonary airway and artery-vein segmentation in CT. IEEE Trans. Med. Imaging **40**(6), 1603–1617 (2021)
11. Ronneberger, O., Fischer, P., Brox, T.: U-net: convolutional networks for biomedical image segmentation. In: Navab, N., Hornegger, J., Wells, W.M., Frangi, A.F. (eds.) MICCAI 2015. LNCS, vol. 9351, pp. 234–241. Springer, Cham (2015). https://doi.org/10.1007/978-3-319-24574-4_28
12. Shit, S., et al.: clDice: a novel topology-preserving loss function for tubular structure segmentation. In: Proceedings of the IEEE/CVF Conference on Computer Vision and Pattern Recognition, pp. 16560–16569 (2021)
13. Tetteh, G., et al.: DeepvesselNet: vessel segmentation, centerline prediction, and bifurcation detection in 3D angiographic volumes. Front. Neurosci. 1285 (2020)
14. Wang, C., et al.: Tubular structure segmentation using spatial fully connected network with radial distance loss for 3D medical images. In: Shen, D., et al. (eds.) MICCAI 2019. LNCS, vol. 11769, pp. 348–356. Springer, Cham (2019). https://doi.org/10.1007/978-3-030-32226-7_39
15. Yu, W., Zheng, H., Zhang, M., Zhang, H., Sun, J., Yang, J.: Break: bronchi reconstruction by geodesic transformation and skeleton embedding. In: 2022 IEEE 19th International Symposium on Biomedical Imaging (ISBI), pp. 1–5. IEEE (2022)
16. Zhang, M., Yang, G.Z., Gu, Y.: Differentiable topology-preserved distance transform for pulmonary airway segmentation. arXiv preprint arXiv:2209.08355 (2022)
17. Zhang, M., Zhang, H., Yang, G.Z., Gu, Y.: CFDA: collaborative feature disentanglement and augmentation for pulmonary airway tree modeling of COVID-19 CTs. In: Wang, L., Dou, Q., Fletcher, P.T., Speidel, S., Li, S. (eds.) MICCAI 2022. LNCS, vol. 13431, pp. 506–516. Springer, Cham (2022). https://doi.org/10.1007/978-3-031-16431-6_48
18. Zhang, X., et al.: Progressive deep segmentation of coronary artery via hierarchical topology learning. In: Wang, L., Dou, Q., Fletcher, P.T., Speidel, S., Li, S. (eds.) MICCAI 2022. LNCS, vol. 13435, pp. 391–400. Springer, Cham (2022). https://doi.org/10.1007/978-3-031-16443-9_38
19. Zheng, H.: BAS dataset download. https://github.com/haozheng-sjtu/3d-airway-segmentation
20. Zheng, H., et al.: Alleviating class-wise gradient imbalance for pulmonary airway segmentation. IEEE Trans. Med. Imaging **40**(9), 2452–2462 (2021)
21. Zhou, Z., Siddiquee, M.M.R., Tajbakhsh, N., Liang, J.: UNet++: redesigning skip connections to exploit multiscale features in image segmentation. IEEE Trans. Med. Imaging **39**(6), 1856–1867 (2019)

Deep Learning

WeightRelay: Efficient Heterogeneous Federated Learning on Time Series

Wensi Tang and Guodong Long[✉]

Australian Artificial Intelligence, Institute Faculty of Engineering and IT,
University of Technology Sydney, Ultimo, Australia
wensi.tang@student.uts.edu.au, guodong.long@uts.edu.au

Abstract. Federated learning for heterogeneous devices aims to obtain models
of various structural configurations in order to fit multiple devices according to
their hardware configurations and external environments. Existing solutions train
those heterogeneous models simultaneously, which requires extra cost (e.g. computation, communication, or data) to transfer knowledge between models. In this
paper, we proposed a method, namely, weight relay (WeightRelay), that could get
heterogeneous models without any extra training cost. Specifically, we find that,
compared with the classic random weight initialization, initializing the weight
of a large neural network with the weight of a well-trained small network could
reduce the training epoch and still maintain a similar performance. Therefore, we
could order models from the smallest and train them one by one. Each model
(except the first one) can be initialized with the prior model's trained weight for
training cost reduction. In the experiment, we evaluate the weight relay on 128-
time series datasets from multiple domains, and the result confirms the effectiveness of WeightRelay. More theoretical analysis and code can be found in (https://
github.com/Wensi-Tang/DPSN/blob/master/AJCAI23_wensi_fedTSC.pdf).

Keywords: Time series classification · Federated learning · Heterogeneous
model

1 Introduction

With the development of smart devices, an increasing amount of time series data can be
collected, such as daily heartbeats, blood oxygen levels, electronic consumption, and
motion signals for smart device control [1,6–8,19,21,35].

With the advent of sophisticated devices and advanced data analysis technologies,
there is potential to bring immense value to society. However, privacy concerns limit the
integration of smart devices with state-of-the-art deep learning [24,59]. Specifically,
smart devices encounter challenges when training large models due to limited power
and computational resources for local training. Concurrently, uploading information
from these devices, which are commonly found in homes, can pose significant privacy
risks, as they often contain sensitive data [14,23].

T. Liu et al. (Eds.): AI 2023, LNAI 14471, pp. 129–140, 2024.
https://doi.org/10.1007/978-981-99-8388-9_11

Federated learning [14] can be used to train deep learning models with privacy protection. However, there are still many unsolved challenges in applying those solutions to smart devices. Specifically, smart devices even for a similar function, typically have varying hardware configurations. The hardware heterogeneous brings a challenge for federated learning. Specifically, how to transfer knowledge between heterogeneous models [14,21,59]. Specifically, this heterogeneous federated learning aims to obtain models of various structural configurations to fit multiple devices according to their hardware configurations and working environment. Under this setting, it is hard for low-capacity devices to contribute their knowledge to big models, for they might have enough memory, bandwidth or computational power to join the big model training via Federate average. Therefore, solutions that could enable big models to get knowledge from the small models are highly desired [14,21].

Existing solutions tackle the problem by adding one or more resources such as computational, communication, and extra data. For example, distillation-based methods require training cost on multiple models and extra cost for knowledge transfer between those models [20,30,31,58,61]. Pruning-based methods [25,29] require an extra cost to pruning the single model onto multiple smaller models. Weight sharing methods, some of them [51] require the computational cost to match the weight of various models iteratively, and some methods [11,55] need a weight scale module to adjust weight before sharing.

Although existing solutions enable knowledge sharing between heterogeneous models, the extra resource consumption makes it hard to implement them on smart devices. This is because most of those devices do not have strong computational, memory or commutations capacity [14,21]. Therefore, solutions with huge training costs bring an embarrassing burden to low-capacity smart devices. When a device's ability limits it from using big models, it may also limit it from contributing its knowledge to big models or getting knowledge from big models via distillations, pruning or weight matching.

Other than capacity neglection, for smart devices, the appropriate 1D-CNN (one-dimensional-convolutional neural networks) model is also seldom mentioned. Specifically, most solutions were tested with 2D-CNN(two-dimensional-convolutional neural networks) models [11,20,30,51,55,58,61] or language models [31]. However, we should notice that most of the data gathered by smart devices are time-series data [33,53] which can mathematically be described as a series of data points recorded in time order [8,9,13,49]. Such as the heartbeat data collected by smartwatches [37,39], the electricity consumption data gathered by energy management devices [15], building structural vibration data recorded by motion sensors [27,50], etc. According to the University of California, Riverside time series archive (UCR archive) [9], the state-of-the-art solutions for time series classification tasks are all 1D-CNNs [10,13,49].

The characteristics of 1D-CNN allow a novel weight relay solution, which does not need any extra resources. Specifically, suppose we have a small 1D-CNN network and a large 1D-CNN network. For the large network training, we could initialize it with a classic random weight initialization or we could initialize it with the weight from a well-trained small network. We find that these two kinds of initialization will be of similar performance, but the second initialization could reduce the training cost of the large network. This training cost reduction could be used to lower the capacity requirement and allows more low-capacity devices to join the big model training. Via ordering those heterogeneous 1D-CNN models from the smallest to the largest, except for the first smallest model, all the other models' training will be benefited.

In experiments, we show the consistently training cost reduction ability of the weight relay on time series datasets from multiple domains i.e., healthcare, human activity recognition, speech recognition, and material analysis. Despite the dynamic patterns of these datasets, weight relay robustly shows its effectiveness.

2   Related Work

2.1   Deep Learning for Time Series Classification

The success of deep learning encourages the exploration of its application on time series data [12,13,28]. Intuitively, the Recurrent Neural Network (RNN), which is designed for temporal sequence, should work on the time series tasks. However, in practice, RNN is rarely applied to TS classification [13]. One widely accepted reason among many is that RNN models suffer from vanishing and exploding gradients when dealing with long sequence data [2,13,38]. Nowadays, 1D-CNN is the most popular deep-learning method TSC tasks. [22,26,40,41,52,60]. According to the University of California, Riverside time series archive (UCR archive) [9], the state-of-the-art solutions for time series classification tasks are all 1D-CNNs [10,22,49,52].

2.2   Federated Learning on Heterogeneous Devices

Based on the method of transferring knowledge between heterogeneous models, the solutions to heterogeneous Federated learning could be divided into three columns. These are distillation-based, weight-sharing-based pruning-based methods and prototype-based methods.

The knowledge distillation [5,18] allows the knowledge sharing between heterogeneous models. It requires extra data or computational resources to enable knowledge sharing between models [20,30,56], which brings multiple challenges under the federated learning setting. For example, the distillation process requires a large amount of computational resources [61]. What's more, the performance of the distillation is highly related to the similarity of the distributions between the training data and distillation data.

The weight sharing method is based on the assumption that some parts of the weight of various structure models are the shareable, and the shareable part could help to transfer the knowledge between various structure models [4,36,42,51]. In practice, finding which parts of various models should be of the same parameter is hard. Therefore, a large number of computation resources have to be taken to find which parts of models should be matched together [51], or taken to calculate the adjustment module for weight re-scale [11,43].

The pruning-based method aims at training a large neural network and pruning it into various small networks [57] according to the hardware configuration and external environment [25,32,54]. One limitation is that it is hard to control the structure of the pruned network, which challenges fitting those small networks according to the configuration of each edge device. [25,29,47]

The prototype-based federated learning [3,16,17,46] can also be viewed as a solution for heterogeneous devices for it requires a very limited resource for classification calculation. However, as the number of classes increases, prototype learning struggles to scale effectively. Put simply, due to the warp characteristic, comparing the distance between two-time series incurs a computational cost of N^2. Where N is the length of the signal. As the number of classes grows, this cost rises substantially. As a result, the computational cost shifts from training to classification, which is not conducive for small devices. While some methods can map the time series into a feature space [45,48], their classification accuracy cannot match that of larger models.

3   Motivation

To train a 1D-CNN model, we could 1) start from a classic random weight initialization or 2) replace parts of the random initialization weight with a well-trained weight from a small network. The second initialization could reduce the training cost of the large network and won't influence the final performance.

Therefore, to train multiple models on a smaller budget, we don't need to train every model from the stretch. We could initialize some of those models by the trained weight from the others for fast convergence. Figure 1 gives an example of the weight relay on the Crop [44] dataset.

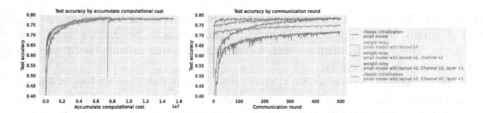

Fig. 1. The left image shows the relationship between the accumulated computational cost and the test accuracy of each model. The accumulated computational cost calculates the computational cost we used to obtain a well-trained model. As the image shows, to obtain a single largest model from classic initialization (purple), we need about $0.8e^7$ computational resources, and we could only get 1 model. However, with weight relay, at the point $0.8e^7$, we have four models(blue, orange, green and red). And the red model, which has the same structure as the purple model, also has the same performance. The right image shows the performance of each model by communication round. We could see that the weight relay model (red) converged much faster than the classic initialization models (purple). (Color figure online)

4   Weight Relay

In Fig. 2, a schematic of weight relay is given. What's more, in this Section, we will explain the weight relay in detail. Specifically, Sect. 4.1 will introduce heterogeneous models for time series classifications. Section 4.2 will introduce how to align those models. Specifically, when a well-trained weight is passed to a large model, which part of the large network should be replaced.

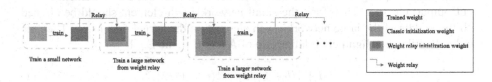

Fig. 2. The schematic shows training multiple models with weight relay. Weight relay starts from the training of the smallest network with classic initialization. The trained weight of the smallest network will be used to replace a part of the classic initialization weight of a large network. When the large network is trained, its weight could be used to accelerate the training of a larger network. Since the weight relay only replaces the random initialization with a well-trained one, it requires almost no cost, and the training cost of each model (except the first one) is also smaller than training those from classic initialization.

4.1   Heterogeneous Models

According to the result statistics on the UCR archive, all state-of-the-art neural network solutions on time series classification tasks are 1D-CNN models [13,34,49,52]. Therefore, this paper mainly talked about 1D-CNN models and heterogeneous could happen on all three main structure configurations: the number of layers, kernel sizes and the number of channels.

4.2   Weight Alignment

The weight alignment defines which part of the large network should be replaced with the weight of the small network. The alignment will have three steps because the neural network has three hierarchies. Specifically, a neural network is composed of layers. Layers are composed of weight sets. And the weight sets are composed of weight tensors. Therefore, we need to pair weight sets and layers before we align weight tensors.

First Step: Pair Weight Sets by Layers. The convolutional layer and the batch normalization layer will be indexed from the input to the output. For example, the first convolutional layer is the convolutional layer closest to the input. The fully connected layer will be indexed by **reverse order**. Therefore, the first fully connected layer will be the layer closest to the output. For each of those three types of layers, the $*$-th layer's tensor set of the small network should be paired with the $*$-th layer's tensor set of the large network

Second Step: Pair Tensors by Paired Weight Sets: According to the definitions of the 1D-CNN, tensors in each set will have different functions, such as weight, bias, and running mean. Therefore, for two paired sets, tensors in the two sets will be paired by their function name.

Third Step: Align Two Paired Tensors. For two paired tensors **A** and **B**, the alignment of the two tensors is to **align the output and input dimensions with the left margin** and **align the kernel dimension with the centre**. Specifically, the small network's *-th element should be aligned with the middle of the large network's *-th element on the input(output) channel. Therefore, the small network's i-th element should be aligned with the middle of the large network's j-th element on the kernel dimension. Then the i and j describe the alignment relationship of **A** and **B** in the Eq. 1

$$j = i + \lfloor (b-1)/2 \rfloor - \lfloor (a-1)/2 \rfloor \tag{1}$$

where the b and a are kernel sizes of large network and small network.

5   Analysis of Weight Relay

This section will have two parts. In Sect. 5.1, we will show that though the alignment method in Eq. 1 is defined in pair, it is reliable for the alignment of multiple models. Secondly, we will give a macro (Sect. 5.2) and a micro (Sect. 5.3) explanation of the coverage acceleration of the weight relay.

5.1   Consistency Proof for the Alignment

This section will show that despite we only define the relationship between two kernels, this operation will keep the **consistency** when we have **multiple** kernels. The consistency of weight alignment can be describe as below: For any three kernel weights $\{\mathbf{A}, \mathbf{B}, \mathbf{C}\}$ and their length relationship are:

$$a < b < c$$

We could use weight alignment to deter the alignment relationship between **C** and the other two kernels as:

$$\mathbf{A}_i \xrightarrow{\text{align with}} \mathbf{C}_k \tag{2}$$

$$\mathbf{B}_j \xrightarrow{\text{align with}} \mathbf{C}_k \tag{3}$$

The signal Phase alignment can be called consistency if, with the same operation on **A** and **B** , we should have

$$\mathbf{A}_i \xrightarrow{\text{align with}} \mathbf{B}_j$$

Here, we give an example to illustrate the consistency of multiple kernels. Supposing we have three kernels of length 3, 5, 8. Via Eq. 1, the $\mathbf{A}_1$ should align with $\mathbf{C}_4$, the $\mathbf{B}_3$ should align with $\mathbf{C}_4$. **Consistency** means that the $\mathbf{A}_1$ should align with $\mathbf{B}_3$.

Proof of Consistency for Eq. 1:

When we combine Eq. 1 and Eq. 2 we know that the a th element of kernel **A** should align with c th element of kernel **C** and their index relationship is:

$$k = i + \lfloor (c-1)/2 \rfloor - \lfloor (a-1)/2 \rfloor \tag{4}$$

With Eq. 1 and Eq. 3 we know the alignment relationship between elements in **B** and **C** is:

$$k = j + \lfloor (c-1)/2 \rfloor - \lfloor (b-1)/2 \rfloor \tag{5}$$

Using the Eq. 5 to subtract Eq. 4 and we have:

$$k - k = \begin{array}{c} i + \lfloor (c-1)/2 \rfloor - \lfloor (a-1)/2 \rfloor \\ + \\ -j - \lfloor (c-1)/2 \rfloor + \lfloor (b-1)/2 \rfloor \end{array} \tag{6}$$

Therefore, we know

$$j \equiv i + \lfloor (b-1)/2 \rfloor - \lfloor (a-1)/2 \rfloor$$

Which means that

$$\mathbf{A}_i \xrightarrow{\text{should align with}} \mathbf{B}_j$$

as it should be.

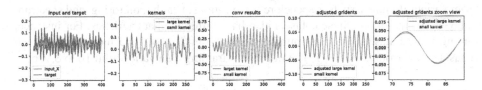

Fig. 3. This example shows that: Given the same input and target, the gradient direction of the small kernel is the same as the gradient direction of the aligned part of the large network. Specifically, images from left to right are: 1) input and target, which are generated by random noise; 2) A large kernel weight which is random noise, and the weight of the small kernel, which is cropped from the large kernel; 3) The convolution results for large kernel and small kernel with the input in image 1); 4) The gradients of kernels which is calculated with the random target in image 1). To demonstrate the overlapping parts of the two kernels are in a similar direction, their magnitude was adjusted; 5). A zoom view of the overlapping gradient parts (indexed 70 to 90).

5.2 Macro Explanation of the Training Acceleration

One explanation of the training acceleration is that compared with the classic initialization weight, the weight relay initialization is closer to the final weight. In Fig. 4 We statistic the distance between the initialization weight and trained weight, and we will see that compared with the random initialization, the weight relay initialization are of a smaller distance to the final value.

Fig. 4. The blue points are results from 5000 tests with random input, targets, and weight. The orange point is the origin point. We could see that when the value of the x-axis approaches 0 (target similar), the value of the y-axis comes to 0 (gradient direction similar). (Color figure online)

5.3   Micro Explanation of the Training Acceleration

The micro explanation to the training acceleration is that: the small network's optimization direction is the optimization direction of the large network's sub-network, where the small network should replace the weight. Therefore, the training acceleration is because the training of the small network makes the sub-network close to the final target.

To explain this, we could start from a simple case study in Fig. 3.

6   Experiment

6.1   Benchmarks

University of California, Riverside (UCR) 128 archive [9] is selected to evaluate the weight relay under the federated setting. This is an archive of 128 univariate TS datasets from various domains, such as speech reorganizations, health monitoring, and spectrum analysis. What's more, those datasets also have different characteristics. For instance, among those datasets, the class number varies from 2 to 60, and the length of each dataset varies from 24 to 2844. The number of training data varies from 16 to 8,926.

6.2   Evaluation Criteria

Following the datasets archive [9], the accuracy score is selected to measure the performance. Following the paper, the communication round multiple model size is selected to measure the training cost.

6.3   Experiment Setup

Following [13,22,49], for all benchmarks, we follow the standard and unify settings [52] for all 128 datasets in the UCR archive. The training will stop when the training loss is less than 1e-3 or reach 5000 epoch. To mimic the federated learning scenario, the client number is 10. More details can be found in the supplementary material.

6.4   Experiment Result

The experiment shows that weight relay has similar performance and fewer computation resources costs than using federated average on all devices to obtain each model. Due to the larger number of datasets, we cannot list the results of all datasets. Therefore, we plot the statistical result of the 128 datasets, and the result is shown in Fig. 5.

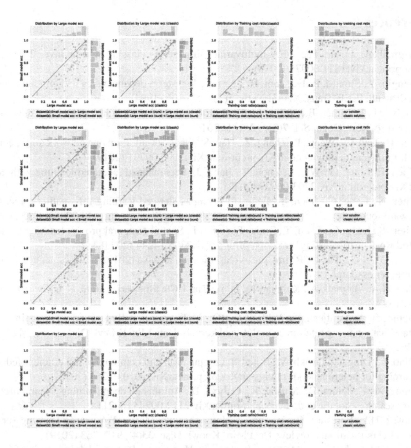

Fig. 5. From top to bottom, each row shows the statistical results when the large model has 2Xkernel size, 2Xchannel number, one extra layer, and all extensions than the smaller model. From the left to the right image, we can see that, for most of the dataset, the large model will perform better than the small model (the first column). The performance of the weight relay is similar to the performance of training from classic (the second column); the weight relay has a lower training cost (the third column); To archive similar performance, the weight relay has a smaller training cost (the fourth column).

7   Conclusion

In this paper, we proposed the weight relay method, which could reduce the training cost for heterogeneous model training. We theoretically analyse the mechanism of weight relay and experimentally verify the effectiveness on multiple datasets from multiple domains.

Acknowledgements. Please place your acknowledgments at the end of the paper, preceded by an unnumbered run-in heading (i.e. 3rd-level heading).

References

1. Bagnall, A., et al.: The UEA multivariate time series classification archive, 2018. arXiv preprint arXiv:1811.00075 (2018)
2. Bengio, Y., Simard, P., Frasconi, P.: Learning long-term dependencies with gradient descent is difficult. IEEE Trans. Neural Netw. **5**(2), 157–166 (1994)
3. Biehl, M., Hammer, B., Villmann, T.: Prototype-based models in machine learning. Wiley Interdisc. Rev.: Cogn. Sci. **7**(2), 92–111 (2016)
4. Cai, H., Gan, C., Wang, T., Zhang, Z., Han, S.: Once-for-all: train one network and specialize it for efficient deployment. arXiv preprint arXiv:1908.09791 (2019)
5. Chen, F., Long, G., Wu, Z., Zhou, T., Jiang, J.: Personalized federated learning with graph. arXiv preprint arXiv:2203.00829 (2022)
6. Chen, S., Long, G., Shen, T., Jiang, J.: Prompt federated learning for weather forecasting: toward foundation models on meteorological data. arXiv preprint arXiv:2301.09152 (2023)
7. Chen, S., Long, G., Shen, T., Zhou, T., Jiang, J.: Spatial-temporal prompt learning for federated weather forecasting. arXiv preprint arXiv:2305.14244 (2023)
8. Chen, Y., et al.: The UCR time series classification archive (2015). www.cs.ucr.edu/eamonn/time_series_data/
9. Dau, H.A., Bagnall, A., Kamgar, K., et al.: The UCR time series archive. arXiv:1810.07758 (2018)
10. Dempster, A., Petitjean, F., Webb, G.I.: ROCKET: exceptionally fast and accurate time series classification using random convolutional kernels. Data Min. Knowl. Disc. **34**(5), 1454–1495 (2020)
11. Diao, E., Ding, J., Tarokh, V.: Heterofl: computation and communication efficient federated learning for heterogeneous clients. arXiv preprint arXiv:2010.01264 (2020)
12. Dong, X., Kedziora, D., Musial, K., Gabrys, B.: Automated deep learning: Neural architecture search is not the end. arXiv preprint arXiv:2112.09245 (2021)
13. Fawaz, H.I., Forestier, G., Weber, J., Idoumghar, L., Muller, P.A.: Deep learning for time series classification: a review. Data Min. Knowl. Disc. **33**(4), 917–963 (2019)
14. Ferrag, M.A., Friha, O., Maglaras, L., Janicke, H., Shu, L.: Federated deep learning for cyber security in the internet of things: concepts, applications, and experimental analysis. IEEE Access **9**, 138509–138542 (2021)
15. Gans, W., Alberini, A., Longo, A.: Smart meter devices and the effect of feedback on residential electricity consumption: evidence from a natural experiment in northern ireland. Energy Econ. **36**, 729–743 (2013)
16. Gee, A.H., Garcia-Olano, D., Ghosh, J., Paydarfar, D.: Explaining deep classification of time-series data with learned prototypes. In: CEUR Workshop Proceedings, vol. 2429, p. 15. NIH Public Access (2019)
17. Ghods, A., Cook, D.J.: PIP: pictorial interpretable prototype learning for time series classification. IEEE Comput. Intell. Mag. **17**(1), 34–45 (2022)
18. Gou, J., Yu, B., Maybank, S.J., Tao, D.: Knowledge distillation: a survey. Int. J. Comput. Vision **129**(6), 1789–1819 (2021)
19. Gu, P., et al.: Multi-head self-attention model for classification of temporal lobe epilepsy subtypes. Front. Physiol. **11**, 1478 (2020)
20. He, C., Annavaram, M., Avestimehr, S.: Group knowledge transfer: federated learning of large CNNs at the edge. Adv. Neural. Inf. Process. Syst. **33**, 14068–14080 (2020)
21. Imteaj, A., Thakker, U., Wang, S., Li, J., Amini, M.H.: A survey on federated learning for resource-constrained IoT devices. IEEE Internet Things J. **9**(1), 1–24 (2021)
22. Ismail Fawaz, H., et al.: InceptionTime: finding AlexNet for time series classification. arXiv e-prints arXiv:1909.04939 (2019)

23. Ji, S., Long, G., Pan, S., Zhu, T., Jiang, J., Wang, S.: Detecting suicidal ideation with data protection in online communities. In: Li, G., Yang, J., Gama, J., Natwichai, J., Tong, Y. (eds.) DASFAA 2019. LNCS, vol. 11448, pp. 225–229. Springer, Cham (2019). https://doi.org/10.1007/978-3-030-18590-9_17

24. Jiang, J., Ji, S., Long, G.: Decentralized knowledge acquisition for mobile internet applications. World Wide Web 23(5), 2653–2669 (2020)

25. Jiang, Y., et al.: Model pruning enables efficient federated learning on edge devices. IEEE Trans. Neural Netw. Learn. Syst. (2022)

26. Kashiparekh, K., Narwariya, J., Malhotra, P., Vig, L., Shroff, G.: Convtimenet: a pre-trained deep convolutional neural network for time series classification. arXiv:1904.12546 (2019)

27. Kavyashree, B., Patil, S., Rao, V.S.: Review on vibration control in tall buildings: from the perspective of devices and applications. Int. J. Dyn. Control 9(3), 1316–1331 (2021)

28. Längkvist, M., Karlsson, L., Loutfi, A.: A review of unsupervised feature learning and deep learning for time-series modeling. Pattern Recogn. Lett. 42, 11–24 (2014)

29. Li, A., Sun, J., Li, P., Pu, Y., Li, H., Chen, Y.: Hermes: an efficient federated learning framework for heterogeneous mobile clients. In: Proceedings of the 27th Annual International Conference on Mobile Computing and Networking, pp. 420–437 (2021)

30. Li, D., Wang, J.: Fedmd: Heterogenous federated learning via model distillation. arXiv preprint arXiv:1910.03581 (2019)

31. Liu, R., et al.: No one left behind: inclusive federated learning over heterogeneous devices. arXiv preprint arXiv:2202.08036 (2022)

32. Liu, S., Yu, G., Yin, R., Yuan, J.: Adaptive network pruning for wireless federated learning. IEEE Wirel. Commun. Lett. 10(7), 1572–1576 (2021)

33. Liu, Y., et al.: Deep anomaly detection for time-series data in industrial IoT: a communication-efficient on-device federated learning approach. IEEE Internet Things J. 8(8), 6348–6358 (2020)

34. Long, G., Shen, T., Tan, Y., Gerrard, L., Clarke, A., Jiang, J.: Federated learning for privacy-preserving open innovation future on digital health. In: Chen, F., Zhou, J. (eds.) Humanity Driven AI, pp. 113–133. Springer, Cham (2022). https://doi.org/10.1007/978-3-030-72188-6_6

35. Long, G., Tan, Y., Jiang, J., Zhang, C.: Federated learning for open banking. In: Yang, Q., Fan, L., Yu, H. (eds.) Federated Learning. LNCS (LNAI), vol. 12500, pp. 240–254. Springer, Cham (2020). https://doi.org/10.1007/978-3-030-63076-8_17

36. Long, G., Xie, M., Shen, T., Zhou, T., Wang, X., Jiang, J.: Multi-center federated learning: clients clustering for better personalization. World Wide Web 26(1), 481–500 (2023)

37. Park, S., Constantinides, M., Aiello, L.M., Quercia, D., Van Gent, P.: Wellbeat: a framework for tracking daily well-being using smartwatches. IEEE Internet Comput. 24(5), 10–17 (2020)

38. Pascanu, R., Mikolov, T., Bengio, Y.: On the difficulty of training recurrent neural networks. In: International Conference on Machine Learning, pp. 1310–1318 (2013)

39. Progonov, D., Sokol, O.: Heartbeat-based authentication on smartwatches in various usage contexts. In: Saracino, A., Mori, P. (eds.) ETAA 2021. LNCS, vol. 13136, pp. 33–49. Springer, Cham (2021). https://doi.org/10.1007/978-3-030-93747-8_3

40. Rajpurkar, P., Hannun, A.Y., Haghpanahi, M., Bourn, C., Ng, A.Y.: Cardiologist-level arrhythmia detection with convolutional neural networks. arXiv:1707.01836 (2017)

41. Serrà, J., Pascual, S., Karatzoglou, A.: Towards a universal neural network encoder for time series. In: CCIA, pp. 120–129 (2018)

42. Singh, A., Vepakomma, P., Gupta, O., Raskar, R.: Detailed comparison of communication efficiency of split learning and federated learning. arXiv preprint arXiv:1909.09145 (2019)

43. Tan, A.Z., Yu, H., Cui, L., Yang, Q.: Towards personalized federated learning. IEEE Trans. Neural Netw. Learn. Syst. (2022)

44. Tan, C.W., Webb, G.I., Petitjean, F.: Indexing and classifying gigabytes of time series under time warping. In: Proceedings of the 2017 SIAM International Conference on Data Mining, pp. 282–290. SIAM (2017)
45. Tan, Y., Liu, Y., Long, G., Jiang, J., Lu, Q., Zhang, C.: Federated learning on non-IID graphs via structural knowledge sharing. In: Proceedings of the AAAI Conference on Artificial Intelligence, vol. 37, pp. 9953–9961 (2023)
46. Tan, Y., et al.: Fedproto: federated prototype learning across heterogeneous clients. In: Proceedings of the AAAI Conference on Artificial Intelligence, vol. 36, pp. 8432–8440 (2022)
47. Tan, Y., Long, G., Ma, J., Liu, L., Zhou, T., Jiang, J.: Federated learning from pre-trained models: a contrastive learning approach. Adv. Neural. Inf. Process. Syst. **35**, 19332–19344 (2022)
48. Tang, W., Liu, L., Long, G.: Interpretable time-series classification on few-shot samples. In: 2020 International Joint Conference on Neural Networks (IJCNN), pp. 1–8. IEEE (2020)
49. Tang, W., Long, G., Liu, L., Zhou, T., Blumenstein, M., Jiang, J.: Omni-scale CNNs: a simple and effective kernel size configuration for time series classification. In: International Conference on Learning Representations (2021)
50. Vidal, F., Navarro, M., Aranda, C., Enomoto, T.: Changes in dynamic characteristics of Lorca RC buildings from pre-and post-earthquake ambient vibration data. Bull. Earthq. Eng. **12**(5), 2095–2110 (2014)
51. Wang, H., Yurochkin, M., Sun, Y., Papailiopoulos, D., Khazaeni, Y.: Federated learning with matched averaging. arXiv preprint arXiv:2002.06440 (2020)
52. Wang, Z., Yan, W., Oates, T.: Time series classification from scratch with deep neural networks: a strong baseline. In: 2017 International Joint Conference on Neural Networks, pp. 1578–1585. IEEE (2017)
53. Xing, L.: Reliability in internet of things: current status and future perspectives. IEEE Internet Things J. **7**(8), 6704–6721 (2020)
54. Xu, W., Fang, W., Ding, Y., Zou, M., Xiong, N.: Accelerating federated learning for IoT in big data analytics with pruning, quantization and selective updating. IEEE Access **9**, 38457–38466 (2021)
55. Xu, Z., Yang, Z., Xiong, J., Yang, J., Chen, X.: Elfish: resource-aware federated learning on heterogeneous edge devices. Ratio **2**(r1), r2 (2019)
56. Yan, P., Long, G.: Personalization disentanglement for federated learning. arXiv preprint arXiv:2306.03570 (2023)
57. Zhang, C., et al.: Dual personalization on federated recommendation. arXiv preprint arXiv:2301.08143 (2023)
58. Zhang, L., Yuan, X.: Fedzkt: zero-shot knowledge transfer towards heterogeneous on-device models in federated learning. arXiv preprint arXiv:2109.03775 (2021)
59. Zhang, T., Gao, L., He, C., Zhang, M., Krishnamachari, B., Avestimehr, A.S.: Federated learning for the internet of things: applications, challenges, and opportunities. IEEE Internet Things Mag. **5**(1), 24–29 (2022)
60. Zheng, Y., Liu, Q., Chen, E., Ge, Y., Zhao, J.L.: Time series classification using multi-channels deep convolutional neural networks. In: Li, F., Li, G., Hwang, S., Yao, B., Zhang, Z. (eds.) WAIM 2014. LNCS, vol. 8485, pp. 298–310. Springer, Cham (2014). https://doi.org/10.1007/978-3-319-08010-9_33
61. Zhu, Z., Hong, J., Zhou, J.: Data-free knowledge distillation for heterogeneous federated learning. In: International Conference on Machine Learning, pp. 12878–12889. PMLR (2021)

Superpixel Attack

Enhancing Black-Box Adversarial Attack with Image-Driven Division Areas

Issa Oe[1]([✉]) [iD], Keiichiro Yamamura[1] [iD], Hiroki Ishikura[1] [iD], Ryo Hamahira[1] [iD], and Katsuki Fujisawa[2] [iD]

[1] Graduate School of Mathematics, Kyushu University, Fukuoka, Japan
issa-oe@kyudai.jp
[2] Institute of Mathematics for Industry, Kyushu University, Fukuoka, Japan

Abstract. Deep learning models are used in safety-critical tasks such as automated driving and face recognition. However, small perturbations in the model input can significantly change the predictions. Adversarial attacks are used to identify small perturbations that can lead to misclassifications. More powerful black-box adversarial attacks are required to develop more effective defenses. A promising approach to black-box adversarial attacks is to repeat the process of extracting a specific image area and changing the perturbations added to it. Existing attacks adopt simple rectangles as the areas where perturbations are changed in a single iteration. We propose applying superpixels instead, which achieve a good balance between color variance and compactness. We also propose a new search method, versatile search, and a novel attack method, Superpixel Attack, which applies superpixels and performs versatile search. Superpixel Attack improves attack success rates by an average of 2.10% compared with existing attacks. Most models used in this study are robust against adversarial attacks, and this improvement is significant for black-box adversarial attacks. The code is available at https://github.com/oe1307/SuperpixelAttack.git.

Keywords: adversarial attack · security for AI · computer vision · deep learning

1 Introduction

Deep learning models have recently found applications in automatic driving and face recognition tasks. These tasks are critical for safety, involving potential risks to life and information privacy. It has been observed that even small perturbations to the model input can significantly alter predictions [26], leading to worst-case scenarios like accidents in automatic driving or information leakage in face recognition. Adversarial attacks are used to identify such perturbations that cause misclassifications. To counter these attacks, defense methods such as adversarial training [28,32] and adversarial detection [3,19] have been explored. However, more potent attacks are needed to develop more effective defenses.

© The Author(s), under exclusive license to Springer Nature Singapore Pte Ltd. 2024
T. Liu et al. (Eds.): AI 2023, LNAI 14471, pp. 141–152, 2024.
https://doi.org/10.1007/978-981-99-8388-9_12

This study targets black-box adversarial attacks, which operate under real-world constraints where only the model's predictions can be accessed. We focus on black-box adversarial attacks that aim to maximize attack success rates within allowed perturbations. A promising approach is to repeat the process of extracting a specific image area and changing the perturbations added to it.

Existing attacks use simple rectangles as the areas where perturbations are changed in a single iteration (Sect. 3.1). However, it is natural to determine the areas based on the image's color information, as it directly influences the perturbation to be added. Therefore, we focus on the color variance of the area where perturbations are changed in a single iteration (Sect. 3.2). Additionally, we focus on the compactness of the area, because existing attacks have adopted rectangles (Sect. 3.3). Through our analysis of the relationship among color variance, compactness, and attack success rates (Sect. 3.4), we discovered that areas that are compact and have a low color variance result in higher attack success rates (Sect. 3.5). Consequently, we propose applying superpixels, which achieve a good balance between color variance and compactness.

Additionally, we introduce versatile search, a new search method that restricts the search to the boundary of perturbation and allows for searches using areas beyond rectangles. With these advancements, we propose Superpixel Attack, a novel attack method that applies superpixels and performs versatile search (Sects. 4.1 and 4.2). To evaluate the performance of Superpixel Attack, we conducted comparison experiments with existing attacks using 19 models trained on the ImageNet dataset [13] and available on RobustBench [8] (Sect. 5). Superpixel Attack significantly enhances attack success rates, resulting in an average improvement of 2.10% compared to existing attacks. Considering that most models used in this study are robust against adversarial attacks, this improvement becomes especially noteworthy for black-box adversarial attacks. Our contributions can be summarized as follows:

1. We analyze the relationship among the color variance, compactness, and attack success rates.
2. We propose applying superpixels to black-box adversarial attacks and a new search method called versatile search.
3. We conducted comparison experiments on Superpixel Attack, which applies superpixel and performs versatile search, and found improvement in attack success rates by an average of 2.10% compared to existing attacks.

2   Preliminaries

2.1   Problem Definition

Let $H \in \mathbb{N}$ be the height, $W \in \mathbb{N}$ be the width, and $C \in \mathbb{N}$ be the number of color channels of the input image. Let $\mathcal{D} = [0, 1]^{H \times W \times C}$ denote the image space, $Y \in \mathbb{N}$ denote the number of classes of the model, and $f : \mathcal{D} \to [0, 1]^Y$ denote the classification model. The output of f is the predicted probability of each class, and we denote $f_i(x) \in [0, 1]$ the predicted probability of class i when image

$x \in \mathcal{D}$ is the input. Adversarial attacks are to find an image $x_{adv} \in \mathcal{D}$ with the predicted label differs from the ground truth label $y \in \{1, \dots, Y\}$ of the original image $x_{org} \in \mathcal{D}$ by adding perturbations that are imperceptible to humans. The inputs generated by adversarial attacks are called adversarial examples. This study focuses on adversarial attacks that maximize attack success rates within the allowed perturbations. We set the allowed perturbation size $\epsilon \in \mathbb{R}^+$ and the loss function $L : [0,1]^Y \times \{1, \dots, Y\} \to \mathbb{R}$, and solve the following constrained nonlinear optimization problem:

$$\max_{x_{adv} \in \mathcal{D}} \quad L\left(f(x_{adv}), y\right)$$
$$\text{s.t.} \quad \|x_{adv} - x_{org}\|_\infty \leq \epsilon \tag{1}$$

2.2   Related Work

Parsimonious attack [20], Square Attack [4], and SignHunter [2] have been proposed as black-box adversarial attacks defined by Eq. (1). Parsimonious attack restricts the search space to the boundaries of allowed perturbations because attacks mostly succeed even on the boundaries. Square Attack achieves high success rates despite its reliance on random sampling. It is a part of AutoAttack [10], a well-known white-box adversarial attack. SignHunter searches for adversarial examples by repeating image division and gradient direction estimation.

Black-box adversarial attacks that minimize perturbations under misclassification [22,27] and those that reduce the number of perturbed pixels [9,11] have also been investigated. Attacks that generate adversarial examples from gradient information of surrogate models have also been proposed [21,31]. These methods are based on transferability, that is, adversarial examples of one model often become those of others. However, training is required to make surrogate models resemble an attacking model and incurs high computational costs.

3   Research on Update Areas

3.1   Update Areas of Existing Methods

The most promising approach for black-box adversarial attacks defined by Eq. (1) involves searching for adversarial examples by repeating the following steps: i. Extract a specific area from the image, ii. Collectively change the perturbation added to the extracted area, iii. Calculate the value of the loss function and update the perturbations when the loss increases. In this paper, we refer to the area where perturbations are changed in a single iteration as *Update Area*. Existing black-box adversarial attacks have adopted simple rectangles as Update Areas. Parsimonious attack sets them using squares that divide the image equally. Square Attack sets them using randomly sampled squares from a uniform distribution. SignHunter sets them using rectangles that divide the image into equal horizontal sections.

3.2   Color Variance of Update Areas

As described in the previous section, Update Areas of the existing attacks are set using simple rectangles. However, it is natural to determine the area by considering the color information of the image because it determines the perturbation to be added. Therefore, we focus on the color variance of Update Areas. As a metric to express the color variance in divided areas of an image, Intra-Cluster Variation (ICV) [5] is proposed. ICV is calculated based on the following equation:

$$
\text{ICV} = \frac{1}{\#\tilde{S}} \sum_{s \in \tilde{S}} \frac{\sqrt{\sum_{p \in s}(I(p) - \mu(s))^2}}{|s|},
\tag{2}
$$

where $\tilde{S}$ is the set of image segmentations. In this paper, it refers to the set of all Update Areas used in an attack. $s \in \tilde{S}$ denotes a single Update Area, and $p \in s$ denotes a pixel. $I(p)$ is the value of the pixel p in the LAB color space[1] and $\mu(s)$ is the average value in the LAB color space within a single Update Area. $\#\tilde{S}$ is the number of Update Areas and $|s|$ is the number of pixels in a single Update Area. Smaller ICV indicates smaller color variance in each Update Area.

3.3   Compactness of Update Areas

Furthermore, considering that existing attacks use rectangles to set Update Areas, we focus on the compactness of Update Areas. The compactness (CO) [24] is a metric calculated by dividing the size of the segments by that of a circle with the same perimeter length. The following equation defines this:

$$
\text{CO} = \frac{\sum_{s \in \tilde{S}} Q(s) \cdot |s|}{\sum_{s \in \tilde{S}} |s|}, \qquad Q(s) = \frac{4\pi|s|}{|R(s)|^2},
\tag{3}
$$

where $|R(s)|$ is the perimeter length of the Update Areas (number of pixels on the boundary). Higher CO indicates more centrally clustered Update Areas. We examined ICV and CO and attack success rates for various Update Areas construction in Sect. 3.5.

3.4   Superpixel Calculated by SLIC

Superpixel is a set of pixels that are close in color and position. They have applications in object detection [30], semantic segmentation [16], and depth estimation [7]. Dong et al. proposed a white-box adversarial attack that adds the same perturbation to each superpixel to avoid disrupting the local smoothness of a natural image [14]. We use superpixels to improve the efficiency of black-box adversarial attacks. To the best of our knowledge, no black-box adversarial attacks that apply superpixels have been proposed. Various methods have been proposed for computing superpixels. We use one of the most popular methods:

---

[1] LAB color spaces in this paper refer to CIELAB (L, a*, b*) color space.

Simple Linear Iterative Clustering (SLIC) algorithm [1]. It places representative points at equal intervals according to the maximum number of segments and clusters pixels based on the k-means method. Let (h_i, w_i) and (h_j, w_j) be the positions in the image, and (l_i, a_i, b_i) and (l_j, a_j, b_j) be the values in the LAB color space. Clustering is performed based on similarity k.

$$k_{color} = \sqrt{(l_i - l_j)^2 + (a_i - a_j)^2 + (b_i - b_j)^2}$$
$$k_{space} = \sqrt{(h_i - h_j)^2 + (w_i - w_j)^2} \qquad (4)$$
$$k = \max(0, \ k_{color} + \alpha \cdot k_{space}),$$

where α is a hyperparameter that weighs the positional distance relative to the color distance. $\alpha = 10$ is generally set to calculate superpixels. We examine the relationship between ICV, CO, and attack success rates for $\alpha = \pm 0.1$, ± 1, ± 10, ± 100, ± 1000 in Sect. 3.5. In addition, the SLIC implementation of scikit-image has the option to force each superpixel to be connected. The experiment in Sect. 3.5 examine both cases. For $\alpha = 1000$, Update Areas are constructed as squares that divide the image equally, regardless of whether they are forced to be connected.

3.5   Analysis of Color Variance and Compactness

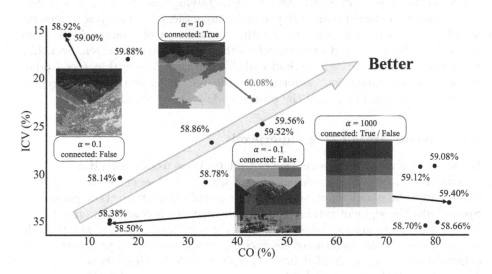

Fig. 1. Relationship between ICV, CO and attack success rates

The experiments use Salman et al. (ResNet-18) [23] trained on the ImageNet dataset and available on RobustBench. According to the RobustBench settings, we use 5,000 images randomly sampled from the ImageNet dataset, and the allowed perturbation size is set to $\epsilon = 4/255$. We adopted versatile search, a new search method proposed in Sect. 4.2. We examine attack success rates at the maximum iterations $T = 500$ for each Update Area construction. The attack success rate is calculated as follows: (number of misclassified images after the attack)/(total number of images), where the higher the attack success rate, the more powerful the attack. The seed value is fixed at 0. We used a CPU: Intel(R) Xeon(R) Gold 5220R CPU@2.20 GHz×2, GPU: Nvidia RTX A6000, RAM:768 GB. The results are shown in Fig. 1.

Each point in Fig. 1 represents the values of CO and ICV for different Update Area construction. The numerical values represent the attack success rate at the point. The horizontal axis represents the value of CO, and the right side indicates that more centrally clustered Update Areas are constructed. The vertical axis represents the value of ICV, where the upper side indicates that Update Areas with lower color variance are constructed. Note that the same Update Areas are constructed for some parameters of α, and the points with equal ICV, CO, and attack success rates coincided with each other. For some representative points, the Update Areas generated by the SLIC algorithm are shown in different colors. This result indicates that it is effective to set Update Areas that are compact and have a low color variance.

4   Superpixel Attack

Based on the analysis in Sect. 3, we consider applying superpixels, which achieve a good balance between color variance and compactness, to black-box adversarial attacks. In this section, we describe the construction of Update Areas using superpixels (Sect. 4.1) and a new search method called versatile search (Sect. 4.2). We propose a novel attack method called *Superpixel Attack* that sets Update Areas using superpixels and performs versatile search. An overview of Superpixel Attack is shown in Fig. 2, and the pseudo-code is shown in Algorithm 1.

4.1   Update Areas Using Superpixels

Below, we describe the construction of Update Areas using superpixels. Inspired by existing attacks, Update Areas are set using a few segments of superpixels at an early stage and many segments of superpixels as the attack progresses. Specifically, the segment ratio r is given and superpixels $\mathcal{S}$ are computed following the maximum number of segmentations $n = r^j$ ($j = 1, 2, \ldots$). Let S be the set of Update Areas constructed for each maximum number of segments n. The original image x_{org} is divided into superpixels $\mathcal{S}$ for each RGB color channel $\{1, \ldots, C\}$, which are set as Update Areas $S = \mathcal{S} \times \{1, \ldots, C\}$. Note that the maximum number of superpixel segments n is not always equal to the number of superpixels computed $\#\mathcal{S}$ in the SLIC algorithm employed in this study. The segment ratio is set to $r = 4$ based on pre-examination. We set $\alpha = 10$ and force the areas to be connected.

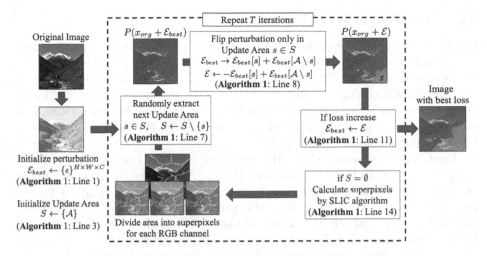

Fig. 2. Flow of proposed method: Superpixel Attack

Algorithm 1. Superpixel Attack

Input: Image height $H \in \mathbb{N}$, Image width $W \in \mathbb{N}$, Number of color channels $C \in \mathbb{N}$,
Allowed perturbation size $\epsilon \in \mathbb{R}^+$, Maximum iterations $T \in \mathbb{N}$,
Original image $x_{org} \in \mathcal{D}$, Ground truth label $y \in \{1, \ldots, Y\}$, Segments ratio $r \in \mathbb{N}$,
Classification model $f : \mathcal{D} \to [0,1]^Y$, Loss function $L : [0,1]^Y \times \{1, \ldots, Y\} \to \mathbb{R}$,
Projection function $P : \mathbb{R}^{H \times W \times C} \to \mathcal{D}$

Output: Image with best loss x_{best}

1: $\mathcal{E}_{best} \leftarrow \{\epsilon\}^{H \times W \times C}$ Initialize perturbation
2: $\mathcal{A} \leftarrow \{(h, w, c) | h \in [1, H], w \in [1, W], c \in [1, C]\}$ Entire area of image
3: $S \leftarrow \{\mathcal{A}\}$ Initialize Update Area
4: $\mathcal{L}_{best} \leftarrow -\infty$ Best loss
5: $n \leftarrow 1$ Maximum number of superpixel
6: **for** $t = 1, 2, \ldots, T$ **do**
7:     $s \in S, \quad S \leftarrow S \setminus \{s\}$ Randomly extract next Update Area
8:     $\mathcal{E}_{best} \to \mathcal{E}_{best}[s] + \mathcal{E}_{best}[\mathcal{A} \setminus s], \quad \mathcal{E} \leftarrow -\mathcal{E}_{best}[s] + \mathcal{E}_{best}[\mathcal{A} \setminus s]$
       Flip perturbation only in Update Area $s \in S$
9:     $\hat{x} \leftarrow P(x_{org} + \mathcal{E}), \quad \mathcal{L} \leftarrow L(f(\hat{x}), y)$
10:     **if** $\mathcal{L} \geq \mathcal{L}_{best}$ Loss increase   **then**
11:        $\mathcal{L}_{best} \leftarrow \mathcal{L}, \quad \mathcal{E}_{best} \leftarrow \mathcal{E}$
12:     **end if**
13:     **if** $S = \emptyset$ All areas are searched   **then**
14:        $n \leftarrow n \times r, \quad S \leftarrow SLIC(x_{org}, n)$ Calculate superpixels by SLIC algorithm
15:        $S \leftarrow S \times \{1, \ldots, C\}$ Divide area into superpixels for each RGB channel
16:     **end if**
17: **end for**
18: $x_{best} \leftarrow P(x_{org} + \mathcal{E}_{best})$

4.2   Procedure of Versatile Search

Below, we describe a new search method called *versatile search*. It searches only the boundaries of the allowed perturbations $\{-\epsilon, \epsilon\}^{H \times W \times C}$ according to the analysis by Moon et al. [20]. At the beginning of the search, the perturbations are initialized with $\mathcal{E}_{best} = \{\epsilon\}^{H \times W \times C}$. Let $\mathcal{A}$ be the entire area of the image and initialize the set of Update Areas with $S = \{\mathcal{A}\}$. The best loss is initialized as $\mathcal{L}_{best} = -\infty$. The following steps are repeated until the number of iterations t reaches the maximum iterations T.

First, the next area where the perturbations are changed is randomly extracted $s \in S$. In the first iteration, Update Area is set to the entire image ($s = \mathcal{A}$). Only the perturbations in the extracted Update Area $\mathcal{E}_{best}[s]$ is flipped to generate new perturbations $\mathcal{E}$. These perturbations $\mathcal{E}$ are added to the original image x_{org}, and the loss $\mathcal{L}$ is calculated. When the calculated loss $\mathcal{L}$ is higher than the best loss $\mathcal{L}_{best}$, the best loss $\mathcal{L}_{best}$ and the perturbation $\mathcal{E}_{best}$ are updated. Superpixels are computed when all Update Areas are searched ($S = \emptyset$), and new Update Areas are set using them.

When the attack is completed, the image with the best loss x_{best} is returned. Superpixel Attack employs CW loss [6] (L_{cw}) as the loss function based on pre-examination. CW loss is calculated as follows:

$$L_{cw}(f(x), y) = \max_{i \neq y} f_i(x) - f_y(x) \tag{5}$$

5   Experiments

In this section, we describe the comparison experiments conducted to confirm the performance of Superpixel Attack. We compare it to Parsimonious attack (Parsimon) [20], Square Attack (Square) [4], SignHunter (SignH) [2], and Accelerated SignHunter (AccSignH) [17] as a baseline. All of these are black-box adversarial attacks with the same problem settings. The experiments use 19 models trained on the ImageNet dataset and available on RobustBench. According to the RobustBench settings, we use 5,000 images randomly sampled from the ImageNet dataset, and the allowed perturbation size is set to $\epsilon = 4/255$. We examine the attack success rates at the maximum iterations $T = 100$ and 1000. The baseline hyperparameters are the same as those in the original paper. The seed value is fixed at 0. We use the same computational environment as in Sect. 3.5. Table 1 presents the results. The highest attack success rate for each iteration is bolded, and the difference between the best baseline method and Superpixel Attack is noted on the right side.

The results in Table 1 show that Superpixel Attack improves the attack success rates by an average of 1.65% for 100 iterations and 2.10% for 1000 iterations compared to existing attacks. Most models used in this study are robust against adversarial attacks, and this improvement is significant for black-box adversarial attacks. In fact, the difference between the second-best and next-best existing attacks averaged 0.67% for 100 iterations and 0.71% for 1000 iterations. For

Table 1. Comparison experiments with baselines

100 iter		Attack Success Rate (%)					
source	Architecture	Parsimon	Square	SignH	AccSignH	**Superpixel**	diff
Wong [29]	ResNet-50	48.32	49.10	50.86	49.48	**53.86**	3.00
Engstrom [15]	ResNet-50	42.40	41.68	42.92	42.08	**45.26**	2.34
Salman [23]	ResNet-50	41.24	40.42	41.98	41.06	**44.44**	2.46
Salman	ResNet-18	52.08	51.50	52.58	52.06	**56.06**	3.48
Salman	WideResNet-50-2	36.84	35.64	37.82	36.54	**39.84**	2.02
PyTorch[a]	ResNet-50	33.92	47.56	**50.08**	38.80	47.52	-2.52
Debenedetti [12]	XCiT-S12	31.72	30.66	32.36	31.64	**33.86**	1.50
Debenedetti	XCiT-M12	30.36	29.38	31.06	30.14	**32.84**	1.78
Debenedetti	XCiT-L12	30.12	29.58	30.66	29.94	**32.32**	1.66
Singh [25]	ViT-S+ConvStem	31.16	30.10	31.40	30.92	**33.48**	2.08
Singh	ViT-B+ConvStem	27.12	26.40	27.56	26.68	**29.22**	1.66
Singh	ConvNeXt-T+ConvStem	30.52	29.76	30.64	30.04	**32.78**	2.14
Singh	ConvNeXt-S+ConvStem	29.26	28.46	29.72	28.98	**31.34**	1.62
Singh	ConvNeXt-B+ConvStem	26.90	26.20	27.38	26.82	**28.86**	1.48
Singh	ConvNeXt-L+ConvStem	25.36	24.82	25.94	25.34	**26.94**	1.00
Liu [18]	ConvNeXt-B	26.48	25.88	26.84	26.44	**28.36**	1.52
Liu	ConvNeXt-L	25.08	24.26	25.78	24.90	**26.88**	1.10
Liu	Swin-B	26.86	26.06	27.20	26.74	**28.88**	1.68
Liu	Swin-L	24.16	23.36	24.62	23.80	**26.06**	1.44
1,000 iter		Attack success rate (%)					
source	Architecture	Parsimon	Square	SignH	AccSignH	**Superpixel**	diff
Wong	ResNet-50	56.62	56.62	52.46	50.34	**59.96**	3.34
Engstrom	ResNet-50	48.92	48.16	45.10	44.12	**51.84**	2.92
Salman	ResNet-50	46.96	46.70	44.06	43.08	**50.16**	3.20
Salman	ResNet-18	58.60	58.72	54.92	54.26	**61.98**	3.26
Salman	WideResNet-50-2	42.94	42.22	39.66	38.32	**44.86**	1.92
PyTorch	ResNet-50	72.04	84.64	80.80	55.80	**87.28**	2.64
Debenedetti	XCiT-S12	37.44	36.48	33.74	32.96	**39.66**	2.22
Debenedetti	XCiT-M12	36.04	35.10	32.70	31.86	**37.64**	1.60
Debenedetti	XCiT-L12	35.32	34.64	32.38	31.52	**37.02**	1.70
Singh	ViT-S+ConvStem	35.50	35.30	32.68	32.44	**37.58**	2.08
Singh	ViT-B+ConvStem	31.02	30.38	28.76	28.20	**32.56**	1.54
Singh	ConvNeXt-T+ConvStem	34.88	34.50	32.24	31.58	**37.12**	2.24
Singh	ConvNeXt-S+ConvStem	33.36	32.80	30.94	30.62	**35.28**	1.92
Singh	ConvNeXt-B+ConvStem	30.78	30.14	28.62	28.24	**32.44**	1.66
Singh	ConvNeXt-L+ConvStem	29.24	28.80	27.30	26.42	**30.64**	1.40
Liu	ConvNeXt-B	30.32	29.76	28.22	27.62	**31.94**	1.62
Liu	ConvNeXt-L	28.92	28.34	26.86	26.40	**30.20**	1.28
Liu	Swin-B	30.88	30.44	28.64	28.08	**32.60**	1.72
Liu	Swin-L	28.18	27.44	25.86	25.32	**29.90**	1.72

[a] https://pytorch.org/vision/stable/models.html

Wong (ResNet-50), PyTorch (ResNet-50), and Singh (ViT-S+ConvStem), we plot the trends of attack success rates per iteration for each attack method in Fig. 3.

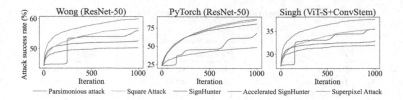

Fig. 3. Transition of attack success rates of each attack method

Figure 3 indicates that Superpixel Attack achieves high success rates in all iterations, including the PyTorch (ResNet-50) model in contrast to SignHunter, which only has high success rates in short iterations. For the other models, each attack method exhibited trends similar to those of Wong (ResNet-50) and Singh (ViT-S+ConvStem). Furthermore, Fig. 4 shows the computational time for superpixels and forward propagation in Superpixel Attack. Although it depends on the computational environment, the computation time of superpixels is less than that of the forward propagation. This indicates that applying superpixels to adversarial attacks is practical in terms of the computation time. For 1000 iterations, the superpixel computation accounts for a very small percentage of the attacks, as indicated by the orange bars.

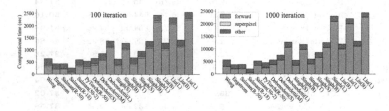

Fig. 4. Computational time of superpixels and forward propagation

6   Conclusion

This study demonstrated that the attack success rates are related to the color variance and compactness of the Update Area. The experimental results suggest that Update Areas with low color variance and high compactness is desirable. Therefore, we propose the Superpixel Attack, which employs superpixels as Update Areas to achieve a good balance between color variance and compactness. The comparison experiments show that the Superpixel Attack improves the attack success rates by an average of 2.10% compared with existing methods for 1000 iterations, which is significant for black-box adversarial attacks. This study indicates that adjusting the Update Areas according to the image can enhance the attack success rates.

Acknowledgements. This research project was supported by the Japan Science and Technology Agency (JST), the Core Research of Evolutionary Science and Technology (CREST), the Center of Innovation Science and Technology based Radical Innovation and Entrepreneurship Program (COI Program), JSPS KAKENHI Grant Number JP16H01707 and JP21H04599, Japan.

References

1. Achanta, R., Shaji, A., Smith, K., Lucchi, A., Fua, P., Süsstrunk, S.: Slic super-pixels. Technical report (2010)
2. Al-Dujaili, A., O'Reilly, U.M.: Sign bits are all you need for black-box attacks. In: International Conference on Learning Representations (2020)
3. Aldahdooh, A., Hamidouche, W., Fezza, S.A., Déforges, O.: Adversarial example detection for DNN models: a review and experimental comparison. Artif. Intell. Rev. **55**(6), 4403–4462 (2022)
4. Andriushchenko, M., Croce, F., Flammarion, N., Hein, M.: Square attack: a query-efficient black-box adversarial attack via random search. In: Vedaldi, A., Bischof, H., Brox, T., Frahm, J.-M. (eds.) ECCV 2020. LNCS, vol. 12368, pp. 484–501. Springer, Cham (2020). https://doi.org/10.1007/978-3-030-58592-1_29
5. Benesova, W., Kottman, M.: Fast superpixel segmentation using morphological processing. In: Conference on Machine Vision and Machine Learning, pp. 67–1 (2014)
6. Carlini, N., Wagner, D.: Towards evaluating the robustness of neural networks. In: 2017 IEEE Symposium on Security and Privacy (SP), pp. 39–57. IEEE (2017)
7. Chen, J., Hou, J., Ni, Y., Chau, L.P.: Accurate light field depth estimation with superpixel regularization over partially occluded regions. IEEE Trans. Image Process. **27**(10), 4889–4900 (2018)
8. Croce, F., et al.: RobustBench: a standardized adversarial robustness benchmark. arXiv preprint arXiv:2010.09670 (2020)
9. Croce, F., Andriushchenko, M., Singh, N.D., Flammarion, N., Hein, M.: Sparse-RS: a versatile framework for query-efficient sparse black-box adversarial attacks. In: Proceedings of the AAAI Conference on Artificial Intelligence, vol. 36, pp. 6437–6445 (2022)
10. Croce, F., Hein, M.: Reliable evaluation of adversarial robustness with an ensemble of diverse parameter-free attacks. In: International Conference on Machine Learning, pp. 2206–2216. PMLR (2020)
11. Dai, Z., Liu, S., Tang, K., Li, Q.: Saliency attack: towards imperceptible black-box adversarial attack. arXiv preprint arXiv:2206.01898 (2022)
12. Debenedetti, E., Sehwag, V., Mittal, P.: A light recipe to train robust vision transformers. arXiv preprint arXiv:2209.07399 (2022)
13. Deng, J., Dong, W., Socher, R., Li, L.J., Li, K., Fei-Fei, L.: ImageNet: a large-scale hierarchical image database. In: 2009 IEEE Conference on Computer Vision and Pattern Recognition, pp. 248–255. IEEE (2009)
14. Dong, X., et al.: Robust superpixel-guided attentional adversarial attack. In: Proceedings of the IEEE/CVF Conference on Computer Vision and Pattern Recognition, pp. 12895–12904 (2020)
15. Engstrom, L., Ilyas, A., Salman, H., Santurkar, S., Tsipras, D.: Robustness (python library) (2019). https://github.com/MadryLab/robustness

16. Kwak, S., Hong, S., Han, B.: Weakly supervised semantic segmentation using super-pixel pooling network. In: Proceedings of the AAAI Conference on Artificial Intelligence, vol. 31 (2017)
17. Li, S., Huang, G., Xu, X., Yang, Y., Shen, F.: Accelerated sign hunter: a sign-based black-box attack via branch-prune strategy and stabilized hierarchical search. In: Proceedings of the 2022 International Conference on Multimedia Retrieval, pp. 462–470 (2022)
18. Liu, C., et al.: A comprehensive study on robustness of image classification models: benchmarking and rethinking. arXiv preprint arXiv:2302.14301 (2023)
19. Metzen, J.H., Genewein, T., Fischer, V., Bischoff, B.: On detecting adversarial perturbations. arXiv preprint arXiv:1702.04267 (2017)
20. Moon, S., An, G., Song, H.O.: Parsimonious black-box adversarial attacks via efficient combinatorial optimization. In: International Conference on Machine Learning, pp. 4636–4645. PMLR (2019)
21. Papernot, N., McDaniel, P., Goodfellow, I., Jha, S., Celik, Z.B., Swami, A.: Practical black-box attacks against machine learning. In: Proceedings of the 2017 ACM on Asia Conference on Computer and Communications Security, pp. 506–519 (2017)
22. Rahmati, A., Moosavi-Dezfooli, S.M., Frossard, P., Dai, H.: GeoDA: a geometric framework for black-box adversarial attacks. In: Proceedings of the IEEE/CVF Conference on Computer Vision and Pattern Recognition, pp. 8446–8455 (2020)
23. Salman, H., Ilyas, A., Engstrom, L., Kapoor, A., Madry, A.: Do adversarially robust ImageNet models transfer better? In: Advances in Neural Information Processing Systems, vol. 33, pp. 3533–3545 (2020)
24. Schick, A., Fischer, M., Stiefelhagen, R.: Measuring and evaluating the compactness of superpixels. In: Proceedings of the 21st International Conference on Pattern Recognition (ICPR2012), pp. 930–934. IEEE (2012)
25. Singh, N.D., Croce, F., Hein, M.: Revisiting adversarial training for ImageNet: architectures, training and generalization across threat models. arXiv preprint arXiv:2303.01870 (2023)
26. Szegedy, C., et al.: Intriguing properties of neural networks. arXiv preprint arXiv:1312.6199 (2013)
27. Wang, X., et al.: Triangle attack: a query-efficient decision-based adversarial attack. In: Avidan, S., Brostow, G., Cissé, M., Farinella, G.M., Hassner, T. (eds.) ECCV 2022. LNCS, vol. 13665, pp. 156–174. Springer, Cham (2022). https://doi.org/10.1007/978-3-031-20065-6_10
28. Wang, Y., Ma, X., Bailey, J., Yi, J., Zhou, B., Gu, Q.: On the convergence and robustness of adversarial training. arXiv preprint arXiv:2112.08304 (2021)
29. Wong, E., Rice, L., Kolter, J.Z.: Fast is better than free: revisiting adversarial training. arXiv preprint arXiv:2001.03994 (2020)
30. Yan, J., Yu, Y., Zhu, X., Lei, Z., Li, S.Z.: Object detection by labeling superpixels. In: Proceedings of the IEEE Conference on Computer Vision and Pattern Recognition, pp. 5107–5116 (2015)
31. Zhang, J., et al.: Towards efficient data free black-box adversarial attack. In: Proceedings of the IEEE/CVF Conference on Computer Vision and Pattern Recognition, pp. 15115–15125 (2022)
32. Zhao, W., Alwidian, S., Mahmoud, Q.H.: Adversarial training methods for deep learning: a systematic review. Algorithms **15**(8), 283 (2022)

Cross Domain Pulmonary Nodule Detection Without Source Data

Rui Xu[1], Yong Luo[1(✉)], and Yan Xu[2]

[1] Wuhan University, Wuhan 430072, China
{rui.xu,luoyong}@whu.edu.cn
[2] The University of Chicago, Chicago, IL 60637, USA
yanx1@uchicago.edu

Abstract. The model performance on cross-domain pulmonary nodule detection usually degrades because of the significant shift in data distributions and the scarcity of annotated medical data in the test scenarios. Current approaches to cross-domain object detection assume that training data from the source domain are freely available; however, such an assumption is implausible in the medical field, as the data are confidential and cannot be shared due to privacy concerns. Thus, this paper introduces source data-free cross-domain pulmonary nodule detection. In this setting, only a pre-trained model from the source domain and a few annotated samples from the target domain are available. We introduce a novel method to tackle this issue, adapting the feature extraction module for the target domain through minimizing the proposed General Entropy (GE). Specifically, we optimize the batch normalization (BN) layers of the model by GE minimization. Thus, the dataset-level statistics of the target domain are utilized for optimization and inference. Furthermore, we tune the detection head of the model using annotated target samples to mitigate the rater difference and improve the accuracy. Extensive experiments on three different pulmonary nodule datasets show the efficacy of our method for source data-absent cross-domain pulmonary nodule detection.

Keywords: Pulmonary Nodule Detection · Domain Adaptation · Source Free · Entropy Minimization · Model Reuse

1 Introduction

There has been much progress in various object detection tasks [13,15,16,28,37] with the prosperity of deep learning. In the medical field, detection algorithms are able to obtain performance comparable to that of clinical experts, e.g. pulmonary nodule detection [18,27,33,34], etc. Nonetheless, most of the approaches are based on the assumption that the training/source and test/target data come from similar distributions. This assumption restricts the application of these approaches in the real world, because there often exists nontrivial domain difference between the training data and the real-world test data; the domain shift causes significant performance degradation of the algorithms in the test/target

T. Liu et al. (Eds.): AI 2023, LNAI 14471, pp. 153–164, 2024.
https://doi.org/10.1007/978-981-99-8388-9_13

domain. Hence, a great deal of effort has been directed towards cross-domain object detection [1,2,5,7,12,23,32,38,39] in recent years to enhance the performance of the source model on the target domain.

However, current approaches for cross-domain object detection still contain an improper assumption for medical applications. They assume that the training samples from the source domain are freely accessible, while in reality, medical data are usually not shareable due to privacy issues and merely a pre-trained source model is accessible. What's more, acquiring and annotating medical data are both time-consuming and costly, resulting in limited training samples of the target domain, making cross-domain object detection in the medical field very challenging. Considering these two aspects, we present a realistic but demanding setting, source data-free cross-domain detection of lung nodule. In this scenario, merely a pre-trained source model and a few annotated samples from the target domain are available. As far as we know, this is the first work that tackles source data-absent cross-domain adaptation in the pulmonary nodule detection task.

The batch normalization (BN) [9] layers of a model normalize and modulate the features, and thus are closely tied to the model performance when there is a shift in data distribution. In cross-domain image classification and semantic segmentation tasks, some studies simply substitute the source batch statistics with the statistics of the current batch of the target domain [14]. Some studies combine the statistics of both source and target [36]. Some other studies [31,35] pay attention to the target statistics, and minimize entropy loss to optimize the affine parameters as well. Nevertheless, these methods are either too weak or not applicable for cross-domain object detection.

In our cross-domain pulmonary nodule detection setting, which does not rely on source data, we propose adapting to the target domain by reducing the entropy of the model predictions. However, the original entropy [25] only supports image classification and segmentation currently. We successfully solve this problem by extending entropy to its detection variant, termed General Entropy (GE). We choose entropy for its ability to quantify uncertainty and shifts, as low entropy predictions are all-in-all more reliable and high entropy predictions represent larger shifts. To better utilize the source information and efficiently adapt, we only optimize the affine parameters and estimate the target dataset-level statistics in the batch normalization layers via entropy minimization. This step enables us to learn a target-specific feature encoding module under the same detection head, without requiring access to the source data or the labels of the target data.

To enhance the detection performance further and alleviate the common problem of rater disagreement in the medical field, we also fine-tune the detection head of the model using annotated samples from the target domain.

Our primary contributions are summarized as follows:

- We establish a source data-free setting for cross-domain lung nodule detection, utilizing merely a well-trained source model and a limited number of labeled target samples.

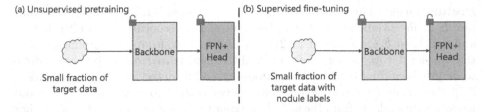

Fig. 1. The pipeline of our proposed method. The source model is composed of a feature encoding module and a detection head module. (a) We keep the detection head frozen, and adapt the batch normalization (BN) layers in the feature extraction module by minimizing our Generalized Entropy (GE) to obtain target dataset-level statistics. (b) The detection head is fine-tuned using a small fraction of target data with labels.

- We propose a novel method, which adapts the model feature extraction module for the target domain via General Entropy (GE) minimization. We further fine-tune the model detection head with labeled target samples to improve the adaptation performance.
- For the purpose of evaluation, we curate a benchmark using four widely used pulmonary nodule datasets.

Experiments on the benchmark show our method can achieve the state-of-the-art results, demonstrating the effectiveness of our method.

2   Method

For a vanilla cross-domain adaptation (DA) task, we have N^s labeled samples $\{x_i^s, y_i^s\}_{i=1}^{N^s}$ from the source domain and also N^t labeled samples $\{x_i^t, y_i^t\}_{i=1}^{N^t}$ from the target domain. The main goal of DA is to address the domain shift between the source domain and the target domain, thus to well predict labels $\{y_i^t\}_{i=1}^{N^t}$ in the target domain. In this work, we assume that we cannot obtain samples from the source domain because of concerns related to privacy. Instead of the source dataset, we are given a well-trained source model $f_\theta(x)$ with parameters θ. Based on this assumption, we present source data-free cross-domain pulmonary nodule detection, and aim to learn a target model with the given well-trained source model $f_\theta(x)$ and target samples $\{x_i^t, y_i^t\}_{i=1}^{N^t}$.

Our method comprises two steps as shown in Fig. 1. First, the feature extraction module of the well-trained source model is adjusted to the target domain using unsupervised learning. To be specific, the batch normalization (BN) layers of the model are optimized by minimizing entropy loss to obtain target dataset-level statistics, where a general form of entropy termed Generalized Entropy (GE) is proposed. Then, using the annotated target samples, we further employ supervised learning to fine-tune the detection head of the model for rater difference mitigation and performance enhancement. In the following, we would like first to revisit two types of the uncertainty of the bounding box, the probability distribution representation and localization quality estimation, and then elaborate on our method in detail.

Preliminaries. There are two conventional representations for the bounding box $\mathcal{B}$ in detection. For instance, the central point coordinates, width, height, and depth, $\{a, b, c, w, h, d\}$ [3,17,21], and the distance from the sampling point to the up, down, top, bottom, left, and right planes, $\{u, d, t, b, l, r\}$ [28] are utilized to denote bounding boxes in the pulmonary nodule detection task. According to [37], there is no performance difference between the two representations. In this work, relative offsets from the sampling point to the six planes of a bounding box $\mathcal{B} = \{u, d, t, b, l, r\}$ are used as the regression targets, since the physical meaning of each variable in $\{u, d, t, b, l, r\}$ is consistent. Given the $\{a, b, c, w, h, d\}$ form, we will convert it to the $\{u, d, t, b, l, r\}$ form.

Yet this form follows the Dirac delta distribution that only concentrates on the ground-truth locations, and is too rigid to reflect the ambiguity of bounding boxes [6,13]. Recently, some works [13,20] adopt the probability distribution representation of the bounding box to learn its localization uncertainty. Let $y \in \mathcal{B}$ be the distance to a certain plane of a bounding box, whose estimated value $\hat{y}$ can be represented as:

$$\hat{y} = \int_{y_{min}}^{y_{max}} s \Pr(s) ds, \tag{1}$$

where s is the regression distance in range of $[y_{min}, y_{max}]$, and $\Pr(s)$ is the corresponding probability. Then, to be congenial with the convolutional neural networks, the continuous regression range $[y_{min}, y_{max}]$ is converted into a uniform discretized representation, $\{y_0, y_1, ..., y_i, y_{i+1}, ..., y_{n-1}, y_n\}$ with even intervals Δ, where $\Delta = y_{i+1} - y_i, \forall i \in [0, n-1]$, $y_0 = y_{min}$, and $y_n = y_{max}$. Thus, the estimated value $\hat{y}$ becomes:

$$\hat{y} = \sum_{i=0}^{n} \Pr(y_i) y_i, \tag{2}$$

where $\sum_{i=0}^{n} \Pr(y_i) = 1$, and the $\Pr(s)$ can be easily implemented using a SoftMax function with $n+1$ outputs. Hereto, the uncertainty of the bounding box offsets are modeled.

There is also another simple way to model the localization uncertainty of the bounding box, i.e. the localization quality estimation in the form of IoU [30] or centerness [28] score. Thereinto, the centerness [28] represents the distance measurement between the center points of the location and its corresponding object. Given the regression targets u^*, d^*, t^*, b^*, l^*, and r^* for a sampling point, the centerness $\hat{y}$ can be defined as:

$$\hat{y} = \sqrt{\frac{\min(u^*, d^*)}{\max(u^*, d^*)} \times \frac{\min(t^*, b^*)}{\max(t^*, b^*)} \times \frac{\min(l^*, r^*)}{\max(l^*, r^*)}}. \tag{3}$$

In our method, we employ the centerness [28] score measurement for its simplicity and good performance in pulmonary nodule detection.

2.1 Feature Extractor Adaptation

Entropy Objective. Our training goal is to reduce the entropy $H(\hat{y})$ of the model detection results $\hat{y} = f_\theta(x^t)$. This is because entropy is an unsupervised objective for uncertainty measurement, while related to the supervised task and model. However, the current Shannon entropy [25] only supports classification. Therefore, we propose Generalized Entropy (GE) that generalizes the Shannon entropy [25] for dense detectors. Assume that a model's final prediction $\hat{y}$ is the linear combination of two variables $\hat{y} = y_l p_{y_l} + y_r p_{y_r}, (y_l \le \hat{y} \le y_r)$, where $p_{y_l}, p_{y_r} (p_{y_l} \ge 0, p_{y_r} \ge 0, p_{y_l} + p_{y_r} = 1)$ are probabilities for these variables estimated by the model respectively. The proposed GE is able to cover the three special cases of the General Focal Loss (GFL) [13] for dense detectors:

When $\beta = \gamma, y_l = 0, y_r = 1, p_{y_r} = p, p_{y_l} = 1 - p$ and $y \in \{1, 0\}$ in GFL [13], GE for focal loss (FL) can be written as:

$$H(p) = -((1 - \alpha)p^\gamma(1 - p)\log(1 - p) + \alpha(1 - p)^\gamma p \log(p)). \quad (4)$$

When $y_l = 0, y_r = 1, p_{y_r} = \sigma$ and $p_{y_l} = 1 - \sigma$ in GFL [13], GE for quality focal loss (QFL) can be written as:

$$H(\sigma) = -(\sigma^\beta(1 - \sigma)\log(1 - \sigma) + (1 - \sigma)^\beta \sigma \log(\sigma)). \quad (5)$$

When $\beta = 0, y_l = y_i, y_r = y_{i+1}, p_{y_l} = \Pr(y_l) = \Pr(y_i) = \mathcal{S}_i$ and $p_{y_r} = \Pr(y_r) = \Pr(y_{i+1}) = \mathcal{S}_{i+1}$ in GFL [13], GE for distribution focal loss (DFL) can be written as:

$$H(\mathcal{S}_i, \mathcal{S}_{i+1}) = -(\mathcal{S}_i \log(\mathcal{S}_i) + \mathcal{S}_{i+1} \log(\mathcal{S}_{i+1})). \quad (6)$$

Modulation Parameters. As shown in Fig. 1, the pulmonary nodule detection network $f_\theta(x)$ is composed of two modules: the feature encoding module $g_\theta : x \rightarrow \mathbb{R}^d$ and the detection head module $h_\theta : \mathbb{R}^d \rightarrow \mathbb{R}^K$; $f_\theta(x) = h_\theta(g_\theta(x))$, d and K are dimensions of the extracted feature and the model output respectively. To keep the same hypothesis h_θ, a natural choice of the modulation parameters is all the feature extractor parameters g_θ; however, altering g_θ may cause the model to diverge from its training, since θ is the only representation of the source data in our setting. Besides, the limited number of training samples from the target domain is not suitable for optimizing the high dimensional θ. Previous works [31,35] find that adapting the batch statistics, especially dataset-level statistics, is effective for domain adaptation. Considering the feature modulation ability and low dimensional computation of the batch normalization (BN) layers, we choose to update the BN layers during training. Inside the BN layer, there are two sets of parameters: the statistics (μ, σ), which normalize the feature, and the affine parameters (β, γ), which modulate the feature. Given a batch of target samples $\{x_i^t\}_{i=1}^B$, where B is the batch size, the outputs of the BN layer $\{x_i^{t'}\}_{i=1}^B$ are calculated as:

$$x_i^{t'} = \gamma \overline{x_i^t} + \beta = \gamma \frac{x_i^t - \mu}{\sigma} + \beta,$$

$$\mu = \mathbb{E}[x_i^t], \sigma^2 = \mathbb{E}[(x_i^t - \mu)^2].$$

In the meantime, a running mean vector μ_r and a running variance vector σ_r are estimated using moving average to derive dataset-level statistics for the target domain:

$$\mu_r = \lambda\mu + (1 - \lambda)\mu_r, \sigma_r^2 = \lambda\sigma^2 + (1 - \lambda)\sigma_r^2. \tag{7}$$

The affine parameters (β, γ) are optimized via minimizing the GE loss.

2.2 Detection Head Adaptation

Transfer learning by fine-tuning is a common way to adjust a well-trained network to a new domain. To enhance the performance of pulmonary nodule detection even further, we tune the detection head of the model h_θ using the training samples from the target domain $\{x_i^t, y_i^t\}_{i=1}^{N^t}$. Meanwhile, this can also alleviate the issue of rater disagreement between different datasets, a common problem in the medical field.

3 Experiments

3.1 Benchmark and Evaluation

We establish a benchmark from PN9 [18] to LUNA16 [24]/tianchi [29]/russia [19] for shifts, as shown in Fig. 2. The specifics of these datasets are listed in Table 1. As seen, the CT scans in these datasets, which are gathered from various sites, have different image sizes and voxel sizes. In Table 2, we display the lung nodule size and quantity distribution of the four datasets.

Recall that vanilla domain adaptation requires the use of the labeled source data, while our setting denies the use of source data PN9 [18] during adaptation. We take into account only those CT scans having publicly available nodule annotations. The annotation files of the four datasets are csv files. Each line of the files holds the information of one nodule, including the CT scan filename it belongs to, and its location. In the three target datasets, the nodule location is indicated by the center coordinates and diameter, whereas in PN9 [18], it is marked by the top-left and bottom-right coordinates.

LUNA16 [24], tianchi [29], and russia [19] are divided into 7/1/2 for training, validation, and testing. In these three datasets, the raw CT data undergoes three pre-processing steps: 1) We use lungmask [8] to extract lung regions from each CT image and mask other regions to minimize irrelevant calculations. In this process, the HU values of the raw CT data are clipped into the range $[-1200, 600]$ and then linearly converted into the range $[0, 255]$, resulting in uint8 values. Then we set a padding value of 170 for regions outside the lung masks. 2) To prevent an excess of unnecessary hyper-parameters, the spacing of all the CT images is resampled to $(1.00, 1.00, 1.00)$ mm, ensuring consistency for the anchor design across all detectors. 3) To further improve the computational efficiency, we crop

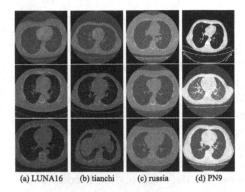

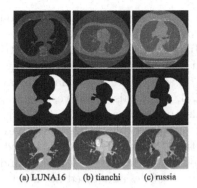

(a) LUNA16   (b) tianchi   (c) russia   (d) PN9

(a) LUNA16   (b) tianchi   (c) russia

Fig. 2. Samples from four lung nodule datasets are shown, with each column corresponding to a dataset as marked. CT images from different datasets exhibit domain discrepancy, for instance, color contrast/saturation, voxel intensity, image spacing, amount of nodules.

Fig. 3. Samples of the pre-processed images in the LUNA16, tianchi, and russia. The 1st row contains the raw images, the 2nd row shows the extracted lung regions, and the 3rd row displays the pre-processed images.

Table 1. Pulmonary nodule datasets. 'Scans' and 'Class' indicates the number of CT scans and the class, respectively. 'Raw' denotes whether the CT images in the dataset are pre-processed. 'Image Size' refers to the CT image matrix size in the direction of the x, y, and z axes. 'Spacing' denotes the voxel sizes (mm) in the direction of the x, y, and z axes.

Dataset	Year	Scans	Class	Raw	Image Size	Spacing
LUNA16 [24]	2016	601	2	Yes	$512 \times 512 \times 95 - 512 \times 512 \times 733$	$(0.86, 0.86, 2.50) - (0.64, 0.64, 0.50)$
tianchi [29]	2017	800	2	Yes	$512 \times 512 \times 114 - 512 \times 512 \times 1034$	$(0.66, 0.66, 2.50) - (0.69, 0.69, 0.30)$
russia [19]	2018	364	2	Yes	$512 \times 512 \times 313 - 512 \times 512 \times 1636$	$(0.62, 0.62, 0.80) - (0.78, 0.78, 0.40)$
PN9 [18]	2021	8796	9	No	$212 \times 212 \times 181 - 455 \times 455 \times 744$	$(1.00, 1.00, 1.00) - (1.00, 1.00, 1.00)$

Table 2. Distribution of the pulmonary nodule size. 'd' indicates the nodule diameter (mm).

Dataset	$d < 3$	$3 \leq d < 5$	$5 \leq d < 10$	$10 \leq d < 30$	$30 \leq d$	All
LUNA16 [24]	-	270	635	279	2	1186
tianchi [29]	1	213	596	423	11	1244
russia [19]	6	552	907	360	25	1850
PN9 [18]	9	4678	29213	6053	483	40436

the CT images according to the extracted lung masks. Figure 3 shows the CT image samples after being pre-processed. For PN9 [18] dataset, the data pre-processing procedure is kept the same as in [18]. In our experiments, the voxel coordinates are utilized. Based on our pre-processing procedures and the voxel coordinates, the nodule locations in the annotation files are recalculated.

In terms of the evaluation metric, the Free-Response Receiver Operating Characteristic (FROC), a commonly used measure for pulmonary nodule detection, is selected. It is calculated by averaging the sensitivities at 0.125, 0.25, 0.5, 1, 2, 4, and 8 false positives per scan. We also use the detection sensitivity at 8 false positives per image for evaluation, since false positives in the medical field are preferable to false negatives. The detected nodule is counted as a true positive if there exists one annotated nodule, and the distance between the center points of the detected nodule and the annotated nodule is smaller than the radius R of the annotated nodule. Otherwise, the detected nodule is considered a false positive.

3.2 Implementation Details

In our experiments, we employ the same backbone as the SANet [18], thus utilizing the weights pre-trained on PN9 [18] for source model training. Concretely, the backbone is U-shaped [22], consisting of a 3D ResNet50 [4] equipped with Slice Grouped Non-local modules [18] and a decoder. Different from [18], the backbone is followed by FPN [15] as neck, and the FCOS-style [28] anchor-free head for classification and localization. The network is optimized using the Stochastic Gradient Descent (SGD). The training batch size of the 3D patches is 16. We implement the patch-based input strategy for training and use the complete 3D volume for inference as in [18]. The learning rate, the momentum, and the weight decay coefficients are respectively fixed at 0.001, 0.9, and 1×10^{-4}. To obtain the source model, the network is set to be trained for a maximum of 30 epochs. For learning in the target domain, we tune the pre-trained source model for 1 epoch. For other training and testing hyper-parameters, we follow the [28], and specialize some hyper-parameters in the task of detecting pulmonary nodules. We use FPN [15] with two levels, a detection head with two classification/regression towers, and a radius of 3. All the experiments are carried out with PyTorch on four NVIDIA GeForce RTX 3090 GPUs, each having 24 GB of memory.

3.3 Results

We evaluate the proposed method by contrasting it with the baseline approach, which simply fine-tunes all the parameters of the source model using the labeled samples from the target domain. Experiments are conducted with 20%, 40%, 60%, 80%, and 100% labeled training samples from the target domains respectively, and the results are reported for the whole target testing sets. Table 3 lists the experimental results of our method and the baseline on target dataset LUNA16 [24] and tianchi [29]. Our method obviously outperforms the baseline. Meanwhile, it adapts more efficiently. It is especially noteworthy that utilizing only the feature extraction module adaptation, the first step of our method without the use of any labeled training samples from the target domain, already brings a good performance. This shows the potential of our method in the more

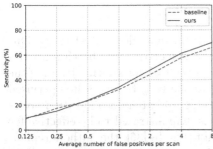
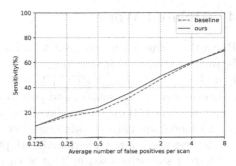

Fig. 4. FROC curves of our method and the baseline on target dataset russia w.r.t 60% percentage of its training set.

Fig. 5. FROC curves of our method and the baseline on target dataset russia w.r.t 80% percentage of its training set.

Table 3. Comparison of our method and the baseline on target dataset LUNA16 and tianchi w.r.t percentage of their training set. The values are pulmonary nodule detection sensitivities (unit: %) at 8 false positives per CT image, with each column indicating the percentage of the training set.

Method	LUNA16					tianchi				
	20%	40%	60%	80%	100%	20%	40%	60%	80%	100%
Fine-tuning	88.53	90.36	91.74	92.66	92.20	91.04	94.02	91.79	90.67	92.91
Ours (Step1)	85.77	89.90	88.99	89.44	88.99	93.28	93.28	93.28	92.91	93.28
Ours (+ Step2)	89.90	86.23	92.20	93.57	94.03	91.41	92.53	92.91	93.65	93.65

Table 4. Comparison of our method and the baseline on target dataset russia w.r.t percentage of its training set. The values are FROCs (unit: %) with each column indicating the percentage of the training set.

Method	russia				
	20%	40%	60%	80%	100%
Fine-tuning	**31.18**	**33.12**	35.69	36.52	38.26
Ours	29.91	33.08	**37.39**	**38.02**	**38.34**

wild and challenging settings. Nonetheless, the performance of our method on target dataset russia [19] is unsatisfactory, probably due to its larger shift with the source. For more adaptation, we tune all the parameters of the model in our second step on russia [19]. As listed in Table 4, our method obtains better FROC scores for lung nodule detection than the baseline, which verifies the effectiveness of our proposed adapting via entropy minimization. The FROC curves illustrated in Fig. 4 and Fig. 5 further confirm the superiority of our method.

4   Related Works

Recently, some works propose to adapt the trained model in test-time. This branch of study originates from the works of recalculating the batch statistics [14]. Test-time training (TTT) [26] relies on a proxy task for altering training the entire model on the source, and then adapts to the target using self-supervised learning. Tent [31] optimizes the affine parameters of batch normalization layers of the model via entropy minimization. This is demonstrated to be effective for robustness and source-free domain adaptation tasks. In [36], the authors replace the target statistics used in Tent with mixed source and target statistics. T3A [10] utilizes centroid-based modification to adapt the classifier in test-time for domain generalization. In [35], the authors revisit the batch normalization in the training process and develop a test-time batch normalization layer design named GpreBN, which is optimized during testing by minimizing entropy loss. This newly designed batch normalization operation preserves the same gradient backpropagation form as training and uses dataset-level statistics for robust optimization and inference. Unfortunately, all these works focus on image classification or semantic segmentation [11], and may not work well on object detection. In contrast, our method revisits the batch statistics for cross-domain pulmonary nodule detection, delving into the model optimization method specific for the detection.

5   Conclusion

In this paper, we present a source data-free setting for cross-domain lung nodule detection and present a method to tackle this issue, requiring only a pre-trained source model and a limited number of annotated samples from the target domain. Specifically, our method adapts the feature extraction module of the model by minimizing the proposed general entropy loss, and tunes the detection head with labeled target samples to enhance the detection performance even more. Experiments on our established benchmark verify that our method is an effective way to solve cross-domain object detection with data privacy issues involved. To the best of our knowledge, this is the first work on cross-domain pulmonary nodule detection without access to the source data. We also hope that this work in the medical field can bring insights into the general object detection field. In the future, we plan to pursue adaptation to more and harder types of shifts.

Acknowledgements. This work was partially supported by the Special Fund of Hubei Luojia Laboratory under Grant 220100014, and the Fundamental Research Funds for the Central Universities (No. 2042023kf1033).

References

1. Cai, Q., Pan, Y., Ngo, C., Tian, X., Duan, L., Yao, T.: Exploring object relation in mean teacher for cross-domain detection. In: CVPR, pp. 11457–11466. Computer Vision Foundation/IEEE (2019)

2. Chen, Y., Li, W., Sakaridis, C., Dai, D., Gool, L.V.: Domain adaptive faster R-CNN for object detection in the wild. In: CVPR, pp. 3339–3348. Computer Vision Foundation/IEEE (2018)
3. Girshick, R.B.: Fast R-CNN, In: ICCV. pp. 1440–1448. IEEE (2015)
4. He, K., Zhang, X., Ren, S., Sun, J.: Deep residual learning for image recognition. In: CVPR, pp. 770–778. IEEE (2016)
5. He, M., et al.: Cross domain object detection by target-perceived dual branch distillation. In: CVPR, pp. 9560–9570. IEEE (2022)
6. He, Y., Zhu, C., Wang, J., Savvides, M., Zhang, X.: Bounding box regression with uncertainty for accurate object detection. In: CVPR, pp. 2888–2897. Computer Vision Foundation/IEEE (2019)
7. He, Z., Zhang, L.: Domain adaptive object detection via asymmetric tri-way faster-RCNN. In: Vedaldi, A., Bischof, H., Brox, T., Frahm, J.-M. (eds.) ECCV 2020. LNCS, vol. 12369, pp. 309–324. Springer, Cham (2020). https://doi.org/10.1007/978-3-030-58586-0_19
8. Hofmanninger, J., Prayer, F., Pan, J., Rohrich, S., Prosch, H., Langs, G.: Automatic lung segmentation in routine imaging is a data diversity problem, not a methodology problem. CoRR abs/2001.11767 (2020)
9. Ioffe, S., Szegedy, C.: Batch normalization: accelerating deep network training by reducing internal covariate shift. In: ICML, vol. 37, pp. 448–456 (2015)
10. Iwasawa, Y., Matsuo, Y.: Test-time classifier adjustment module for model-agnostic domain generalization. In: NeurIPS, pp. 2427–2440 (2021)
11. Jiang, Y., et al.: A novel negative-transfer-resistant fuzzy clustering model with a shared cross-domain transfer latent space and its application to brain CT image segmentation. IEEE ACM Trans. Comput. Biol. Bioinform. **18**(1), 40–52 (2021)
12. Khodabandeh, M., Vahdat, A., Ranjbar, M., Macready, W.G.: A robust learning approach to domain adaptive object detection. In: ICCV, pp. 480–490. IEEE (2019)
13. Li, X., et al.: Generalized focal loss: learning qualified and distributed bounding boxes for dense object detection. In: NeurIPS (2020)
14. Li, Y., Wang, N., Shi, J., Liu, J., Hou, X.: Revisiting batch normalization for practical domain adaptation. In: ICLR (2017)
15. Lin, T., Dollár, P., Girshick, R.B., He, K., Hariharan, B., Belongie, S.J.: Feature pyramid networks for object detection. In: CVPR, pp. 936–944. IEEE (2017)
16. Lin, T., Goyal, P., Girshick, R.B., He, K., Dollár, P.: Focal loss for dense object detection. In: ICCV, pp. 2999–3007. IEEE (2017)
17. Liu, W., et al.: SSD: single shot multibox detector. In: Leibe, B., Matas, J., Sebe, N., Welling, M. (eds.) ECCV 2016. LNCS, vol. 9905, pp. 21–37. Springer, Cham (2016). https://doi.org/10.1007/978-3-319-46448-0_2
18. Mei, J., Cheng, M.M., Xu, G., Wan, L.R., Zhang, H.: SANet: a slice-aware network for pulmonary nodule detection. IEEE Trans. Pattern Anal. Mach. Intell. **44**, 4374–4387 (2021)
19. Morosov, S., et al.: Tagged results of lung computed tomography scans (RU 2018620500) (2018)
20. Qiu, H., Li, H., Wu, Q., Shi, H.: Offset bin classification network for accurate object detection. In: CVPR, pp. 13185–13194. Computer Vision Foundation/IEEE (2020)
21. Redmon, J., Divvala, S.K., Girshick, R.B., Farhadi, A.: You only look once: unified, real-time object detection. In: CVPR, pp. 779–788. IEEE (2016)
22. Ronneberger, O., Fischer, P., Brox, T.: U-Net: convolutional networks for biomedical image segmentation. In: Navab, N., Hornegger, J., Wells, W.M., Frangi, A.F. (eds.) MICCAI 2015. LNCS, vol. 9351, pp. 234–241. Springer, Cham (2015). https://doi.org/10.1007/978-3-319-24574-4_28

23. Saito, K., Ushiku, Y., Harada, T., Saenko, K.: Strong-weak distribution alignment for adaptive object detection. In: CVPR, pp. 6956–6965. Computer Vision Foundation/IEEE (2019)
24. Setio, A.A.A., et al.: Validation, comparison, and combination of algorithms for automatic detection of pulmonary nodules in computed tomography images: the LUNA16 challenge. Med. Image Anal. **42**, 1–13 (2017)
25. Shannon, C.E.: A mathematical theory of communication. Bell Syst. Tech. J. **27**(3), 379–423 (1948)
26. Sun, Y., Wang, X., Liu, Z., Miller, J., Efros, A.A., Hardt, M.: Test-time training with self-supervision for generalization under distribution shifts. In: ICML, vol. 119, pp. 9229–9248. PMLR (2020)
27. Tang, H., Zhang, C., Xie, X.: NoduleNet: decoupled false positive reduction for pulmonary nodule detection and segmentation. In: Shen, D., et al. (eds.) MICCAI 2019. LNCS, vol. 11769, pp. 266–274. Springer, Cham (2019). https://doi.org/10.1007/978-3-030-32226-7_30
28. Tian, Z., Shen, C., Chen, H., He, T.: FCOS: fully convolutional one-stage object detection. In: ICCV, pp. 9626–9635. IEEE (2019)
29. Tianchi: Tianchi medical AI competition: Intelligent diagnosis of pulmonary nodules (2017). https://tianchi.aliyun.com/competition/entrance/231601/introduction
30. Tychsen-Smith, L., Petersson, L.: Improving object localization with fitness NMS and bounded IOU loss. In: CVPR, pp. 6877–6885. Computer Vision Foundation/IEEE (2018)
31. Wang, D., Shelhamer, E., Liu, S., Olshausen, B.A., Darrell, T.: TENT: fully test-time adaptation by entropy minimization. In: ICLR (2021)
32. Xu, C., Zhao, X., Jin, X., Wei, X.: Exploring categorical regularization for domain adaptive object detection. In: CVPR, pp. 11721–11730. Computer Vision Foundation/IEEE (2020)
33. Xu, R., et al.: SGDA: towards 3D universal pulmonary nodule detection via slice grouped domain attention. IEEE/ACM Trans. Comput. Biol. Bioinform. 1–13 (2023). https://doi.org/10.1109/TCBB.2023.3253713
34. Xu, R., Luo, Y., Du, B., Kuang, K., Yang, J.: LSSANet: a long short slice-aware network for pulmonary nodule detection. In: Wang, L., Dou, Q., Fletcher, P.T., Speidel, S., Li, S. (eds.) MICCAI 2022. LNCS, vol. 13431, pp. 664–674. Springer, Cham (2022). https://doi.org/10.1007/978-3-031-16431-6_63
35. Yang, T., Zhou, S., Wang, Y., Lu, Y., Zheng, N.: Test-time batch normalization. CoRR abs/2205.10210 (2022)
36. You, F., Li, J., Zhao, Z.: Test-time batch statistics calibration for covariate shift. CoRR abs/2110.04065 (2021)
37. Zhang, S., Chi, C., Yao, Y., Lei, Z., Li, S.Z.: Bridging the gap between anchor-based and anchor-free detection via adaptive training sample selection. In: CVPR, pp. 9756–9765. IEEE (2020)
38. Zhang, Y., Wang, Z., Mao, Y.: RPN prototype alignment for domain adaptive object detector. In: CVPR, pp. 12425–12434. Computer Vision Foundation/IEEE (2021)
39. Zhao, G., Li, G., Xu, R., Lin, L.: Collaborative training between region proposal localization and classification for domain adaptive object detection. In: Vedaldi, A., Bischof, H., Brox, T., Frahm, J.-M. (eds.) ECCV 2020. LNCS, vol. 12363, pp. 86–102. Springer, Cham (2020). https://doi.org/10.1007/978-3-030-58523-5_6

3RE-Net: Joint Loss-REcovery and Super-REsolution Neural Network for REal-Time Video

Liming Ge[1(✉)], David Zhaochen Jiang[2], and Wei Bao[1]

[1] Faculty of Engineering, The University of Sydney, Sydney, Australia
{liming.ge,wei.bao}@sydney.edu.au
[2] Department of Electrical and Computer Engineering, University of Western Ontario, London, ON, Canada
djiang72@uwo.ca

Abstract. Real-time video over the Internet suffers from packet loss and low network bandwidth. The receiving side may receive down-sampled video with damaged frames. In this work, we are motivated to enhance the quality of video by joint loss recovery and super-resolution. We propose Joint Loss-REcovery and Super-REsolution Neural Network for REal-time Video (3RE-Net), to recover the loss and super-resolve a damaged frame. 3RE-Net has two unprecedented advantages: (1) It only utilizes preceding frames and the current frame as input, as waiting for future frames causes additional delay, which is not suitable for real-time video streaming. (2) 3RE-Net induces small inference delay, which is applaudable for real-time videos. To mitigate computational workload, we jointly process the loss recover and super-resolve by reusing motions and features beneficial for both super-resolution and loss recovery. The design of 3RE-Net can be summarized as follows: It first extracts motions and features from the frames, and propagates the extracted motions and features through warping, synthesis, feature-level alignment, and deep detail refinement modules. Through this way, we can first obtain a set of warped candidate frames, which are later used to generate spatio-temporal consistent feature maps through synthesis and alignment. The output frame can be reconstructed by the feature maps in high resolution and loss free. We conduct experiments to compare 3RE-Net with state-of-the-art benchmark schemes. Results demonstrate that 3RE-Net outperforms all existing benchmarks in terms of both quality and delay.

Keywords: Video super-resolution · Video loss recovery · Motion estimation · Video enhancement

1 Introduction

Real-time video traffic over the Internet has experienced a tremendous growth with the rise of video conference applications such as Zoom and Skype. With the outbreak of the COVID-19 pandemic, classes, meetings, and conferences have gone virtual. The demand for real-time video streaming is unprecedented.

© The Author(s), under exclusive license to Springer Nature Singapore Pte Ltd. 2024
T. Liu et al. (Eds.): AI 2023, LNAI 14471, pp. 165–177, 2024.
https://doi.org/10.1007/978-981-99-8388-9_14

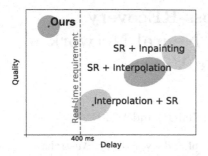

Fig. 1. Output quality vs. inference delay.

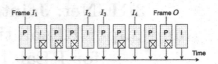

Fig. 2. Demonstration of the selection of supporting frames, where I and P denotes I-frame and P-frame respectively. The loss region bounded by solid lines are caused by packet loss directly, while the loss region bounded by dashed lines are caused by loss propagation.

Different from stored video streaming (e.g., Youtube and Netflix) where up to 10 s of delay is tolerable, real-time video streaming requires a much smaller delay (e.g., < 400 ms [27]) to satisfy the interactive experience, while consuming large network bandwidth. User Datagram Protocol (UDP), rather than Transmission Control Protocol (TCP) is widely employed as the transport-layer protocol to avoid retransmission delay. There are two obstacles to realizing real-time video delivery with high quality of experience (QoE). First, UDP works for the best-effort packet transmission and does not guarantee error-free packet delivery. Packet loss causes damaged frames in the video application and brings negative visual impacts. Second, due to the complexity of the Internet, the end-to-end throughput may fluctuate. Unlike stored video playback, where a buffer can be used to smooth the video playout at a cost of additional delay, real-time video has to be down-sampled to adapt to the network dynamics, as delaying packet transmission is not allowed.

Recent advances in deep neural networks (DNNs) have strong potential to address the aforementioned issues. Video inpainting networks, e.g. [12], can achieve video loss recovery by reconstructing a missing region in a sequence of frames. Video super-resolution networks, e.g. [11], use the low-resolution input frames to reconstruct high-resolution output frames. These DNNs can be combined to enhance the received real-time video. However, it is not very practical to directly use them at the receiver side of real-time video applications. There are two challenges. (1) Existing DNNs [2,8,12,16,29] are not designed for real-time video streaming; they introduce seconds of inference delay (see Fig. 1), which is intolerable by real-time video streaming. (2) Existing DNNs require both the preceding frames and the succeeding frames for reconstruction. To meet the stringent delay requirement of real-time video streaming, we cannot use a future (succeeding) frame as this will introduce an additional delay (to wait for the arrival of the future frame). In fact, we can only utilize the current frame and its preceding frames to enhance real-time video frames.

In this paper, we design a new network, namely 3RE-Net, to effectively realize super-resolution and loss recovery for real-time video. 3RE-Net takes the partially missed (damaged) current frame and the preceding frames as input, all in low-resolution. It recovers the missed region and performs super-resolve to output a high-resolution and complete frame with a small delay.

To address the aforementioned Challenge (1), 3RE-Net re-uses the similarities in video inpainting and video super-resolution approaches, to reduce duplicate processes and to improve effectiveness. In particular, the extracted optical flows and features of the frames can be regarded as the semantics of the video content, and the extractors can be tuned towards a generic direction to facilitate both loss recovery and super-resolution. By sharing the extractors, we not only reduce redundant computation, but also achieve even better performance.

To address the aforementioned Challenge (2), 3RE-Net takes only the partially missed current frame and the preceding frames as input. We capture multiple optical flows (motions) using the reformative motion extraction DNN. By cross-checking the optical flows, we are able to achieve comparable performance with the DNNs which utilize both the preceding frames and the succeeding frames for reconstruction.

We have conducted comprehensive experiments to evaluate the performance of 3RE-Net. We compare 3RE-Net with state-of-the-art benchmark schemes. In terms of video quality improvement (both the correctness of the reconstructed region, and the accuracy of the super-resolved details), 3RE-Net substantially outperforms other benchmark schemes. In terms of quality-delay trade-off, 3RE-Net substantially improves the video quality at a cost of a small amount of delay, which is more advantageous compared with all other benchmarks. (Other benchmarks either cannot realize as much performance gain as 3RE-Net, or cause too much delay, intolerable by real-time video applications.) A rough comparison is shown in Fig. 1. Detailed comparison will be shown later in the experiments.

2   Related Work

Video Enhancement Using DNNs. There are three types of video enhancement DNNs related to our work. Video super-resolution (VSR) [2,17] refers to the task of restoring high-resolution frames from multiple low-resolution observations of the same scene. Different from video super-resolution, we do not use succeeding frames as input to avoid additional delay, and simultaneously recover the loss. Video inpainting [12,29] fills the missing regions of a given video sequence with contents that are both spatially and temporally coherent. Similar to inpainting, one of our objectives is to recover the missing regions. However, we do not use succeeding frames as input and the inference is lightweight without incurring a long delay. Video interpolation [6,8] increases the temporal resolution of a video by synthesizing non-existent frames between two original frames. Their optical flow extractors have fewer parameters and cause less delay among the three types of DNNs.

Video Delivery and Packet Loss. UDP is widely deployed to carry real-time [25] and interactive [13] video applications [22] (e.g., WebRTC, Zoom, and Teamviewer), to guarantee conversational behaviors. UDP does not recover packet loss. The packet loss [19] usually brings a missing block after decoding

[5]. Frames in a video are put together to form a Group of Pictures (GOP) [18], which has an intra-coded frame (I-frame) and predicted frames (P-frames), but no bi-directional predicted frames (B-frames) are used in real-time video as they cause delay [28]. A lost frame in a GOP affects subsequent P-frames, and recovery may require earlier error-free frames instead of succeeding ones. These issues are expected to be addressed by our solution.

3   Model Design

Let $O \in \mathbb{R}^{H \times W \times C}$ be the damaged (partial missing) video frame in low-resolution (LR), and $\overline{O} \in \mathbb{R}^{sH \times sW \times C}$ be the corresponding original (ground-truth) video frame in high-resolution (HR), where H, W, and C denote the height, weight, and number of channels of the frame respectively, and s is the upscaling factor. Please note that $C = 3$ is a constant, and s is an integer where $s > 1$. We utilize the N complete frames in LR prior to frame O to aid the reconstruction, these frames are known as the supporting frames, where $N \in \{2, 3, 4\}$. (The larger number of N brings higher accuracy, but is more time-consuming. This trade-off is discussed in the ablation studies of the supplementary.) The proposed network takes the reference frame O and N supporting frames $\{I_1, ..., I_N\}$ in LR as input, and outputs the recovered and super-resolved frame $\hat{O} \in \mathbb{R}^{sH \times sW \times C}$ in HR:

$$\hat{O} = f_{DNN}(\{I_1, ..., I_N\}, O), \tag{1}$$

with the objective to minimize the difference between the ground-truth frame $\overline{O}$ and the reconstructed frame $\hat{O}$.

Figure 3 is an overview of the proposed network architecture. First, we extract two groups of motions using the motion extraction network (Step A). We also extract the corresponding feature map of each input frame (Step B). Then, we warp the extracted features towards the extracted motions (Step C), and obtain a set of candidate frames along with their feature maps. After that, we synthesize the candidate frames (Step D), and align the synthesized frame with spatio-temporal neighboring frames (Step E). Finally, we apply deep detail

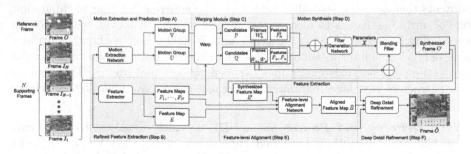

Fig. 3. DNN model architecture.

refinement to obtain the recovered and super-resolved frame (Step F). Note that the motions extracted in Step A and the features extracted in Step B benefit both loss recovery and super-resolution processes. By extracting them only once, we alleviate the redundant computation and reduce delay.

Please also note that the supporting frames $\{I_1, ..., I_N\}$ may not be consecutive frames due to the existence of loss propagation caused by GOPs. Let t_o denote the timestamp of frame O, and $\{t_1, ..., t_N\}$ be the timestamps of frames $\{I_1, ..., I_N\}$ respectively. Figure 2 shows an example, where the immediately preceding frame of the reference frame O also contains loss. In this case, we iterate backwards through the preceding frames until $N = 4$ error-free frames $\{I_1, I_2, I_3, I_4\}$ are found. In addition, we do not use a recovered frame to reconstruct O. The reason is twofold: (1) This avoids error propagation among the sequence of frames. (2) This avoids the processing delay to wait for a previous frame to be reconstructed. We assume that the network loss is not significant and only a small percentage of frames need to recover [4].

Motion Extraction and Prediction (Step A). We establish a motion extraction DNN (PWC-Net [23]). Given two input frames, `frame x` and `frame y`, the DNN will output the estimated motions from `frame y` to `frame x`, with respect to all pixels in `frame x`. The output depends on the order of input.

By feeding different pairs of inputs into the DNN, we can extract two groups of motions using direct and indirect approaches to better estimate true motions, which is detailed as follows.

First Group of Motions. For each supporting frame I_n, where $n \in \{1, \cdots, N\}$, we feed the reference frame O and the supporting frame I_n into the motion extraction network. The motion extraction network outputs the extracted motion $V_{n \to o}$. The same is done for each supporting frame, resulting in a group of N motions $\{V_{1 \to o}, ..., V_{N \to o}\}$. We denote the first group of motions as $\mathbb{V}$, where $\mathbb{V} = \{V_{1 \to o}, ..., V_{N \to o}\}$ and $|\mathbb{V}| = N$.

Second Group of Motions. We estimate the second group of motions using the supporting frames only. As for each pair of frames in the supporting frames, we estimate a set of 4 motions. Let frame I_m and I_n be a pair of two non-repetitive supporting frames, where $m, n \in \{1, \cdots, N\}$, and $m \neq n$. We obtain a set of 4 estimated motions using this pair of frames.

Given frames I_n and I_m as input, the motion extraction network outputs the estimated motion $V_{m \to n}$ for all pixels in I_n. By reversing the order of input, we obtain another motion $V_{n \to m}$ for all pixels in I_m. Given timestamps t_m, t_n, and t_o, we expand the motion vector $V_{m \to n}$ and obtain the estimated motion vector, denoted by $U_{m \to o | m \to n}$ and $U_{n \to o | m \to n}$. Formally, we have

$$U_{m \to o | m \to n} = \frac{t_o - t_m}{t_n - t_m} \times V_{m \to n}, \tag{2}$$

$$U_{n \to o | m \to n} = \frac{t_o - t_n}{t_n - t_m} \times V_{m \to n},$$

$$\forall m, n \in \{1, \cdots, N\}, m \neq n. \tag{3}$$

Similarly, as for the reversed motion $V_{n \to m}$, we have

$$U_{m \to o|n \to m} = -\frac{t_o - t_m}{t_n - t_m} \times V_{n \to m}, \tag{4}$$

$$U_{n \to o|n \to m} = -\frac{t_o - t_n}{t_n - t_m} \times V_{n \to m},$$

$$\forall m, n \in \{1, \cdots, N\}, m \neq n. \tag{5}$$

We apply the same procedure to each pair of frames in the supporting frames. We denote the second group of motions as $\mathbb{U}$. As a result, we obtain $|\mathbb{U}| = \binom{N}{2} \times 4$ motions, where

$$\mathbb{U} = \left\{ U | U_{m \to o|m \to n}, U_{n \to o|m \to n}, U_{m \to o|n \to m}, U_{n \to o|n \to m} \right\},$$

$$\forall m, n \in \{1, \cdots, N\}, m \neq n. \tag{6}$$

On the one hand, the first group of motions are extracted using the reference frame O and one of the supporting frames. The reference frame contains loss. Therefore, the first group of motions does not include the motions of the pixels in the missing region. Despite the missing region, all other pixels in the reference frame are accurate. On the other hand, the second group of motions are estimated using the complete supporting frames. The motions missing in the first group of motions are likely to be in the second group of motions. However, the motions in the second group are less accurate since the estimation does not utilize any information in the reference frame itself. The two groups of motions complement each other. Through this way, we are able to obtain more accurate motions for the subsequent steps.

Refined Feature Extraction (Step B). Given frame I_n as input, the feature extraction module extracts the feature map F_n of the input frame I_n. The feature extraction module uses the first convolutional layer of ResNet [7]. Since ResNet is designed for image classification tasks, and the rest of the layers narrow down the parameters and lose valuable contextual information. The module takes a 3-channel frame as input, and outputs a 64-channel feature map, which contains rich and generic contextual information to serve as a backbone for the alignment and up-sampling modules.

For each supporting frame I_n, where $n \in \{1, \cdots, N\}$, the corresponding feature map F_n is obtained using this module. As a result, we obtain feature maps $\{F_1, ..., F_N\}$. Similarly, we apply feature extraction to the input reference frame O and obtain feature map E. The contextual information in the extracted feature map works alongside the extracted motions, to perform loss recovery and super-resolution. Let $\texttt{Conv}$ denote the feature extraction module, we have

$$E = \texttt{Conv}(O), F_n = \texttt{Conv}(I_n), \forall n \in \{1, \cdots, N\}. \tag{7}$$

Warping Module (Step C). This module warps frames and feature maps based on the given motion vector. We adopt and modify the warping layer in the spatial transformer networks [9], which applies a spatial transformation to a pair of frames and feature maps.

As for each supporting frame I_n, where $n \in \{1, \cdots, N\}$, we warp the frame itself and its corresponding feature map F_n towards motion $V_{n \to o} \in \mathbb{V}$ (the first group of motions derived). We obtain a warped reference frame W'_n along with its feature map F'_n. We denote the aggregated pair of warped frame and its feature map as $P_n = [W'_n, F'_n]$, and let $\mathbb{P}$ denote the set of all P_n.

$$P_n \triangleq [W'_n, F'_n] = \mathtt{Warp}(I_n, F_n, V_{n \to o}), \forall n \in \{1, \cdots, N\}. \tag{8}$$

As for each pair of frames I_m and I_n in the supporting frames, where $m, n \in \{1, \cdots, N\}$, and $m \neq n$, we warp the supporting frame and its feature map towards the motions in $\mathbb{U}$ (the second group of motions derived) as stated above.

$$Q_m \triangleq [\bar{W}_m, \bar{F}_m] = \mathtt{Warp}(I_m, F_m, U_{m \to o|m \to n}), \tag{9}$$

$$\tilde{Q}_m \triangleq [\tilde{W}_m, \tilde{F}_m] = \mathtt{Warp}(I_m, F_m, U_{m \to o|n \to m}), \tag{10}$$

$$Q_n \triangleq [\bar{W}_n, \bar{F}_n] = \mathtt{Warp}(I_n, F_n, U_{n \to o|m \to n}), \tag{11}$$

$$\tilde{Q}_n \triangleq [\tilde{W}_n, \tilde{F}_n] = \mathtt{Warp}(I_n, F_n, U_{n \to o|n \to m}),$$
$$\forall m, n \in \{1, \cdots, N\}, m \neq n. \tag{12}$$

Let $\mathbb{Q}$ denote the set of all $Q_m, \tilde{Q}_m, Q_n, \tilde{Q}_n$.

$$\mathbb{Q} = \left\{ Q | Q_m, \tilde{Q}_m, Q_n, \tilde{Q}_n \right\}, \forall m, n \in \{1, \cdots, N\}, m \neq n. \tag{13}$$

Since $|\mathbb{P}| = |\mathbb{V}| = N$ and $|\mathbb{Q}| = |\mathbb{U}| = \binom{N}{2} \times 4$, we obtain $|\mathbb{P}| + |\mathbb{Q}| = N + \binom{N}{2} \times 4$ pairs of warped frames and feature maps. These frames, along with their corresponding feature maps, then become candidates for the synthesis module.

Motion Synthesis (Step D). We do not use a simple average on the candidate warped frames, as it leads to blurry results caused by disputed estimation of fast-moving and occluded pixels. To address this issue, we feed the missing and unreliable pixels into a DNN (dynamic filter network [10]), which takes spatio-temporal factors into consideration to soundly resolve the disputes.

We first adopt the filter generation network to cross check the $|\mathbb{P}| + |\mathbb{Q}|$ candidate warped frames on the image level. As for each candidate warped frame, the network generates a set of $5 \times 5 \times \left[N + \binom{N}{2} \times 4\right]$ parameters for the blending filter denoted as X, where 5×5 is the size of the blending filter. We then apply convolution using the blending filter on each warped candidate frame, and the results of each candidate frame are aggregated to generate the synthesized frame O'. We extract the feature map E' of frame O' using the aforementioned refined feature extractor (Step B).

Feature-Level Alignment (Step E). Spatial and temporal alignment are crucial for video enhancement. Due to the varying motions of cameras or objects, the synthesized reference frame and the supporting frames are not aligned. Misaligned frames can hinder aggregation and reduce performance, especially with a larger number of supporting frames N. Alignment modules align frames and features to enable subsequent aggregation. Traditional image-level alignment methods result in artifacts when combined with the deep detail refinement module (Step F), as they suffer from artifacts around image structures being propagated into final reconstructed frames. To avoid this, we resort to deformable convolution [3], which provides additional offsets to allow the convolutional network to obtain information from beyond its regular local neighborhood [24,26], and redevelop the feature-level deformable alignment module in our design.

The module inputs the synthesized frame E', aligns the supporting feature maps $F_1, \cdots, F_N$ and E with E', and outputs the aligned feature map $\bar{E}$.

Note that the synthesis module handles the disputed estimation of the occluded or missing pixels at the same timestamp t_o. It performs spatial aggregation on the image level. The alignment module, on the other hand, performs feature-level adjustments to align the temporal neighboring frames (at timestamps $\{t_1, ..., t_N\}$) with the reference frame. It effectively handles complex motions and large parallax problems.

Deep Detail Refinement (Step F). Since we have the synthesized reference frame O' along with its aligned feature map $\bar{E}$, we aggregate the reference frame and the supporting frames cross the time-space (since each of the supporting frames may contain different detail), and then up-sample to obtain the output in HR. We directly concatenate the $N+1$ frames and feed them into a convolutional layer to output the fused feature map. Then, we feed the fused feature map and the aligned feature map derived in Step E into a nonlinear mapping module, which utilizes the fused features to predict deep features. After extracting deep features, we utilize an up-sampling module composed of an up-scaling layer [14] to increase the resolution of the feature map, with a sub-pixel convolution [21]. The final output frame in HR is obtained by a convolutional layer from the zoomed feature map.

4   Experiments

Datasets and Metrics. We adopt Vimeo90k [30] as the training and testing dataset. It is a public dataset consists of 448×256 video for video enhancement tasks, including super-resolution. We further use Vid4 [15] as another testing set, since its results are more visually comparable. Vid4 consists of multiple scenes with various motions and occlusions in 720p. We apply $4\times$ down-sampling using bicubic degradation. We stream the down-sampled H.264 encoded video over the Internet using UDP-based RTSP protocol [20] via FFmpeg. We employ Peak Signal to Noise Ratio (PSNR), measured in decibels (dB) where higher values indicate superior quality, and the Structural Similarity Index (SSIM),

Table 1. Average PSNR (in dB) and SSIM along with the delay (in ms) under all schemes. B1–B12 denote the benchmark schemes.

Schemes	Vid4 Dataset		Vimeo90k Dataset		Delay
	PSNR	SSIM	PSNR	SSIM	
RAW (Bicubic)	17.7606	0.4889	26.7701	0.7559	None
(B1) DAIN + RRN	18.3051	0.4712	27.4539	0.7432	488
(B2) RRN + DAIN	18.0939	0.5000	26.8427	0.7649	1339
(B3) DAIN + RBPN	18.8285	0.5141	27.9267	0.7916	1414
(B4) RBPN + DAIN	18.8222	0.5442	27.6205	0.8174	2276
(B5) BMBC + RRN	18.0185	0.5093	26.7970	0.7767	302
(B6) RRN + BMBC	19.4391	0.5688	28.5432	0.8460	1260
(B7) BMBC + RBPN	18.0883	0.5370	27.6172	0.8046	1228
(B8) RBPN + BMBC	19.2415	0.5477	27.5649	0.8053	2389
(B9) RRN + DFC-Net	18.1021	0.5074	26.9107	0.7719	1454
(B10) RBPN + DFC-Net	18.6556	0.5318	27.3627	0.7833	3380
(B11) RRN + VI-Net	19.5097	0.5460	28.9774	0.8304	1053
(B12) RBPN + VI-Net	19.5979	0.5376	29.2534	0.8108	2975
3RE-Net (ours)	**21.2680**	**0.6435**	**31.2192**	**0.9064**	**187**

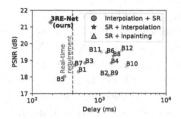

Fig. 4. Quality vs. delay under different schemes.

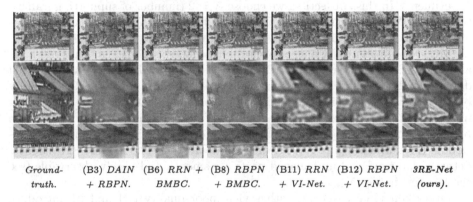

| Ground-truth. | (B3) DAIN + RBPN. | (B6) RRN + BMBC. | (B8) RBPN + BMBC. | (B11) RRN + VI-Net. | (B12) RBPN + VI-Net. | 3RE-Net (ours). |

Fig. 5. Qualitative comparison of the reconstructed loss region. Two red boxes are zoomed in. The second row shows one of the reconstructed regions. The third row shows the border between the damaged and undamaged parts. (Color figure online)

which ranges from 0 to 1 with higher values denoting greater similarity with the ground-truth, as quantitative evaluation metrics for comparing the quality of reconstructions.

Benchmarks. We conduct comprehensive experiments by comparing our 3RE-Net with the following benchmarks. For the loss recovery part, we consider two video interpolation networks (DAIN [2] and BMBC [17]) and two video inpainting networks (DFC-Net [29] and VI-Net [12]). For the super-resolution part, we consider two video super-resolution networks (RRN [8] and RBPN [6]). Then, we run interpolation/inpainting and super-resolution in tandem to compare with ours. Since the above schemes are not designed for real-time loss recovery, we modify them slightly for a fair comparison. The modified schemes take as input two previous (supporting) frames and the current damaged (reference) frame, identical to 3RE-Net. The timestamps in the optical flow manipulation are changed to ensure that the benchmarks synthesize the damaged frame as desired. Please note that we can swap the order of interpolation and super-resolution and treat them as different benchmarks. For example, $DAIN + RRN$ means we run DAIN first and then RRN; while $RRN + DAIN$ means we run RRN first and then DAIN. Inpainting (DFC-Net and VI-Net) can only be run after super-resolution, because these networks contain large convolution kernels and cannot handle input frames in low resolution. Note that for video prediction, its performance is worse than interpolation and thus it is not necessary to compare with benchmarks utilizing prediction. In sum, we consider 12 benchmarks (labeled by B1–B12) in Table 1.

Results. The quantitative and qualitative results are summarized in Table 1 and Fig. 5, respectively. As for the visualized result, we only show our results and top 5 benchmarks due to the space limit. The quality and delay comparison using Vid4 is illustrated in Fig. 4. Note that the delay is averaged after 1,000 rounds of experiments. In this subsection, we choose $N = 2$ (number of supporting frames) by default. We will further discuss the choice of different N in the ablation studies of the supplementary material.

3RE-Net substantially outperforms all benchmark schemes in terms of PSNR and SSIM, and achieves 21.26 dB in PSNR (resp. 0.6435 in SSIM) for Vid4 dataset, and 31.21 dB in PSNR (resp. 0.9064 in SSIM) for Vimeo90k. This is because 3RE-Net effectively exploits the motions and features in the supporting frames. With packet loss, the damaged part of the frame is still observable in the preceding (supporting) frames. 3RE-Net aggregates the information and reconstructs the damaged part in a spatio-temporal consistent manner. The border between damaged and undamaged parts are smooth and barely observable (see the last row in Fig. 5).

Other benchmark schemes either yield poor quality (B1 and B5 introduce merely acceptable $\sim$ 400 ms delay, but low PSNR (18 dB)), or cause too much delay ($>$ 1000 ms delay for the rest of the benchmarks). 3RE-Net causes the least delay (187 ms) and yields satisfactory results. In our joint design, the extracted motions and features are used by multiple modules (including motion synthesis, feature-level alignment, and deep detail refinement), which significantly reduces redundancy.

We observe a trade-off when we swap the order of interpolation and super-resolution. Performing interpolation before super-resolution (B1, B3, B5, and B7) yields weaker quality but less delay, compared with the opposite order (B2, B4, B6, and B8). This is because the load of the motion and feature extractors is smaller on low-resolution inputs, while the results are less accurate due to the limited resolution of inputs. 3RE-Net can provide high quality without causing large delays. All modules have access to accurate motions and features.

In terms of quality-delay trade-off, 3RE-Net substantially improves the video quality at a cost of a small amount of delay. It is more advantageous compared with all other benchmarks, as they either cannot realize as much performance gain as ours, or cause too much delay, intolerable by real-time video applications.

Other Studies. Results with confidence intervals, ablation study, inference delay breakdown, and more visualized examples are shown in the supplementary materials [1].

5   Conclusion

In this work, we propose the 3RE-Net (Joint Loss-REcovery and Super-REsolution Neural Network for REal-time Video). Our pipeline utilizes the reference (damaged) frame and preceding frames. It exploits the motions and features of the input frames, and propagates them through warping and synthesis modules. The output frame is then reconstructed using deep features in high resolution and loss free, with small inference delay. We run benchmark interpolation/inpainting and super-resolution schemes, and results show that 3RE-Net outperforms all other benchmarks with substantial improvements in both quality and delay.

References

1. Supplementary materials for 3RE-Net: joint loss-recovery and super-resolution neural network for real-time video (2023). https://github.com/greg308/3RE-Net/blob/main/3RE_Net_AJCAI_Supplementary.pdf
2. Bao, W., Lai, W.S., Ma, C., Zhang, X., Gao, Z., Yang, M.H.: Depth-aware video frame interpolation. In: IEEE Conference on Computer Vision and Pattern Recognition, pp. 3703–3712 (2019)
3. Dai, J., et al.: Deformable convolutional networks. In: IEEE International Conference on Computer Vision, pp. 764–773 (2017)
4. Gu, Y., Grossman, R.L.: UDT: UDP-based data transfer for high-speed wide area networks. Comput. Netw. **51**(7), 1777–1799 (2007)
5. Han, J., et al.: A technical overview of AV1. Proc. IEEE **109**, 1435–1462 (2021)
6. Haris, M., Shakhnarovich, G., Ukita, N.: Recurrent back-projection network for video super-resolution. In: IEEE Conference on Computer Vision and Pattern Recognition (CVPR) (2019)
7. He, K., Zhang, X., Ren, S., Sun, J.: Deep residual learning for image recognition. In: IEEE Conference on Computer Vision and Pattern Recognition. pp. 770–778 (2016)
8. Isobe, T., Zhu, F., Jia, X., Wang, S.: Revisiting temporal modeling for video super-resolution. arXiv preprint arXiv:2008.05765 (2020)
9. Jaderberg, M., Simonyan, K., Zisserman, A.: Spatial transformer networks. In: Advances in Neural Information Processing Systems, vol. 28, pp. 2017–2025 (2015)
10. Jia, X., De Brabandere, B., Tuytelaars, T., Gool, L.V.: Dynamic filter networks. In: Advances in Neural Information Processing Systems, vol. 29, pp. 667–675. Curran Associates, Inc. (2016)
11. Kappeler, A., Yoo, S., Dai, Q., Katsaggelos, A.K.: Video super-resolution with convolutional neural networks. IEEE Trans. Comput. Imaging **2**(2), 109–122 (2016)

12. Kim, D., Woo, S., Lee, J.Y., Kweon, I.S.: Deep video inpainting. In: IEEE/CVF Conference on Computer Vision and Pattern Recognition, pp. 5792–5801 (2019)
13. Kirmizioglu, R.A., Tekalp, A.M.: Multi-party WebRTC services using delay and bandwidth aware SDN-assisted IP multicasting of scalable video over 5G networks. IEEE Trans. Multimedia 22(4), 1005–1015 (2019)
14. Lim, B., Son, S., Kim, H., Nah, S., Mu Lee, K.: Enhanced deep residual networks for single image super-resolution. In: IEEE Conference on Computer Vision and Pattern Recognition Workshops, pp. 136–144 (2017)
15. Liu, C., Sun, D.: On Bayesian adaptive video super resolution. IEEE Trans. Pattern Anal. Mach. Intell. 36(2), 346–360 (2013)
16. Liu, C., Yang, H., Fu, J., Qian, X.: Learning trajectory-aware transformer for video super-resolution. In: IEEE/CVF Conference on Computer Vision and Pattern Recognition, pp. 5687–5696 (2022)
17. Park, J., Ko, K., Lee, C., Kim, C.-S.: BMBC: bilateral motion estimation with bilateral cost volume for video interpolation. In: Vedaldi, A., Bischof, H., Brox, T., Frahm, J.-M. (eds.) ECCV 2020. LNCS, vol. 12359, pp. 109–125. Springer, Cham (2020). https://doi.org/10.1007/978-3-030-58568-6_7
18. Porter, T., Peng, X.H.: Hybrid TCP/UDP video transport for H. 264/AVC content delivery in burst loss networks. In: 2011 IEEE International Conference on Multimedia and Expo, pp. 1–5. IEEE (2011)
19. Rane, S.D., Sapiro, G., Bertalmio, M.: Structure and texture filling-in of missing image blocks in wireless transmission and compression applications. IEEE Trans. Image Process. 12(3), 296–303 (2003)
20. Schulzrinne, H., Rao, A., Lanphier, R.: Real time streaming protocol (RTSP) (1998)
21. Shi, W., et al.: Real-time single image and video super-resolution using an efficient sub-pixel convolutional neural network. In: IEEE Conference on Computer Vision and Pattern Recognition, pp. 1874–1883 (2016)
22. Suciu, G., Stefanescu, S., Beceanu, C., Ceaparu, M.: WebRTC role in real-time communication and video conferencing. In: Global Internet of Things Summit (GIoTS), pp. 1–6. IEEE (2020)
23. Sun, D., Yang, X., Liu, M.Y., Kautz, J.: PWC-Net: CNNs for optical flow using pyramid, warping, and cost volume. In: IEEE Conference on Computer Vision and Pattern Recognition, pp. 8934–8943 (2018)
24. Tian, Y., Zhang, Y., Fu, Y., Xu, C.: TDAN: temporally-deformable alignment network for video super-resolution. In: IEEE/CVF Conference on Computer Vision and Pattern Recognition, pp. 3360–3369 (2020)
25. Vucic, D., Skorin-Kapov, L.: The impact of packet loss and google congestion control on QoE for WebRTC-based mobile multiparty audiovisual telemeetings. In: Kompatsiaris, I., Huet, B., Mezaris, V., Gurrin, C., Cheng, W.-H., Vrochidis, S. (eds.) MMM 2019. LNCS, vol. 11295, pp. 459–470. Springer, Cham (2019). https://doi.org/10.1007/978-3-030-05710-7_38
26. Wang, X., Chan, K.C., Yu, K., Dong, C., Change Loy, C.: EDVR: video restoration with enhanced deformable convolutional networks. In: IEEE/CVF Conference on Computer Vision and Pattern Recognition Workshops (2019)
27. Wang, Z., Cui, Y., Hu, X., Wang, X., Ooi, W.T., Li, Y.: MultiLive: adaptive bitrate control for low-delay multi-party interactive live streaming. In: IEEE Conference on Computer Communications, pp. 1093–1102. IEEE (2020)
28. Wenger, S.: H. 264/AVC over IP. IEEE Trans. Circ. Syst. Video Technol. 13(7), 645–656 (2003)

29. Xu, R., Li, X., Zhou, B., Loy, C.C.: Deep flow-guided video inpainting. In: IEEE/CVF Conference on Computer Vision and Pattern Recognition, pp. 3723–3732 (2019)
30. Xue, T., Chen, B., Wu, J., Wei, D., Freeman, W.T.: Video enhancement with task-oriented flow. Int. J. Comput. Vis. (IJCV) **127**(8), 1106–1125 (2019)

Neural Networks in Forecasting Financial Volatility

Wenbo Ge[1]($\boxtimes$)(iD), Pooia Lalbakhsh[2](iD), Leigh Isai[3](iD), and Hanna Suominen[1,4](iD)

[1] School of Computing, The Australian National University, 145 Science Road, Acton, Canberra, ACT 2601, Australia
wenbo.ge@anu.edu.au

[2] Department of Data Science and Artificial Intelligence, Monash University, 20 Exhibition Walk, Clayton, VIC 3168, Australia

[3] Euler Capital, 63-67 Cemetery Road, Drysdale, VIC 3222, Australia

[4] Department of Computing, University of Turku, 20014 Turku, Finland

Abstract. In 2020s, the state of the art (SOTA) in financial volatility forecasting is underpinned by deep learning (DL). Despite this, forecasting methods in practice tend to be dominated by their more traditional counterparts (e.g., Generalised Auto-Regressive Conditional Heteroscedasticity (GARCH) models) or relatively simple neural networks (NN), leaving much of DL unexplored. Hence, this study experimented the power of DL in forecasting financial volatility and expedited further progress in such multidisciplinary DL applications to quantitative finance by releasing open-source software and proposing a shared task. We compared the financial forecasting ability of the SOTA methods used to more recent DL work, proceeding from simpler or shallower to deeper and more complex models. Specifically, the volatility of five assets (i.e., S&P500, NASDAQ100, gold, silver, and oil) was forecast with the GARCH models, multi-layer perceptrons, recurrent NNs, temporal convolutional networks, and Temporal Fusion Transformer. The results indicated that in almost all cases, DL models forecast volatility with less error than the SOTA models in financial volatility research. These experiments were repeated and the difference between competing models was shown to be statistically significant, therefore encouraging their use in practice.

Keywords: Applications · Deep learning · Economics · Financial volatility · Multidisciplinary AI · Neural networks · Time series analysis

Supported by Euler Capital Pty Ltd, under APR Intern Agreement (INT - 0804). We would also like to thank the APR.Intern and *Australian National University* (ANU) for their support to this research through the Australian Postgraduate Research Internship Program (APR.Intern) program. We also gratefully acknowledge funding from the *Australian Government Research Training Program* (AGRTP) Domestic Scholarship for the first author's PhD studies.

T. Liu et al. (Eds.): AI 2023, LNAI 14471, pp. 178–189, 2024.
https://doi.org/10.1007/978-981-99-8388-9_15

1   Introduction

One of the most important tasks for a financial institution is to monitor the *volatility* of its portfolio and other market variables. However, there are many different ways to quantify this latent and unobservable variable, such as *historical volatility* (HV, a.k.a. close-to-close, the standard deviation of log-returns over a time window) [27][1], *realised volatility* (RV, the square root of the sum of squared log-returns over a time window) [1], *implied volatility* (IV, backwards calculated from options prices via an option pricing model, such as Black-Scholes) [20], and many more [26]. Because volatility is a key factor in security valuation, risk management, and options pricing, as well as affecting investment choice and valuation of public and corporate liabilities, sophisticated computational models are studied for financial volatility forecasting to support practitioners' judgment and decision-making in quantitative finance [6,19,27,30]. In the 2020s, such computer-assisted forecasting methods are dominated by *Generalised Auto-Regressive Conditional Heteroscedasticity* (GARCH) models and relatively simple *Neural Networks* (NN), leaving much of *Machine Learning* (ML) and *Deep Learning* (DL) unexplored [10].

Hence, this multidisciplinary paper will exemplify the power of ML/DL in forecasting financial volatility to practitioners in quantitative finance. We will compare the financial forecasting ability of a range of methods by proceeding from simpler or shallower models (i.e., the GARCH models and *Multi-Layer Perceptrons* (MLP)) to deeper and more complex NNs (i.e., the *Recurrent NNs* (RNN), *Convolutional NNs* (CNN), *Temporal Convolutional Networks* (TCN), and *Temporal Fusion Transformer* (TFT)). These performance evaluations and statistical analyses on five assets (i.e., S&P500, NASDAQ100, gold, silver, and oil), completed by releasing our Python code[2] under the MIT license should encourage practitioners to apply DL as a way to reduce error in forecasting financial volatility.

2   Related Work

A convenient property of financial price data is the efficient market hypothesis, which stipulates that all publicly available information is reflected in the market prices of assets at a given time point [30]. At the finest resolution, market prices are a list of prices of all the buy and sell orders that were matched, which can then be aggregated over time (e.g., 1-h or 1-day intervals) to create more coarse-grained views and can be further described by its highest, lowest, opening, and closing price for that interval, as well as the total number of assets traded, known as volume; however, introducing additional data tends to be helpful in

---

[1] Despite the name containing the word 'historical', it is not defined exclusively for historical data. This can still be forecast in the same way that realized volatility can forecast.

[2] All results, tables, figures, and analysis methods can be found at https://github.com/xyz, along with extended results.

this predictive modeling task [21,35]. In addition to the definition of volatility and financial price data, the volatility forecasting model should consider the time period for which the data is useful: If the goal was to forecast for the next 15 min, using data from the previous 50 years might be wasteful but with 1 week, information from past market regimes that could repeat might be missed. Moreover, the amount of information provided at inference time is important as it impacts the computation time, as well as may dilute the useful information; when inferring the volatility of the next 30 days, all data could be useful, but the most recent entries are likely to carry more insights than earlier ones. Finally, the timing of the data and the window of time that the volatility captures must also be considered, keeping in mind that the further into the future we are trying to forecast, the more uncertain any forecast will be. Although this part of modeling should depend on the reason for forecasting volatility[3], asking the model to forecast volatility in a wide range of time frames may be beneficial[4].

Of the many types of models that can be used to understand and forecast volatility, none are as widespread as the *auto-regressive* (AR) models: The seminal *Auto-Regressive Conditional Heteroscedasticity* (ARCH) models future volatility conditioned on previous observations [7] and its adaptation as the well-known GARCH model includes an *Auto-Regressive Moving Average* (ARMA) component [2]. Since these models from the 1980s, there have been many advancements that attempt to address the models' inability to capture several stylized facts of volatility [8][5]. Despite the countless variants of the GARCH model, several experiments have found that the simple ARCH(1) and GARCH(1, 1) forecasting models perform the best [11,24].

ML and DL models have also shown much success and are rising in popularity [4,5,12]. NN-based models are commonly used, and although they do not have the same theoretical underpinnings as the GARCH models, they are flexible, possessing the ability to learn any arbitrary mapping f from input $\mathbf{X}$ to output y; $y = f(\mathbf{X})$. In the context of time series analysis, a *Nonlinear Auto-Regressive* (NAR) framework is often adopted with the MLP, enforcing an AR property to the nonlinear mapping (e.g., $\widehat{y}_{t+1} = f([y_t, y_{t-1}, ..., y_{t-m}]^T)$ with t referring to a given time point) [15]. This can be extended into a NARX framework by including exogenous variables (such as those derived from several indices,

---

[3] E.g., market makers and day traders may want to monitor short-term volatility in the span of minutes to watch for entry/exit signals.

[4] E.g., (1) if using daily prices in forecasting financial volatility, some assets are not on the market every day, and as a result, when using multiple data streams, a mismatch between the date and time of each point in the time series is likely to be present, calling for interpolation to fill in the missing data between points, or (2) when using ML/DL methods for modeling, having multiple learning signals aggregated can keep the learning on track.

[5] E.g., *Exponential, Threshold,* and *Glosten-Jagannathan-Runkle* versions (EGARCH, TGARCH, GJR-GARCH) which allow for asymmetric dependencies in volatility, and the *Integrated* and *Fractionally Integrated* versions (IGARCH, FIGARCH) which address volatility persistence, where an observed shock in the volatility series seems to impact future volatility over a long horizon.

exchange rates, and outputs of GARCH models), thus providing more information to the model [3] which has been beneficial for forecasting performance [14]. Other NN architectures (e.g., RNNs, CNNs, and *Long Short Term Memory* (LSTM) models) have also been used in volatility forecasting. For instance, LSTM and GARCH models have been combined to forecast HV [13] and gold prices can be converted into a 3-channel RGB image and then processed with a pre-trained vgg16 (a well-known and high performing CNN model) [32].

Whilst RNNs, LSTMs, and CNNs are deep models, they are not considered as the *state-of-the-art* (SOTA) for time series processing in DL, and models used in financial volatility forecasting tend to be even shallower and simpler, a distinct gap highlighted in a recent systematic literature review [10]. This is reserved for recent models that have the extremely deep capacity and use complex models, often adapted from other fields such as TCNs, which have been successful in music generation, speech enhancement, and many other areas involving time series [16,23,25]. The TCN is a CNN adaptation, consisting of 1-dimensional convolutional blocks structured in a way that does not violate the temporal ordering of data (i.e., only past data can be seen when forecasting), known as a causal convolution [23]. In conjunction with a progressively increasing dilation size, the receptive field can be increased exponentially as layers increase, thus allowing the exploitation of long-term relationships. These blocks also often use residual connections, layer normalization, gradient clipping, and dropout, all of which have been shown to improve learning and performance [34]. Another recently developed SOTA model that handles sequential data well is the Transformer [31]. Its TFT variant deploys a gating mechanism to skip unused components of the network, variable selection networks to select relevant input variables at each time step, static co-variate encoders to provide context to the model, temporal processing to learn long and short-term relationships, and quantile predictions to forecast with a corresponding confidence [17].

3   Experimental Comparison of Forecasting Models

Our experimental study of forecasting models will next exemplify through comparative performance evaluation and statistical significance testing the power of DL in forecasting financial volatility. The volatility of five assets will be forecast with the SOTA methods; simpler or shallower DL models; and recent deeper and more complex models. The results will indicate that in almost all cases, DL models forecast volatility with less error than the SOTA models in financial volatility research. These experiments will be repeated to give evidence that the difference between competing models is statistically significant, therefore encouraging their use in practice and further study as a shared task.

3.1   Posing the Problem as a Shared Task

Volatility was forecast for five assets: S&P500, NASDAQ-100 (NDX), gold, silver, and oil. The data for each, as well as the corresponding volatility indices, were

Table 1. Description of data

Asset	Start date	End date
S&P500	22/Sep/2003	31/Dec/2018
NASDAQ100	01/Jan/2003	31/Dec/2018
Gold	03/Jun/2008	31/Dec/2018
Silver	16/Mar/2011	31/Dec/2018
WTI Crude Oil	10/May/2007	31/Dec/2018

retrieved from Global Financial Data[6] (Table 1). The proper permissions to use the data for the purposes of this study and its reporting were obtained from Global Financial Data[7]. The data consisted of the daily closing prices, as well as the open, high, and low prices for S&P500, NDX, gold, and oil. Volume was available only for S&P500 and NDX. Each asset was restricted to a starting date that corresponded with when the volatility index was introduced, except for S&P500 and NDX. This was because the volatility index for S&P500 was originally for the S&P100 and later changed on 22 September 2003 and because the volatility index for NDX began earlier than one of the exogenous variables used. Additionally, the ending date was restricted to 31 December 2018.

Exogenous variables were also retrieved from Global Financial Data, consisting of several other indices (SZSE, BSE SENSEX, FTSE100, and DJIA), exchange rates (US-YEN, US-EURO, and the US dollar trade weighted index), and United States fundamentals (Federal Reserve primary credit rate, mean and median duration of unemployment, consumer price index inflation rate, Government debt per *Gross Domestic Product* (GDP), gross Federal debt, and currency in circulation). All variables were date matched with the underlying assets by bringing forward the nearest historical value.

The task was to forecast the month-long HV and IV (Fig. 1), starting from 1 day ahead, for S&P500, NDX, gold, silver, and oil. The ground truth for HV was the standard deviation of log returns starting from 1 trading day ahead to 22 trading days ahead (= one calendar month). In other words, with t referring to the current time, we defined HV over a certain period $[\tau_1, \tau_2] = [t+1, t+22]$ as the standard deviation $(\text{std}(\cdot))$ of log-returns as follows:

$$
\text{HV} = \sqrt{\frac{1}{N}\sum_{t=\tau_1}^{\tau_2}\left(r_t - \frac{1}{N}\sum_{t=\tau_1}^{\tau_2} r_t\right)^2} = \text{std}\left(\begin{bmatrix} r_{\tau_2} \\ r_{\tau_2-1} \\ r_{\tau_2-2} \\ \vdots \\ r_{\tau_1} \end{bmatrix}\right) \tag{1}
$$

where $N = \tau_2 - \tau_1 = 21$ is the number of samples between the time steps, P_t is price at time t, and $r_t = \log(P_t/P_{t-1}) \cdot 100$. For IV, this meant the ground

---

[6] https://globalfinancialdata.com/.

[7] Due to the underlying data use agreement, the data or their derivatives cannot be released as part of this paper.

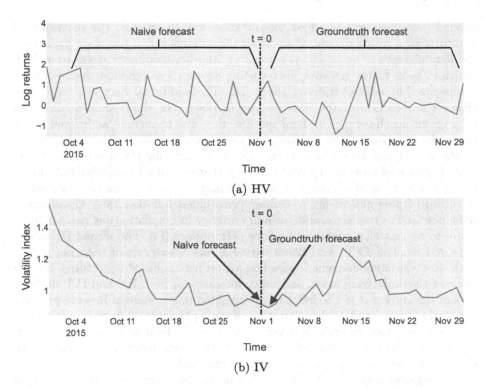

Fig. 1. Groundtruth and naïve forecasts for HV (a) and IV (b).

truth was simply the value of the volatility index for the next trading day, as the volatility index was already defined for the next calendar month. The values of the volatility indices were also adjusted by a factor of $1/\sqrt{252}$, de-annualizing the value to be on the same scale as HV.

3.2   Methods

Five methods were used to represent the SOTA financial volatility forecasting performance, the combination of which will be the benchmark for comparison. These five methods were: a naïve model[8], a GARCH model, an MLP model, and two models from literature: ANN-GARCH [14], and CNN-LSTM [32].

Two models from DL were investigated to represent the experimental fore-casting performance. The first model was the TCN, as well as the TCN with

---

[8] The naïve model simply repeated the most recent known value of volatility (Fig. 1). For IV, this was the value of the corresponding volatility index at the current time. For HV, this was the standard deviation of log returns for the current day and previous 20 days, that is, Eq. (1) with $\tau_1 = t - 20$ and $\tau_2 = t$. This assumed that at the time of forecasting, the current trading day is over and observed, an assumption maintained for all forecasting models.

several modifications. The first modification was to leverage the naïve model and forecast a residual, defined as either the difference $y_{\text{groundtruth},t} - y_{\text{naïve},t}$, or the log difference $log(y_{\text{groundtruth},t}/y_{\text{naïve},t})$. Another modification was to include multiple tasks to the network, introducing another loss function that will have a separate but related training effect [28]. The additional task was to predict either the direction of the forecast (up or down), or the change in direction (change or no change). The final modification was to include additional input channels, introducing new information to the network [33], such as descriptors of the underlying asset (log returns, naïve forecasts for HV and IV, and current direction of movement), and variables that describe the market (US dollar trade-weighted index, Federal Reserve primary credit rate, mean and median duration of unemployment, consumer price index inflation rate, Government debt per GDP, gross Federal debt, and currency in circulation) as there is literature to suggest that this may improve performance [14]. The second DL model explored was the TFT. Additional variables like descriptors of the asset (open, high, low, close, volume where possible, log returns, squared log returns, inverse price of the underlying asset, and naïve forecasts for both HV and IV), descriptors of the market (the US dollar trade-weighted index, Federal Reserve primary credit rate, mean and median duration of unemployment, consumer price index inflation rate, Government debt per GDP, gross Federal debt, and currency in circulation), and descriptors of time (day of the week, month, and a number of days since previous observation) were also included.

To engineer and evaluate the forecasting models using these five methods, a 70-15-15 train-validation-test split of the data was used because it did not violate the temporal aspect of the data[9]. All performances were quantified with the *Mean Squared Error* (MSE) with statistical significance testing to distinguish if competing models were statistically significantly different from each other. After the hyperparameters of a model were chosen and the testing phase was completed, the model was reinitialized with a random seed, re-trained, and re-tested. This was repeated until ten MSE values were obtained for each model. These values were then tested across different models in a pair-wise fashion to determine if they were from the same distribution[10]. The *Shapiro-Wilk* (SW) test was first applied to assess the normality of the distribution with significance level $\alpha = 0.05$[11]. If both distributions were normal then Student's t-test was used, otherwise the *Kruskal-Wallis* (KW) test was employed.

---

[9] The training set was used to train the models using different hyperparameters, which were then evaluated against the validation set to determine the performance with that given set of hyperparameters. Different combinations of hyperparameters were searched, and the best performing set of hyperparameters proceeded to the test phase. Here, the model was re-initialized and trained again with the union of both the training and validation set, then evaluated once using the test set.

[10] If so, the models were assumed to be equivalent, if not, then the model with the smaller mean MSE was assumed to be superior.

[11] The SW test was justified by the applicability of the test to the data with unspecified mean and variance, as well as its appropriateness for small sample sizes.

Table 2. Smallest error models (in bold), and models for which no statistically significant difference could be found from the smallest error models.

Task	Model	S&P500	NDX	Gold	Silver	Oil
HV	Smallest error	**TFT**	**TFT**	**ANN-GARCH**	**TCN var.**	**TFT**
	Equivalent	–	–	TFT	TCN	–
IV	Smallest error	**TCN var.**	**TCN var.**	**TCN var.**	**TCN var.**	**TCN var.**
	Equivalent	–	–	–	–	–

Table 3. Performance (MSE) of forecasting models on the test set. Smaller is better, bold is the best. HS refers to Hyperparameter Search.

Model Family and Task	Method	S&P500	NDX	Gold	Silver	Oil
Benchmark Models						
HV	Naïve	0.1281	0.1887	0.0216	0.0745	0.2799
	GARCH	0.2662	0.9430	0.0191	0.0611	7.9448
	MLP	0.2935	0.3542	0.0146	0.0666	**0.2080**
	ANN-GARCH	**0.1136**	**0.1735**	**0.0105**	0.0563	0.2191
	CNN-LSTM	0.1550	0.1869	0.0554	**0.0488**	0.3870
IV	Naïve	**0.0098**	**0.0084**	**0.0013**	**0.0023**	**0.0139**
	GARCH	0.2296	0.5859	0.0263	0.0279	7.6927
	MLP	0.0172	0.0151	0.0021	0.0032	0.0278
	ANN-GARCH	0.0248	0.0159	0.0022	0.0036	0.0312
	CNN-LSTM	0.0141	0.0104	0.0064	0.0104	0.0246
Experimental Models						
HV	Naïve	0.1281	0.1887	0.0216	0.0745	0.2799
	Best benchmark	0.1136	0.1735	**0.0105**	0.0488	0.2080
	TCN	0.1467	0.1912	0.0331	0.0364	0.2614
	TCN variants	0.1300	0.1728	0.0153	**0.0315**	0.2050
	TFT	**0.0294**	**0.0513**	0.0116	0.0341	**0.0864**
IV	Naïve	0.0098	0.0084	0.0013	0.0023	0.0139
	Best benchmark	0.0141	0.0104	0.0021	0.0032	0.0246
	TCN	0.0098	0.0083	0.0029	0.0032	0.0136
	TCN variants	**0.0093**	**0.0081**	**0.0013**	**0.0022**	**0.0127**
	TFT	0.0094	0.0083	0.0013	0.0025	0.0134
HS for TCN						
HV	Grid search	**0.1467**	**0.1912**	**0.0331**	**0.0364**	0.2614
	BOHB	0.1591	0.2010	0.0616	0.0800	**0.2107**
IV	Grid search	**0.0098**	**0.0083**	0.0029	0.0032	**0.0136**
	BOHB	0.0126	0.0093	**0.0017**	**0.0025**	0.0139

3.3   Result Evaluation and Analysis

A comparison of the benchmark models and experimental models gave evidence of a clear trend. Across almost all volatility forecasting tasks and assets investigated, the experimental models outperformed the benchmark models, with statistical significance. Based on 10 repetitions, for almost all assets and tasks, the performance values from the experimental models were superior and found to be statistically significant (Table 2).

Of the benchmark models, the ANN-GARCH model from literature performed best overall in forecasting HV, but for IV forecasting, the naïve model performed the best overall, achieving the smallest errors for all five assets (Table 3). However, a comparison of the traditional grid search hyperparameter optimization method against the more recent *Bayesian Optimisation HyperBand* (BOHB) search [9] indicated no clear trend; BOHB only produced a better forecasting model for the HV of oil, and the IV of gold and silver . Both methods were given roughly the same wall time and were both tested using the un-modified TCN. Though it is difficult to say if one method is superior to the other, the continued use of grid search is justified and was the primary hyperparameter optimization method for the remaining experimental TCN models.

Of the experimental models, the TFT performed best overall for HV forecasting, achieving the smallest errors for S&P500, NDX, and oil (Table 3). An encoder length of 21 days was optimal for all assets, with no set of input variables that were consistently best. S&P500 and NDX performed best with the addition of variables that describe time and the underlying asset, gold and silver performed best with the addition of variables that describe time, and oil performed best with the addition of market and time descriptors. The inclusion of exogenous variables only increased the performance for forecasting gold HV. The smallest error for gold was achieved by a benchmark model, specifically the ANN-GARCH.

The TCN variants were the best performing experimental model for IV forecasting, achieving the smallest errors for all assets (Table 3). The optimal modification was to use a secondary task of predicting the direction, as well as forecasting the residuals, consistent amongst all assets. S&P500, NDX, and gold also benefited from the inclusion of the volatility index value and previous direction of movements. For the TFT model, an encoder length of 10 days was preferred for all assets, except for S&P500 which preferred a length of 126.

4   Discussion

This experimental study exemplified the value of DL in forecasting financial volatility and expedited further progress in such DL applications by releasing open-source software and proposing a shared task. It created a benchmark of experimental evaluation results that consisted of the SOTA in NN-based financial volatility forecasting, several traditional models, and a naïve baseline model. This was then compared to several DL methods, representing the competing experimental models.

These results, however, come with some limitations. The main limitation is that the implementation of several models (GARCH and TFT) was open source and thus not necessarily under the same strict control as the other models used.

This study differs from prior publications by presenting a multidisciplinary approach to DL experimentation in forecasting financial volatility. While some other studies on financial volatility forecasting exist, they tend to be limited to literature reviews [10, 27, 29] or expert systems in economics [5, 14]. Our results imply that DL may offer better volatility forecasting performance than traditional methods, and hence, our code release and proposed shared task should expedite this future work. The most obvious is an investigation into other DL models that have not yet been used for volatility forecasting. Combined with the larger capacity of deeper models, another avenue to enhance the models is to make use of multi-modal data (e.g., extend from numeric data to text [18, 22]).

Moving forward, the most vital work is not further exploration of DL models and methods, but rather, the establishment of the proposed shared task that could include, for example, sharing of relevant resources (e.g., code to train models and/or the resulting trained models) and tracks for studying models on a given data modality or expanding them across modalities. This would allow easy and direct comparisons without the need to implement competing models, enabling the synthesis of publications, and propelling the field of financial volatility forecasting further and faster. This task should help gain a deeper understanding of the factors and mechanisms that may affect the economic feasibility of a statistical result. In conclusion, harvesting the diversity of thought and other community effects is likely to accelerate knowledge discovery and methodological innovations required to proceed from statistical significance to economic impact.

References

1. Andersen, T.G., Bollerslev, T.: Answering the skeptics: yes, standard volatility models do provide accurate forecasts. Int. Econ. Rev. **39**(4), 885 (1998). https://doi.org/10.2307/2527343
2. Bollerslev, T.: Generalized autoregressive conditional heteroskedasticity. J. Econometrics **31**(3), 307–327 (1986). https://doi.org/10.1016/0304-4076(86)90063-1
3. Bucci, A.: Cholesky-ANN models for predicting multivariate realized volatility. J. Forecast. **39**(6), 865–876 (2020). https://doi.org/10.1002/for.2664
4. Cavalcante, R.C., Brasileiro, R.C., Souza, V.L.F., Nobrega, J.P., Oliveira, A.L.I.: Computational intelligence and financial markets: a survey and future directions. Expert Syst. Appl. **55**, 194–211 (2016). https://doi.org/10.1016/j.eswa.2016.02.006
5. Chen, W.J., Yao, J.J., Shao, Y.H.: Volatility forecasting using deep neural network with time-series feature embedding. Econ. Res.-Ekonomska Istraživanja 1–25 (2022). https://doi.org/10.1080/1331677X.2022.2089192
6. Edwards, S., Biscarri, J.G., Pérez de Gracia, F.: Stock market cycles, financial liberalization and volatility. J. Int. Money Finance **22**(7), 925–955 (2003). https://doi.org/10.1016/j.jimonfin.2003.09.011
7. Engle, R.F.: Autoregressive conditional heteroscedasticity with estimates of the variance of United Kingdom inflation. Econometrica **50**(4), 987–1008 (1982). https://doi.org/10.2307/1912773

8. Engle, R.F., Patton, A.J.: 2 - What good is a volatility model?*. In: Knight, J., Satchell, S. (eds.) Forecasting Volatility in the Financial Markets, 3rd edn, pp. 47–63. Quantitative Finance, Butterworth-Heinemann, Oxford (2007). https://doi.org/10.1016/B978-075066942-9.50004-2

9. Falkner, S., Klein, A., Hutter, F.: BOHB: robust and efficient hyperparameter optimization at scale. In: Dy, J., Krause, A. (eds.) Proceedings of the 35th International Conference on Machine Learning. Proceedings of Machine Learning Research, vol. 80, pp. 1437–1446. PMLR (2018). https://proceedings.mlr.press/v80/falkner18a.html

10. Ge, W., Lalbakhsh, P., Isai, L., Lenskiy, A., Suominen, H.: Neural network based financial volatility forecasting: a systematic review. ACM Comput. Surv. **55**(1), 14:1–14:30 (2022). https://doi.org/10.1145/3483596

11. Hansen, P.R., Lunde, A.: A forecast comparison of volatility models: does anything beat a GARCH(1,1)? J. Appl. Economet. **20**(7), 873–889 (2005). https://doi.org/10.1002/jae.800

12. Ismail Fawaz, H., Forestier, G., Weber, J., Idoumghar, L., Muller, P.A.: Deep learning for time series classification: a review. Data Min. Knowl. Disc. **33**(4), 917–963 (2019). https://doi.org/10.1007/s10618-019-00619-1

13. Kim, H.Y., Won, C.H.: Forecasting the volatility of stock price index: a hybrid model integrating LSTM with multiple GARCH-type models. Expert Syst. Appl. **103**, 25–37 (2018). https://doi.org/10.1016/j.eswa.2018.03.002

14. Kristjanpoller, W., Hernández, E.: Volatility of main metals forecasted by a hybrid ANN-GARCH model with regressors. Expert Syst. Appl. **84**, 290–300 (2017). https://doi.org/10.1016/j.eswa.2017.05.024

15. Kumar, P.H., Patil, S.B.: Estimation forecasting of volatility using ARIMA, ARFIMA and neural network based techniques. In: 2015 IEEE International Advance Computing Conference (IACC), pp. 992–997. IEEE (2015). https://doi.org/10.1109/IADCC.2015.7154853

16. Lea, C., Flynn, M.D., Vidal, R., Reiter, A., Hager, G.D.: Temporal convolutional networks for action segmentation and detection. In: 2017 IEEE Conference on Computer Vision and Pattern Recognition (CVPR), Honolulu, HI, pp. 1003–1012. IEEE (2017). https://doi.org/10.1109/CVPR.2017.113

17. Lim, B., Arik, S.O., Loeff, N., Pfister, T.: Temporal fusion transformers for interpretable multi-horizon time series forecasting. arXiv:1912.09363 [cs, stat] (2020)

18. Malik, F.: Estimating the impact of good news on stock market volatility. Appl. Financ. Econ. **21**(8), 545–554 (2011). https://doi.org/10.1080/09603107.2010.534063

19. Masset, P.: Volatility Stylized Facts. SSRN Scholarly Paper ID 1804070, Social Science Research Network, Rochester (2011). https://doi.org/10.2139/ssrn.1804070

20. Mayhew, S.: Implied volatility. Financ. Anal. J. **51**(4), 8–20 (1995). https://doi.org/10.2469/faj.v51.n4.1916

21. Neupane, B., Thapa, C., Marshall, A., Neupane, S.: Mimicking insider trades. J. Corp. Finan. **68**, 101940 (2021). https://doi.org/10.1016/j.jcorpfin.2021.101940

22. Oliveira, N., Cortez, P., Areal, N.: The impact of microblogging data for stock market prediction: using Twitter to predict returns, volatility, trading volume and survey sentiment indices. Expert Syst. Appl. **73**, 125–144 (2017). https://doi.org/10.1016/j.eswa.2016.12.036

23. van den Oord, A., et al.: WaveNet: a generative model for raw audio. arXiv:1609.03499 [cs] (2016). https://arxiv.org/abs/1609.03499

24. Orhan, M., Köksal, B.: A comparison of GARCH models for VaR estimation. Expert Syst. Appl. **39**(3), 3582–3592 (2012). https://doi.org/10.1016/j.eswa.2011. 09.048
25. Pandey, A., Wang, D.: TCNN: temporal convolutional neural network for real-time speech enhancement in the time domain. In: 2019 IEEE International Conference on Acoustics, Speech and Signal Processing (ICASSP), ICASSP 2019, pp. 6875–6879 (2019). https://doi.org/10.1109/ICASSP.2019.8683634
26. Patton, A.J.: Volatility forecast comparison using imperfect volatility proxies. J. Econometrics **160**(1), 246–256 (2011). https://doi.org/10.1016/j.jeconom.2010.03. 034
27. Poon, S.H., Granger, C.W.J.: Forecasting volatility in financial markets: a review. J. Econ. Lit. **41**(2), 478–539 (2003). https://doi.org/10.1257/002205103765762743
28. Ruder, S.: An overview of gradient descent optimization algorithms. arXiv:1609.04747 [cs] (2016). https://arxiv.org/abs/1609.04747
29. Sezer, O.B., Gudelek, M.U., Ozbayoglu, A.M.: Financial time series forecasting with deep learning?: a systematic literature review: 2005–2019. Appl. Soft Comput. **90**, 106181 (2020). https://doi.org/10.1016/j.asoc.2020.106181
30. Timmermann, A., Granger, C.W.J.: Efficient market hypothesis and forecasting. Int. J. Forecast. **20**(1), 15–27 (2004). https://doi.org/10.1016/S0169-2070(03)00012-8
31. Vaswani, A., et al.: Attention is all you need. In: Proceedings of the 31st International Conference on Neural Information Processing Systems, NIPS 2017, pp. 6000–6010. Curran Associates Inc., Red Hook (2017). https://papers.nips.cc/paper_files/paper/2017/file/3f5ee243547dee91fbd053c1c4a845aa-Paper.pdf
32. Vidal, A., Kristjanpoller, W.: Gold volatility prediction using a CNN-LSTM approach. Expert Syst. Appl. **157**, 113481 (2020). https://doi.org/10.1016/j.eswa. 2020.113481
33. Wan, R., Mei, S., Wang, J., Liu, M., Yang, F.: Multivariate temporal convolutional network: a deep neural networks approach for multivariate time series forecasting. Electronics **8**(8), 876 (2019). https://doi.org/10.3390/electronics8080876
34. Zhang, J., He, T., Sra, S., Jadbabaie, A.: Why gradient clipping accelerates training: a theoretical justification for adaptivity. In: International Conference on Learning Representations, p. 21 (2020). https://openreview.net/forum? id=BJgnXpVYwS
35. Zhang, X., Shi, J., Wang, D., Fang, B.: Exploiting investors social network for stock prediction in China's market. J. Comput. Sci. **28**, 294–303 (2018). https:// doi.org/10.1016/j.jocs.2017.10.013

CLIP-Based Composed Image Retrieval with Comprehensive Fusion and Data Augmentation

Haoqiang Lin[1] , Haokun Wen[2] , Xiaolin Chen[1] , and Xuemeng Song[1(✉)]

[1] Shandong University, Shandong, China
zichaohq@gmail.com, cxlicd@gmail.com, sxmustc@gmail.com
[2] Harbin Institute of Technology (Shenzhen), Guangdong, China
whenhaokun@gmail.com

Abstract. Composed image retrieval (CIR) is a challenging task where the input query consists of a reference image and its corresponding modification text. Recent methodologies harness the prowess of visual-language pre-training models, *i.e.*, CLIP, yielding commendable performance in CIR. Despite their promise, several shortcomings linger. First, a salient domain discrepancy between the CLIP's pre-training data and the CIR's training data leads to suboptimal feature representation. Second, the existing multimodal fusion mechanisms solely rely on weighted summing and feature concatenation, neglecting the intricate higher-order interactions inherent in the multimodal query. This oversight poses challenges in modeling complex modification intents. Additionally, the paucity of data impedes model generalization. To address these issues, we propose a CLIP-based composed image retrieval model with comprehensive fusion and data augmentation (CLIP-CD), consisting of two training stages. In the first stage, we fine-tune both the image and text encoders of CLIP to alleviate the aforementioned domain discrepancy. In the second stage, we propose a comprehensive multimodal fusion module that enables the model to discern complex modification intentions. Furthermore, we propose a similarity-based data augmentation method for CIR, ameliorating data scarcity and enhancing the model's generalization ability. Experimental results on the Fashion-IQ dataset demonstrate the effectiveness of our method.

Keywords: Image retrieval · Vision-Language pre-training model · Multimodal fusion

1 Introduction

Image retrieval [7] stands as a cornerstone within the computer vision field, playing pivotal roles in diverse domains ranging from face recognition [19] to fashion retrieval [27]. Traditional image retrieval has predominantly centered on single-modal queries, including text-based image retrieval [22] and content-based image retrieval [17]. However, in many cases, expressing precise search intent via a single-modal query often poses formidable challenges for users.

ⓒ The Author(s), under exclusive license to Springer Nature Singapore Pte Ltd. 2024
T. Liu et al. (Eds.): AI 2023, LNAI 14471, pp. 190–202, 2024.
https://doi.org/10.1007/978-981-99-8388-9_16

To address the limitations of conventional image retrieval, composed image retrieval (CIR) [21] has been proposed and gained increasing research attention. In this task, the input is a multimodal query, i.e., a reference image plus a modification text. The reference image reflects the user's overarching retrieval demands, while the modification text delineates the user's specific unsatisfactory features in the reference image and his/her desired modifications. This multimodal query enables users to express their retrieval intents more flexibly and accurately.

Recent efforts have been increasingly devoted toward CIR. Predominantly, these works apply conventional frameworks like CNNs [12] and LSTM [8] to cultivate representations for the multimodal query. Yet, with the burgeoning prowess of vision-language pre-training models in feature extraction, the innovative method Clip4Cir [2] has integrated CLIP [4] and achieved impressive results. Nevertheless, there are still some limitations that need to be addressed. 1) Clip4Cir overlooks the substantial domain discrepancy between the CLIP's pre-training image data and the CIR's image data, which results in suboptimal feature extraction. 2) Clip4Cir employs a simple combiner network reliant on mere weighted summing and feature concatenation. It neglects the higher-order interaction between the multimodal query and potentially fails to model complex modification intents. And 3) the laborious data annotation limits the scale of most CIR's datasets. Like other existing works, Clip4Cir overlooks this issue, resulting in insufficient model generalization.

To address the above limitations, we present a CLIP-based composed image retrieval model with comprehensive fusion and data augmentation (CLIP-CD), as illustrated in Fig. 1, which comprises two stages. In the first stage, we fine-tune the CLIP's image and text encoders to alleviate the domain discrepancy problem. In the second stage, we design a multimodal fusion module, which incorporates weighted summing, feature concatenation, and bilinear pooling [20], to enhance the model's multimodal fusion capabilities. Moreover, we propose a similarity-based data augmentation method to expand the dataset size and enhance the model's generalization capabilities. Extensive experiments on the Fashion-IQ dataset corroborate the superiority of our method.

Our main contributions can be summarized as follows:

- We present a novel CLIP-based method for CIR. Our approach not only incorporates a fine-tuning strategy specifically designed to address the domain discrepancy problem but also integrates a comprehensive multimodal fusion module to enhance the effectiveness of multimodal fusion.
- To the best of our knowledge, we are the first to introduce a similarity-based data augmentation mechanism in CIR, which alleviates the insufficient training data problem.
- Extensive experiments conducted on the real-world Fashion-IQ dataset validate the superiority of our model.

2   Related Work

Our work is closely related to composed image retrieval (CIR) and vision-language pre-training (VLP).

2.1 Composed Image Retrieval

Recently, there have been numerous works aiming to solve this problem. For example, Vo *et al.* [21] employed gate mechanisms coupled with residual modules to fuse the multimodal query features. Later, Lee *et al.* [13] handled changes in both content and style conveyed by modification text through the designed content modulator and the style modulator. Meanwhile, Wen *et al.* [23] harnessed the mutual learning strategy to unify both local-wise multimodal fusion and global-wise multimodal fusion. Baldrati *et al.* [2] pioneered the integration of CLIP into this task and achieved remarkable performance. However, they overlooked the domain discrepancy between the CLIP's pre-training image data and the CIR's image data. The combiner network employed in their approach also exhibits limitations in effectively modeling complex modification intents within multimodal queries. Furthermore, like other works, they fail to address the issue of the limited scale of most CIR's datasets, resulting in constrained model generalization. In light of this, we fine-tuned both the image and text encoders to alleviate the domain discrepancy, and also introduced a comprehensive multimodal fusion module to effectively capture complex modification intents. Besides, we proposed a similarity-based data augmentation method to expand the dataset size and enhance the model's generalization capabilities.

2.2 Vision-Language Pre-training

Vision-language pre-training models leverage vast data for pre-training and generalize well on numerous downstream tasks through fine-tuning. Examples of vision-language pre-training models include ViLBERT [16], Oscar [14], and CLIP. Among them, CLIP stands out due to its contrastive learning based on 400 million image-text pairs. It can handle both text and visual inputs and model the relationship between them. This capability has led to advancements in multimodal areas like fine-grained classification [4], zero-shot retrieval [5], and visual commonsense reasoning [22]. Drawing from these insights, we proposed an effective fine-tuning strategy to bridge the gap between the CLIP's pre-training data and the CIR's data.

3 Methodology

In this section, we first formulate the problem, and then detail the proposed CLIP-CD.

3.1 Problem Formulation

In this work, we aim to tackle the CIR task, which can be formally defined as that given a multimodal query comprising a reference image and its modification text, the goal is to retrieve the optimal target image from a set of gallery images. Suppose we have a set of triples denoted as $\mathcal{D} = \{(I_r, T_m, I_t)_i\}_{i=1}^{N}$, where I_r is the reference image, T_m is the modification text, I_t signifies the target image, and N

is the total number of triplets. Based on $\mathcal{D}$, our goal is to train a model that can effectively fuse the multimodal query (I_r, T_m) to be close to the representation of the target image I_t. This can be formalized as follows,

$$f(I_r, T_m) \rightarrow h(I_t), \tag{1}$$

where $f(\cdot)$ represents the multimodal fusion function mapping the multimodal query to the latent space, while $h(\cdot)$ denotes the feature embedding function for the target image.

3.2   CLIP-CD

As illustrated in Fig. 1, we propose a CLIP-based composed image retrieval model with comprehensive fusion and data augmentation, which consists of two training stages. In the first stage (encoder fine-tuning stage), we simultaneously fine-tune the text and image encoders of CLIP, which helps alleviate the problem of domain discrepancy. In the second stage (multimodal fusion stage), we freeze the parameters of the CLIP's encoder fine-tuned in the first stage and focus on learning a multimodal fusion module. Additionally, to further expand the dataset size and improve the model's generalization, we generate pseudo triplets by replacing the reference/target image with another similar one.

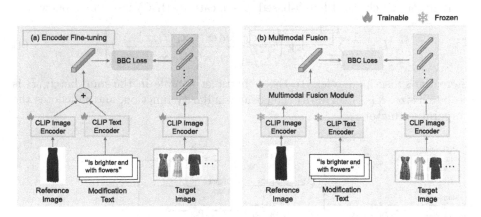

Fig. 1. The overall architecture of the proposed framework consists of two training stages: (a) Encoder Fine-tuning and (b) Multimodal Fusion. The parameters of the image encoder are shared by the reference image and the target image.

3.2.1   Encoder Fine-Tuning

The first step is to address the domain discrepancy between the data of the CLIP's pre-training and that of the CIR's training.

Regarding text data, CLIP is trained on descriptive texts directly related to images, such as "A photo of a dog". However, CIR's modification texts highlight differences between the reference and target images, such as "has more colors and is purple". In addition, as for image data, CLIP's pre-training data contain images from diverse domains in the open domain, like objects, landscapes, and humans. Conversely, CIR's training image data are usually domain-specific, such as fashion-related items. Hence, there are significant domain discrepancies between these two tasks in both image and text training data.

Based on the above analysis, in this stage, we focus on fine-tuning both the image and text encoders of CLIP to address the domain discrepancy problem. Figure 1(a) illustrates the overview of the first stage. Specifically, we first utilize CLIP to extract image and text features from the training triplet (I_r, T_m, I_t), which can be formulated as follows,

$$\begin{cases} \mathbf{x_r} = \text{IE}\,(I_r)\,, \\ \mathbf{t_m} = \text{TE}\,(T_m)\,, \\ \mathbf{x_t} = \text{IE}\,(I_t)\,, \end{cases} \tag{2}$$

where $\text{IE}(\cdot)$ and $\text{TE}(\cdot)$ represent the image and text encoders of CLIP, respectively. $\mathbf{x_r}, \mathbf{t_m}, \mathbf{x_t} \in \mathbb{R}^D$ represent the encoded reference image feature, modification text feature, and target image feature, respectively. Then we fuse $\mathbf{x_r}$ and $\mathbf{t_m}$ with an element-wise summation followed by $L2$-normalization as follows,

$$\phi = L2\,(\mathbf{x_r} \oplus \mathbf{t_m})\,, \tag{3}$$

where ϕ represents the combined features of $\mathbf{x_r}$ and $\mathbf{t_m}$. $\oplus$ serves as element-wise summation. Finally, to fine-tune the CLIP's image and text encoders, we leverage the widely-used batch-based classification (BBC) loss [2] as follows,

$$L = \frac{1}{B} \sum_{i=1}^{B} \left[-\log \left(\frac{exp\{\kappa(\phi^{(i)}, \mathbf{x_t}^{(i)})/\tau\}}{\sum_{j=1}^{B} exp\{\kappa(\phi^{(j)}, \mathbf{x_t}^{(j)})/\tau\}} \right) \right], \tag{4}$$

where the subscript i refers to the i-th triplet sample in the mini-batch, B is the batch size, $\kappa\,(\cdot, \cdot)$ serves as the cosine similarity function, and τ denotes the temperature factor.

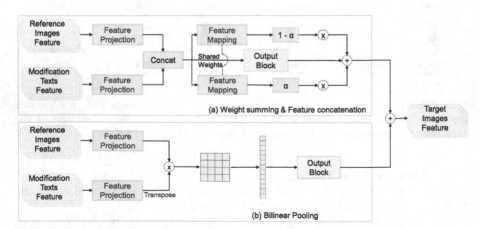

Fig. 2. The architecture of the multimodal fusion module. It takes the reference image feature and the modification texts feature as inputs and outputs a fused representation.

3.2.2 Multimodal Fusion

The overview of the second training stage is depicted in Fig. 1(b). In this stage, we freeze the parameters of CLIP and focus on training a comprehensive multi-modal fusion module.

The details of the proposed multimodal fusion module are illustrated in Fig. 2. It employs three strategies for multimodal feature fusion: weighted summing, feature concatenation, and bilinear pooling. Notably, the former two strategies have been explored in Clip4Cir. However, we argue that merely relying on weighted summing and feature concatenation neglects the intricate higher-order interactions inherent in the multimodal query, posing challenges in modeling complex modification intents. To address this limitation, we additionally introduce bilinear pooling as depicted in Fig. 2(b). By leveraging bilinear pooling, the model can comprehensively capture the multimodal query features. This enhancement allows for capturing higher-order interactions between the reference image and modification text features and hence boosts the understanding of the multimodal query.

The bilinear pooling consists of three essential components: the feature projection layer, the bilinear pooling layer, and the output block. First, the features of the reference image and the modification text are processed through the feature projection layers, respectively. Formally, we have,

$$\begin{cases} \mathbf{f}_I = \xi \left(\mathrm{FC} \left(\mathbf{x_r} \right) \right), \\ \mathbf{f}_T = \xi \left(\mathrm{FC} \left(\mathbf{x_m} \right) \right), \end{cases} \tag{5}$$

where $\mathbf{f}_I \in \mathbb{R}^K$ and $\mathbf{f}_T \in \mathbb{R}^K$ refer to the output vectors through the feature projection layer of the reference image and the modification text, respectively. ξ is the RELU activation function and $\mathrm{FC}(\cdot)$ denotes the fully-connected layer. Note that considering the memory cost of storing high dimensional features, we set $K < D$.

In the subsequent bilinear pooling layer, an outer product computation is performed between $\mathbf{f}_I$ and $\mathbf{f}_T$ to obtain a feature map matrix, and this matrix is then flattened into a vector $\mathbf{f}_{bil} \in \mathbb{R}^{K^2}$. It can be formulated as follows,

$$\mathbf{f}_{bil} = \mathrm{Flatten} \left(\mathbf{f}_I \otimes \mathbf{f}_T \right), \tag{6}$$

where $\otimes$ represents the outer product operation. Next, we feed $\mathbf{f}_{bil}$ to the output block, which is as follows,

$$\mathbf{f}_{out} = \mathrm{FC} \left(\xi \left[\mathrm{FC} \left(\mathbf{f}_{bil} \right) \right] \right). \tag{7}$$

Similar to the first stage, we add $\mathbf{f}_{out} \in \mathbb{R}^D$ to the features obtained from the weighted summing and the feature concatenation as the final output of the multimodal fusion module. And the parameters of the multimodal fusion module are optimized by the BBC loss the same as Eq. 4.

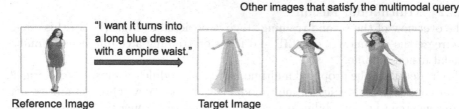

Fig. 3. Examples that multiple images satisfy the same multimodal query.

3.2.3   Similarity-Based Data Augmentation

Another major novelty of our work is that we propose a similarity-based data augmentation method, to alleviate the overfitting phenomenon caused by the limited size of training data.

We observe that a single multimodal query could correspond to multiple images that meet the retrieval requirements. Intuitively, these images that align with the same multimodal query tend to exhibit a high similarity degree to the desired target image. As illustrated in Fig. 3, the multimodal query specifies a red dress and intends to modify it into a long blue one with an empire waist. However, apart from the target image, two additional blue dresses meet the multimodal query.

This observation suggests the potential of creating a pseudo triplet by substituting the target image with another similar one. Likewise, replacing the reference image with a similar one can also yield a new pseudo triplet. Specifically, we first utilize the CLIP's image encoder trained from the first stage to extract features of all images in the training dataset, which can be represented as follows,

$$\mathbf{x}_k = \mathrm{IE}\left(I_k\right), k = 1, \cdots, n, \tag{8}$$

where I_k is the image in the training dataset, n denotes the total number of training images, and $\mathbf{x}_k \in \mathbb{R}^D$ is the feature vector corresponding to I_k. Subsequently, we employ the cosine similarity function to calculate the similarity score between the reference/target image and the training images based on the extracted features. To ensure the validity of the pseudo triplets upon image replacement, we design specific constraints from two perspectives: 1) image similarity constraint and 2) triplet matching constraint.

In the image similarity constraint, to ensure the quality of pseudo triplets and distinguish them from the original triplets, we establish the lower similarity threshold ε_l and upper similarity threshold ε_u. Besides, to control the number of pseudo triplets, we follow a method where for each image, only the first s_{max} images with the highest similarity that meet the similarity threshold are selected as its substitute set. The image similarity constraint is as follows,

$$\mathcal{M}_k = \{I_1^{'}, I_2^{'}, ..., I_{s_{max}}^{'}\}, \forall I_i^{'} \in \mathcal{M}_k, \varepsilon_l \leq \kappa(\mathbf{x}_k, \mathbf{x}_i^{'}) \leq \varepsilon_u, \tag{9}$$

where $I_i^{'}$ is a suitable substitute image for I_k, $\mathcal{M}_k$ is the substitute image set for I_k, $\mathbf{x}_i^{'}$ is the feature vector corresponding to $I_i^{'}$. Following this constraint, we can construct the pseudo triplet by replacing the reference image or the target image in the original triplet from their substitute image set. Specifically, the pseudo triplets are denoted as $\tilde{\mathcal{D}}$.

Considering that obtaining similar images directly based on the visual feature outputted by the image encoder could introduce certain noise similar images, we additionally incorporate the triplet matching constraint where we further evaluate the matching degree between the multimodal query and the target image within the pseudo triplets. Specifically, we employ the model derived from the first stage to compute the matching score. If the score surpasses the pre-defined threshold α, the pseudo triplet is retained; otherwise, it is discarded. Finally, the pseudo triplets are a subset of $\tilde{\mathcal{D}}$, denoted as $\tilde{\mathcal{D}}_{sub}$.

Although we have introduced two constraints to ensure the quality of the pseudo triplets, their quality may be still inferior compared to the original dataset. Therefore, during the second training stage, we adopt an iterative approach utilizing the two types of data. Specifically, we train the model on the original training data $\mathcal{D}$ for k epochs, followed by one training epoch using pseudo data $\tilde{\mathcal{D}}_{sub}$. This training strategy can not only augment the training data to improve the model's generalization ability but also mitigate the adverse effects posed by the lower quality of the pseudo triplets.

4 Experiments

In this section, we first present the experimental settings and then detail the experiments conducted on the Fashion-IQ dataset.

4.1 Datasets and Metrics

Fashion-IQ [24] is a fashion image retrieval dataset based on natural language descriptions. It comprises $77,648$ clothing images and is divided into three sub-training sets: dress, shirt, and top&tee. The training data includes over $18,000$ triplets, where each triplet includes a reference image, a target image, and a modification text. As the test set of Fashion-IQ is not publicly available, we followed the experimental setup of other related works [2,26] in terms of dividing the dataset into training and testing sets.

Following previous efforts [2,23], we adopted the Recall at rank k ($R@k$) as the evaluation metric. It measures the fraction of queries for which the ground truth target is retrieved among the top k results. For all three subsets of the Fashion-IQ dataset, k is set to 10 and 50.

4.2 Implementation Details

Similar to Clip4Cir, we used RN50×4 CLIP as the feature encoder for our model. In the similarity-based data augmentation section, to ensure the validity of the obtained pseudo triplets, we set the lower similarity threshold ε_l and the upper similarity threshold ε_u in Eq. (9) to 90% and 99.5%, respectively. The max number of substitute images s_{max} in Eqn. (9) is set to 4 and the threshold α for the triplet matching constraint is set to 0.6. In the first stage, we employed AdamW optimizer [15] with a learning rate of $2e - 6$ and a weight decay coefficient of $1e - 2$ to optimize the model. The batch size is set to 64. Additionally, following [18], the temperature factor τ in Eq. (4) is set to 100 to ensure an adequately wide dynamic range of similarity probabilities without interfering with the normal training process. In the second stage, we frozen the CLIP's image and text encoders from the first stage and focused on training the multimodal fusion module. We adopted Adam [11] optimizer with a learning rate of $2e - 5$. The batch size is set to 1,024. All the experiments are implemented by PyTorch, and we fixed the random seeds to ensure reproducibility.

4.3 Performance Comparison

To validate the effectiveness of our method in CIR, we chose the following baselines: TIRG [21], ComAE [1], VAL [3], DATIR [6], CosMo [13], Heteroge [25], SAC [9], DCNet [10], CLVC-Net [23], Clip4Cir [2].

Table 1 summarizes the performance comparison on the Fashion-IQ dataset. From this table, we obtained the following observations. 1) Our proposed method outperforms all baselines over the Fashion-IQ dataset. This confirms the advantages of leveraging a two-stage training approach with a novel fine-tuning strategy and incorporating pseudo triplets. And 2) both our proposed method and Clip4Cir perform better than others without employing visual-language pre-training models, i.e., CLIP. This highlights the crucial role of visual-language pre-training models in this task.

Table 1. Performance comparison on Fashion-IQ. The best results are in boldface, while the second best are underlined.

Method	Dress		Shirt		Top&Tee		Avg	
	R@10	R@50	R@10	R@50	R@10	R@50	R@10	R@50
TIRG [21]	14.87	34.66	18.26	37.79	19.08	39.62	17.40	37.39
ComAE [1]	14.03	35.10	13.88	34.59	15.80	39.26	19.89	36.31
VAL [3]	21.12	42.19	21.03	43.44	25.64	49.49	22.60	45.04
DATIR [6]	21.90	43.80	21.90	43.70	27.20	51.60	23.70	46.40
CosMo [13]	25.64	50.30	24.90	49.18	29.21	57.46	26.58	52.31
Heteroge [25]	26.20	51.20	22.40	46.00	29.30	56.40	26.10	51.20
SAC [9]	26.52	51.01	28.02	51.86	32.70	61.23	29.08	54.70
DCNet [10]	28.95	56.07	23.95	47.30	30.44	58.29	27.78	53.89
CLVC-Net [23]	29.85	56.47	28.75	54.76	33.50	64.00	30.70	58.41
Clip4Cir [2]	<u>33.81</u>	<u>59.40</u>	<u>39.99</u>	<u>60.45</u>	<u>41.41</u>	<u>65.37</u>	<u>38.52</u>	<u>61.74</u>
CLIP-CD	**37.68**	**62.62**	**42.44**	**63.74**	**45.33**	**67.72**	**41.82**	**64.79**

Table 2. Ablation study on Fashion-IQ. The results are the average result of the three categories on Fashion-IQ's three sub-training sets.

	Method	R@10	R@50
Stage1	w/o FT	23.44	43.15
	w/ FT-Text	32.71	54.59
	w/ FT-Image	34.02	55.74
	w/ FT-Both	**39.45**	**62.88**
Stage2	w/o Pseudo Data	41.55	64.49
	w/o Bilinear	41.51	64.40
	CLIP-CD	**41.82**	**64.79**

4.4 Ablation Study

To verify the importance of each part of our model, we conducted ablation experiments in two parts: 1) Fine-tune strategy (Stage1) and 2) Component ablation (Stage2).

In the first part, to demonstrate the importance of our fine-tuning strategy, we devised four different fine-tuning experiments which are conducted during the first training stage. In the second part, we compared our proposed method with two other variants of our model to investigate the effectiveness of our key components. Both of them are conducted in the second training stage.

- w/o FT, w/ FT-Text, w/ FT-Image, and w/ FT-Both: To investigate the effectiveness of our fine-tune strategy, we conducted four fine-tuning related variants, including without any CLIP's encoder fine-tuning and with fine-tuning text/ image/ both encoders of CLIP, respectively.
- w/o Pseudo Data: To verify the effectiveness of the proposed similarity-based data augmentation method in the retrieval process, we removed the generated pseudo data and only used the original dataset to train.
- w/o Bilinear: To check the importance of bilinear pooling, we removed bilinear pooling from the multimodal fusion module.

Table 2 shows the ablation results of our proposed method. As can be seen from this table, we gained the following observations. 1) w/ FT-Both achieves better performance than other fine-tune related variants, which confirms the importance of alleviating the domain discrepancy of both text and image simultaneously. 2) Our method surpasses w/o Pseudo Data, indicating that the proposed similarity-based data augmentation method helps improve the model's generalization performance and alleviates the overfitting phenomenon. And 3) w/o Bilinear performs worse compared to our method, which demonstrates that bilinear pooling is useful for boosting the model's multimodal fusion capability.

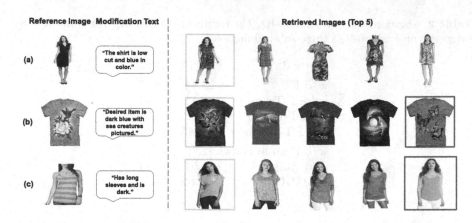

Fig. 4. Retrieval examples obtained by our CLIP-CD on Fashion-IQ.

4.5 Case Study

Figure 4 illustrates CIR examples from the three sub-datasets of Fashion-IQ. The top 5 retrieved images are listed, where the green boxes indicate the ground-truth target images labeled in the dataset, whereas the red boxes signify images that fail to meet the retrieval requirements based on our evaluation. It can be observed that the proposed method ranks the ground-truth target image in the first place for all three examples. Specifically, in Fig. 4(a), all the retrieved images can align with the multimodal query. This confirms the effectiveness of our method. Meanwhile, in Fig. 4(b), the 5-th image doesn't exhibit the "sea creatures" trait, even though it shows high similarity to the target image. A similar case can be observed in Fig. 4(c). This may be due to that although we designed two constraints for filtering pseudo samples, a few low-quality triplets might still mislead the optimization of the method. Nevertheless, these observations show that our method succeeds in retrieving the desired target image and most of the retrieved images can meet the multimodal query requirements. This confirms the effectiveness and robustness of our method.

5 Conclusions

In this work, we present a CLIP-based composed image retrieval model with comprehensive fusion and data augmentation, which consists of two training stages. In the first stage, the focus is fine-tuning the image and text encoders of CLIP to alleviate the issue related to domain discrepancy. In the second stage, the emphasis is placed on training a multimodal fusion module to fully integrate the features extracted from the reference image and the modification text. Furthermore, we propose a similarity-based data augmentation method to overcome the problem of insufficient training triplets in the dataset used for this task. Extensive experiments have been conducted on the public Fashion-IQ dataset, and the results demonstrate the effectiveness of our method.

Acknowledgment. The work is supported by the Shandong Provincial Natural Science Foundation (No.:ZR2022YQ59).

References

1. Anwaar, M.U., Labintcev, E., Kleinsteuber, M.: Compositional learning of image-text query for image retrieval. In: Proceedings of the IEEE/CVF Winter conference on Applications of Computer Vision, pp. 1140–1149 (2021)
2. Baldrati, A., Bertini, M., Uricchio, T., Del Bimbo, A.: Conditioned and composed image retrieval combining and partially fine-tuning clip-based features. In: Proceedings of the IEEE/CVF Conference on Computer Vision and Pattern Recognition, pp. 4959–4968 (2022)
3. Chen, Y., Gong, S., Bazzani, L.: Image search with text feedback by visiolinguistic attention learning. In: Proceedings of the IEEE/CVF Conference on Computer Vision and Pattern Recognition, pp. 3001–3011 (2020)
4. Conde, M.V., Turgutlu, K.: Clip-art: contrastive pre-training for fine-grained art classification. In: Proceedings of the IEEE/CVF Conference on Computer Vision and Pattern Recognition, pp. 3956–3960 (2021)
5. Fang, H., Xiong, P., Xu, L., Chen, Y.: Clip2video: mastering video-text retrieval via image clip. arXiv preprint arXiv:2106.11097 (2021)
6. Gu, C., Bu, J., Zhang, Z., Yu, Z., Ma, D., Wang, W.: Image search with text feedback by deep hierarchical attention mutual information maximization. In: Proceedings of the ACM International Conference on Multimedia, pp. 4600–4609 (2021)
7. Guo, J., et al.: HGAN: hierarchical graph alignment network for image-text retrieval. IEEE Trans. Multimedia (2023)
8. Hochreiter, S., Schmidhuber, J.: Long short-term memory. Neural Comput. $9(8)$, 1735–1780 (1997)
9. Jandial, S., Badjatiya, P., Chawla, P., Chopra, A., Sarkar, M., Krishnamurthy, B.: SAC: semantic attention composition for text-conditioned image retrieval. In: Proceedings of the IEEE/CVF Winter Conference on Applications of Computer Vision, pp. 4021–4030 (2022)
10. Kim, J., Yu, Y., Kim, H., Kim, G.: Dual compositional learning in interactive image retrieval. In: Proceedings of the AAAI Conference on Artificial Intelligence, vol. 35, pp. 1771–1779 (2021)
11. Kingma, D.P., Ba, J.: Adam: a method for stochastic optimization. arXiv preprint arXiv:1412.6980 (2014)
12. Krizhevsky, A., Sutskever, I., Hinton, G.E.: ImageNet classification with deep convolutional neural networks. In: Advances in Neural Information Processing Systems, vol. 25 (2012)
13. Lee, S., Kim, D., Han, B.: Cosmo: Content-style modulation for image retrieval with text feedback. In: Proceedings of the IEEE/CVF Conference on Computer Vision and Pattern Recognition, pp. 802–812 (2021)
14. Li, X., et al.: OSCAR: object-semantics aligned pre-training for vision-language tasks. In: Vedaldi, A., Bischof, H., Brox, T., Frahm, J.-M. (eds.) ECCV 2020. LNCS, vol. 12375, pp. 121–137. Springer, Cham (2020). https://doi.org/10.1007/978-3-030-58577-8_8
15. Loshchilov, I., Hutter, F.: Decoupled weight decay regularization. arXiv preprint arXiv:1711.05101 (2017)

16. Lu, J., Batra, D., Parikh, D., Lee, S.: ViLBERT: pretraining task-agnostic visiolinguistic representations for vision-and-language tasks. In: Advances in Neural Information Processing Systems, vol. 32 (2019)

17. Philbin, J., Chum, O., Isard, M., Sivic, J., Zisserman, A.: Object retrieval with large vocabularies and fast spatial matching. In: IEEE Conference on Computer Vision and Pattern Recognition, pp. 1–8. IEEE (2007)

18. Radford, A., et al.: Learning transferable visual models from natural language supervision. In: International Conference on Machine Learning, pp. 8748–8763 (2021)

19. Sun, Y., Wang, X., Tang, X.: Deep learning face representation from predicting 10,000 classes. In: Proceedings of the IEEE Conference on Computer Vision and Pattern Recognition, pp. 1891–1898 (2014)

20. Tenenbaum, J.B., Freeman, W.T.: Separating style and content with bilinear models. Neural Comput. **12**(6), 1247–1283 (2000)

21. Vo, N., et al.: Composing text and image for image retrieval-an empirical odyssey. In: Proceedings of the IEEE/CVF Conference on Computer Vision and Pattern Recognition, pp. 6439–6448 (2019)

22. Wang, L., Li, Y., Lazebnik, S.: Learning deep structure-preserving image-text embeddings. In: Proceedings of the IEEE Conference on Computer Vision and Pattern Recognition, pp. 5005–5013 (2016)

23. Wen, H., Song, X., Yang, X., Zhan, Y., Nie, L.: Comprehensive linguistic-visual composition network for image retrieval. In: Proceedings of the International ACM SIGIR Conference on Research and Development in Information Retrieval, pp. 1369–1378 (2021)

24. Wu, H., et al.: Fashion IQ: a new dataset towards retrieving images by natural language feedback. In: Proceedings of the IEEE/CVF Conference on Computer Vision and Pattern Recognition, pp. 11307–11317 (2021)

25. Zhang, G., Wei, S., Pang, H., Zhao, Y.: Heterogeneous feature fusion and cross-modal alignment for composed image retrieval. In: Proceedings of the ACM International Conference on Multimedia, pp. 5353–5362 (2021)

26. Zhao, Y., Song, Y., Jin, Q.: Progressive learning for image retrieval with hybrid-modality queries. In: Proceedings of the International ACM SIGIR Conference on Research and Development in Information Retrieval, pp. 1012–1021 (2022)

27. Zhou, Y., Guo, J., Sun, H., Song, B., Yu, F.R.: Attention-guided multi-step fusion: a hierarchical fusion network for multimodal recommendation. arXiv preprint arXiv:2304.11979 (2023)

LiDAR Inpainting of UAV Based 3D Point Cloud Using Supervised Learning

Muhammad Talha$^{(\boxtimes)}$ (iD), Aya Hussein (iD), and Mohammed Hossny (iD)

School of Engineering and Information Technology, UNSW, Canberra, Australia
m.talha@student.unsw.edu.au

Abstract. Unmanned Aerial Vehicles (UAV) can quickly scan unknown environments to support a wide range of operations from intelligence gathering to search and rescue. LiDAR point clouds can give a detailed and accurate 3D representation of such unknown environments. However, LiDAR point clouds are often sparse and miss important information due to occlusions and limited sensor resolution. Several studies used inpainting techniques on LiDAR point clouds to complete the missing regions. However, these studies have three main limitations that hinder their use in UAV-based environment 3D reconstruction. First, existing studies focused only on synthetic data. Second, while the point clouds obtained from a UAV flying at moderate to high speeds can be severely distorted, none of the existing studies applied inpainting to UAV-based LiDAR point clouds. Third, all existing techniques considered inpainting isolated objects and did not generalise to inpainting complete environments. This paper aims to address these gaps by proposing an algorithm for inpainting point clouds of complete 3D environments obtained from a UAV. We use a supervised learning encoder-decoder model for point cloud inpainting and environment reconstruction. We tested the proposed approach for different LiDAR parameters and different environmental settings. The results demonstrate the ability of the system to inpaint the objects with a minimum average Chamfer Distance (CD) loss of 0.028 at a UAV speed of 5 ms^{-1}. We present the results of the 3D reconstruction for a few test environments.

Keywords: LiDAR inpainting · 3D Reconstruction · Point clouds

1 Introduction

The highly delicate presenting ability of point clouds led to their increased popularity in many robotics and autonomous vehicle applications [21–23]. Light Detection and Ranging (LiDAR) sensors are used to get point clouds of an environment. The data collected from LiDAR sensors mounted on UAVs can be used to reconstruct a 3D map of the environment. This can be very helpful in applications like aerial mapping, forestry, disaster recovery, safety and survey inspections, search and rescue (SAR), and underground mapping. The main advantage of using UAVs is their capacity to rapidly survey vast, unfamiliar areas, thereby expediting the scanning process.

T. Liu et al. (Eds.): AI 2023, LNAI 14471, pp. 203–214, 2024.
https://doi.org/10.1007/978-981-99-8388-9_17

LiDAR sensors have the ability to form high-definition 3D graphics from real-time sensing of the surrounding environment. Unlike cameras, LiDAR captures extremely accurate 3D depth measurements. Depth cameras are sensitive to light conditions and more suitable for indoor environments. On the other hand, LiDAR has the advantages of high accuracy and long-range capabilities. According to Wang [16], among the known environmental measurement solutions, LiDAR is the one with the highest measurement accuracy. Thus, LiDAR is preferred over cameras in 3D mapping tasks.

When used in real-world environments, both LiDAR and depth cameras often generate incomplete 3D data due to limited sensor resolution and occlusions. This is exacerbated when the 3D data is collected from a UAV flying at a moderate or high speed leading to a further drop in resolution and more missing information. One way to improve the quality is to use a high-quality LiDAR with more channels and high resolution, but it will be expensive. Also, the detailed LiDAR point cloud will contain a large number of data points which will require more processing power, time and storage which imposes heavier setups impractical for deployment on UAVs. This is a limitation in applications like SAR and disaster recovery. Therefore, we aim to use a low-cost LiDAR sensor to get a sparse point cloud and complete it using inpainting.

Inpainting is a technique used in image processing for filling in missing or corrupted parts of an image using the surrounding information. It has also been extended to point clouds to estimate missing information from a given point cloud. The main contribution of this work is the inpainting of sparse UAV-based 3D point clouds for the 3D reconstruction of complete environments. We use low-cost LiDARs mounted on a UAV flying at a high speed to scan an unknown environment.

The rest of this paper is organised as follows: Section 2 provides an overview of the previous work related to our topic. Section 3 provides background information about some important concepts used in this paper. Section 4 provides a concrete description of the problem that we are trying to address. Section 5 talks about the methodology and components of the system. Finally, we provide the experimental results and evaluation in Sect. 6 and the conclusion and future work in Sect. 7.

2   Related Work

Inpainting is a technique that has been used to complete missing or corrupted parts of images [8,12,14,18,19]. Some of the previous works have also investigated inpainting in the context of point clouds. Chen et al. [3] proposed a self-adaptive method in which the input and output sizes of the point clouds were kept flexible. They applied the axis-aligned bounding box (AABB) to project a point cloud into a 2D structured representation and made use of hierarchical depth image painting to complete the task. Another efficient point cloud inpainting method was proposed by Fu et al. [6]. Their method leveraged graph signal processing and was based on the observation of non-local self-similarity in 3D point clouds. They filled holes in a point cloud by using similar geometry. They used fixed-size cubes from the point cloud as the processing unit and globally searched for the cube most similar to the target cube with holes inside. A limitation of their approach is that it can only inpaint small missing regions (holes) in the point clouds.

Most of the existing methods for 3D shape completion make use of supervised learning. These methods use complete ground-truth shape labels for training models. Huang et al. [9] proposed a learning-based approach, the Point Fractal Network, for precise point cloud completion. They used a feature-points-based multi-scale generating network to estimate the missing point cloud hierarchically. They used a combination of multi-stage completion loss and adversarial loss to generate more realistic missing regions. A main feature of their network was that it could preserve the spatial arrangements of the incomplete point cloud in the output. However, a key limitation was that it required the input incomplete point clouds to be of high density. Therefore, their model might fail if a sparse incomplete point cloud is presented.

Yuan et al. [20] presented an encoder-decoder style architecture that operates directly on the raw point clouds without any assumption about the structure of the underlying shape. The architecture generated dense point clouds in a coarse-to-fine fashion with a higher resolution. An advantage of their decoder design was that it was able to produce dense point clouds with much fewer parameters than previous methods. Another method was proposed by Tchapmi et al. [15] in which they presented a novel decoder that was combined with encoders from existing frameworks. Their decoder generated a structured point cloud using a hierarchical rooted tree structure. Each node of the tree represented a subset of the point cloud. The embedded point cloud structure was shown by visualizing nodes in the decoder, as a collection of all its descendant points. A limitation of this work is that the tree needs to be sufficiently large which is intractable. Therefore, the capacity of the decoder can limit the possible topologies learned.

Some previous works have used weakly-supervised methods for point cloud completion. Gu et al. [7] proposed a weakly-supervised approach for 3D shape completion from unaligned partial point clouds without shape and pose supervision. The proposed network jointly optimized the pose of the input partial point cloud and 3D canonical shapes. However, their model was not robust to view alignment errors and generated coarse shapes and missed details due to the noise of pose estimation. Insafutdinov & Dosovitskiy [10] proposed a Differentiable Point Cloud (DPC) method which, given two image views of the same object, jointly predicted camera poses and a 3D shape representation. An end-to-end differentiable point cloud projection was used to enforce the geometric consistency between the estimated 3D shape and the input images.

All of these studies used synthetic data from ShapeNet for training their models. Some of them tested their approach on the real-world LiDAR dataset (Semantic KITTI [2]) but none used LiDAR data for training. A common limitation was that all of these works focused on the inpainting of individual objects, no one did a complete environment inpainting. Also, none of the above works used a UAV-collected point cloud for inpainting which is very important in aerial applications as mentioned in the Introduction.

In contrast to all aforementioned studies, our method uses inpainting directly on the sparse point clouds produced by LiDAR sensors mounted on a UAV. The main advantage of this is that our training data is similar to real-world LiDAR data, therefore our approach can be easily implemented in the real world. Also, we do a complete environment inpainting, so the output from our model can be used in various applications like search-and-rescue, aerial mapping, underground mapping etc.

3   Preliminaries

LiDAR is a remote sensing method used to provide a 3D representation of a target. The working principle of the LiDAR sensor is based on the reflection of light. It consists of laser light that is projected onto the target's surface and then the reflected light is measured to see the variation in wavelength and arrival time of the light. The LiDAR system can calculate the distance from these measurements which is used to draw the digital representation of the target. Please refer to [13] for further details on this topic.

The distance result for a single laser scan gives a very accurate estimate of the 3D coordinates of a point in space from where the beam was reflected. By repeating this over time in varying directions, we can get different coordinates for various different points, also known as a point cloud. So basically, the point cloud is a set of data points in 3D space that provides a 3D representation of the LiDAR's surroundings, with great precision and detail.

4   Problem Definition

Consider a UAV Q with a position $P_Q^t = (x_Q^t, y_Q^t, z_Q^t)$ at time t deployed in an unknown environment with N static objects at initially unknown positions $\{P_1, P_2, \ldots P_N\}$, $P_i = (x_i, y_i, z_i)$. The velocity of the UAV can range between V_{min} and V_{max}. The UAV is equipped with an accurate positioning system used to calculate its position and with LiDAR sensors for scanning the environment.

Let X be a sparse point cloud of an object O obtained from LiDAR sensors with the UAV flying at a high speed V_{Q_h}. Let Y_{GT} be the corresponding ground truth dense point cloud obtained from the UAV LiDARs when flying at a low speed V_{Q_l}. Given an unknown environment, the objective is to quickly scan the environment by flying the UAV at a high speed and use the partial point clouds X collected during the scan to predict the corresponding dense point cloud Y_{GT} such that an accurate 3D reconstruction of the environment as a whole can be generated.

5   Methodology

The proposed approach contains offline and online operations, as shown in Fig. 1. The offline operation is concerned with generating a machine-learning model for inpainting partial point clouds of individual objects. The offline operation involves model development, data collection, data pre-processing and augmentation, and model training and evaluation. On the other hand, the online operation involves using the trained model to reconstruct a detailed 3D representation of an unknown environment by flying a UAV above the environment in question at moderate to high speeds. The rest of this section gives details about the processes used in the online and offline operations of the proposed approach.

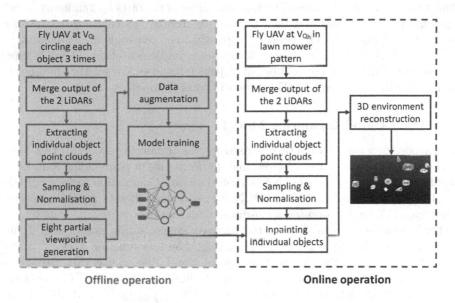

Fig. 1. The proposed approach. Left: data collection and model training. Right: reconstructing unknown 3D environments.

5.1   Simulator

Airsim is a photo-realistic simulator for drones and ground vehicles developed on top of Epic's Unreal Engine 4 (UE4) by Microsoft Research. Airsim is selected for simulating the UAV behaviours during environment scanning because of its high fidelity. The UAV model used is AR Quadcopter 2.0. The UAV has two Velodyne LiDAR sensors (one horizontally oriented and the other vertically oriented) to collect point clouds of the objects in the environment along with a GPS device for position information. The decision to use two LiDARs is based on some initial experiments which showed that one LiDAR is not enough for capturing meaningful 3D information from the environment. The point clouds collected from both LiDARs are combined and sent to the next stages for processing.

5.2   Extracting Individual Point Clouds

The LiDAR scan of an environment contains point clouds of the objects along with the ground points. The ground point clouds are not needed for inpainting, therefore we remove them by deleting all points with a z-value equal to zero (ground value) from the merged LiDAR file.

As the point cloud of a complete environment contains multiple objects, k-means clustering is used to separate these objects. The k-means clustering algorithm searches an unlabeled multidimensional dataset for a pre-determined number of clusters. It accomplishes this by meeting two criteria of optimal clustering. First, the centre of the cluster is the arithmetic mean of all the points belonging to that cluster. Second, each point is closer to its own cluster centre than to other cluster centres. The k-means algorithm starts by randomly selecting a group of centroids and then performing iterative calculations to optimize the centroid's position. The process stops when the centroids are stabilised.

We used the *scikit-learn* library for k-means implementation. The only input required for k-means is the number of objects (clusters). We get that number by simply viewing the merged LiDAR file in Cloud Compare software.

5.3   Point Cloud Inpainting Model

An encoder-decoder architecture is used for the inpainting model in this work, as these architectures have achieved high performance in point cloud completion problems [9,15,20]. Our model is based on the Point Completion Network (PCN) architecture proposed in [20]. In PCN, the encoder takes as input the partial point cloud X and generates an n-dimensional feature vector f. This feature vector is then used by the decoder to produce a coarse output point cloud Y_c and a detailed (inpainted) point cloud $Y_{inpainted}$. The loss function L is calculated using $Y_{inpainted}$ and the ground truth point cloud Y_{GT}, and is then utilised to train the network using backpropagation. Unlike autoencoders, PCN does not enforce the network to keep the points from X in $Y_{inpainted}$. Instead, PCN learns a projection from the space of partial to the space of complete point clouds [20].

Encoder. The encoder summarises the geometric information in X as a feature vector $\mathbf{f} \in \mathbb{R}^n$ where $n = 1024$. The encoder consists of two stacked Point Net layers. The first layer consumes a $p \times 3$ matrix M that consists of p input points. Each row of M is the 3D coordinate of a point $\mathbf{m}_i = (x, y, z)$. A shared multi-layer perceptron (MLP) consisting of two linear layers with ReLU activation transforms each $\mathbf{m}_i$ into a point feature vector $\mathbf{v}_i$. The output of MLP is a feature matrix V that consists of learned point features $\mathbf{v}_i$ as the row entries. The matrix V then passes through a point-wise max-pool to obtain an n-dimensional global feature $\mathbf{g}$, where $\mathbf{g}_j = max_{i=1,...,m}\{V_{ij}\}$ for $j = 1,...,n$. The second Point Net layer takes V and $\mathbf{g}$ as input. It produces an augmented point feature matrix $\tilde{V}$ by concatenating $\mathbf{g}$ to each $\mathbf{v}_i$. The rows of this matrix are the concatenated feature vectors $[\mathbf{v}_i \ \mathbf{g}]$. The matrix $\tilde{V}$ then passes through another shared MLP and point-wise max-pool to produce the feature vector $\mathbf{f}$.

Decoder. The decoder is in charge of generating the output point cloud $Y_{inpainted}$ from the feature vector **f**. The decoder of PCN combines the advantages of the folding-based decoder [17] and the fully-connected decoder [1] in a multistage point generation pipeline. The PCN decoder divides the generation of the output point cloud into two stages to allow for generating a detailed output point cloud with fewer parameters. The first stage consists of generating a coarse output Y_c of k points by passing f through a fully-connected network with $3k$ output units and reshaping the output into a $k \times 3$ matrix. The second stage uses the folding operation (details can be found in [20]) on each point q_i in Y_c. This operation generates a patch of $t = u^2$ points in the local coordinates centred at q_i. The local coordinates are then transformed into the global coordinates by adding q_i to the output. Finally, the combination of all k patches gives the detailed output $Y_{inpainted}$ consisting of $s = kt$ points.

Loss Function. The loss function is responsible for measuring the difference between the ground truth and the detailed output point cloud. The loss must be invariant to the permutations of the points because both point clouds are unordered. Fan et al. [4] introduced a permutation invariant function called Chamfer Distance (CD) which finds the average closest distance for each point between the output point cloud S_1 and the ground truth point cloud S_2. We use the symmetric version of CD as calculated in Eq. 1. The first term of Eq. 1 forces the output points to lie as close as possible to the ground truth points, and the second term makes sure that the output point cloud covers the majority of the ground truth point cloud.

$$CD(S1, S2) = \frac{1}{|S_1|} \sum_{x \in S_1} \min_{y \in S_2} ||x - y||_2 + \frac{1}{|S_2|} \sum_{y \in S_2} \min_{x \in S_1} ||y - x||_2 \qquad (1)$$

Since we are using PCN as our baseline architecture, our loss function is the same as proposed in [20]. The loss function consists of two terms given by Eq. 2. The first term gives the CD between Y_c and the sub-sampled ground truth $\tilde{Y}_{GT}$. Note that Y_c and $\tilde{Y}_{GT}$ have the same size. The second term, weighted by hyperparameter α, gives the CD between the inpainted point cloud $Y_{inpainted}$ and the ground truth point cloud Y_{GT}.

$$L(Y_c, Y_{inpainted}, Y_{GT}) = CD(Y_c, \tilde{Y}_{GT}) + \alpha \; CD(Y_{inpainted}, Y_{GT}) \qquad (2)$$

Training Data Collection and Augmentation. Here, we describe the creation of the dataset used to train our model via supervised learning. We use two LiDARs mounted on an AR Quadcopter 2.0 to collect the partial (X) and complete (Y) point clouds of the objects in the *Blocks* environment. Specifically, we collect point clouds for 11 different objects: ball, box, rectangular prism, cone, straight stairs, curved stairs, diamond, octagon prism, cylinder, stair-step cube, and square prism.

The complete point clouds are obtained by flying the UAV in circles (3 times) around an object at a speed of $V_{Q_l} = 1$ ms^{-1}. These point clouds are used as ground truth Y_{GT}. The partial point clouds for the training dataset are generated in a different way than the validation and testing datasets. For the training dataset, 8 partial point clouds are generated for each Y_{GT}. Each partial point cloud is generated from a random viewpoint of the object. We use k-means clustering to divide the complete point cloud into eight clusters where multiple clusters can share points from other clusters. For the validation dataset, partial point clouds are generated from a single UAV flight above an object at 5 ms^{-1}. The partial point clouds for testing are generated in the same way but the UAV speed is increased to 7 ms^{-1}. The LiDAR parameters setting used for data collection is shown in Table 1. The same parameters were used for both LiDAR sensors.

Table 1. LiDAR parameters for data collection.

Channels	Rotations/sec	Range	Points/sec	Horizontal FOV	Vertical FOV
64	10	30	300000	0 to 359	-15 to $+15$

After collecting the point clouds, the floor points are removed from the two LiDAR output files and the files are merged into a single .ply file. Since the complete point clouds from the two LiDAR sensors contain a large number of points, random sampling is done to make the point clouds less computationally expensive. All complete point clouds are sampled to contain 16384 points only. Sampling is followed by normalization to standardise all point clouds so that their coordinates become in the range $[-0.5, 0.5]$ before being used as input for the inpainting model.

Data augmentation is done to increase the size of the training and validation datasets and to avoid overfitting. We use MATLAB Computer Vision toolbox for augmenting data through x and y-axis translation, rotation and scaling of the collected point clouds. We increase the dataset to 341 dense point clouds (31 for each object), out of which 253 are used for training and 88 for validation. The total partial point clouds for training are 2024 (8 partials for one dense) and for validation are 88 (one partial for one dense). The collected data can be found here, **Point cloud data**.

5.4   Inpainting Complete Environments

Constructing detailed 3D representations of unknown environments can be achieved using the model described in the previous section. First, a partial scan of a given environment is performed by flying the UAV in a lawn mower path at $V_{Q_h} = 7\,\text{ms}^{-1}$. The width of the lawn mower path is set to 15 m. Note that the maximum horizontal speed of the AR Quadcopter 2.0 is 10 ms^{-1} [5]. The point cloud data from the two LiDAR sensors are merged and the point clouds of individual objects are extracted, as described in Sect. 5.2. Then, the pre-trained inpainting model is used to inpaint missing parts of each point cloud. After that, the inpainted objects are re-projected into the environment using the min-max scaling given by Eq. 3. It is a commonly used technique in the field of data preprocessing and normalization.

$$x' = a + \frac{(x - \min(x))(b - a)}{\max(x) - \min(x)} \tag{3}$$

such that, x' is the re-scaled value, a and b are the minimum and maximum values of x, y or z in the original point cloud before normalization, x is the normalized value, and $\min(x)$ and $\max(x)$ are the minimum and maximum values of the range used for normalization. By applying this formula to each dimension (x, y, z) of the inpainted point clouds, we denormalize the values and regenerate the complete environment.

6   Experimental Results

Our approach is implemented in Python using the Pytorch library. Data collection is done on Windows 10. The training and evaluation were done on a Linux system (1080 Nvidia GPU). We trained the model using Adam optimizer [11] for 600 epochs (10.8k time steps) with an initial learning rate of 0.0005 which decays by 0.7 every 50 epochs. The batch size was 16. The best model is the one with the minimum CD in validation data. The training loss averaged from five training runs is shown in Fig. 2. The total loss decreases exponentially until it becomes almost constant after 8k time steps.

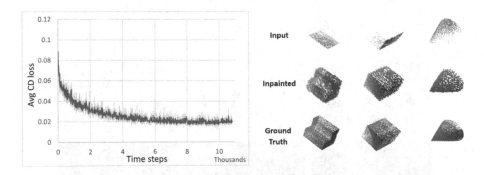

Fig. 2. Results of the inpainting model. Left: training loss over 5 runs. Right: examples of inpainting individual objects.

The inpainting results for 3 out of 11 objects are shown in Fig. 2. It is evident from the visual results that the model learnt to inpaint the objects even when very little information is available in the input point clouds. The average CD loss achieved is $27.5 * 10^{-3}$ (std $= 1.3 * 10^{-3}$) indicating the effectiveness of the inpainting model.

The trained model was tested for different LiDAR parameters and different environment settings. Various partial scans have been performed using different UAV speeds and varying LiDAR parameters as shown in Table 2. Although the quality of the partial scans was reduced by decreasing the LiDAR capabilities (number of channels and points per second) and increasing the UAV speed, our model successfully managed to inpaint the environment under different evaluation settings. It can be seen that the average CD loss is slightly higher for low-quality scans (high speeds, fewer channels and fewer points per second) than high-quality scans (low speeds, more channels and more points per second), demonstrating the ability of the proposed algorithm to generate accurate 3D representations even when low-quality partial scans are available.

Figure 3 shows the inpainting result of three test environments under the following parameters: UAV speed = 7 ms^{-1} and LiDAR parameters: 32 channels, 200000 points per second. This setting gives the least quality partial point cloud with the given Quadcopter and LiDAR sensors.

Table 2. CD loss for different partial scans.

UAV speed (ms^{-1})	Channels	Points/sec	Avg CD Loss ($\times 10^{-3}$)
5	32	200000	32.1206
5	32	300000	31.7038
5	64	200000	30.1912
5	64	300000	**28.5745**
7	32	200000	**39.4098**
7	32	300000	32.5207
7	64	200000	31.9208
7	64	300000	31.8370

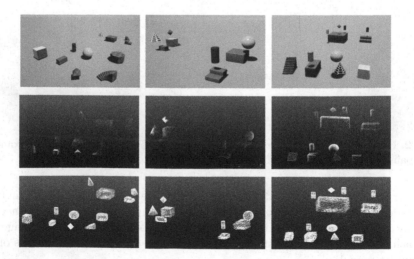

Fig. 3. Test environments.

7   Conclusion and Future Work

LiDAR point clouds can be used in 3D environment reconstruction. However, 3D data from LiDAR is often incomplete due to occlusions and limitations of sensor resolution. Inpainting can be done to complete sparse point clouds. Previous studies on LiDAR inpainting used synthetic data for training and none of them studied inpainting of point clouds obtained from UAV. In this paper, we used a supervised learning encoder-decoder architecture to inpaint LiDAR point clouds collected by a UAV flying at speeds ranging from 5–7 ms^{-1}. This is the first study to propose LiDAR inpainting for point clouds collected from a UAV. Also, this is the first work to perform inpainting of whole environments as previous studies focused on single objects. The visual representations and the values of average CD loss in Sect. 6 demonstrate the effectiveness and accuracy of our model. The minimum CD loss recorded was 0.028 for the UAV speed of 5 ms^{-1} and sensor parameters: 64 channels, 300000 points per second.

Future studies can extend this work in several directions. First, the current work used a limited data set (11 different types of objects) for training, however in the general case, there will be many other complex objects in the environment. Second, extending the system to a swarm of UAVs flying at high speed will further reduce the scan time and will allow for covering large areas. This can benefit from federated learning to enable the collection of large amounts of training data from different environments.

Acknowledgement. This work is funded by the Australian Research Council Grant DP200101211.

References

1. Achlioptas, P., Diamanti, O., Mitliagkas, I., Guibas, L.: Learning representations and generative models for 3D point clouds. In: International Conference on Machine Learning, pp. 40–49. PMLR (2018)
2. Behley, J., et al.: Semantickitti: a dataset for semantic scene understanding of lidar sequences. In: Proceedings of the IEEE/CVF International Conference on Computer Vision, pp. 9297–9307 (2019)
3. Chen, J., Yi, J.S.K., Kahoush, M., Cho, E.S., Cho, Y.K.: Point cloud scene completion of obstructed building facades with generative adversarial inpainting. Sensors **20**(18), 5029 (2020)
4. Fan, H., Su, H., Guibas, L.J.: A point set generation network for 3d object reconstruction from a single image. In: Proceedings of the IEEE Conference on Computer Vision and Pattern Recognition, pp. 605–613 (2017)
5. Foley, K.: Parrot AR drone 2.0 elite edition (Jun 2022)
6. Fu, Z., Hu, W., Guo, Z.: Point cloud inpainting on graphs from non-local self-similarity. In: 2018 25th IEEE International Conference on Image Processing (ICIP), pp. 2137–2141. IEEE (2018)
7. Gu, J., et al.: Weakly-supervised 3D shape completion in the wild. In: Vedaldi, A., Bischof, H., Brox, T., Frahm, J.-M. (eds.) ECCV 2020. LNCS, vol. 12350, pp. 283–299. Springer, Cham (2020). https://doi.org/10.1007/978-3-030-58558-7_17

8. Hong, X., Xiong, P., Ji, R., Fan, H.: Deep fusion network for image completion. In: Proceedings of the 27th ACM International Conference on Multimedia, pp. 2033–2042 (2019)

9. Huang, Z., Yu, Y., Xu, J., Ni, F., Le, X.: PF-Net: point fractal network for 3D point cloud completion. In: Proceedings of the IEEE/CVF Conference on Computer Vision and Pattern Recognition, pp. 7662–7670 (2020)

10. Insafutdinov, E., Dosovitskiy, A.: Unsupervised learning of shape and pose with differentiable point clouds. In: Advances in Neural Information Processing Systems, vol. 31 (2018)

11. Kingma, D.P., Ba, J.: Adam: a method for stochastic optimization. arXiv preprint arXiv:1412.6980 (2014)

12. Liu, H., Jiang, B., Song, Y., Huang, W., Yang, C.: Rethinking image inpainting via a mutual encoder-decoder with feature equalizations. In: Vedaldi, A., Bischof, H., Brox, T., Frahm, J.-M. (eds.) ECCV 2020. LNCS, vol. 12347, pp. 725–741. Springer, Cham (2020). https://doi.org/10.1007/978-3-030-58536-5_43

13. Mehendale, N., Neoge, S.: Review on lidar technology. Available at SSRN 3604309 (2020)

14. Pathak, D., Krahenbuhl, P., Donahue, J., Darrell, T., Efros, A.A.: Context encoders: feature learning by inpainting. In: Proceedings of the IEEE Conference on Computer Vision and Pattern Recognition, pp. 2536–2544 (2016)

15. Tchapmi, L.P., Kosaraju, V., Rezatofighi, H., Reid, I., Savarese, S.: Topnet: structural point cloud decoder. In: Proceedings of the IEEE/CVF Conference on Computer Vision and Pattern Recognition, pp. 383–392 (2019)

16. Wang, P.: Research on comparison of lidar and camera in autonomous driving. In: Journal of Physics: Conference Series, vol. 2093, p. 012032. IOP Publishing (2021)

17. Yang, Y., Feng, C., Shen, Y., Tian, D.: Foldingnet: interpretable unsupervised learning on 3d point clouds. arXiv preprint arXiv:1712.07262 2(3), 5 (2017)

18. Yeh, R.A., Chen, C., Yian Lim, T., Schwing, A.G., Hasegawa-Johnson, M., Do, M.N.: Semantic image inpainting with deep generative models. In: Proceedings of the IEEE Conference on Computer Vision and Pattern Recognition, pp. 5485–5493 (2017)

19. Yu, J., Lin, Z., Yang, J., Shen, X., Lu, X., Huang, T.S.: Generative image inpainting with contextual attention. In: Proceedings of the IEEE Conference on Computer Vision and Pattern Recognition, pp. 5505–5514 (2018)

20. Yuan, W., Khot, T., Held, D., Mertz, C., Hebert, M.: PCN: point completion network. In: 2018 International Conference on 3D Vision (3DV), pp. 728–737. IEEE (2018)

21. Zhang, X., Le, X., Panotopoulou, A., Whiting, E., Wang, C.C.: Perceptual models of preference in 3D printing direction. ACM Trans. Graph. (TOG) 34(6), 1–12 (2015)

22. Zhang, X., Le, X., Wu, Z., Whiting, E., Wang, C.C.: Data-driven bending elasticity design by shell thickness. In: Computer Graphics Forum, vol. 35, pp. 157–166. Wiley Online Library (2016)

23. Zhao, B., Le, X., Xi, J.: A novel SDASS descriptor for fully encoding the information of a 3D local surface. Inf. Sci. **483**, 363–382 (2019)

A Sampling Method for Performance Predictor Based on Contrastive Learning

Jingrong Xie, Yuqi Feng, and Yanan Sun[✉]

School of Computer Science, Sichuan University, Chengdu 610065, China
xiejingrong@stu.scu.edu.cn, ysun@scu.edu.cn

Abstract. Performance predictors are commonly dedicated to mitigating the substantial resource consumption of neural architecture search. Nevertheless, existing performance predictors are typically constructed based on the randomly sampled training data. Such a sampling method will not only lead to unnecessary computation budget caused by the sampled similar architectures, but also induce performance deterioration resulting from the poor spanning of search space. In this paper, we propose a contrastive learning-based sampling method to address the aforementioned issues. Specifically, we first encode the architectures as directed acyclic graphs, based on which a large number of architectures are augmented to learn invariant knowledge of architectures. After that, we maximize agreement based on augmented architectures to express similar architectures to analogous representations. Consequently, representative architectures are selected through clustering similar architectures to improve the spanning of the search space. We conduct extensive experiments on NAS-Bench-101 and NAS-Bench-201. The experimental results show that the proposed method can improve the predictive ability of performance predictors compared with the random sampling-based ones and can help search superior architectures when integrating with neural architecture search. In addition, an ablation study shows the effectiveness of contrastive learning and the clustering method used in the proposed sampling method.

Keywords: Neural Architecture Search · Performance Predictor · Contrastive Learning

1 Introduction

Neural Architecture Search (NAS) can automatically design high-performance architectures and has demonstrated promising performance in practice. Generally, a typical NAS algorithm consists of three parts: search space, search strategy, and performance evaluation [10]. In particular, the search space encompasses the entire set of potential architectures. The search strategy is leveraged to explore and identify promising architectures from the search space. In addition, the performance evaluation serves as a criterion to determine whether a particular architecture satisfies the given requirements or not, which determines the

© The Author(s), under exclusive license to Springer Nature Singapore Pte Ltd. 2024
T. Liu et al. (Eds.): AI 2023, LNAI 14471, pp. 215–226, 2024.
https://doi.org/10.1007/978-981-99-8388-9_18

efficiency of NAS. In practice, the performance evaluation often involves training all the architectures generated in the search process, which is prohibitively expensive and has become the most significant bottleneck in NAS [25].

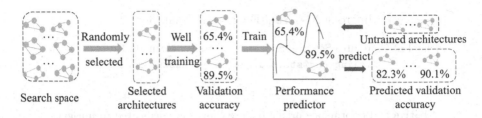

Search space · · · Selected architectures · · · Validation accuracy · · · Performance predictor · · · Predicted validation accuracy

Randomly selected · · · Well training 65.4% 89.5% · · · Train 65.4% 89.5% predict · · · Untrained architectures 82.3% 90.1%

Fig. 1. Construction of performance predictor

To significantly alleviate the huge consumption of resources, performance predictors have emerged as the mainstream solution [16]. As shown in Fig. 1, the construction process of a performance predictor contains three parts. First, the architectures are randomly selected from the search space, and then they are well-trained to get validation accuracy. Second, the obtained architecture-accuracy pairs (i.e., training data) are used to train the performance predictor which is essentially a regression model. Third, the performance predictor is used to predict the validation accuracy of the untrained architectures.

In practice, a large amount of such training data is often required to get an accurate performance predictor. However, too much training data requires training a large number of architectures, which is computationally expensive. To solve this problem, the random sampling method is often used to get a set of architectures as the training dataset [16,22]. Unfortunately, this will inevitably lead to two limitations. First, the random sampling method will lead to additional computation costs. This is because such a method may sample similar architectures. These similar architectures usually have similar accuracy [26], so training each of these similar architectures is not necessary and incurs additional computational resources. Second, the random sampling method will cause the performance degradation of performance predictors. This is because randomly sampled architectures are not representative. Specifically, the sampled architectures can only represent a small region of the whole search space most of the time, and have gaps with other regions of the search space. This makes the trained performance predictors cannot accurately predict the architectures in the whole search space.

To tackle these issues, we propose an architecture **S**ampling method for **P**erformance predictor based on **C**ontrastive **L**earning (SPCL). This is because architectures can be denoted as Directed Acyclic Graphs (DAGs), and contrastive learning has been proven to be highly effective in graph data learning [24]. By leveraging contrastive learning, similar architectures can be easily selected to avoid unnecessary training. In addition, by running SPCL multiple times, a set of representative architectures that have the potential to span the whole search space can be obtained. In summary, the proposed SPCL method has the following contributions:

- SPCL can avoid the generation of pseudo-random numbers in the random sampling method by sampling representative samples, thereby preventing the occurrence of similar architectures sampled in the search space. Therefore, SPCL can avoid unnecessary wastage of computational resources compared with the random sampling method.
- SPCL can obtain representative samples, alleviating the unrepresentative problem brought by the random sampling method when constructing the training dataset. Therefore, the training dataset constructed by SPCL can be spanned to the entire search space, and the trained performance predictors can be more accurate.
- SPCL can achieve promising performance in extensive experiments. Specifically, SPCL can improve the predictive accuracy of many performance predictors compared with the random sampling method. Furthermore, the performance predictors constructed based on SPCL can help NAS algorithms find better architectures.

The remainder of this paper is organized as follows. First, the background is provided in Sect. 2. Then, the details of the proposed SPCL method are presented in Sect. 3. After that, the experiments are conducted to demonstrate the effectiveness of SPCL in Sect. 4. Finally, the paper is concluded in Sect. 5.

2 Background

2.1 Contrastive Learning

Contrastive learning is a self-supervised learning technique to learn the common characteristics of data [3]. Generally, such characteristics are learned by minimizing the distance between input samples of the same categories, and maximizing the distances between samples of different categories at the same time. In particular, the main process of contrastive learning has the following three steps [15]. First, the data is augmented to learn invariant knowledge of data, i.e., essential features of the data. Second, the augmented data are entered into the encoder to map the data to the characteristic space. Third, the agreement of the encoded data is maximized by minimizing the contrastive loss function to express analogous architectures similarly.

Because contrastive learning can learn characteristics of the data to measure similarities between them, it can also express similar architectures to analogous representations. In other words, contrastive learning can effectively cluster similar architectures together to avoid unnecessary training. Therefore contrastive learning is used in this work to represent the architectures.

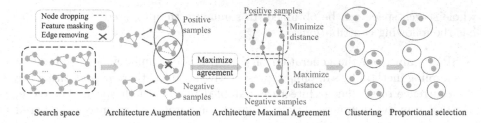

Fig. 2. The framework of SPCL contains three parts. First, the architectures in the search space are augmented, and distinguished into positive and negative samples according to the sources of augmented architectures. Second, the augmented architectures are maximized agreement. Third, these architectures are clustered and proportionally selected to get representative architectures.

2.2 Graph Data Sampling Methods

The construction of performance predictors needs to sample architectures which are often represented by DAGs. However, such architectures cannot be sampled out by traditional graph data sampling methods. This is because traditional graph data sampling only selects a portion of subgraphs or nodes from large-scale graph data. For example, Graph learn [29] uses a common method of traditional graph sampling called neighbor sampling, which includes a random sampling of neighbor nodes, sampling according to the probability weight of edges.

To address the above problem, plenty of work uses the random sampling method to get training datasets in performance predictors. For example, Wei *et al.* [25] randomly selected a small subset of architectures and constructed a performance predictor based on them. Similarly, HOP [4] obtained a small proportion of architectures by the random sampling method, then redistributed the weight of the operations in these architectures, and subsequently constructed a performance predictor. However, because of the similar and unrepresentative architectures sampled, the random sampling method will bring additional computational costs and reduce the predictive accuracy of performance predictors.

To address these limitations, we propose SPCL to sample representative architectures and then construct the training dataset of performance predictors. Because the proposed sampling method can learn the characteristics of architectures, SPCL can avoid the additional computational consumption brought by sampling similar architectures, and alleviate the degradation of performance predictors caused by the unrepresentative architectures sampled.

3   Approach

On the basis of the contrastive learning method, the overall framework of SPCL is shown in Algorithm 1. Given the architecture dataset of the whole search space, the two augmentation functions, and the clustering method, through the three sequential parts of SPCL, the training dataset for performance predictor is obtained. Specifically, in the first part (lines 2–4), the architectures based on DAGs are represented and augmented for learning invariant knowledge of architectures. In the second part (lines 5–6), analogous architectures are expressed similarly by the maximal agreement. In the third part (lines 8–9), these architectures are clustered to gather similar architectures together. Then architectures are selected proportionally from each cluster. After SPCL, the selected representative architectures will be well-trained and used as the training dataset for the performance predictor. For a better understanding, an overall framework of SPCL is presented in Fig. 2. The details about the key components including architecture augmentation and maximal agreement are presented in the following two paragraphs.

Algorithm 1: Procedure of SPCL

---

input : The architecture dataset of the whole search space $\mathcal{G}_{\text{source}}$,
augmentation functions t and t', clustering method DBSCAN
output: The training dataset for performance predictor $\mathcal{D}_{\text{train}}$

1 **for** $epoch \leftarrow 1,2\cdots$ **do**
2 $\quad$ Generate two augmented architectures $\bar{\mathcal{G}}_i = t(\mathcal{G})$ and $\bar{\mathcal{G}}_j = t'(\mathcal{G})$;
3 $\quad$ Obtain representations g_i of $\bar{\mathcal{G}}_i$ using GCN;
4 $\quad$ Obtain representations g_j of $\bar{\mathcal{G}}_j$ using GCN;
5 $\quad$ Compute the contrastive objective $\mathcal{L}$ with Eq. 1;
6 $\quad$ Update parameters by applying gradient ascent to minimize $\mathcal{L}$;
7 **end**
8 Cluster representations $g_{ni}, g_{nj}, \cdots$ using DBSCAN;
9 Sample from each cluster get $\mathcal{G}_{\text{train}}$

3.1   Architecture Augmentation

To learn invariant knowledge of architectures, we propose the architecture augmentation method which is detailed in Algorithm 2. Specifically, based on an architecture and four different augmentation functions, through the architecture augmentation method, two augmented architectures are obtained. In the architecture augmentation method, several commonly used graph data enhancement methods are integrated into this data enhancement process. In practice, for the first enhancement method, the architecture is obtained by data augmentation with feature masking and edge removing (lines 3–4). For the second one, the architecture is randomly obtained from two of the four methods (lines 5–6): feature masking, node dropping, edge removing, and random wandering [8]. Specifically, feature masking is the operation of partially masking or obscuring the node features in the graph. Node dropping is the operation of randomly selecting some nodes and removing them from the graph. Edge dropping is the operation of randomly selecting some edges and removing them from the graph. Random wandering is the architecture after the edge indexes and edge weights are wandered and newly enhanced. On the basis of these above four enhancement methods, the second augmented architecture is obtained through the two randomly selected methods from them (lines 7–8).

In this architecture augmentation method, we use two different data enhancement methods on the same architecture to obtain two different architectures. That is because different enhancement methods are more conducive to the representation of contrastive learning [30]. In the four different augmentation methods, feature masking and node or edge dropping can simulate the missing information in real applications, aiming to generate a more robust representation of architectures. For random wandering, different connectivity patterns can be explored, thus adapting to potential uncertainties in architecture. After applying data augmentation, a new dataset is obtained. This dataset serves as a solid foundation for subsequent maximization of the agreement among architectures through positive and negative samples.

Algorithm 2: Architecture Augmentation

---

    input   : An architecture $\mathcal{G}$(graph data and node features), function $\mathcal{F}_1$
                feature_masking, function $\mathcal{F}_2$ node_dropping, function $\mathcal{F}_3$
                edge_dropping, function $\mathcal{F}_4$ random_wandering
    output: Two augmented architecture $\mathcal{G}_1$, $\mathcal{G}_2$

1 $\mathcal{G}_1 = copy(\mathcal{G})$;
2 $\mathcal{G}_2 = copy(\mathcal{G})$;
3 $\mathcal{G}_1 = \mathcal{F}_1(\mathcal{G}_1.node\_features)$;
4 $\mathcal{G}_1 = \mathcal{F}_2(\mathcal{G}_1.graph\_data)$;
5 augmentation_methods $= [\mathcal{F}_1, \mathcal{F}_2, \mathcal{F}_3, \mathcal{F}_4]$;
6 $\mathcal{F}_{selected\_1}$, $\mathcal{F}_{selected_2} = $ random_choice(augmentation_methods, 2, replace=False);
7 $\mathcal{G}_2 = \mathcal{F}_{selected\_1}(\mathcal{G}_2.node\_features)$;
8 $\mathcal{G}_2 = \mathcal{F}_{selected\_2}(\mathcal{G}_2.graph\_data)$;

3.2 Architecture Maximal Agreement

In order to express analogous architectures similarly by maximizing agreement, we propose the architecture maximal agreement method which is detailed in Algorithm 3. Specifically, given the architecture augmented dataset, through the architecture maximal agreement, the training architecture dataset is obtained. Specifically, in the first part, the feature of graph data is extracted by the method of Graph Convolution Network [23] (lines 2–3). That is because it has achieved promising results in graph feature extraction. In practice, for the n_{th} architecture in a batch, the augmented architectures $\mathcal{G}_i, \mathcal{G}_j$ are encoded as g_{ni}, g_{nj}. In the second part, augmented architectures are maximized agreement through positive and negative samples (lines 4–6). In other words, the augmented architectures derived from the same original architecture are treated as positive samples of each other, while different ones are treated as negative samples. Based on these positive and negative samples, the similarity of the two encoded architectures is defined as $sim(g_{ni}, g_{nj}) = g\top_{n,i}g_{n,j}/\|g_{n,i}\|\|g_{n,j}\|$. Then the Normalized Temperature-Scaled Cross-Entropy Loss [21] is employed as the contrastive loss, which can be represented as Eq. 1:

$$l_n = -log\frac{exp(sim(g_{ni}, g_{nj})\backslash\tau)}{\sum_{n'=1,n'\neq n}^{N} exp(sim(g_{ni}, g_{n'j})\backslash\tau)} \tag{1}$$

In Eq. 1, sim denotes the cosine similarity function, and τ denotes the temperature parameter. The numerator finds the distance between the two positive pairs, while the denominator finds the distance between the positive pair and all the negative pairs in the batch.

By optimizing the distance between positive and negative samples, this architecture maximal agreement method can distinguish between similar and dissimilar architectures, thus ensuring that similar structures have analogous representations.

Algorithm 3: Architecture Maximal Agreement

input   : The architecture augmented dataset $\bar{\mathcal{G}}$
output: The training architecture dataset $\mathcal{G}_{train}$

1 for $epoch \leftarrow 1, 2, ...$ do
2 | Obtain representations g_1 of $\bar{\mathcal{G}}_1$ using GCN;
3 | Obtain representations g_2 of $\bar{\mathcal{G}}_2$ using GCN;
4 | Compute the contrastive objective $\mathcal{L}$ with Eq. 1;
5 | Update parameters by applying gradient ascent to minimize $\mathcal{L}$
6 end

---

4   Experiments

In this section, we conduct experiments to demonstrate the effectiveness of the SPCL method. The experiments contain two parts. First, SPCL is combined with different regression models and other open-source performance predictors, aiming to demonstrate an increase in predictive ability. Second, SPCL is integrated into performance predictor to search superior architectures in the search space of NAS, aiming to prove the enhancement to performance predictors.

For a fair comparison with other methods, we choose NAS-Bench-101 [27] and NAS-Bench-201 [9] which are two widely used benchmark datasets as the search space to perform experiments. NAS-Bench-101 includes 423,624 different DNN architectures and their performance on the CIFAR-10 dataset [14]. The search space of NAS-Bench-201 results in 15,625 neural cell candidates in total. This benchmark dataset includes the performance of three datasets, which are CIFAR-10, CIFAR-100, and Imagenet-16-120.

4.1 Overall Performance

In order to demonstrate the generalizability and portability of SPCL, experiments are conducted first on regression models and second on well-known open-source performance predictors for experimental validation. Specifically, we choose some regression models for our experiments, including Random Forest (RF) [1], Decision Tree (DT) [17], Support Vector Regression (SVR) [2], K-Nearest Neighbors (KNN) [5], and Gradient Boosted Regression Tree (GBRT) [13]. Afterward, we apply SPCL to the construction of performance predictors including Peephole [7], E2EPP [22], and HAAP [16] to replace the sampling method.

To quantify the performance gains from SPCL, we compare two parameters: Kendall's Tau (KTau) [20] and mean square error (MSE). Specifically, KTau measures the correlation between the predicted performance and the actual performance. The larger the value is closer to one, the higher the performance of the predictor. MSE is another commonly used performance evaluation metric that measures the average of the sum of squares of the errors between the predicted and actual values. The smaller the MSE value, the higher the performance of the performance predictor.

Table 1. Performance comparison of various regression models and performance predictors

Indicators	Method	Regression Model					Performance Predictor		
		RF	DT	SVR	KNN	GBRT	Peephole	E2EPP	HAAP
KTau	RS	0.4187	0.3071	0.3180	0.3462	0.3721	0.4373	0.5705	0.7010
	SPCL	**0.6619**	**0.3208**	**0.3312**	**0.3519**	**0.5674**	**0.4402**	**0.5831**	**0.7021**
MSE	RS	2.0284	0.0066	0.0051	0.0037	0.0038	0.0071	0.0042	0.0024
	SPCL	**0.0028**	**0.0055**	**0.0046**	**0.0028**	**0.0029**	**0.0053**	**0.0031**	**0.0020**

RS denotes random sampling.

The specific results of the experiment are shown in Table 1, each column represents a regression model or performance predictor. For each metric of each regression model or performance predictor, two methods are used to obtain its dataset, one using the random sampling method and the other using the SPCL method to compare the improvement in prediction performance resulting. It can be seen that some of the regression models have a significant improvement after using SPCL, such as the KTau of GBRT improves by 0.1953, while RF improves by 0.2432, etc. Therefore, the sampling method of SPCL has some improvement over the random sampling method. Meanwhile, replacing the random sampling method with SPCL in performance predictors also has a certain improvement in predictive ability. In practice, KTau is 0.0029 higher than Peephole, 0.0126 higher than E2EPP, and 0.001 higher than HAAP, while MSE is 0.0018 lower than Peephole, 0.0011 lower than E2EPP, and 0.0004 lower than HAAP.

4.2   Performance Evaluation in NAS Datasets

Combined with different regression models and performance predictors, it can be concluded that SPCL appears to have a significant improvement in predictive ability compared with the random sampling method. However, we still need to compare the performance predictors constructed based on SPCL in NAS to validate the improvement of the sampling method of the predictive ability of the predictors. Therefore we compare the search for the optimal architecture on two search spaces based on the evolutionary algorithm.

Specifically, we construct a performance predictor with SPCL and RF. This is because RF is a relatively simple regression model [1] that can handle high dimensional data and is used well in performance predictors [16, 22]. In NAS-Bench-101, we compare our performance predictor with Peephole [7] and E2EPP [22]. Please note that among these three performance predictors, Peephole uses a well-designed regression model approach, and both E2EPP and our proposed SPCL sampling approach use a simpler regression model of RF. Similarly, in NAS-Bench-201, we compare the accuracy of three algorithms on CIFAR-10. Among them, BOHB [12] is a super elaborated optimized search strategy that achieves good results on CIFAR-10 through the NAS search method. To ensure the accuracy of the searched-out architecture, we compared not only the test accuracy but also the verification accuracy. After repeating the experiment 20 times, we take the best results and write them in Table 2.

For detailed results, as shown in Table 2, we can conclude that the architecture obtained by combining the sampling method of SPCL with RF is better than the other methods. Specifically, in NAS-Bench-101, our predictor achieves a KTau of 0.6316, which is 0.1999 higher than Peephole, and 0.0611 higher than E2EPP. At the same time, our performance predictor achieves the MSE of 0.0028, which is 0.0043 lower than Peephole, and 0.0014 lower than E2EPP. Similarly, in NAS-Bench-201, we can conclude that the architecture obtained by our predictor. Specifically, the validation dataset is higher than E2EPP 0.4 and higher than BOHB 0.19, and the test dataset is higher than E2EPP 0.39 and higher than BOHB 0.17. This improvement in performance demonstrates the effectiveness of SPCL.

4.3   Ablation Study

In this section, we demonstrate the effectiveness of the proposed SPCL method through a series of comparative experiments. Same as previous subsection, we combine SPCL with RF to construct a performance predictor. First, we compare SPCL with a sampling method that only uses clustering to demonstrate the need for contrastive learning. Specifically, we perform the clustering process directly after extracting features of architectures, aiming to confirm that only the clustering method cannot sample representative architectures. Second, we leverage different clustering methods to obtain datasets for constructing performance predictors, aiming to demonstrate the verification of the clustering

Table 2. Classification accuracies on CIFAR-10 of the best searched-out architecture in the NAS-Bench-101 and NAS-Bench-201 search space using different performance predictor algorithms.

predictor	NAS-Bench-101		NAS-Bench-201	
	Top-1 Accuracy(%)	Ranking(%)	Validation(%)	Test(%)
Peephole	93.41±0.34	1.64	—	—
E2EPP	93.77±0.33	0.15	90.61±0.89	93.39±0.75
BOHB	—	—	90.82±0.53	93.61±0.52
SPCL+RF	**93.94±0.12**	**0.02**	**91.01±0.34**	**93.78±0.34**

method. Generally, clustering methods consist of the following categories hierarchical clustering, partition-based clustering, density-based clustering, and model-based clustering [19]. Therefore we select four clustering methods from the four categories to compare the impact of different clustering methods on dataset construction. Specifically, Balanced Iterative Reducing [28] is a type of hierarchical clustering, Spectral Clustering (SC) [18] is a type of partition-based clustering, DBSCAN [11] is a type of density-based clustering, and Self-Origanizing Maps (SOM) [6] is a type of model-based clustering.

Table 3 presents the results of the above two experiments. First, we can conclude that only the clustering method is not stable and the result will drop. Second, compared with the predictive ability and stability of the performance predictors, we believe that DBSCAN has the best clustering effect.

Table 3. Ablation experiments on comparative learning and clustering methods

Method	Contrastive Learning	Clustering	Ktau	MSE
SPCL+RF	✗	DBSCAN	0.4684±0.0205	0.0689±0.0062
	✓	BRICH	0.4204±0.0605	1.8420±0.6785
	✓	SOM	0.4421±0.0046	0.0096±0.0005
	✓	SC	0.4659±0.0214	1.9960±0.0364
	✓	DBSCAN	**0.6619±0.0031**	**0.0028±0.0003**

5   Conclusion

The goal of this paper is to propose an efficient sampling method for performance predictors. This goal is achieved by augmenting the architectures, maximizing their agreement, expressing similar architectures to analogous representations, and then obtaining representative architectures using the DBSCAN clustering method. The dataset of performance predictors obtained by this method can avoid additional computational costs and improve the performance of predictors. Experiments were performed on many regression models and performance predictors, and the effectiveness of SPCL has been proven. Meanwhile, the optimal architecture is derived when the performance predictor is constructed based on SPCL and leveraged to NAS algorithms. In addition, ablation studies also confirm the superiority of the contrastive learning and the clustering method in the SPCL method. Our future work will attempt to encode the architecture in a better encoding way, providing all the features of the architecture as simply as possible.

References

1. Breiman, L.: Random forests. Mach. Learn. **45**(1), 5–32 (2001)
2. Chang, C.C., Lin, C.J.: LIBSVM: a library for support vector machines. ACM Trans. Intell. Syst. Technol. **2**(3), 1–27 (2011)
3. Chen, T., Kornblith, S., Norouzi, M., Hinton, G.: A simple framework for contrastive learning of visual representations. In: Proceedings of the 37th International Conference on Machine Learning (2020)
4. Chen, Z., Zhan, Y., Yu, B., Gong, M., Du, B.: Not all operations contribute equally: hierarchical operation-adaptive predictor for neural architecture search. In: 2021 IEEE/CVF International Conference on Computer Vision, pp. 10488–10497 (2021)
5. Cover, T., Hart, P.: Nearest neighbor pattern classification. IEEE Trans. Inf. Theory **13**(1), 21–27 (1967)
6. Crespo, R., Alvarez, C., Hernandez, I., Garcia, C.: A spatially explicit analysis of chronic diseases in small areas: a case study of diabetes in Santiago, Chile. Int. J. Health Geograph. **19**(1), 1–13 (2020)
7. Deng, B., Yan, J., Lin, D.: Peephole: Predicting Network Performance Before Training. arXiv e-prints arXiv:1712.03351 (2017)
8. Ding, K., Xu, Z., Tong, H., Liu, H.: Data augmentation for deep graph learning: a survey. ACM SIGKDD Explor. Newsl **24**(2), 61–77 (2022)
9. Dong, X., Yang, Y.: Nas-bench-201: Extending the scope of reproducible neural architecture search. arXiv preprint arXiv:2001.00326 (2020)
10. Elsken, T., Hendrik Metzen, J., Hutter, F.: Neural architecture search: a survey. arXiv e-prints arXiv:1808.05377 (2018)
11. Ester, M., Kriegel, H.P., Sander, J., Xu, X.: A density-based algorithm for discovering clusters in large spatial databases with noise, pp. 226–231. AAAI Press (1996)
12. Falkner, S., Klein, A., Hutter, F.: BOHB: robust and efficient hyperparameter optimization at scale. In: International Conference on Machine Learning, pp. 1437–1446. PMLR (2018)
13. Friedman, J.H.: Greedy function approximation: a gradient boosting machine. Ann. Stat. **29**, 1189–1232 (2001)
14. Krizhevsky, A.: Learning multiple layers of features from tiny images (2009)
15. Liu, Y., et al.: Graph self-supervised learning: a survey. IEEE Trans. Knowl. Data Eng. 1–1 (2022). https://doi.org/10.1109/TKDE.2022.3172903
16. Liu, Y., Tang, Y., Sun, Y.: Homogeneous architecture augmentation for neural predictor. In: 2021 IEEE/CVF International Conference on Computer Vision, pp. 12229–12238 (2021)
17. Loh, W.Y.: Classification and regression trees. Wiley Interdisciplinary Rev. Data Mining Knowl. Discov. **1**(1), 14–23 (2011)

18. Luxburg, U.: A tutorial on spectral clustering. Stat. Comput. **17**(4), 395–416 (2007)
19. Milligan, G.W., Cooper, M.: Methodology review: clustering methods. Appl. Psychol. Meas. **11**, 329–354 (1987). https://api.semanticscholar.org/CorpusID: 121335572
20. Sen, P.K.: Estimates of the regression coefficient based on Kendall's tau. J. Am. Stat. Assoc. **63**(324), 1379–1389 (1968)
21. Sohn, K.: Improved deep metric learning with multi-class n-pair loss objective. In: Advances in Neural Information Processing Systems, vol. 29 (2016)
22. Sun, Y., Wang, H., Xue, B., Jin, Y., Yen, G.G., Zhang, M.: Surrogate-assisted evolutionary deep learning using an end-to-end random forest-based performance predictor. IEEE Trans. Evol. Comput. **24**(2), 350–364 (2020)
23. Veličković, P., Cucurull, G., Casanova, A., Romero, A., Liò, P., Bengio, Y.: Graph attention networks. In: International Conference on Learning Representations (2018)
24. Verma, V., Qu, M., Lamb, A., Bengio, Y., Kannala, J., Tang, J.: Graphmix: regularized training of graph neural networks for semi-supervised learning. arxiv e-prints, art. arXiv preprint arXiv:1909.11715 (2019)
25. Wen, W., Liu, H., Chen, Y., Li, H., Bender, G., Kindermans, P.-J.: Neural predictor for neural architecture search. In: Vedaldi, A., Bischof, H., Brox, T., Frahm, J.-M. (eds.) ECCV 2020. LNCS, vol. 12374, pp. 660–676. Springer, Cham (2020). https://doi.org/10.1007/978-3-030-58526-6_39
26. Wu, B., et al.: Fbnet: hardware-aware efficient convnet design via differentiable neural architecture search. In: Proceedings of the IEEE/CVF Conference on Computer Vision and Pattern Recognition, pp. 10734–10742 (2019)
27. Ying, C., Klein, A., Christiansen, E., Real, E., Murphy, K., Hutter, F.: NAS-bench-101: towards reproducible neural architecture search. In: Proceedings of the 36th International Conference on Machine Learning, vol. 97, pp. 7105–7114 (2019)
28. Zhang, T., Ramakrishnan, R., Livny, M.: Birch: an efficient data clustering method for very large databases. In: Proceedings of the 1996 ACM SIGMOD International Conference on Management of Data, pp. 103–114 (1996)
29. Zhu, R., et al.: Aligraph: a comprehensive graph neural network platform. In: Proceedings of the VLDB Endowment, vol. 12. no. 12, pp. 2094–2105 (2019)
30. Zhu, Y., Xu, Y., Yu, F., Liu, Q., Wu, S., Wang, L.: Graph contrastive learning with adaptive augmentation. In: Proceedings of the Web Conference 2021, pp. 2069–2080 (2021)

AdaptMatch: Adaptive Consistency Regularization for Semi-supervised Learning with Top-k Pseudo-labeling and Contrastive Learning

Nan Yang(ID), Fan Huang(ID), and Dong Yuan(✉)

The University of Sydney, Camperdown, NSW 2006, Australia
dong.yuan@sydney.edu.au

Abstract. Semi-supervised learning has been established as a very effective paradigm for utilizing unlabeled data in order to reduce dependency on large labeled datasets. Most of the semi-supervised learning (SSL) methods proposed recently rely on a predefined and extremely high threshold to select unlabeled data that contribute to the training, thus failing to consider different learning statuses of the model and feature learning from unlabeled data. To address this issue, we propose AdaptMatch, an adaptive learning approach to leverage unlabeled data using Top-k pseudo-labeling and contrastive learning according to the model's learning status. The core of AdaptMatch is to adaptively adjust rules for different learning phases to allow informative unlabeled data and their pseudo-labels. If we cannot get high-confidence pseudo-labels from unlabeled data, contrastive learning can help the model learn more common features within the class. AdaptMatch outperforms or equals the state-of-the-art performance on a range of SSL benchmarks, exceptionally superior when the labeled data are extremely limited or imbalanced. For example, AdaptMatch reaches 91.56% and 97.44% accuracy with 4 labeled examples per class on CIFAR-10 and SVHN respectively, substantially improving over the previously best 88.70% and 96.66% accuracy achieved by FixMatch and ReMixMatch. Meanwhile, AdaptMatch also improves the accuracy of FixMatch in CIFAR10-LT with a performance gain of up to 2.3%.

Keywords: Semi-Supervised Learning · Pseudo-labeling · Contrastive Learning

1 Introduction

Training deep neural network models through supervised learning will produce better performance based on the empirical observation [17,28]. However, the collection of labeled data is extremely expensive for many learning tasks, as it requires human labor and expert knowledge. In contrast, unsupervised learning

T. Liu et al. (Eds.): AI 2023, LNAI 14471, pp. 227–238, 2024.
https://doi.org/10.1007/978-981-99-8388-9_19

and semi-supervised learning (SSL) are effective methods for training models on huge amounts of data without the need for a large number of labels. Since unsupervised learning does not learn the precise output in advance, the prediction may be inaccurate, making it difficult to apply in the industry. Therefore, semi-supervised learning (SSL) becomes the most promising approach to address this weakness by leveraging a small number of labeled and a large amount of unlabeled data. The performance boost conferred by SSL typically comes with a low cost on acquiring unlabeled data and labeling a small number of data. Many current methods to SSL have added a loss term calculated on unlabeled data and encouraged the model to generalize more effectively to the unseen data. The recent state-of-the-art loss term falls into three categories [4], which are entropy minimization, consistency regularization, and generic regularization. Generating a pseudo-label for unlabeled images with the loss term and training the model to predict the artificial labels when feeding unlabeled images as input is a popular SSL method. The pseudo-labels for unlabeled data can be obtained by model's class prediction [28] or classes' predicted distribution.

However, a drawback of current SSL methods such as Pseudo-Labeling [16], Unsupervised Data Augmentation (UDA) [28], and FixMatch [22] is that they calculate the unsupervised loss using only unlabeled data with prediction confidence higher than the fixed threshold. While this technique ensures that only high-quality unlabeled data are used to train the model, it misses a large amount of additional unlabeled data, particularly in the early training stages, when only a few unlabeled data have a prediction confidence higher than the threshold. Throughout the training process, even in the late stages, some unlabeled data will still be discouraged due to the high-confidence threshold. Therefore, the model should prioritize not just acquiring more unlabeled data with high-confidence labels but also additional unlabeled data. When there are no high-confidence labels, the model can still learn some feature information from unlabeled data to assist the learning process.

In this paper, we propose AdaptMatch, an SSL algorithm that introduces Top-k pseudo-labeling and an additional contrastive loss, which gracefully unifies with consistency regularization. Top-k pseudo-labeling is when the model ensures consistency of the first-k class prediction of the same input across multiple perturbations, and the model will take the highest predicted class as the pseudo-label of this image. Previous pseudo-label methods only set a fixed way to extract high-confidence labels throughout the whole training process, while our Top-k pseudo-labeling can dynamically adjust the k value to adapt to the current learning status of the model.

Due to the limited number of labeled data samples, only a small number of pseudo-labels can be generated at the early stage of model training, which consumes considerable computational resources. Inspired by recent contrastive learning algorithms, AdaptMatch learns representations by maximizing agreement between two differently augmented views of the same data sample via a contrastive loss in the latent space. Representation learning with contrastive cross-entropy loss will improve the model's learning of the same objects with different augmentations if the model does not produce reliable "guessed labels". By combining these two critical techniques, our AdaptMatch method yields better SSL with more effective and stable consistency regularization.

The main contributions of this paper can be summarized as follows:

- We propose a Top-k algorithm to measure the label consistency, which shows that, by using the dynamic k value, we can focus more on high-confidence instances in different training epochs in an adaptive and flexible fashion.
- To the best of our knowledge, we are the first to apply contrastive loss in semi-supervised learning for deep neural networks. By incorporating contrastive learning, our work can help the model learn unlabeled information effectively without the guidance of pseudo-labels. Previous empirical studies always focus on how to find pseudo-labels while ignoring the mutual information between different versions.
- We conducted extensive experiments and showed that our AdaptMatch outperforms or equals the state-of-the-art performance on a range of SSL benchmarks. Furthermore, our strategy significantly increases the learning speed at the early stage of model training, while simultaneously minimizing the label bias in the imbalanced dataset to identify a superior final classifier.

2 Related Work

2.1 Consistency Regularization

Consistency regularization is a widely used technique for learning from unlabeled data in semi-supervised learning. SSL models rely on consistency regularization to generate consistent predictions between given data and its meaningfully distorted variants, which means that a sample and its close neighbors are expected to have the same label. This method, first suggested in [2,15], is referred to as "Regularization With Stochastic Transformations and Perturbations" and the "Π-Model" respectively. The most common regularization technique is to apply domain-specific data augmentation [4,28], which can be used to artificially enlarge the size of a training dataset artificially by producing a new, modified, and nearly infinite data stream. Consistency is usually quantified by the mean squared error or cross-entropy [28] between the model output for perturbed and unperturbed inputs. This procedure makes the model noise-insensitive and therefore smoother as the input (or hidden) space changes. From another aspect, reducing consistency loss allows label information to be progressively propagated from labeled to unlabeled data.

2.2 Contrastive Learning

The methods of contrastive visual representation learning can be dated all the way back to [10], which can learn representations by comparing positive and negative data information. Following these lines, [8] suggests that each instance should be treated as a class denoted by a feature vector. [26] suggests storing the

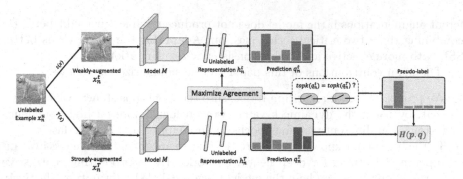

Fig. 1. The framework of AdaptMatch for semi-supervised learning.

instance class representation vector in a memory bank, a technique that has been adopted and expanded in a number of subsequent studies [11,25]. Additionally, other research utilizes the in-batch samples for negative sampling rather than a memory bank [29]. Recent literature has tried to connect the effectiveness of their techniques to maximize mutual information between latent representations [12,13,20]. Our AdaptMatch uses contrastive learning to find more of the same information from the same class of different versions when the model cannot find high-confidence labels.

Unlike previous work, this paper introduces a novel framework for general semisupervised learning, in which we intend to use the combination of pseudo-labeling and contrastive learning to develop more targeted strategies.

3   Our Approach: AdaptMatch

In this section, we introduce AdaptMatch shown in Fig. 1, our proposed semi-supervised learning approach. AdaptMatch is a "holistic" method that incorporates ideas and components from consistency regularization, pseudo-labeling, and contrastive learning. We also use a separate strong and weak augmentation for unlabeled data. We first set up the problem of semi-supervised learning. For a L class classification problem, there is a training dataset B, let a batch of labeled data $X = \{(x^s_m, y^s_m) : m \in (1, ..., M)\}$, where $x^s_m \in \mathbb{R}^n$ are labeled examples, $y^s_m \in \{1, ..., L\}$ are the corresponding one-hot labels, and s denotes "supervised" learning. In addition to the labeled batch X, there is a batch of unlabeled data $U = \{x^u_n : n \in (1, ..., N)\}$, where u denotes "unsupervised" learning. We define $\eta = \frac{M}{N}$ as a hyperparameter that specifies the relative sizes of X and U in the training dataset B. Our goal is to learn a classifier $f :\rightarrow \{1, ..., L\}$ that generalizes well from labeled data to unlabeled data.

3.1   Data Augmentation

As a common practice in various SSL methods, we use labeled data with weak augmentation and unlabeled data with weak and strong augmentation. Weak augmentation is used to randomly flip images horizontally with a probability of 50% and randomly crop images with the padding mode of reflect except SVHN. For strong augmentation, we use RandAugment [7], which is inspired by AutoAugment [6], which is a simple procedure to automatically provide more data augmentation policies. AutoAugment employs a reinforcement learning search algorithm to combine all image processing transformations available in the Python Image Library (PIL) in order to determine an optimal augmentation approach. However, RandAugment uniformly samples from the same set of augmentation transformations in PIL in fixed global probabilities and magnitudes without using search. Compared to AutoAugment, RandAugment is simpler and requires no labeled data as there is no need to search for optimal policies, which greatly reduces the sample space generated by data augmentation, thereby integrating the process of data augmentation and the training process of deep learning models.

3.2   Top-k Label Guessing

For each unlabeled sample in U, AdaptMatch produces a "guess" process for the unlabeled data's pseudo-label using a Top-k comparison, which relies on a dynamic way to estimate the model's learning status based on the test dataset. As is believed in current SSL methods, a high threshold that filters out noisy pseudo-labels and leaves only high-quality ones can considerably reduce the confirmation bias. However, not-so-high thresholds can also provide high-confidence labels as the performance of the model gradually improves. As a result, an unfixed k value may support the model in learning more high-confidence labels more efficiently.

In this work, the same unlabeled example x_n^u undergoes weak augmentation t and strong augmentation T will generate two versions of the input x_n^t, x_n^T into the same Convolutional Neural Network (CNN), which will later give two kinds of representation h_n^t, h_n^T, where

$$h_n^t = p_{model}(y|t(x_n^u)) \tag{1}$$

$$h_n^T = p_{model}(y|T(x_n^u)) \tag{2}$$

AdaptMatch computes an artificial label for each unlabeled example which is then used in a standard CrossEntropy loss H. The model generates predicted class distribution for a weakly-augmented version of a given unlabeled image:

$$q_n^t = Softmax(h_n^t) \tag{3}$$

Similarly, the model also provides prediction distribution of the same input with strong augmentation:

$$q_n^T = Softmax(h_n^T) \tag{4}$$

Then, we compare $topk(q_n^t, k)$ and $topk(q_n^T, k)$. To obtain high accurate artificial labels, AdaptMatch will set a larger value of k throughout the initial stages of training. If $topk(q_n^t, k) = topk(q_n^T, k)$, we will use

$$\hat{q}_n = argmax(q_n^t) \tag{5}$$

as a pseudo-label of the input data with strong augmentation. The final stage is to enforce the CrossEntropy loss against the model output for a strongly augmented version of x_n^u:

$$l^{ue} = \frac{1}{N} \sum_{n=1}^{N} \mathbb{1}(topk(q_n^t, k) == topk(q_n^T, k))H(\hat{q}_n, q_n^t) \tag{6}$$

This loss is minimized by AdaptMatch is simply λl^{ue} where λ is a fixed scalar hyperparameter representing the relative weight of the pseudo-labeling loss.

While current SSL algorithms generate pseudo-labels for only high-confidence unlabeled data cut off by a predefined and fixed threshold, AdaptMatch generates the pseudo-labels based on dynamic rules and at different time steps. This method is accomplished by adjusting the k values to alter the rules in response to the model's learning status.

However, it is non-trivial to determine the rule according to the learning status dynamically. The ideal approach would be to decrease the value of k to obtain more artificial labels with the improvement of the model performance, so Γ is used to indicate the reliability of the model, which can be summarized as:

$$\Gamma(M) = \mathbb{1}(Acc^{td} \geq \tau) \tag{7}$$

where τ is a scalar hyperparameter, denoting accuracy threshold, and Acc^{td} represents the model M's performance on the test dataset (td). In this way, lower accuracy that indicates a less satisfactory learning status of the model will lead to a more strict k value that encourages more trustworthy labels to be learned, while higher accuracy represents that models can learn more reliable labels from a not very large value of k.

3.3 Contrastive Learning

In this part, we introduce a new form of consistency regularization, which is crucial to AdaptMatch's success, by which we intend to tackle the problem of over-fitting the insufficient labeled target samples. Motivated by the fact that unlabeled samples contain potential information about the data structure, we use unlabeled target samples to assist labeled samples in learning representations with a higher capacity for generalization. Unlike Top-k pseudo-labeling, which incorporates the alignment at the distribution level, the representation consistency affects training at the sample level. Specifically, we use the classical contrastive loss to measure the distance between the representations of two versions of unlabeled data.

At the early stage of training, the model cannot find highly reliable pseudo-labels for backpropagation most of the time, that is, when $topk(q_n^t, k) \neq topk(q_n^T, k)$. AdaptMatch learns representations by maximizing agreement between differently augmented views of the same example via a contrastive loss in the latent space.

For each unlabeled sample in U, the x_n^t and x_n^T obtained after data augmentation should be very similar, so the output h_n^t and h_n^T after model M should also be similar to each other. We use the Euclidean distance here to express the similarity, which has been generalized to abstract metric spaces. As shown in Fig. 1, the more similar the feature values of h_n^t and h_n^T, the closer they are in the space. Define the parameterized distance function D to be learned between x_n^t and x_n^T as the Euclidean distance between the outputs of the model M. That is,

$$D(x_n^t, x_n^T) = \parallel h_n^t - h_n^T \parallel_2 \tag{8}$$

To shorten notation, $D(x_n^t, x_n^T)$ is written D.

Due to the particularity of our method, the model is only generated in positive pairs without negative pairs, then the loss function in its most general form is

$$l^{uc} = \frac{1}{2N} \sum_{n=1}^{N} (1 - \mathbb{1}(topk(q_n^t) == topk(q_n^T)))D^2 \tag{9}$$

The loss that AdaptMatch minimizes is μl^{uc}, where μ is a fixed scalar hyperparameter indicating the contrastive loss's relative weight.

Note that the sample selection result is adaptively changing as the target model progressively fits more training examples. Considering that the number of selected unlabeled samples with pseudo-labels in the earlier training may not be sufficient for model training, we impose a buffer of contrastive learning to assist the model in acquiring feature information from unlabeled data. This buffer enables us to calculate contrastive loss with more unlabeled input, which is helpful for rapidly improving the performance of the model.

3.4   Summarization of the Framework

We finally present the process of AdaptMatch in the Supplementary Material. The Cross-Entropy loss will be used for supervised learning in the labeled batch. We applied weak augmentation and strong augmentation on the unlabeled data, and these processed data will be sent to the same CNN model used in supervised training. For each unlabeled sample, the model predicts the same first k classes among the prediction distributions of the example with weakly-augmented and with strongly-augmented, we will use the CrossEntropy loss to compare the output from the model using strong augmentation with the class of the highest probability. Otherwise, we will use the contrastive loss to maximize the agreement between feature values from the outputs of the same CNN with different augmentations. Therefore, the total loss L can be summarized as follows:

$$L = l^s + \lambda l^{ue} + \mu l^{uc} \tag{10}$$

where λ and μ are weighted factors for semi-supervised consistency loss.

4 Experiments

This section performs an empirical analysis to demonstrate the effectiveness and benefits of our SSL algorithm over other state-of-the-art approaches. The first part introduces the used datasets and experimental setup in the following tasks. The second part evaluates the efficacy of AdaptMatch on various standards of image classification of SSL following standard SSL evaluation protocols [3,4,19]. In the supplementary material, we conduct experiments with fewer labeled data than previously considered, indicating that AdaptMatch shows better generalization in extremely label-scarce settings. In addition, we perform experiments on imbalanced datasets, such as long-tail datasets, since AdaptMatch shows a better capability to minimize label bias. Results on large-scale datasets will also be shown in the supplementary material, such as STL-10 and ImageNet.

4.1 Datasets and Experimental Setup

Datasets. We use several widely used benchmark SSL datasets to evaluate AdaptMatch performance, including the well-known datasets CIFAR-10 [14], CIFAR-100 [14], and SVHN [18]. For CIFAR-10, CIFAR-100 and SVHN, we apply 4 labeled examples per class, 25 labeled examples per class, and 100 examples per class in our comparison experiments. In addition, we assess AdaptMatch on ImageNet, which is shown in the supplementary material, to verify whether it works well on a larger and more complicated dataset. Following [54], we label 10% of the training data and consider the remainder as unlabeled examples. For research of imbalanced learning, we evaluate the ability of AdaptMatch to learn from imbalanced datasets of long-tailed CIFAR10 (CIFAR10-LT) and long-tailed CIFAR100 (CIFAR100-LT) [5].

Experimental Setup. To provide a fair comparison environment, we produce labeled and unlabeled data divisions using the same settings and random seeds as in FixMatch. We use a Wide ResNet-28-2 [30] for CIFAR-10 and SVHN, and Wide ResNet-28-8 for CIFAR-100, and train it for 1024 epochs using SGD with momentum factor $\beta = 0.9$ [23]. For a learning rate schedule, we use a cosine learning rate decay which sets the learning rate at $\theta\cos\frac{7\pi i}{16I}$ where θ is the initial learning rate, i is the current training step and I is the total number of training steps. As introduced in Sect. 3.1, we apply RandAugment for strong augmentation. For our method, we set $\lambda = 1$ and $\mu = 0.6$ by default and investigate the effect of λ and μ in Sect. 5. To maintain a comparable training calculation cost to FixMatch, we set the default number $n = 1$ of augmentations per image.

For imbalanced training in the CIFAR10-LT and CIFAR100-LT datasets, training images are randomly selected for each class to maintain a predefined imbalance rate γ. The number of training data in B of class L is denoted as B_l, i.e., $\sum_{l=1}^{L} B_l = B$ We assume that the classes are arranged in descending order by cardinality, i.e., $B_1 \geq B_2 \geq \cdots \geq B_L$. Therefore, we use the imbalance ratio to measure the degree of class imbalance, $\gamma = \frac{B_1}{B_L}$, and the label fraction is used to measure the percentage of labeled data, $\beta = \frac{M}{M+N}$. The results are shown in the supplementary material.

Table 1. Test accuracy for CIFAR-10, CIFAR-100 and SVHN on 5 different folds. All baseline models (Π-Model [21], Pseudo-Labeling [16], Mean Teacher [24], MixMatch [4], UDA [27], and ReMixMatch [3], FixMatch [22]) are tested using the same codebase.

Method	CIFAR-10		CIFAR-100		SVHN	
	250 labels	4000 labels	2500 labels	10000 labels	250 labels	1000 labels
Fully-Supervised	95.55		80.93		97.86	
Π-Model	52.39	85.99	41.49	62.12	81.04	92.46
Pseudo-Labeling	52.51	84.76	40.38	63.79	79.79	90.06
Mean Teacher	64.68	90.40	46.09	64.17	96.43	96.58
MixMatch	88.95	93.58	60.06	71.69	96.02	96.50
UDA	91.12	94.18	66.87	75.50	94.31	97.54
ReMixMatch	94.56	95.28	**72.57**	76.97	**97.08**	97.35
FixMatch	**94.93**	**95.74**	71.71	**77.40**	96.04	**97.52**
AdaptMatch	**95.28**	**96.05**	**72.60**	**78.36**	**97.21**	97.50

4.2 Main Results

We summarize the performance of all baselines along with AdaptMatch in Table 1. We compute the average accuracy for fully-supervised learning and different models with the same training times. When the labeled data is less, the performance of Π-Model, Mean Teacher, and Pseudo-Labeling were poor. While MixMatch, ReMixMatch, and UDA perform reasonably well with 250 or more labels, we show that AdaptMatch provides outstanding performance while maintaining a simple design.

As a reference, we train fully-supervised baselines, which entails training the models with all available training labels. The test accuracy is 95.55%, 80.93%, and 97.86% on datasets of CIFAR-10, CIFAR-100, and SVHN, respectively.

CIFAR-10 Results. The results of AdaptMatch and baseline methods on the CIFAR-10 dataset are listed in Table 1. By utilizing Top-k pseudo-labeling and contrastive loss in AdaptMatch, the method of AdaptMatch achieves the best performance in both cases of 250 and 4000 labeled samples. For example, Adapt-Match reached an average accuracy of 95.28% on CIFAR-10 with 25 labels per class. As a point of reference, the highest accuracy achieved on CIFAR-10 with 400 labels per class is 96.05% among the methods using the same network architecture. Our results outperform not only the best FixMatch, but also ReMix-Match when self-supervised loss is used.

CIFAR-100 Results. For CIFAR-100, ReMixMatch performs better than Fix-Match due to the use of Distribution Alignment (DA), which encourages the model predictions to have the same class distribution as the labeled set. Although we do not use DA, we still exceed all algorithm results when using the same CNN architecture. For instance, AdaptMatch achieved an average accuracy of 72.60% with 25 labels per class on CIFAR-100 and improved by nearly 1% on CIFAR-100 with 100 labels per class. As can be observed, the addition of contrastive loss enhances the model's capacity to learn features from unlabeled data and increases the model's generalizability.

SVHN Results. For SVHN, AdaptMatch once again achieved results comparable to state-of-the-art methods with 25 labeled data or more per class. These experiments illustrate that AdaptMatch achieved a best average accuracy of 97.21% on SVHN with 250 labeled data. Although our results did not reach the top one in the scenario of 1000 labeled samples, the difference between our result and the first MixMatch result is only 0.02%. All experimental-labeled data are randomly selected, and minor differences in results are within the acceptable range.

5   Ablation Study

Since AdaptMatch comprises a simple combination of three existing techniques, we performed an ablation study to better understand why it is able to obtain state-of-the-art results. The previous FixMatch and other algorithms that used high-confidence labels set a fixed scalar hyperparameter before the pseudo-label to 1, so we also set λ to 1 here. Due to the number of experiments in our ablation study, we focus on studying with a single 4000 label split from CIFAR-10 and only report results for the chosen of μ shown in Table 2. When contrastive loss is added to semi-supervised learning, test accuracy has reached a very good level, no matter how much we choose μ equals. Note that AdaptMatch with $\mu = 0.6$ achieves the best accuracy, so we use the best μ in our algorithm.

Table 2. Test accuracy for CIFAR-10 under different values of the scalar hyperparameter μ.

μ	0.1	0.2	0.3	0.4	0.5	0.6	0.7	0.8	0.9	1.0
Acc	95.94	95.94	95.78	95.79	95.93	**96.05**	95.77	95.83	95.75	95.80

6   Conclusion

SSL has made a remarkable advance in recent years. Regrettably, most of the gain has been made at the expense of increasingly complicated learning algorithms [9] with sophisticated loss terms or transfer learning models [1]. Adapt-Match provides state-of-the-art performance over a wide variety of datasets, which combines Top-k pseudo-labeling and the contrastive loss with supervised learning to provide a simple, effective and stable SSL method. Our adaptive consistency regularization method could be further combined with unsupervised algorithms, transfer learning, or flexible algorithms that adjust the threshold of each class [31]. On the whole, we believe that this simple, yet performant semi-supervised learning method will enable deep learning to be applied in an expanding number of practical domains where labels are either prohibitively costly or difficult to obtain.

References

1. Abuduweili, A., Li, X., Shi, H., Xu, C.Z., Dou, D.: Adaptive consistency regularization for semi-supervised transfer learning. In: Proceedings of the IEEE/CVF Conference on Computer Vision and Pattern Recognition, pp. 6923–6932 (2021)
2. Bachman, P., Alsharif, O., Precup, D.: Learning with pseudo-ensembles. In: Advance in Neural Information Processing System, vol. 27, pp. 3365–3373 (2014)
3. Berthelot, D., et al.: Remixmatch: semi-supervised learning with distribution matching and augmentation anchoring. In: International Conference on Learning Representations (2019)
4. Berthelot, D., Carlini, N., Goodfellow, I., Papernot, N., Oliver, A., Raffel, C.A.: Mixmatch: a holistic approach to semi-supervised learning. In: Advances in Neural Information Processing Systems, vol. 32 (2019)
5. Cao, K., Wei, C., Gaidon, A., Arechiga, N., Ma, T.: Learning imbalanced datasets with label-distribution-aware margin loss. In: Advance in Neural Information Processing System, vol. 32, pp.p 1567–1578 (2019)
6. Cubuk, E.D., Zoph, B., Mane, D., Vasudevan, V., Le, Q.V.: Autoaugment: learning augmentation strategies from data. In: Proceedings of the IEEE/CVF Conference on Computer Vision and Pattern Recognition, pp. 113–123 (2019)
7. Cubuk, E.D., Zoph, B., Shlens, J., Le, Q.V.: Randaugment: practical automated data augmentation with a reduced search space. In: Proceedings of the IEEE/CVF Conference on Computer Vision and Pattern Recognition Workshops, pp. 702–703 (2020)
8. Dosovitskiy, A., Springenberg, J.T., Riedmiller, M., Brox, T.: Discriminative unsupervised feature learning with convolutional neural networks. In: Advances in Neural Information Processing Systems, vol. 27 (2014)
9. Gong, C., Wang, D., Liu, Q.: Alphamatch: improving consistency for semi-supervised learning with alpha-divergence. In: Proceedings of the IEEE/CVF Conference on Computer Vision and Pattern Recognition, pp. 13683–13692 (2021)
10. Hadsell, R., Chopra, S., LeCun, Y.: Dimensionality reduction by learning an invariant mapping. In: 2006 IEEE Computer Society Conference on Computer Vision and Pattern Recognition (CVPR 2006), vol. 2, pp. 1735–1742. IEEE (2006)
11. He, K., Fan, H., Wu, Y., Xie, S., Girshick, R.: Momentum contrast for unsupervised visual representation learning. In: Proceedings of the IEEE/CVF Conference on Computer Vision and Pattern Recognition, pp. 9729–9738 (2020)
12. Henaff, O.: Data-efficient image recognition with contrastive predictive coding. In: International Conference on Machine Learning, pp. 4182–4192. PMLR (2020)
13. Hjelm, R.D., et al.: Learning deep representations by mutual information estimation and maximization. In: International Conference on Learning Representations (2019)

14. Krizhevsky, A., Hinton, G., et al.: Learning multiple layers of features from tiny images (2009)
15. Laine, S., Aila, T.: Temporal ensembling for semi-supervised learning. arXiv preprint arXiv:1610.02242 (2016)
16. Lee, D.H., et al.: Pseudo-label: the simple and efficient semi-supervised learning method for deep neural networks. In: Workshop on Challenges in Representation Learning, ICML, vol. 3, p. 896 (2013)
17. Mahajan, D., et al.: Exploring the limits of weakly supervised pretraining. In: Proceedings of the European Conference on Computer Vision (ECCV), pp. 181–196 (2018)
18. Netzer, Y., Wang, T., Coates, A., Bissacco, A., Wu, B., Ng, A.Y.: Reading digits in natural images with unsupervised feature learning (2011). https://api.semanticscholar.org/CorpusID:16852518
19. Oliver, A., Odena, A., Raffel, C.A., Cubuk, E.D., Goodfellow, I.J.: Realistic evaluation of deep semi-supervised learning algorithms. In: NeurIPS (2018)
20. Oord, A.V.D., Li, Y., Vinyals, O.: Representation learning with contrastive predictive coding. arXiv preprint arXiv:1807.03748 (2018)
21. Rosenberg, C., Hebert, M., Schneiderman, H.: Semi-supervised self-training of object detection models (2005)
22. Sohn, K., Berthelot, D., Carlini, N., Zhang, Z., Zhang, H., Raffel, C.A., Cubuk, E.D., Kurakin, A., Li, C.L.: Fixmatch: simplifying semi-supervised learning with consistency and confidence. In: Advance in Neural Information Processing System, vol. 33, pp. 596–608 (2020)
23. Sutskever, I., Martens, J., Dahl, G., Hinton, G.: On the importance of initialization and momentum in deep learning. In: International Conference on Machine Learning, pp. 1139–1147. PMLR (2013)
24. Tarvainen, A., Valpola, H.: Mean teachers are better role models: weight-averaged consistency targets improve semi-supervised deep learning results. In: Advances in Neural Information Processing Systems, vol. 30 (2017)
25. Tian, Y., Krishnan, D., Isola, P.: Contrastive multiview coding. In: Vedaldi, A., Bischof, H., Brox, T., Frahm, J.-M. (eds.) ECCV 2020 Part XI. LNCS, vol. 12356, pp. 776–794. Springer, Cham (2020). https://doi.org/10.1007/978-3-030-58621-8_45
26. Wu, Z., Xiong, Y., Yu, S.X., Lin, D.: Unsupervised feature learning via nonparametric instance discrimination. In: Proceedings of the IEEE Conference on Computer Vision and Pattern Recognition, pp. 3733–3742 (2018)
27. Xie, Q., Dai, Z., Hovy, E., Luong, T., Le, Q.: Unsupervised data augmentation for consistency training. In: Advance in Neural Information Processing System, vol. 33, pp. 6256–6268 (2020)
28. Xie, Q., Luong, M.T., Hovy, E., Le, Q.V.: Self-training with noisy student improves imagenet classification. In: Proceedings of the IEEE/CVF Conference on Computer Vision and Pattern Recognition, pp. 10687–10698 (2020)
29. Ye, M., Zhang, X., Yuen, P.C., Chang, S.F.: Unsupervised embedding learning via invariant and spreading instance feature. In: Proceedings of the IEEE/CVF Conference on Computer Vision and Pattern Recognition, pp. 6210–6219 (2019)
30. Zagoruyko, S., Komodakis, N.: Wide residual networks. In: British Machine Vision Conference 2016. British Machine Vision Association (2016)
31. Zhang, B., et al.: Flexmatch: boosting semi-supervised learning with curriculum pseudo labeling. arXiv preprint arXiv:2110.08263 (2021)

Estimation of Unmasked Face Images Based on Voice and 3DMM

Tetsumaru Akatsuka^(✉), Ryohei Orihara, Yuichi Sei,
Yasuyuki Tahara, and Akihiko Ohsuga

The University of Electro-Communications, Tokyo, Japan
akatsuka.tetsumaru@ohsuga.lab.uec.ac.jp, orihara@acm.org,
{seiuny,tahara,ohsuga}@uec.ac.jp

Abstract. Facemasks have become common due to the COVID-19 pandemic. They have begun to affect security and identification systems because they cover almost half of the face. Current state-of-the-art methods have been applied to estimate unmasked faces from masked face images. They are successful in improving the quality of the face texture by 3D Morphable Model (3DMM) as intermediate representations. However, their performance in restoring the face shapes is insufficient, and some of generated faces lack identities. In this study, we focus on voice, which has a particularly high correlation with the shape of the mouth and nose, which are obscured by masks. We propose a multimodal method using 3DMM and voice for face shape estimation under masks. Experimental results show that the proposed method qualitatively and quantitatively improves the quality of shape restoration of a face compared to the baseline method without considering voice.

Keywords: Mask Removal · Face Inpainting · 3DMM · Voice Embedding · Multimodal

1 Introduction

With the outbreak of COVID-19, facemasks became common, and many people still wear them in Japan even now that the requirement has been lifted. Given this experience, people in many nations may continue the habit of wearing masks during future pandemics. Wearing masks is very important to prevent the spread of the infection, however they have the problem of hiding a part of the faces. One of the negative effects is the impact on security and identification systems [3, 11].

With this background in mind, research aiming at highly accurate reconstruction by removing masks from masked face images has been attracting attention, especially in the field of inpainting tasks in deep learning. Among them, the method of Yin et al. [27], which uses a 3D mesh called 3DMM as an intermediate representation, was shown to outperform other methods both qualitatively and quantitatively. It is also superior in that the parameters of the 3DMM are controllable and the generated results can be edited later. However, the performance of the method is insufficient in terms of face shape restoration, and some of generated faces suffered from impaired identities.

Therefore, in this study, we focus on the inpainting of face shape and consider the use of voice, which is physiologically correlated with face shape.

In the study by Wu et al. [26], which investigated the relationship between face shape and voice, they conducted experiments in the Speech2Face task, in particular by shifting the reconstruction target from a 2D face image to 3DMM (without texture) for the first time, and reported that there was a rough relationship between the voice and the shape of 3DMM.

Based on these studies, we hypothesize that the use of the person's voice as additional information can recover the face shape more accurately in the regions covered by the mask, and propose a multimodal method using voice (Fig. 1.)

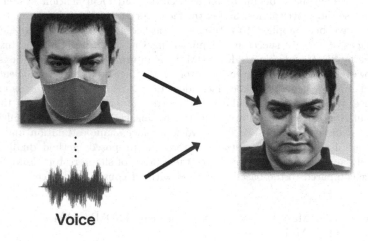

Voice

Fig. 1. Goal of our research. Estimation and generation of unmasked face images from voice and masked face images.

The main contributions of this study are as follows.

- We proposed a multimodal method for estimating unmasked face images from masked face images using voice and 3DMM.
- In order to improve the reconstruction quality of 3DMM, we designed a new loss function that employs the Skin Attention Mask.
- By considering voice features, we confirmed that the proposed method improves the quality both qualitatively and quantitatively compared to the method of Yin et al.

As an example of application of this research, it is expected to be used in the security aspect. For example, suppose a security camera has video and voice recordings of a crime scene and the criminal wears a mask. It may be possible to create a montage of the criminal's maskless face, which may lead to quick identification of the criminal. In addition, it is an interesting topic to be able to estimate the maskless face of an acquaintance whose face with mask only is known by people.

This paper is organized as follows. Section 2 introduces related works, Sect. 3 describes the details of the proposed method, Sect. 4 describes experiments and evaluations using the proposed method, Sect. 5 discusses the results, and finally Sect. 6 summarizes and discusses future prospects.

2   Related Research

2.1   Studies on Mask Removal

Among the inpainting tasks for natural completion of missing regions in images, especially since the COVID-19 outbreak, there has been an increase in the number of research cases that specialize in removing masks and completing faces from masked face images. Such face completion is more difficult than inpainting of landscape photographs, or abstract paintings, because the face has the unique structures. In addition, since the slightest difference can cause a large perceptual discrepancy, accurate inpainting is required.

Din et al. [9] employed a GAN (Generative Adversarial Networks)-based network that employs two types of discriminators, one that monitors global consistency and the other that monitors missing parts in more detail, in stages on the reconstructed face images. They were successful in generating realistic and plausible faces. In a follow-up work, Hosen et al. [13] adopted the Residual Attention U-Net structure as a generator and dealt with distorted facial structures and color mismatch.

Despite the improved inpainting quality, the above methods produce only deterministic results. Yin et al. [27] proposed a face reconstruction architecture with 3DMM as an intermediate representation to address this problem. Using the 3DMM reconstruction results obtained from masked face images as a precondition, they reconstructed face images and were successful in recovering facial textures. Moreover, by changing the parameters of the 3DMM, it is possible to control the shape and facial expression of the reconstructed face. Furthermore, the quality of the reconstructed results by this method is qualitatively and quantitatively superior to that of the conventional reconstruction method based only on 2D images. Our study is based on the architecture proposed in this method.

2.2   Studies on Estimating Facial Shape from Voice

Voice contains many physiological characteristics that influence physical features. For example, voice is produced by articulatory structures such as vocal cords, facial muscles, facial skeleton, and nasal cavity, all of which are closely connected [6,12]. In addition, intonation are related to age, gender, and ethnicity (regionality). These facts indicate that there may be a correlation between voice and facial shape. Several attempts have been made to inpaint face images from voice alone based on this speculation.

Wen et al. [25] employed GAN trained on a large amount of paired data consisting of face images and short voice clips, aiming to generate face images from voice embeddings extracted from the voice clips. In addition, Koseki et al. [17] and Oh et al. [21] employ an encoder-decoder structure, and aim to eventually generate face images from voice alone by training the voice embedding to be close to the features obtained from face images. However, it is difficult to accurately predict the face images because these 2D face images contain elements such as pose, hair style, beard, skin texture and background that are unrelated to the face shape and the voice features. Even if the images are of the same person, these factors may cause unstable identities in the face reconstruction results. Therefore, Wu et al. [26] focus on the prediction of 3DMM with a more physiological basis quantitatively. Specifically, they have created a large paired dataset of 3DMMs reconstructed from face images and voice. They also proposed a new model to reconstruct the embeddings extracted from voice on the parameters of the 3DMMs. According to their experiments, the results is possible to roughly reconstruct the 3D shape of a face from its voice. In this study, we focus on this fact and consider using the embedding extracted from the voice as additional information to the architecture of [27].

2.3   3D Morphable Model (3DMM)

Various models of 3DMM have been published [4,5,10,14,18]. The most widely used one is the Basel Face Model (BFM) [23]. BFM is a 3D face modeling method using Principal Component Analysis (PCA). A 3D face can be constructed by [8] from a 257-dimensional vector of $c_i \in \mathbb{R}^{80}, c_e \in \mathbb{R}^{64}, c_t \in \mathbb{R}^{80}, p \in \mathbb{R}^6,$ and $l \in \mathbb{R}^{27}$, the coefficients representing identity, expression, texture, pose, and illumination, respectively. In this study, we use the BFM as a 3DMM model as well as [27].

3   Proposed Method

3.1   Overview of the Proposal Method

In this section, we propose a method that takes a masked face image and voice as input. It produces a face image with the mask removed. Figure 2 shows a schematic diagram of the proposed model. The model is based on the architecture employed by [27]. We modified the architecture to combine 64-dimensional embeddings extracted from voice with intermediate features of the network and to consider voice embeddings in the 3D face shape represented by 3DMM estimation. The proposed model consists of three modules: (1) a speech encoder that extracts 64-dimensional embeddings from speech, (2) a multitasking module that takes the masked face image and speech embedding information as input and performs mask region prediction in the lower layer of the network and overall 3DMM reconstruction in the upper layer, (3) a generator, which produces a maskless face images, whose inputs are (a) the predicted mask region, (b) the face image with the mask regions replaced by random noise and (c) the reconstructed 3DMM. During the training, the parameters of the layer before the one where the voice embedding and intermediate features are combined, and of the Generator, are fixed at the state of the pre-trained model of [27]. The training is concentrated on the 3DMM predictions by the voice. Section 3.2 describes the extraction of voice embeddings, Sect. 3.3 describes the how to combine the voice embeddings and the intermediate features, and Sect. 3.3 details the training of the multitasking module.

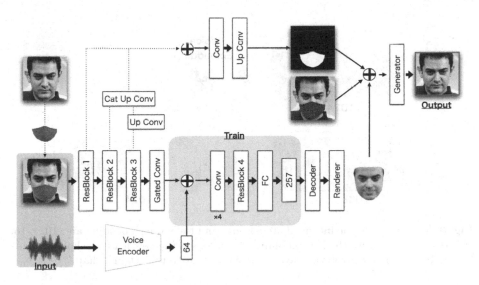

Fig. 2. Schematic diagram of the proposed model. The 3DMM is reconstructed from the input masked face image and voice. Then, the unmasked face image is inpainted by the Generator.

3.2   Extraction of Voice Embedding

The processed voice data provided in [25] is given as input to the trained voice encoder used in the supervised learning framework in [26], and 64-dimensional voice embeddings are extracted. The processed voice data (log mel-spectrograms) is prepared as follows; Speech-bearing regions of the recordings are isolated using a voice activity detector interface from the WebRTC project. Subsequently, log mel-spectrograms are extracted using an analysis window of 25 ms, with a hop of 10 ms between frames.

3.3   Combining Voice Embedding and Intermediate Features

The reconstruction of the 3DMM is performed by a multitasking module that takes the face images and the voice embeddings described in Sect. 3.1 as input.

The voice embedding and the intermediate features are combined by duplicating the voice embedding to produce a new feature. The voice embedding is represented by a 64-dimensional vector. The intermediate features are represented by a tensor whose dimensionality is (1024, 16, 16). It is produced by the gated convolution layer. The resulted new feature is represented by a tensor whose dimensionality is (1088, 16, 16), as shown in Fig. 3. After that, it is passed through the convolution layer, which consists of four layers, and the shape of the feature is returned to (1024, 16, 16). Then, it is merged back into the original pipeline.

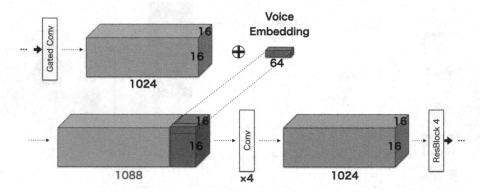

Fig. 3. The shapes of the intermediate features and the voice embedding are (1024, 16, 16), (64, 1, 1) respectively. By combining them, we get the new feature whose shape is (1088, 16, 16). The convolution layer then converts it to the original shape.

3.4   Training of Multitasking Module

Coefficient Loss. The multitasking module produces 257-dimensional coefficients $\hat{c}$ including identity, expression, texture, pose and illumination parameters that constitute the 3DMM. For this output, the 3D coefficients c predicted from the original unmasked face images by the pre-trained model of [8] are used as a ground truth, and the loss is calculated directly from the coefficients as in Eq. (1).

$$L_{coef} = \frac{1}{257}||\hat{c} - c||_1 \tag{1}$$

Photo Loss. Although the above coefficient-level loss treats all dimensional discrepancies equally, it is unreasonable because some dimensions affect the reconstruction results much more than others (e.g., poses v.s. illuminations). Therefore, as in [27], we introduce a photo loss (Eq. (3)) that compares the inpainted result I_{3D} of the rendered 3D face with the unmasked face image I that is the ground truth at the image level. The difference is that, in addition to the cropping of the face region to ignore the background, we also introduce the Skin Attention algorithm [8], as shown in Fig. 4. Skin Attention calculates the skin color probability P_i for each pixel i, then calculates the probability of skin color P_i for each pixel i as follows.

$$A_i = \begin{cases} 1 & \text{if } P_i > 0.5 \\ P_i & otherwise \end{cases} \tag{2}$$

We identified the region M of a face from an unmasked face image. In order to compare the images ignoring the effects of noise such as beard and heavy makeup, we introduce the Skin Attention Mask (M_{skin}), which is the element-wise product of A and M. The photo loss is defined as Eq. (3) using each pixel $M_{skin i}$, $I_{3D i}$ and I_i.

$$L_{photo} = \frac{\sum_{i \subset M} M_{skin i} \odot ||I_{3D i} - I_i||_2}{\sum_{i \in M} M_{skin i}} \tag{3}$$

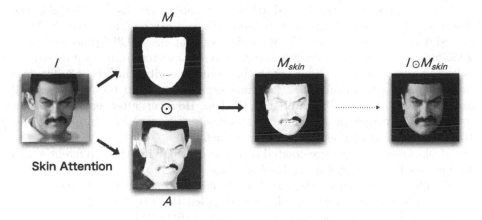

Fig. 4. Skin Attention Mask. $\odot$ represents element-wise product.

Perceptual Loss. Furthermore, we introduced the loss of perceptual level as in Eq. (4). Using the pre-trained Arcface [7] model, we perform the feature extraction operation $F(\cdot)$ on I_{3D} and the unmasked face image I, and calculate the cosine distance.

$$L_{per} = 1 - \frac{F(I_{3D})F(I)}{||F(I_{3D})|| \cdot ||F(I)||} \tag{4}$$

Landmark Loss. Finally, we loosely constrain the shape and pose of the inpainted face by a landmark loss [8] as Eq. (5) where $\hat{q}_i$ and q_i denotes 68 facial landmarks extracted from the predicted 3DMM and the 3DMM reconstructed from the unmasked face images respectively. Also, ω_i is the weight corresponding to the i-th landmark, set to 20 for the points in the nose and mouth, and 1 for the others.

$$L_{lm} = \frac{1}{68} \sum_{i=1}^{68} \omega_i ||\hat{q}_i - q_i||^2 \tag{5}$$

The overall loss function is as follows.

$$L_{3D} = L_{coef} + L_{photo} + \lambda_{per} L_{per} + \lambda_{lm} L_{lm} \tag{6}$$

$$\text{where} \quad \lambda_{per} = 0.1, \; \lambda_{lm} = 0.001$$

4 Experiments and Results

4.1 Dataset

In order to train and test the proposed model, we need a large dataset cosists of masked face images, unmasked face images (ground truth) and voices of the same person.

For the voice and face image dataset, we extended one used in the [25]. The voices and face images have corresponding IDs and can be intersected to obtain paired data for 1225 people. The voice dataset contains 149,354 processed data, which are 64-dimensional log-mel spectrograms converted from voice recordings collected from VoxCeleb [20] using the procedure described in Sect. 3.2. The face image dataset contains 139,572 face images collected from the VGGFace [22] and cropped to 128 × 128, centered on the face. However, after removing images with multiple people in the frame and portraits that were not suitable for this training, the final number of images was 136,698.

Masked face images were created by randomly applying three different masks to unmasked face images using MaskTheFace [2].

Finally, in order to fit the face images and the masked face images to the input size of the proposed model, we upsampled them to 256 × 256 resolution using an open source super-resolution system waifu2x [19].

The generated data were split for train/validation/test as shown in table 1. The split method is based on [20], where individuals whose names begin with A and B are assigned to the validation, those whose names begin with C, D, and E are assigned to the test, and the rest are assigned to the train.

Table 1. Splitting of the dataset

	train	validation	test	total
Face & masked face	104,469	12,239	19,990	136,698
Voice	113,322	14,182	21,850	149,354
No. of person IDs	924	112	189	1,225

4.2   Training Details

Adam with exponential decay rates for the moment estimates $\beta 1 = 0.9$, $\beta 2 = 0.999$ was used to optimize the proposed network. The learning rate was set to 1×10^{-4}, batch size to 10, and 500K steps of training were performed on NVIDIA DGX-1. Since there are multiple face images and voice data for each person, which do not correspond one-to-one, we randomly select voice data for each batch and associate them with a pair of face images.

4.3   Qualitative Evaluation

Figure 5 shows a comparison of the results of face image inpainted by the proposed method and the method of Yin et al. when a masked face image is given as an input. In particular, focusing on the inpainted results of women with sharp contours (columns 3 and 5), we can confirm that the proposed method inpaints the contours more sharply than the method by Yin et al. In addition, as for the male subjects in the columns 2 and 6, we can confirm that the position and relative size of the nose, the mouth and the thickness of the lips are closer to the ground truth. As for the nose shape, there is a particular difference in the restoration performance between the male in column 4 and the female in column 7. The ground truth of the male has a characteristic high nasal head, and the proposed method reconstructs the shape more closely. Similarly, the ground truth of the female has a sharp nasal bridge, and the proposed method inpaints it more closely.

Fig. 5. Face images inpainted by the proposed method and Yin et al.'s when a masked face images are given as an input

Figure 6 shows a comparison of the reconstructed results of the 3DMM, which is an intermediate representation. It can be seen that the 3DMM reconstructed by the proposed method has blackened areas corresponding to sunglasses and hidden areas due to the effect of Skin Attention Mask. This is a result of the loss function of Eq. (3), in which the region that are not skin-colored (i.e., not related to the reconstructed shape) are ignored. Comparing the shapes, it can be seen that the proposed method reconstructs the identity of the ground truth more accurately.

Fig. 6. 3DMM reconstructed by the proposed method and Yin et al.'s

4.4　Quantitative Evaluation

The 19,990 masked face images for testing and the face images inpainted from them were evaluated by the metrics used in [27]. They were L1 loss, PSNR score [15], SSIM score [24], and the cosine similarity of features extracted by Arcface, and the average has been calculated. L1 loss is the absolute error per pixel with ground truth, PSNR and SSIM are objective metrics of image quality, and cosine similarity is the closeness of facial features. As shown in Table 2, the proposed method outperformed Yin et al.'s method in all indices.

Table 2. Quantitative Evaluation

model	L1($\downarrow$)	PSNR($\uparrow$)	SSIM($\uparrow$)	Arcface($\uparrow$)
Yin et al.	0.034	14.710	0.819	0.374
Ours	**0.032**	**14.994**	**0.831**	**0.401**

5 Discussion

5.1 On Qualitative Evaluation

In Sect. 4.3, we showed that the proposed method can inpaint the shape and position of parts such as contours, mouth and nose with higher accuracy than the method of Yin et al. These results suggest that the combination of voice features and 3DMM is effective in inpainting face images with higher fidelity.

5.2 On Quantitative Evaluation

In Sect. 4.3, we showed that the quality of the inpainted images by the proposed method is quantitatively superior in all evaluation metrics. These results show that the face inpainted by considering voice feature improves not only the perceptual quality but also the quantitative quality of the image.

6 Conclusion and Future Works

In this study, we proposed a new multimodal method for the task of inpainting unmasked face images from masked face images, which utilizes voice in addition to the use of 3DMM. Experimental results show that the proposed method improves the quality both qualitatively and quantitatively compared to the method without voice. The dataset used in this study is a combination of two different datasets, VGGFace and VoxCeleb, so that there is no one-to-one correspondence between face images and voices. Therefore, it cannot be guaranteed that there is a correspondence between a voice and a face image during the utterance of the voice. Furthermore, it is not able to cope with the fact that the face shape and voice quality change with age within the same person. As a countermeasure, there are methods to create paired data by directly extracting face images and voices from video clips such as those on YouTube [1, 16]. Despite we used a simple network to combine the voices with images, the accuracy of the recovery was improved. In order to further improve the accuracy, we will investigate studies that are working on more sophisticated multimodal approaches to voice and images, and aim to optimize the network. The estimation of facial expressions is out of scope of the both of Yin et al.'s and our method. Further investigation into the correspondence between voice and facial expressions will be conducted, and in conjunction with the above perspectives, reconstruction of facial images including facial expressions will be a future goal.

Acknowledgements. This work was supported by JSPS KAKENHI Grant Numbers JP21H03496, JP22K12157, JP23H03688 and SEI Group CSR Foundation.

References

1. ISCA (2018). https://doi.org/10.21437/interspeech.2018-1929
2. Anwar, A., Raychowdhury, A.: Masked face recognition for secure authentication (2020)
3. Barragan, D., Howard, J.J., Rabbitt, L.R., Sirotin, Y.B.: Covid-19 masks increase the influence of face recognition algorithm decisions on human decisions in unfamiliar face matching. PLoS ONE **17**(11), e0277625 (2022)
4. Booth, J., Roussos, A., Ponniah, A., Dunaway, D., Zafeiriou, S.: Large scale 3d morphable models. IJCV **126**(2), 233–254 (2018)
5. Cao, C., Weng, Y., Zhou, S., Tong, Y., Zhou, K.: Facewarehouse: a 3D facial expression database for visual computing. IEEE Trans. Visual Comput. Graphics **20**(3), 413–425 (2013)
6. Denes, P.B., Pinson, E.: The speech chain (1993)
7. Deng, J., Guo, J., Xue, N., Zafeiriou, S.: Arcface: additive angular margin loss for deep face recognition. In: CVPR, pp. 4690–4699 (2019)
8. Deng, Y., Yang, J., Xu, S., Chen, D., Jia, Y., Tong, X.: Accurate 3d face reconstruction with weakly-supervised learning: from single image to image set (2020)
9. Din, N.U., Javed, K., Bae, S., Yi, J.: A novel GAN-based network for unmasking of masked face. IEEE Access **8**, 44276–44287 (2020)
10. Gerig, T., et al.: Morphable face models-an open framework. In: IEEE FG, pp. 75–82 (2018)
11. Guo, Y.: Impact on biometric identification systems of COVID-19. Sci. Program. **2021**, 1–7 (2021)
12. Harrington, J.: Acoustic Phonetics. The Handbook of Phonetic Sciences, pp. 81–129 (2010)
13. Hosen, M.I., Islam, M.B.: Masked face inpainting through residual attention UNet. In: ASYU, pp. 1–5 (2022)
14. Huber, P., et al.: A multiresolution 3D morphable face model and fitting framework. In: International Conference on Computer Vision Theory and Applications, vol. 5, pp. 79–86 (2016)
15. Huynh-Thu, Q., Ghanbari, M.: Scope of validity of PSNR in image/video quality assessment. Electron. Lett. **44**(13), 800–801 (2008)
16. Khalid, H., Tariq, S., Kim, M., Woo, S.S.: FakeAVCeleb: a novel audio-video multimodal deepfake dataset (2022)
17. Koseki, K., Sei, Y., Tahara, Y., Ohsuga, A.: Generation of facial images reflecting speaker attributes and emotions based on voice input. In: ICAART (2), pp. 99–105 (2023)
18. Li, T., Bolkart, T., Black, M.J., Li, H., Romero, J.: Learning a model of facial shape and expression from 4D scans. ACM Trans. Graph. **36**(6), 194–1 (2017)
19. nagadomi: waifu2x (2022). https://github.com/nagadomi/nunif
20. Nagrani, A., Albanie, S., Zisserman, A.: Seeing voices and hearing faces: cross-modal biometric matching (2018)
21. Oh, T.H., et al.: Speech2face: learning the face behind a voice. In: CVPR (2019)
22. Parkhi, O., Vedaldi, A., Zisserman, A.: Deep face recognition. In: BMVC (2015)
23. Paysan, P., Knothe, R., Amberg, B., Romdhani, S., Vetter, T.: A 3D face model for pose and illumination invariant face recognition. In: IEEE, pp. 296–301 (2009)
24. Wang, Z., Bovik, A., Sheikh, H., Simoncelli, E.: Image quality assessment: from error visibility to structural similarity. IEEE Trans. Image Process. **13**(4), 600–612 (2004). https://doi.org/10.1109/TIP.2003.819861

25. Wen, Y., Singh, R., Raj, B.: Reconstructing faces from voices (2019)
26. Wu, C.Y., Hsu, C.C., Neumann, U.: Cross-modal perceptionist: can face geometry be gleaned from voices? In: CVPR, pp. 10452–10461 (2022)
27. Yin, X., Huang, D., Chen, L.: Non-deterministic face mask removal based on 3D priors. In: IEEE ICIP, pp. 2137–2141 (2022)

Aging Contrast: A Contrastive Learning Framework for Fish Re-identification Across Seasons and Years

Weili Shi[1]([✉]) [iD], Zhongliang Zhou[2] [iD], Benjamin H. Letcher[3] [iD],
Nathaniel Hitt[3] [iD], Yoichiro Kanno[4] [iD], Ryo Futamura[5] [iD], Osamu Kishida[7] [iD],
Kentaro Morita[6] [iD], and Sheng Li[1] [iD]

[1] University of Virginia, Charlottesville, VA, USA
{rhs2rr,shengli}@virginia.edu
[2] University of Georgia, Athens, GA, USA
Zhongliang.Zhou@uga.edu
[3] United States Geological Survey, Eastern Ecological Science Center, Kearneysville,
USA
{bletcher,nhitt}@usgs.gov
[4] Colorado State University, Fort Collins, CO, USA
yoichiro.kanno@colostate.edu
[5] Hokkaido University, Tomakomai, Hokkaido, Japan
kishida@fsc.hokudai.ac.jp
[6] The University of Tokyo, Kashiwa, Chiba, Japan
moriken@g.ecc.u-tokyo.ac.jp
[7] Hokkaido University, Kozagawa, Wakayama, Japan

Abstract. The fields of biology, ecology, and fisheries management are witnessing a growing demand for distinguishing individual fish. In recent years, deep learning methods have emerged as a promising tool for image-based fish recognition. Our study is focused on the re-identification of masu salmon from Japan, wherein fish were individually marked and photographed to evaluate discriminative body characteristics. Unlike previous studies where fish were sampled during the same time period, we evaluated individual re-identification across seasons and years to address challenges due to aging, seasonal variation, and other factors. In this paper, we propose a new contrastive learning framework called Aging Contrast (AgCo) and evaluate its performance on the masu salmon dataset. Our analysis indicates that, unlike large changes in body size over time, the pattern of parr marks on the lateral line of the fish body

W. Shi and Z. Zhou—Contributed equally to this work.
The views and conclusions contained in this document are those of the authors and should not be interpreted as representing the opinions or policies of the U.S. Geological Survey. Mention of trade names or commercial products does not constitute their endorsement by the U.S. Geological Survey.

Supplementary Information The online version contains supplementary material available at https://doi.org/10.1007/978-981-99-8388-9_21.

remains relatively stable, despite some change in coloration across seasons. AgCo accounts for such seasonally-invariant features and performs re-identification based on the cosine similarity of these features. Extensive experiments show that our AgCo method outperforms other state-of-the-art methods.

Keywords: Fish Re-identification · Contrastive Learning · Seasonally-invariant Features

1 Introduction

With the global human population now exceeding 8 billion, there is a growing need for sustainable food sources. Fish, being a vital source of protein, comprise a significant portion of the diet for many across the globe, and increased harvest pressures underscore the importance of effective approaches for fish production and conservation. Moreover, the United Nations' Food and Agriculture Organization reported that approximately 80 percent of marine fish production is from wild populations [14] where monitoring and assessment can be difficult and expensive. In this paper, we propose a new analytical framework that could potentially revolutionize fish cultivation and conservation in natural environments as well as aquaculture systems.

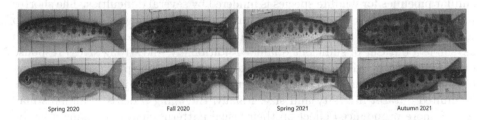

Spring 2020   Fall 2020   Spring 2021   Autumn 2021

Fig. 1. Samples of two fish identities in four different seasons. Compared to the dramatic change of the size, color and dots, the pattern of parr marks located on the lateral line of the fish body remains relatively consistent across seasons. In our study we filter out noisy information to focus on learning the seasonally-invariant features from parr marks.

Distinguishing between individuals within a species is a fundamental step for understanding demographic processes in animal populations [1,4,10,11,18,30, 32,33,35,37]. Currently, most individual-identification systems can be grouped into either invasive or non-invasive processes. For invasive methods, tagging, altering, or coloring specific body parts have been implemented. Among these, tagging is especially popular for its accuracy, and many tagging techniques have been developed. Despite its utility, such tagging procedures have certain inherent drawbacks: mortality rates can increase after tagging, stress caused by tagging could impact the recapture rate, and the time and cost for this process can limit the survey areas and sample sizes available for research. Among non-invasive methods, DNA samples have shown potential in animal individual recognition, but such methods often require operators with advanced professional skills and technical equipment that is not readily available.

Recent advances have enabled imagery-based individual identification for various wildlife species [6,10,13,19,25,27,35,37] based on convolutional neural network analysis of individually-diagnostic pigmentation patterns. However, such applications for wildlife management and conservation are often limited by uncertainty associated with organismal growth and development that may change pigmentation patterns and thus decrease re-identification accuracy. In this paper, we develop and evaluate a new approach to identify *Oncorhynchus masou* (masu salmon) across seasons and years.

Masu salmon is a species of great economic, cultural, and ecological importance in northern Japan (see supplementary material). Typically, masu salmon have a darkened back with small pigmentation spots and larger oval-shaped parr marks on their sides (Fig. 1). The species exhibits a migratory life-history strategy whereby spawning and juvenile development occurs in freshwater environments, a subset of sub-adults migrate into marine environments, and adults return to freshwater environments for spawning [15]. An understanding of individual variation in fish movement patterns and growth rates is necessary to quantify population dynamics and extirpation risks for this valuable species [23].

Compared to recognition tasks based on human datasets, the creation of similar pipelines for wildlife species is hindered by several difficulties. The shorter life spans of animals and fewer opportunities to collect images, as compared to humans, often result in a scarcity of data per individual, making it challenging to assemble a dataset of sufficient quality. Additionally, the strong stress response often exhibited by wild animals can make recapture for the purpose of collecting data difficult. In addition, the time-consuming nature of manual labeling results in limited dataset diversity and size. Finally, the aging process in animals may have a more pronounced effect on their visual patterns than in humans, thereby potentially affecting recognition accuracy.

To mitigate some of the previously mentioned constraints, the advent of self-supervised learning has demonstrated its efficacy as a means to automatically discern patterns from vast quantities of unlabeled data. As an unsupervised representation learning model, contrastive learning seeks to learn representations that bring similar objects together while separating unrelated objects. In our study, we propose the use of individual fish identities and temporal information within the dataset to improve the representation of the same individual over time, while pushing the representations of different individuals farther apart. Compared to simple, random augmentations of the same images in training data, we demonstrate that our method yields more accurate re-identification of individuals over time.

In this paper, we propose a novel framework called Aging Contrast (AgCo) for fish re-identification across multiple seasons and years. The primary challenge in this task is the misalignment of features of the same fish identity, caused by significant changes in fish appearance over time. However, our examination of masu salmon samples revealed that, despite considerable changes in the dots, size of the salmon and background coloration, the pattern of parr marks located on the lateral line of the fish body remains relatively consistent over time. Compared to other unsupervised contrastive learning methods [7,24], our AgCo framework measures the similarity of the query and key from two domains (i.e., different seasons) and perform the data augmentation on the feature level to obtain the transitional features between two seasons. As a result, AgCo can learn seasonally-invariant features from the pattern of parr marks for each fish identity because the change in the pattern of parr marks is more predictable. The major contributions of this paper are summarized as follows:

- Compared to a previous study [10], which performed fish re-identification over a maximum period of six months, our study focuses on a longer time span, with a time gap between sampled fish that can exceed one year. The greater time span can encompass more significant changes in fish appearance and therefore poses a more substantial challenge for fish re-identification. A new dataset for fish re-identification across ages is collected.
- Based on the analysis of masu salmon images, we propose the Aging Contrast (AgCo) framework, which learns seasonally-invariant features from the pattern of the parr marks of fish sampled at different times.
- Extensive experimental results demonstrate that, in the settings with two or more seasos, our AgCo framework significantly outperforms state-of-the-art contrastive learning methods such as MoCo [16] and SimCLR [7].

2 Related Work

2.1 Deep Learning for Fish Recognition

Until recently, fish recognition by images emerged as an appealing area due to its theoretical and applied significance to aquaculture and marine biology, and it has gained great interest from researchers around the world. This task poses great challenges since the collected images of the fish might be of low quality (e.g., noisy or distorted), which heavily affects the recognition. Recently, Alsmadi et al. [2] obtained the distinct features through distance and geometrical measurements. The obtained features were fed into a neural network to distinguish 20 different fish families. To improve image resolution, Sun et al. [34] generated high resolution images from raw images. Then the discriminative features could be obtained from the refined images, and the support vector machine (SVM) was used for fish recognition. Ding et al. [12] proposed several convolutional neural networks (CNN) architectures to identify the fish from four species. Hridayami et al. [20] fine-tuned the VGG16 on four different types of datasets, and their results showed that blending image with an RGB image trained model exhibited the best performance for recognition of fish species.

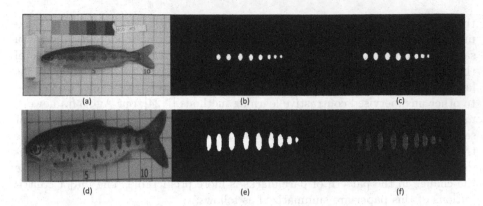

Fig. 2. Illustrations of: (a) original fish image; (b) predicted parr marks from segmentation pipeline; (c) ground truth parr marks from manual labelling; (d) ROI image of fish; (e) ROI image of predicted parr marks; and (f) parr mark pigmentation images generated by the fusion of (d) and (e).

Another branch of work focuses on fish recognition at the individual level. Cisar et al. [10] performed individual identification on Atlantic salmon based on the dot patterns on the skin of the body. Mcinness et al. [28] showed the ability to discriminate the individual freckled hawkfish by the natural markings on their body. [4] further demonstrated that the visible patterns such as the stripes on the fish body could be effective for fish individual identification. Apart from using the visible patterns, other recent studies [3,21,29] explored an alternative method that adopted scale patterns on the body as discriminative features for fish individual identification. For instance, Zhou et al. [37] used the dots patterns from brook trout as the biomarker and trained a CNN based model to learn discriminative features for fish individual recognition.

2.2 Contrastive Learning

Contrastive learning methods provide a powerful tool to pre-train the model without the need for a large number of labels. The core idea behind contrastive learning is to aggregate the positive sample pairs and repulse the negative sample pairs. Most contrastive learning methods [7,8,16,22,24] adopt a contrastive loss to maximize the similarity of positive pairs and enlarge the gap between the negative pairs. Some new mechanisms, such as momentum encoder [16] and cluster alignment [5,24], are introduced in latest contrastive learning framework. Apart from the InfoNCE [31] loss adopted in many contrastive learning methods, ProtoNCE [24] is proposed to estimate the concentration for the feature distribution around each prototype. Although promising, the aforementioned contrastive learning methods do not consider temporal changes in data (e.g., fish across ages), and thus they cannot effectively deal with the fish re-identification task across seasons and years.

3   Dataset

The fish dataset used in our study was acquired from the Horonai River in Japan, as part of a long-term research project that marked masu salmon with Passive Integrated Transponder (PIT) tags. Four capture-mark-recapture surveys were conducted across the survey area in the following seasons: Spring 2020, Autumn 2020, Spring 2021, and Autumn 2021. The samples of the fish from Spring 2020, Autumn 2020, Spring 2021, and Autumn 2021 are illustrated in Fig. 1. More details of our dataset can be found in supplementary material.

4   Proposed Method

We present a novel approach for (1) isolating the defining characteristics of individual masu salmon and (2) employing a contrastive learning framework to extract representations in a self-supervised manner. Our methodology takes into account previous findings that parr marks on a masu salmon's body can serve as a distinct biomarker for individuals. Thus, the first step of our pipeline is to automatically isolate the parr marks from their surroundings and create a more distinctive pattern for analysis. Subsequently, we utilize the proposed AgCo framework to discern aging-resistant features from the image of each fish.

4.1   Segmentation and Feature Extraction

In this stage, we adopt we utilized a segmentation network with a Feature Pyramid Network [26] architecture to extract the parr marks for all images of fish, which served as the primary features for subsequent steps. The results from the segmentation are illustrated in Fig. 2. More details can be found in supplementary material. To further take advantage of information from the parr marks such as the texture and color, the predicted parr marks are fused with the original images to generate parr mark images for analysis as shown in Fig. 2.

4.2   Aging Contrast Framework

Our Aging Contrast (AgCo) framework aims to learn seasonally-invariant features for each fish from the parr mark pigmentation images. Though the overall appearance of each fish may change dramatically over time, the parr mark patterns from each fish are relatively consistent across different seasons. Thus, parr marks are the key for successful re-identification of masu salmon from different ages.

Since there is only a minor shift in the pattern of the parr marks from the observed fish samples in different seasons, we assume that a transitional feature would exist, which represents the potential fish sample between season A and season B. The intermediate feature can be approximated by the linear interpolation of the fish images from season A and season B due to the minor and regular change of the pattern of parr marks. In our study, we adopt mixup [36]

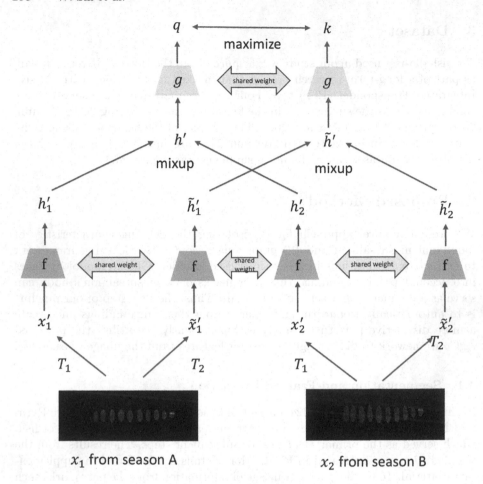

Fig. 3. The diagram depicts our proposed Aging Contrast (AgCo) framework. The features extracted from the same individual in two different seasons are fed into deep neural networks to produce their respective representations. Mixup operation is applied in the representation space to create two augmented representations. We aim to obtain seasonally-invariant features for the same fish identity by maximizing the query and key generated from the augmented latent features h' and $\tilde{h}'$.

to approximate the transitional features. The transitional features contain both temporal information of the parr marks of the fish from both season A and season B. By maximizing the similarity of the transitional features, the season-invariant features can be obtained, and fish re-identification can be performed by measuring the similarity of such seasonally-invariant features.

The AgCo framework is illustrated in Fig. 3. Compared to conventional contrastive learning methods, we take advantage of the fish identities information, and the parr mark pigmentation images (x_1 and x_2) of same identity from two seasons are adopted as the input. T_1 and T_2 are sets of commonly used data augmentation methods which includes rotation, color jittering, and flipping. For each image, two views are generated T_1 and T_2. A neural network encoder f extracts the latent features from all the views of the input images. In our study, we adopt the commonly used ResNet-152 [17] as the encoder f. Different from existing contrastive learning methods, we perform the data augmentation on the features from different views to form the transitional features. To model the transitional features from parr marks of the potential fish between two seasons, transitional features h', $\tilde{h}' \in \mathbb{R}^d$ are generated by combination of the features from different seasons.

$$h' = \lambda h_1' + (1 - \lambda)h_2' \qquad \tilde{h}' = \lambda \tilde{h}_2' + (1 - \lambda)\tilde{h}_1', \tag{1}$$

where h_1', $\tilde{h}_1'$, h_2', and $\tilde{h}_2'$ are the representative features generated from the views of the images from season A and season B. The transitional features h' and $\tilde{h}'$ are joint combination of the features from different seasons. λ denotes the hyperparameter that regulates the proportion of the component features to the targeted features. As shown in Eq. (1), transitional features h' and $\tilde{h}'$ are constructed by the features from the images views augmented by T_1 and T_2, respectively. To make the transitional features distinctive for contrastive learning, the proportion of component features in h' is different to that in $\tilde{h}'$.

Next, we map the transitional features to query q and key k by the projection head g. g is implemented as the two-layered multi-layer perception (MLP). For each query q, k^+ is the key from the same fish identity as query while k^- comes from different identities. By maximizing the similarity of positive pair (q, k^+) and the dissimilarity of negative pair (q, k^-) via a contrastive loss function, the neural network encoder f can learn the seasonally-invariant features for each fish identity. In most contrastive learning framework, the InfoNCE [31] loss is adopted.

$$\mathcal{L}_{\text{InfoNCE}} = -\log \frac{\exp\left(q \cdot k^+ / \tau\right)}{\sum_i \exp\left(q \cdot k_i / \tau\right)},$$

where τ is the hyperparameter that scales the distribution of the similarity distance. q is the query and k is key. k^+ is the key from the positive pair. The query and key are generated from the same domain. In our study, the query and key are combination of the features of fish samples from different seasons. And the loss function is modified accordingly, as shown in Eq (2):

$$\mathcal{L} = -\frac{1}{N} \sum_i \log \frac{\exp\left(q_i \cdot k_i^+ / \tau\right)}{\sum_j \exp\left(q_i \cdot k_j / \tau\right)} = -\frac{1}{N} \sum_i \log \frac{\exp\left(g(h_i') \cdot g(\tilde{h}_i') / \tau\right)}{\sum_j \exp\left(g(h_i') \cdot g(\tilde{h}_j') / \tau\right)}, \tag{2}$$

where N is the total number of fish identities. Details of the AgCo framework can be found in supplementary material. In general, the AgCo framework aims to capture the common traits of the parr marks that exist in the fish samples in different seasons and generate the season-invariant features that can be aligned on different seasons for each fish identity.

5   Experiments

In this section, we introduce the experimental settings for our study. Our proposed method is evaluated for fish observed in all surveys (Spring 2020, Autumn 2020, Spring 2021, Autumn 2021). The results of the AgCo is compared against several representative contrastive learning methods. Ablation studies and visualizations of outcomes can be found in supplementary material.

Experimental Settings. In the experiments, we performed the fish re-identification task by measuring the cosine similarity of the latent features generated by the neural network encoder f shown in Algorithm 1. We compare the results of AgCo against that of latest contrastive learning frameworks including SimSiam [9], PCL [24], MoCo [16] and SimCLR [7] as well as pretrained ResNet-152 [17] without any contrastive learning fine-tuning. More details of the experimental settings can be found in supplementary material.

Table 1. The performance of our AgCo and the baseline methods on S1 → SX (X=2,3,4). Top-1/3/5 accuracy for each model is reported. The bold represents the best results.

Method	Task		
	S1 → S2	S1→ S3	S1→ S4
Pretained Only	29.2/43.9/53.6	14.6/36.5/46.3	12.1/31.7/46.3
SimSiam	34.1/51.2/61.0	17.1/39.0/53.7	17.1/31.7/46.3
PCL	34.1/51.2/61.0	19.5/43.9/48.7/	14.6/29.2/34.1
MoCo	48.8/58.5/70.7	39.0/63.4/73.2	14.6/36.5/44.0
SimCLR	80.4/87.8/90.2	63.4/80.4/95.1	41.4/60.9/80.4
AgCo (Ours)	**100.0/100.0/100.0**	**90.2/100.0/100.0**	**73.2/90.2/92.7**

Results and Analysis. The results of our AgCo and the baseline methods on the tasks of S1 → SX (X = 2, 3, 4) and S2 → SX (X = 3, 4), S3 → S4 are shown in Table 1 and Table 2, respectively. We find that the pretrained model which is not fine-tuned by contrastive learning yields the worst performance on all tasks. The poor performance originated from the inability of the model to generate the discriminative features for each fish identity. When the individual fish grows and the characteristics of the fish are changed, the feature generated by the model for one fish identity in the subsequent season is prone to be aligned with the feature of another fish identity in the previous season. We also observed that for the pretrained-only method, the model can achieve 29.2% top-1 accuracy on S1 → S2. While on S1 → S3 and S1 → S4, this value is decreased to 14.6% and 12.1%. This suggests that the misalignment of the features generated by the model from two different seasons would be exacerbated if the time span of two seasons is larger. It is understandable that since the time span is larger, the

Table 2. The performance of our AgCo and the baseline methods on S2 → SX (X = 3, 4) and S3 → S4. Top-1/3/5 accuracy for each model is reported. The bold represents the best results.

Method	Task		
	S2 → S3	S2 → S4	S3 → S4
Pretained Only	24.3/43.9/60.9	19.5/36.5/46.3	24.3/41.5/51.2
SimSiam	29.2/29.3/39.0	7.3/14.6/26.8	24.4/36.6/68.3
PCL	65.8/83.0/90.2	39.0/56.1/70.7	60.9/73.2/78.0
MoCo	80.5/92.7/97.6	56.1/73.2/85.3	39.1/56.1/68.3
SimCLR	97.5/97.5/97.5	75.6/85.3/90.2	70.7/87.8/92.6
AgCo (Ours)	**100.0/100.0/100.0**	**87.8/95.2/97.6**	**92.7/95.1/97.5**

change in the characteristics of the fish would be more significant. Therefore, it is more difficult for the model to generate the aligned features for the same fish over longer periods of time.

Compared to the pretrained only method, The model fine-tuned by Sim-Siam [9] only gains very limited improvement performance on all the tasks. Compared to other contrastive learning methods [7,16,24] which take the similarity and dissimilarity into account for queries and keys, SimSiam [9] only consider similarity of the queries and keys in the loss function. We assume that the low performance is mainly caused by lack of discrimination of the features from different fish identities.

As shown from the experimental results, there is prominent improvement in the performance of the models for the baseline methods which adopts InfoNCE [31] as the loss function. For instance, on the task S2 → S3, the top-1 accuracy of MoCo [16] and SimCLR [7] can reach 80.5% and 97.5%, respectively. The experiments suggest that the discriminative features learned by contrastive methods is the key to the improved performance. Even if the model can only 'see' the fish in one season in the training stage, the features of same fish identity from different seasons can still be aligned to some extent since features of different fish identities are assumed to be pushed away from each other.

When the pretrained model is further fine-tuned with our proposed AgCo, the performance of the model is further improved. For instance, the top-1 accuracy of the model can reach 100.0% on S1 → S2, almost 20% higher than that of SimCLR [7]. It also is observed that the performance of other contrastive baseline methods is lower than that of AgCo on all tasks. Though the model trained by the these baseline methods is able to learn the discriminative features for each fish identity, the discrimination of the features is confined to a single season and the temporal information of the parr marks on the fish body is neglected. We notice that on S1 → S4, the top-1 accuracy of the model is degraded to 41.4% for SimCLR [7]. It indicates that there is still a large chance that the features learned from different seasons are misaligned for a single fish identity. In contrast, our proposed AgCo framework is not only able to learn the discriminative features for each fish but also to account for possible changes in parr marks. As mentioned, the seasonally-invariant features learned by the AgCo framework can capture common traits of the parr marks that exist in the fish samples from different seasons. Thus, even if the appearance of an individual fish has been dramatically changed, the features of the fish from different seasons can still be aligned. For the most difficult task S1 $\rightarrow$ S4, the top-1 accuracy of the model trained by AgCo can still achieve 73.2%.

6   Conclusion

Recognition of individual fish is necessary for many aspects of fisheries management and conservation, and our study contributes a novel contrastive learning framework called Aging Contrast (AgCo) for this purpose that outperforms other approaches. Our results also highlight the specific importance of parr marks as the key for the fish re-identification task. Our proposed method can learn the seasonally-invariant features that enable accurate re-identification of individual fish as they develop over time. Further applications of our framework could improve fisheries management and conservation globally.

Acknowledgement. This material is based upon work supported by the U.S. Geological Survey under Grant/Cooperative Agreement No. G22AC00372.

References

1. Al-Jubouri, Q., Al-Azawi, R., Al-Taee, M., Young, I.: Efficient individual identification of zebrafish using hue/saturation/value color model. Egypt. J. Aquat. Res. **44**(4), 271–277 (2018)
2. Alsmadi, M.K., Omar, K.B., Noah, S.A., Almarashdeh, I.: Fish recognition based on robust features extraction from size and shape measurements using neural network. J. Comput. Sci. **6**(10), 1088 (2010)
3. Bekkozhayeva, D., Cisar, P.: Image-based automatic individual identification of fish without obvious patterns on the body (scale pattern). Appl. Sci. **12**(11), 5401 (2022)
4. Bekkozhayeva, D., Saberioon, M., Cisar, P.: Automatic individual non-invasive photo-identification of fish (sumatra barb puntigrus tetrazona) using visible patterns on a body. Aquacult. Int. **29**(4), 1481–1493 (2021)
5. Caron, M., Misra, I., Mairal, J., Goyal, P., Bojanowski, P., Joulin, A.: Unsupervised learning of visual features by contrasting cluster assignments. Adv. Neural. Inf. Process. Syst. **33**, 9912–9924 (2020)
6. Chen, P., et al.: A study on giant panda recognition based on images of a large proportion of captive pandas. Ecol. Evol. **10**(7), 3561–3573 (2020)
7. Chen, T., Kornblith, S., Norouzi, M., Hinton, G.: A simple framework for contrastive learning of visual representations. In: International Conference on Machine Learning, pp. 1597–1607. PMLR (2020)
8. Chen, T., Kornblith, S., Swersky, K., Norouzi, M., Hinton, G.E.: Big self-supervised models are strong semi-supervised learners. Adv. Neural. Inf. Process. Syst. **33**, 22243–22255 (2020)

9. Chen, X., He, K.: Exploring simple siamese representation learning. In: Proceedings of the IEEE/CVF Conference on Computer Vision and Pattern Recognition, pp. 15750–15758 (2021)

10. Cisar, P., Bekkozhayeva, D., Movchan, O., Saberioon, M., Schraml, R.: Computer vision based individual fish identification using skin dot pattern. Sci. Rep. **11**(1), 1–12 (2021)

11. Delcourt, J., et al.: Individual identification and marking techniques for zebrafish. Rev. Fish Biol. Fish. **28**(4), 839–864 (2018). https://doi.org/10.1007/s11160-018-9537-y

12. Ding, G., et al.: Fish recognition using convolutional neural network. In: OCEANS 2017-Anchorage, pp. 1–4. IEEE (2017)

13. Ding, R., Wang, L., Zhang, Q., Niu, Z., Zheng, N., Hud, G.: Fine-grained giant panda identification. In: ICASSP 2020–2020 IEEE International Conference on Acoustics, Speech and Signal Processing (ICASSP), pp. 2108–2112. IEEE (2020)

14. Food, Organization, A.: The state of world fisheries and aquaculture. Technical report, Food and Agriculture Organization of the United Nations (2022)

15. Futamura, R., et al.: Size-dependent growth tactics of a partially migratory fish before migration. Oecologia **198**, 371–379 (2022)

16. He, K., Fan, H., Wu, Y., Xie, S., Girshick, R.: Momentum contrast for unsupervised visual representation learning. In: Proceedings of the IEEE/CVF Conference on Computer Vision and Pattern Recognition, pp. 9729–9738 (2020)

17. He, K., Zhang, X., Ren, S., Sun, J.: Deep residual learning for image recognition. In: Proceedings of the IEEE Conference on Computer Vision and Pattern Recognition, pp. 770–778 (2016)

18. Hirsch, P.E., Eckmann, R.: Individual identification of Eurasian perch perca fluviatilis by means of their stripe patterns. Limnologica **54**, 1–4 (2015)

19. Hou, J., et al.: Identification of animal individuals using deep learning: a case study of giant panda. Biol. Cons. **242**, 108414 (2020)

20. Hridayami, P., Putra, I.K.G.D., Wibawa, K.S.: Fish species recognition using vgg16 deep convolutional neural network. J. Comput. Sci. Eng. **13**(3), 124–130 (2019)

21. Huntingford, F., Borçato, F., Mesquita, F.: Identifying individual common carp cyprinus carpio using scale pattern. J. Fish Biol. **83**(5), 1453–1458 (2013)

22. Kalantidis, Y., Sariyildiz, M.B., Pion, N., Weinzaepfel, P., Larlus, D.: Hard negative mixing for contrastive learning. Adv. Neural. Inf. Process. Syst. **33**, 21798–21809 (2020)

23. Kanno, Y., Harris, A., Kishida, O., Utumi, S., Uno, H.: Complex effects of body length and condition on within-tributary movement and emigration in stream salmonids. Ecol. Freshw. Fish **31**, 317–329 (2021)

24. Li, J., Zhou, P., Xiong, C., Hoi, S.C.: Prototypical contrastive learning of unsupervised representations. arXiv preprint arXiv:2005.04966 (2020)

25. Li, W., Ji, Z., Wang, L., Sun, C., Yang, X.: Automatic individual identification of Holstein dairy cows using tailhead images. Comput. Electron. Agric. **142**, 622–631 (2017)

26. Lin, T.Y., Dollár, P., Girshick, R., He, K., Hariharan, B., Belongie, S.: Feature pyramid networks for object detection. In: Proceedings of the IEEE Conference on Computer Vision and Pattern Recognition, pp. 2117–2125 (2017)

27. Matkowski, W.M., Kong, A.W.K., Su, H., Chen, P., Hou, R., Zhang, Z.: Giant panda face recognition using small dataset. In: 2019 IEEE International Conference on Image Processing (ICIP), pp. 1680–1684. IEEE (2019)

28. McInnes, M.G., Burns, N.M., Hopkins, C.R., Henderson, G.P., McNeill, D.C., Bailey, D.M.: A new model study species: high accuracy of discrimination between individual freckled hawkfish (paracirrhites forsteri) using natural markings. J. Fish Biol. **96**(3), 831–834 (2020)

29. Morgado-Santos, M., Matos, I., Vicente, L., Collares-Pereira, M.: Scaleprinting: individual identification based on scale patterns. J. Fish Biol. **76**(5), 1228–1232 (2010)

30. Navarro, J., Perezgrueso, A., Barría, C., Coll, M.: Photo-identification as a tool to study small-spotted catshark scyliorhinus canicula. J. Fish Biol. **92**(5), 1657–1662 (2018)

31. Oord, A.V.d., Li, Y., Vinyals, O.: Representation learning with contrastive predictive coding. arXiv preprint arXiv:1807.03748 (2018)

32. Sandford, M., Castillo, G., Hung, T.C.: A review of fish identification methods applied on small fish. Rev. Aquac. **12**(2), 542–554 (2020)

33. Stien, L.H., et al.: Consistent melanophore spot patterns allow long-term individual recognition of Atlantic salmon salmo salar. J. Fish Biol. **91**(6), 1699–1712 (2017)

34. Sun, X., Shi, J., Dong, J., Wang, X.: Fish recognition from low-resolution underwater images. In: 2016 9th International Congress on Image and Signal Processing, BioMedical Engineering and Informatics (CISP-BMEI), pp. 471–476. IEEE (2016)

35. Whooley, P., Berrow, S., Barnes, C.: Photo-identification of fin whales (balaenoptera physalus L.) off the south coast of Ireland. Mar. Biodivers. Rec. **4** (2011)

36. Zhang, H., Cisse, M., Dauphin, Y.N., Lopez-Paz, D.: mixup: beyond empirical risk minimization. In: International Conference on Learning Representations (2018)

37. Zhou, Z., Hitt, N.P., Letcher, B.H., Shi, W., Li, S.: Pigmentation-based visual learning for salvelinus fontinalis individual re-identification. In: 2022 IEEE International Conference on Big Data (Big Data), pp. 6850–6852. IEEE (2022)

Spatial Bottleneck Transformer for Cellular Traffic Prediction in the Urban City

Hexuan Weng[1]($\boxtimes$) , Yanbin Liu[2] , and Ling Chen[1]

[1] University of Technology Sydney, Ultimo, NSW 2007, Australia
hexuan.weng@student.uts.edu.au, ling.chen@uts.edu.au
[2] Australian National University, Canberra, ACT 2601, Australia

Abstract. Due to the widespread use of portable devices and the advancement of 5G technology, we have received a significant amount of mobile data, which requires prediction models for cellular traffic data. However, accurately forecasting mobile traffic data is challenging due to the complex spatial and temporal correlations, especially when the mobile data comes from a large geographical area. To tackle this challenge, we propose a new model, called *ST-InducedTrans*, to dynamically explore the large geographical correlations (spatial) and periodic variations (temporal). Specifically, a Spatial Bottleneck Transformer is devised to obtain spatial correlations from the most relevant grids in the geographical area at the cost of linear complexity. For the temporal blocks, we embed the elaborately selected temporal clues into a temporal Transformer to offer useful temporal prompts for cellular prediction. Finally, several spatial and temporal blocks are effectively stitched into a whole model for complementary cellular traffic prediction. We conducted comprehensive experiments on the public real-world cellular data from Milan. Results show that our model outperforms the state-of-the-art methods on three metrics (MAE, NRMSE, and R^2) at the cost of lower time complexity.

Keywords: Cellular Traffic Prediction · Spatial-Temporal Data · Deep Learning · Transformer

1 Introduction

In recent years, cellular traffic prediction has become a prominent area of focus due to the advancement of 5G technology. With the development of portable devices and the internet, mobile phones become an essential part of our daily life. Since cellular traffic data normally contains spatial and temporal information as well as their interactions, efficiently identifying the complex spatial and temporal dependence is crucial for accurate prediction. Previous studies have employed statistical methods, machine learning methods and deep learning methods to improve both the accuracy and efficiency of cellular traffic prediction.

T. Liu et al. (Eds.): AI 2023, LNAI 14471, pp. 265–276, 2024.
https://doi.org/10.1007/978-981-99-8388-9_22

Statistical methods extract specific correlation measures from the individual time series. For example, Zhao et al. [6] used Anselin Local Moran's I statistic measure, while Zhang et al. [5] and Liu et al. [7] used the Pearson correlation coefficient to manually get the spatial correlation between the target grid and its neighboring cells. These methods used pre-defined statistics and shallow models, so they struggled to capture the complex non-linear spatial-temporal correlations in real-world cellular traffic data [12,13]. *Machine learning methods* utilize various traditional algorithms such as Gradient Boosting Decision Tree (GBDT) [14], Gaussian mixture model (GMM) [15], and support vector regression [2] to improve the prediction performance. However, similar to statistical methods, they are also constrained by shallow models.

Deep learning methods leverage the representation capability of modern deep networks to solve the above problems, including Recurrent Neural Networks (RNNs), Convolutional Neural Networks (CNNs), Graph Neural Networks (GNNs), and Transformers. Qiu et al. [3] proposed a model using RNNs in the temporal block in their research. However, its performance deteriorates quickly in predicting long-range time-series data. CNNs [1] are used to extract the spatial correlation in cellular traffic prediction. For example, in [5], cellular data of the whole geographical area is treated as an image that can be processed by CNNs to obtain spatial correlations. The disadvantage of CNNs is that the local receptive fields can only model adjacent spatial information within a small region range. GNNs are employed to model the spatial correlations by a graph structure. For example, CNN&GNN model was proposed in [6] to process spatial features in cellular traffic prediction. However, its graph structure requires an adjacent matrix with fine-grained data, which is difficult to obtain due to privacy issues.

The emergence of Transformer advanced research in various fields such as Natural Language Processing (NLP) and computer vision. Hence, it can also serve as a good network architecture for mobile traffic prediction. On the one hand, Transformer can resolve the long-range prediction inability of RNNs by escaping the gradient vanishing problem. On the other hand, Transformer has a global receptive field using the self-attention mechanism, which relaxes the local limitations of CNNs and is free from the adjacent matrix of GNNs. Therefore, the vanilla Transformer has been used in [7] to improve the performance of cellular traffic prediction. Since the complexity of the vanilla Transformer is quadratic w.r.t the input length, modeling the spatial grids of real-world cellular traffic (e.g., 100×100 in Milan) is computationally expensive and infeasible. In this context, a novel, dynamic and more efficient model variant is required to reduce the heavy computation load of Transformer.

Motivated by this, we propose a new model, called **ST-InducedTrans**, to efficiently capture the complex spatial-temporal dependencies of a large mobile grid for accurate cellular traffic prediction. The whole model is composed of several spatial and temporal blocks. Each spatial block is a well-devised Spatial Bottleneck Transformer, which introduces a smaller-size query (i.e. **inducing point**) to only focus on the K most relevant spatial correlations and recovers the input length by an original-size query. This design significantly reduces the

Transformer complexity and also manages to automatically and dynamically select the most relevant region correlations, even for far-away regions. Each temporal block is a vanilla Transformer augmented with elaborately selected temporal clues (e.g., day of the week and holidays). These clues offer important prompts for extracting certain cellular traffic patterns, e.g., the difference between weekdays and weekends. The overall structure is built by stitching the spatial and temporal blocks with several fusion layers. We verify the effectiveness of our model design on a widely used cellular prediction benchmark: Milan. The state-of-the-art comparison, parameter analysis and visualization corroborate the superiority and efficiency of our model design.

The contributions of this paper are summarized as follows:

- We propose ST-InduscedTrans, a novel and efficient model for cellular traffic prediction on a real-world large dataset.
- We design a Spatial Bottleneck Transformer in ST-InducedTrans to capture the spatial dependencies, which can improve the prediction accuracy and reduce the time complexity from quadratic to linear.
- Informative temporal clues (e.g., day of the week and holidays) are embedded in the temporal Transformer to provide useful temporal prompts beneficial for cellular prediction.
- Comprehensive experiments are conducted on the real-world benchmark dataset Milan, which verifies the superiority and efficiency of our method.

2 Related Work

RNNs have been used widely in temporal dependency discoveries in cellular traffic forecasting. In Kuber's research [0], LSTMs are mainly used to build temporal-dependent models. As to spatial components, the auxiliary information (e.g. the location of airports, banks, or restaurants) is used for modeling. However, RNNs may encounter gradient explosion and have limitations when dealing with long-term sequence data. Due to the properties of mobile traffic data, cellular traffic data is in the form of grids, which can be seen as images. Thus, previous works have combined CNNs and RNNs to predict mobile traffic data. Zhang et al. [4] proposed a model of Milan data into a 3D-ConvNet while incorporating LSTMs for the prediction. CNNs and LSTMs can effectively get a temporal and spatial dependency. Zhang et al. [10] proposed STCNet, which used cross-domain data as the information in CNNs and employed the CNN-RNN model to combine the complex spatiotemporal traffic variability in mobile traffic prediction through temporal and spatial perspectives.

In mobile traffic data, GNNs play an increasingly important role. Spatial-based graph convolutional networks have proliferated since spectral methods process the entire graph simultaneously. Wang et al. [11] mainly used GNN-D to predict the mobility data of metropolises. It considers other factors, including land use, population, holidays, and social activities. The graph neural network model can reasonably predict the moving position or the area with complicated traffic and control the flow to avoid accidents, but it requires fine-grained data, which is not accessible from a public data set. In the encoder-decoder framework, CNNs and RNNs can be hybridized, and whichever acts as an encoder and decoder can be flexibly combined for various tasks. Transformer is of a parallel encoder-decoder structure, and Xu et al. [16] proposed Spatial-Temporal Transformer Networks (STNNs), which jointly leverage dynamically directed spatial dependencies and long-range temporal dependencies to improve the accuracy of long-term traffic flow forecasting. Traffic transformer, raised by [18], can learn dynamic and hierarchical features in sequential data, and Cai et al. [17] apply Traffic transformer to capture the continuity and periodicity of time series and to model spatial dependency. Liu et al. [7] were the first to apply the transformer architecture to predict cellular traffic.

3   Problem Formulation

We conduct cellular traffic research on the large-scale, real-world public telecommunication dataset from a well-known telecommunication provider: Telecom Italia. The dataset, collected from Milan, divides the large geographical areas into a $H \times W$ grid. Various cellular traffic activities are recorded, including Received SMS, Sent SMS, Incoming Call, Outgoing Call and Internet usage.

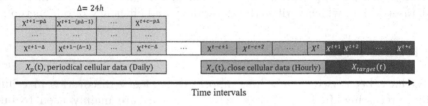

Fig. 1. $X_p(t)$, $X_c(t)$, $X_{target}(t)$ at the $X(t)$ interval.

In this paper, we deal with the cellular traffic forecasting problem. In particular, we study the behavior of Call data prediction. Cellular traffic prediction is performed on the $H \times W$ grid representing the whole geographical location. Each cell of the grid reflects the cellular traffic in a certain location. As a time-series prediction problem, we make use of two types of historical data: *close* neighboring cellular data and *periodical* cellular data. Denote the current time step of traffic data as t in Fig. 1; we want to predict future data $X_{target}(t) \in \mathbb{R}^{N \times c}$ of the future c steps. Then, we define close neighboring data $X_c(t) \in \mathbb{R}^{N \times c}$ as the previous c steps just before future data on all N locations. We define periodic data $X_p(t) \in \mathbb{R}^{N \times p \times c}$ as historical data having the same multiplicative intervals relative to future steps.

To summarize, the cellular traffic forecasting problem is defined as follows: ***Given*** cellular traffic data from N geographical grids: (1) Close neighboring cellular data $X_c(t) = (X^{t-c+1}, X^{t-c+2}, \ldots, X^t) \in \mathbb{R}^{N \times c}$, (2) Periodical cellular data $X_p(t) = (X^{t+1-p\Delta}, \ldots, X^{t+c-p\Delta}, \ldots, X^{t+1-\Delta}, \ldots, X^{t+c-\Delta}) \in \mathbb{R}^{N \times p \times c}$, ***Forecast*** future traffic data $X_{target}(t) = (X^{t+1}, X^{t+2}, \ldots, X^{t+c}) \in \mathbb{R}^{N \times c}$.

4   Methodology

For real-world cellular traffic prediction problems, data usually comes from a large geographical area (e.g., 100×100 in Milan). We propose a novel *ST-InducedTrans* model for cellular traffic prediction, which contains a well-devised *Spatial Bottleneck Transformer* to obtain the spatial correlations from all grids by inducing points (i.e., bottleneck). Moreover, we embed informative temporal clues in the temporal Transformer to augment the temporal pattern extraction.

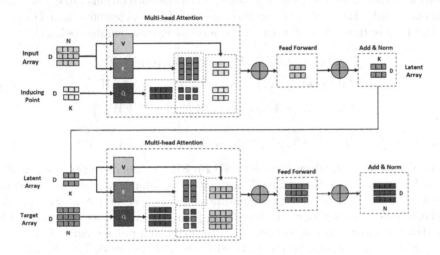

Fig. 2. Spatial Bottleneck Transformer

4.1   Spatial Bottleneck Transformer

Spatial Bottleneck Transformer (SBT) is the major contribution in this paper, which eliminates the quadratic scaling problem of all-to-all attention of a vanilla Transformer and decouples the network depth from the input's size, allowing us to construct very deep models. Concretely, SBT is a Transformer-based neural network architecture. Unlike the vanilla Transformer, each SBT block is composed of two Multihead Attention Blocks (MABs), as shown in Fig. 2. We design a bottleneck structure for the two MABs. The first MAB (*encoder*) takes as input a small-size query to extract the K most relevant features with the cross-attention mechanism. Then, the second MAB (*decoder*) retrieves the spatial correlations from the extracted K relevant features with another cross-attention.

The bottleneck refers to mapping from N input elements to a much smaller $K(K \ll N)$ features and then recovering the original length size N. There are several advantages of the bottleneck design over the vanilla transformer design: (1) the computation complexity is significantly reduced with this design; (2) the bottleneck design has the feature selection effect by only selecting K relevant features. This is consistent with cellular traffic properties that only a small number of locations sharing similar patterns are highly related in cellular traffic prediction; (3) K can be adjusted to reflect different circumstances, such as different cities and different grid-scale grained.

The main components of the SBT are as follows.

Multi-head Attention. This module is similar to the Multi-Head Attention in the vanilla Transformer. It splits the vectors in the input sequence into multiple subsets and then performs a self-attentive computation on the vectors in each subset. With an input tensor of shape (N, D), the Multi-Head Attention outputs a tensor of the same shape (N, D), but embeds the correlations among all input elements inside. Here, N refers to the number of input elements, and D stands for the feature dimension. Formally, the module can be denoted below:

$$
\begin{aligned}
\text{Attention}(Q, K, V) &= \text{softmax}\left(QK^T\right)V, \\
\text{Multihead}(Q, K, V) &= \text{concat}\left(O_1, \cdots, O_h\right)W^O, \\
\text{where } O_j &= \text{Attention}\left(QW_i^Q, KW_i^K, VW_i^V\right).
\end{aligned}
\tag{1}
$$

Here, $W^O, QW_i^Q, KW_i^K, VW_i^V$ are the relevant fully connected layer weights.

Spatial Bottleneck Transformer (SBT). SBT model comprises an Encoder and a Decoder. Unlike the vanilla Transformer, we adopt cross-attention mechanisms in the Multi-Head Attention module instead of self-attention. Specifically, in the *Encoder*, we introduce the **inducing point** $I \in \mathbb{R}^{K \times D}$ as the query probe to extract K most relevant features from the original input sequence $X \in \mathbb{R}^{N \times D}$ ($N = H \times W$ in a cellular traffic grid). Since $K \ll N$ in our design, the Encoder has the advantages of information compression and feature extraction, which only keeps the most related features beneficial for future cellular traffic prediction. I serves as a query (Q), and X serves as key (K) and value (V) in Eq. 1. Taking the Feed-Forward Network and layer normalization into consideration, we can represent the structure as:

$$
\begin{aligned}
\text{CrossAttention}(I, X) &= \text{LayerNorm}(H + \text{FeedForward}(H)), \\
\text{where } H &= \text{LayerNorm}(X + \text{Multihead}(I, X, X)).
\end{aligned}
\tag{2}
$$

Similarly, the *Decoder* adopts a cross-attention structure but designs different query, key, and value. We take the original input sequence X as query (Q), and the output from the Encoder as key (K) and value (V). This way, the output of Decoder has the same sequence length as the input.

The overall structure of the SBT can be represented as:

$$
\begin{aligned}
\text{SBT}_K(X) &= \text{CrossAttention}(X, H) \in \mathbb{R}^{N \times D}, \\
\text{where } H &= \text{CrossAttention}(I, X) \in \mathbb{R}^{K \times D}.
\end{aligned}
\tag{3}
$$

Here, K stands for the bottleneck size for choosing only K relevant features.

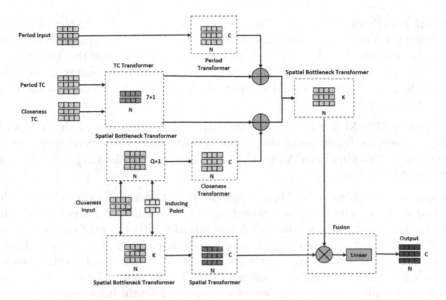

Fig. 3. ST-InducedTran Model

4.2 ST-InducedTran Model

We now describe the detailed structure of the ST-InducedTrans model, as shown in Fig. 3. Overall, ST-InducedTrans includes the temporal clues, spatial block and temporal block, devised to capture the spatial correlations and temporal patterns, respectively.

Temporal Clues. Before going into the details of the model architecture, we first describe how to leverage indicative temporal clues to help cellular traffic prediction. Our observation is that the cellular patterns vary with certain temporal events, e.g. weekdays and weekends exhibit different patterns in using mobile devices, thus leading to different prediction modes. Similarly, holidays are an important factor. Bearing this in mind, we define the one-hot indicative clues as 'day-of-week' (7-dimension) and 'holidays' (1-dimension). Then, we concatenate them into the temporal clues $X_{tc} \in \mathbb{R}^{n \times (7+1)}$. With an additional projection, the clues can be mapped to $X_{tc} \in \mathbb{R}^{N \times D}$ (D is the same as all the Transformers) and used as temporal clues in the later Transformers.

Spatial Block. There are two types of Transformer in the Spatial Block: Spatial Bottleneck Transformer and Spatial Transformer.

Spatial Bottleneck Transformer can extract possible correlations among all locations in the cellular traffic grid. Since in a real-world grid, not all grid locations are prediction-relevant, the bottleneck structure can efficiently and effectively select only the relevant spatial locations helpful for cellular traffic prediction. As mentioned before, the parameter K of the inducing points in the bottleneck can be chosen to reflect real-world scenarios, such as cities, grid-scale, etc.

Spatial Transformer. The input of this block is $X_c \in \mathbb{R}^{N \times c}$. The top K similar elements are selected and concatenated to $X_s \in \mathbb{R}^{N \times K \times c}$ via Spatial Bottleneck Transformer, which is a learning-based feature selecting module. The Spatial Transformer takes the vanilla Transformer structure and contains an encoder and a decoder. X_s is the encoder input, X_c is the decoder input, and the final output is $X_{sp} \in \mathbb{R}^{N \times 1 \times c}$.

Temporal Block. There are also two types of Transformer in the Temporal Block: Closeness Transformer and Period Transformer. These two types process the close neighboring data $X_c \in \mathbb{R}^{N \times c}$ and the periodic data $X_p \in \mathbb{R}^{N \times p \times c}$ as described in Sect. 3.

Closeness Transformer. We first augment the close neighboring data in the spatial context. The Q most related spatial locations are selected with the Spatial Bottleneck Transformer and concatenate with the grid's own data as $X_c^{aug} = \mathbb{R}^{N \times (Q+1) \times c}$. Then X_c^{aug} is input to the encoder of the Closeness Transformer. The period data X_p is averaged to get X_p^{avg} and input to the decoder of the Closeness Transformer. Closeness Transformer mainly processes the close data while it also refers to the period average for relevant information.

Period Transformer. The period data X_p is directly input to the encoder. The close data X_c is input to the decoder as reference information. Since the period data already contains various period patterns, there is no need to apply extra augmentations. Period Transformer processes the period data while it also refers to the close data for relevant information.

Stitching and Fusion. After getting all the modules and components, we stitch them in an overall model ST-InducedTrans (Fig. 3). Specifically, the model input includes period data, close data, and temporal clues. The temporal block and the Spatial block are put in a parallel layout to process the data. Finally, the Fusion module is proposed (containing several MLP layers) to combine both the temporal and spatial processed information for better prediction.

5   Experiments

5.1   Dataset

We use the common benchmark dataset Milan [8], whose data is collected from Telecom Italia. It contains aggregated cellular traffic, including SMS, call service and internet in the city of Milan. The time duration for data collection is from 2013/11/01 00:00 to 2014/01/01 23:00 for two months (62 days with around 300 million records), with an interval of 10 min. We further aggregated the data into an interval of 60 min (1 h). The geographical area of Milan has been divided into a size of $H \times W$ grids, where H and W refer to the number of rows and columns and $H = W = 100$ (i.e. 10,000 grid cells). Each grid has an approximate area of 235×235 square meters. Following previous studies, we take the center 20×20 grid from the whole 100×100 grid, mainly to predict the cellular traffic in urban areas. In this paper, we use call data, and our approach also fits the other data sets, including SMS and Internet usage.

5.2   Baseline

We compare with several classical and state-of-the-art methods.

- **Historical Average & ARIMA** : Traditional methods that uses one day of mobile phone traffic data from historical data to predict future traffic.
- **STDenseNet** [5]: It jointly models the spatial and temporal dependence of traffic in different cells. Each time, the incoming and outgoing traffic is represented as an image-like two-channel tensor matrix, with a sliding window to generate training and test datasets.
- **STACN(w/o E)** [6]: This model proposes a STACN that considers the dynamic spatiotemporal correlation of cellular traffic. A GCN+CNN convolutional module is investigated to capture mobile traffic's spatial and temporal characteristics through spatial and temporal attention.
- **ConvLSTM** [10]: STCNet can efficiently capture complex patterns in grid data. It actively models three cross-domain datasets to capture the external factors influencing traffic generation.
- **ST-Trans** [7]: This is the first cellular prediction model that implements all modules as Transformer. The model designs the spatial branch and temporal branch composed of different Transformers and finally gathers the information from the two branches.

5.3   Implementation Details

Hyperparameters. In the experiments, the number of Transformer layers was set to $N = 6$. The input and feature dimension D varies among different Transformers. Here, $D = 64$ for the Spatial Transformer, $D = 256$ for Closeness Transformer, and $D = 128$ for the Period Transformer. $D = 128$ in other Transformer modules. In all experiments, a polynomial learning rate decay strategy was used. The initial learning rate was set to 0.001, and the training batch size for the entire model was 16, with 500 epochs of training (Table 1).

Table 1. Input and dimension of different Transformers.

Transformer	D	encoder input	decoder input
Spatial	64	$X_s(t)$	$X_c(t)$
Closeness	256	$X_c(t)$	$X_p(t)$
Period	128	$X_p(t)$	$X_c(t)$

5.4   Evaluation Metrics

To comprehensively evaluate the performance of our model, we utilized three metrics that are widely used in time-series and cellular traffic prediction: MAE (Mean Absolute Error), $NRMSE$ (Normalized Root Mean Squared Error), and R^2 (Coefficient of determination). The closer the MAE and $NRMSE$ are to zero, the better the result is. The higher the R^2 value is, the better the model.

Table 2. Comparison with the state-of-the-art methods.

Methods	MAE	NRMSE	R^2
HA	18.7226	0.9687	0.4419
ARIMA	17.1895	0.8813	0.6564
LSTM	13.9438	0.6079	0.7802
STDenseNet	12.3168	0.6442	0.7842
STACN (w/o E)	12.6450	0.6210	0.7207
ConvLSTM	11.2308	0.5652	0.8097
ST-Trans	10.0244	0.5388	0.8273
ST-InducedTrans (w/o Fusion)	**9.7215**	0.5319	0.8317
ST-InducedTrans	9.7273	**0.5035**	**0.8493**

6 Results and Discussion

6.1 Time Complexity Analysis

One major contribution of this research is to reduce the quadratic scaling problem of Transformer to a linear complexity problem. Denote N as the total number of grid cells (i.e. $H \times W$) we have, and we want to predict through top K spatially relevant data from N grid cells with Spatial Bottleneck Transformer. Our design reduces the number of calculation operations from N^2 of the vanilla Transformer to NK ($K \ll N$). This means that the complexity reduces from quadratic ($O(N^2D)$, vanilla Transformer) to linear ($O(KND)$, Ours). Here, D is the dimension of the features in the intermediate layers. The significant reduction in time complexity makes the computation faster and possible to predict cellular traffic data in a larger-scale geographical grid.

6.2 Comparison with the State-of-the-Art Methods

In Table 2, we compare our method with various baselines, including the state-of-the-art method ST-Trans. Our model variant "w/o Fusion" means the Spatial Bottleneck Transformer is removed before the Fusion module (Fig. 3 left-right). Our method outperforms all the baselines with a large margin w.r.t. three metrics. Note that ST-Tran employed vanilla Transformer architecture. Our method is not only more efficient in time complexity but also much better in performance. The comparison with the "w/o Fusion" variant shows that the Spatial Bottleneck Transformer is effective and generally applicable in our model.

6.3 Parameter Analysis and Visualization

Parameter K. For too small and too large K, the performance deteriorates quickly, while in a reasonable range (15 to 20), the performance is stable. The best K is 20 according to the MAE measure, which is only 5% compared with the overall 400 grid cells.

In this module, the choice of K is significant. K can be selected in the range of $[0, 400]$; when K is chosen as 0, the grid selection function is removed. To enhance the prediction performance, we experimented with the value of K between 0 and 50; the MAE value was the lowest for $K = 20$ when fusion was not implemented (Fig. 4).

Fig. 4. Performance with respect to varying values of K

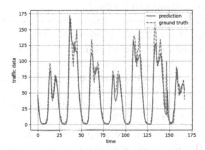

Fig. 5. Prediction without Fusion   **Fig. 6.** Prediction with Fusion

Prediction Visualization. We show the comparison between the predicted values and the ground truth of the 224th grid in Fig. 5 and Fig. 6 for the model "w/o Fusion" and our final model. Both models have a relatively good prediction trend. However, our model performs much better in detail, such as the time range between 75 and 100, verifying the effectiveness of the Spatial Bottleneck Transformer in the Fusion module.

7   Conclusion

This paper proposes a new ST-InducedTrans model to improve forecasting accuracy and computation complexity. We devise a new spatial transformer block, namely Spatial Bottleneck Transformer (SBT), to explore the spatial dependencies. In SBT design, we introduce a learnable inducing point to reduce the algorithmic complexity of the transformer from quadratic to $O(nm)$ linear. Moreover, we explore the informative temporal clues and incorporate them into the temporal embedding of the Transformer. Extensive experiments are conducted on the real-world cellular traffic dataset, which corroborates the efficiency and effectiveness of our ST-InducedTrans model.

Acknowledgement. This work is supported by the Australian Research Council under Grant ARC DP210101347. Yanbin Liu is supported by the Google Cloud Research Credits program with the award GCP19980904.

References

1. Kim, P.: Convolutional neural network. In: MATLAB Deep Learning, pp. 121–147. Apress, Berkeley, CA (2017). https://doi.org/10.1007/978-1-4842-2845-6_6
2. Sapankevych, N.I., Sankar, R.: Time series prediction using support vector machines: a survey. IEEE Comput. Intell. Mag. **4**(2), 24–38 (2009)
3. Qiu, C., et al.: Spatio-temporal wireless traffic prediction with recurrent neural network. IEEE Wireless Commun. Lett. **7**(4), 554–557 (2018)
4. Zhang, C., Patras, P.: Long-term mobile traffic forecasting using deep spatio-temporal neural networks. In: Proceedings of the Eighteenth ACM International Symposium on Mobile Ad Hoc Networking and Computing (2018)
5. Zhang, C., et al.: Citywide cellular traffic prediction based on densely connected convolutional neural networks. IEEE Commun. Lett. **22**(8), 1656–1659 (2018)
6. Zhao, N., et al.: Spatial-temporal attention-convolution network for citywide cellular traffic prediction. IEEE Commun. Lett. **24**(11), 2532–2536 (2020)
7. Liu, Q., Li, J., Zhaoming, L.: ST-Tran: spatial-temporal transformer for cellular traffic prediction. IEEE Commun. Lett. **25**(10), 3325–3329 (2021)
8. Barlacchi, G., et al.: A multi-source dataset of urban life in the city of Milan and the Province of Trentino. Sci. Data **2**(1), 1–15 (2015)
9. Kuber, T., Seskar, I., Mandayam, N.: Traffic prediction by augmenting cellular data with non-cellular attributes. In: 2021 IEEE Wireless Communications and Networking Conference (WCNC). IEEE (2021)
10. Zhang, C., Patras, P., Haddadi, H.: Deep learning in mobile and wireless networking: a survey. IEEE Commun. Surv. Tutor. **21**(3), 2224–2287 (2019)
11. Wang, X., et al.: Spatio-temporal analysis and prediction of cellular traffic in metropolis. IEEE Trans. Mob. Comput. **18**(9), 2190–2202 (2018)
12. Box, G.E.P., et al.: Time Series Analysis: Forecasting and Control. John Wiley & Sons, Hoboken (2015)
13. Snyder, R.D.: Recursive estimation of dynamic linear models. J. Roy. Stat. Soc. Ser. B (Methodol.), 272–276 (1985)
14. Ke, G., et al.: Lightgbm: a highly efficient gradient boosting decision tree. In: Advances in Neural Information Processing Systems, vol. 30 (2017)
15. Zhang, Q., et al.: Machine learning for predictive on-demand deployment of UAVs for wireless communications. In: 2018 IEEE Global Communications Conference (GLOBECOM). IEEE (2018)
16. Xu, M., et al.: Spatial-temporal transformer networks for traffic flow forecasting. arXiv preprint arXiv:2001.02908 (2020)
17. Cai, L., et al.: Traffic transformer: capturing the continuity and periodicity of time series for traffic forecasting. Trans. GIS **24**(3), 736–755 (2020)
18. Yan, H., Ma, X., Ziyuan, P.: Learning dynamic and hierarchical traffic spatiotemporal features with transformer. IEEE Trans. Intell. Transp. Syst. **23**(11), 22386–22399 (2021)

MIDGET: Music Conditioned 3D Dance Generation

Jinwu Wang[✉]ⓘ, Wei Maoⓘ, and Miaomiao Liuⓘ

Australian National University, Canberra, ACT 2601, Australia
jinwu.wang@alumni.anu.edu.au

Abstract. In this paper, we introduce a MusIc conditioned 3D Dance GEneraTion model, named **MIDGET** based on Dance motion Vector Quantised Variational AutoEncoder (VQ-VAE) model and Motion Generative Pre-Training (GPT) model to generate vibrant and high-quality dances that match the music rhythm. To tackle challenges in the field, we introduce three new components: 1) a pre-trained memory codebook based on the Motion VQ-VAE model to store different human pose codes, 2) employing Motion GPT model to generate pose codes with music and motion Encoders, 3) a simple framework for music feature extraction. We compare with existing state-of-the-art models and perform ablation experiments on AIST++, the largest publicly available music-dance dataset. Experiments demonstrate that our proposed framework achieves state-of-the-art performance on motion quality and its alignment with the music.

Keywords: 3D Dance Generation · Music Condition · Deep Learning · Auto-Regressive

1 Introduction

Dance is an art form that expresses emotions and culture through-composed motion patterns. [25]. Moreover, music as the carrier of dance is an indispensable element in dance performance. All dances always have some strong correlations with the music rhythm [4,25]. Music-based human motion generation drives many applications like choreograph and virtual character animation [24] (Fig. 1).

One of the main challenges of this task is to ensure the consistency between the music rhythm and the generated motion e.g., the alignment of music beats and motion. Previous methods e.g., EDGE [26] learn it implicitly from the data which cannot ensure the motion-music beat alignment leading to artifacts like freezing motion. Although other methods such as Bailando [24] try to explicitly encourage the music-motion beat alignment, their actor-critic learning strategy is very unstable during train. By contrast, in this paper, we propose a gradient copying strategy which enable us to directly train the motion model with beat alignment loss in a gradient descent manner. Furthermore, other than Bailando [24] which directly downsample the music feature to match the human

T. Liu et al. (Eds.): AI 2023, LNAI 14471, pp. 277–288, 2024.
https://doi.org/10.1007/978-981-99-8388-9_23

motion, we propose to learn a music feature extractor which is proved to be more effective. In particular, building upon the Bailando [24], we introduce MusIc conditioned 3D Dance GEneraTion (MIDGET), an end-to-end generative model for generating motion that match the beats of the music.

Fig. 1. Dance examples generated by our proposed method. Qualitative human motion generation samples based on our MIDGET model can be found at YouTube.

The main contributions of our work can be summarized as follows:

- We introduce a gradient copying strategy which enables us to train the motion generator with music alignment score directly.
- We propose a simple yet effective music feature extractor improves recognition and analysis performed on music information with few additional parameters.

2   Related Work

The field of music to dance generation research generates dance motions for the traditional music of a specific style or genre through machine learning ideas [14,19,23]. This area has been studied for many years, and various techniques have been used to implement dance motions generation.

Traditional Music-to-Dance Models. Lee *et al.* [14] used a clustering model to group musically similar segments into a cluster and then used the dance motion sequences corresponding to the same clusters as the generated results. This method determines dance motions by classifying music only and needs more diversity and accuracy of dance motions. Ofli *et al.* [19] proposed using four statistical machine learning models to obtain the mapping relationship from music to capture the diversity in dance performances and the dependence on musical segments, and dance figure models are used to model the motion of each dance character to capture the changes in the performance style of a particular dance character.

In addition, Shiratori *et al.* [23] uses a graph-based approach to cut motion clips from existing data into individual nodes and stitch them together to synthesize new motions based on appropriate musical features. However, this clip-based scheme unitizes the motions. So that only a fixed rhythm, tempo, and variety of motions can exist.

Music-Conditioned 2D Human Dance Generation. Most currently available works have studied the 2D choreographic generation of music, Lee *et al.*[13] realized that dance is multi-modal and that various subsequent pose motions are equally possible at any moment. The Generative model of music generative dance is proposed for the first time to decompose dance into a series of basic dance units. In the synthesis phase, the model learns to choreograph a dance by seamlessly organizing multiple basic dance motions based on the input music. Li *et al.*[18] connects the predicted output of the network itself with its future input stream by using the training mechanism of an automatic conditional recurrent neural network [10]. Qi *et al.*[20] proposed a sequence-to-sequence learning architecture that leverages LSTM [7] and Self-Attention mechanism for dance generation based on music.

Huang *et al.*[9] introduces Transformer to realize 2D dance generation under music conditions and uses the local self-attention to efficiently handle long sequences of music features. In addition, Huang *et al.* [9] proposes a dynamic automatic conditional training method to mitigate the error accumulation of auto-regressive models in long motion sequence generation.

Music-Conditioned 3D Human Dance Generation. The application scenario of 2D dance motion is minimal. It only contains flat information and thus needs 3D spatial three-dimensional details. Research in the Music-Conditioned 3D Human Dance Generation field has emerged. Li *et al.* [17] and Li *et al.* [15] allows the models to extract and learn audio features by stacking different forms of Transformers. Siyao *et al.* [24] used the idea of VQ-VAE [22,27] to encode and quantize the spatial standard dance motions into a limited codebook, which can store different dance motions in a codebook so that a more professional, reasonable, and coherent dance motion can be used when generating a sequence. Each sequence inside the codebook represents a unique dance pose, and these dances have contextual semantic information.

Recently, diffusion-based approaches [6,16,26] have excelled in generating musically conditioned motion to create realistic and physically plausible dance motions based on input music. Tseng *et al.* [26] proposed Editable Dance Generation (EDGE), a transformer-based diffusion model paired with the powerful music feature extractor Jukebox [3,26]. It was demonstrated that this unique diffusion-based approach gives powerful editing capabilities well suited to dance, including joint modulation and intermediate processing.

3   Method

An overview of the proposed music-to-dance generation framework, "Music Conditioned 3D Dance Generation (MIDGE)", is shown in Fig. 2. We first quantize the dance motions using the VQ-VAE [27] resulting a finite-length codebook $\mathcal{Z} = \{z_i\}_{i=1}^{K}$ where each code $z_i \in \mathbb{R}^C$ is a vector of size C. Each motion sequence

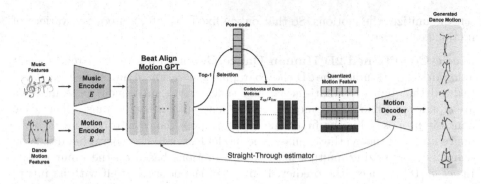

Fig. 2. Overview of the MIDGET Model. Given a piece of music and its corresponding dance instructions, MIDGET can generate corresponding high-quality and smooth dance sequences.

is then represented as a sequence of codes from the codebook. Instead of directly downsampling the music feature as in [24], we adopt a 1-D convolutional network to learn such downsampling process which is proved to be more effective. Conditioning on the music feature and a seed motion, we leverage the motion GPT [24] to obtain the code sequences for both upper body and lower body. Such code sequences can be easily decoded as human motion via the pretrained VQ-VAE. We provide the details for each module below.

3.1   3D Dance Motion VQ-VAE

In order to effectively capture and extract the unique style of dance motion codes and enable them to be reconstructed into corresponding dance motion sequences with actual physical meaning. Therefore, we use 1-D temporal convolutional network to encode dance motion sequences $P \in \mathbb{R}^{T \times (J \times 3)}$ into pose codes $Q \in \mathbb{R}^{T' \times C}$, where T denotes the time duration of the original dance motion sequence and J means the number of joint, and C is the feature dimension, and $T' = T/d$ is the motion code length of the pose codes, and d is the temporal sampling rate (Fig. 3).

Codebook of Dance Poses. The VQ-VAE codebook [27] is a vector collection that maps continuous high-dimensional data into discrete ones. Each of these vectors is called a code. By transforming continuous data into discrete representations, the codebook is able to store and process data in lower dimensions, enabling efficient data compression. In this paper, we mapped the output of the motion encoder module to the nearest neighbour vector in the codebook to discretize the continuous high-dimensional data into a low-dimensional code. In this scheme, we analyze the output $e_i \in \mathbb{R}^C$ of the encoder and map it to the nearest discrete pose code in the codebook:

$$q_i = \arg \min_{z_j \in \mathcal{Z}} \|e_i - z_j\|_2 \,, \tag{1}$$

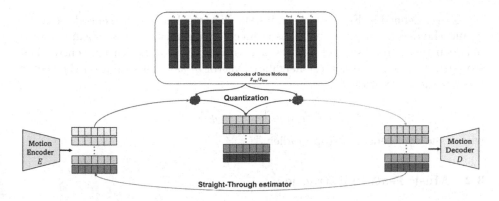

Fig. 3. 3D Dance Motion VQ-VAE. The main purpose of the VQ-VAE model is to obtain codebooks containing diverse quantified dance motion sequences. Learnable encoder and decoder to quantify features and reconstruct target poses.

where $q_i \in \mathbb{R}^C$ is the i-th row of Q and $i \in \{1, 2, \cdots, T'\}$. Finally, The motion decoder is defined as a 1-D temporal deconvolutional network to reconstruct the corresponding dance pose sequences $\hat{P} \in \mathbb{R}^{T \times J \times 3}$ from the latent codes.

The Training Process of Motion VQ-VAE. With this gradient conduction, we can achieve the training of the Motion Encoder E, Decoder D and codebook $\mathcal{Z}$ simultaneously with the following loss function:

$$\mathcal{L} = \mathcal{L}_{REC} + \mathcal{L}_{VQ} + \beta \cdot \mathcal{L}_{COM} \tag{2}$$

The loss function consists of three terms, where $\mathcal{L}_{REC}$ is the *reconstruction loss* used to assess the quality of dance motions. It measures the difference between the generated and target pose sequences. In particular, we propose to define the reconstruction loss function by focusing on the original 3D coordinate points of the body joints and the velocity and acceleration of the motion sequences. The detailed definition is presented below:

$$\mathcal{L}_{REC} = \left\| P - \hat{P} \right\|_2 + \alpha_1 \left\| V - \hat{V} \right\|_2 + \alpha_2 \left\| A - \hat{A} \right\|_2 \tag{3}$$

where $P, \hat{P}$ are the ground truth and generated joints of the original 3D points, respectively. V and A represent the motion velocity and acceleration partial derivatives of the 3D joint sequence on time to learn the time-dependent and spatial relationships in the dance motion sequences, thus learning the transition patterns between different motions and the structure of the motion sequences.

The second and third part of $\mathcal{L}$ is the quantification loss with $\mathcal{L}_{VQ}$ and $\mathcal{L}_{COM}$ to constrain the difference between the discretized code and the continuous code. $\mathcal{L}_{VQ}$ using exponential moving averages function to learn latent embedding vectors in the codebook. We use Exponential Moving Averages (EMA) [12,27] to update the vectors in the codebook of the VQ-VAE model.

$$\mathcal{L}_{VQ} = (e_i, q_i)_{EMA} \tag{4}$$

$\mathcal{L}_{COM}$ defined in Eq. (5) constrains the distance between the encoder output e_i and the decoder input q_i which is reduced to minimise the Euclidean distance between the encoder output and its nearest neighbour quantization centre. The stop gradient [1,2] the nearest neighbour vector q_i as a constraint on the vector quantization operation.

$$\mathcal{L}_{COM} = \|e_i - \text{sg}[q_i]\|_2 \tag{5}$$

where the $sg[\cdot]$ denotes "stop gradient".

3.2 Music Feature Extraction

The Music Feature Extractor is presented in Fig. 4. The learned music features enable the model to ensure less feature information loss with little additional parameters. The purpose of such feature extractor is to downsample the musical features so as to match the quantized motion codes.

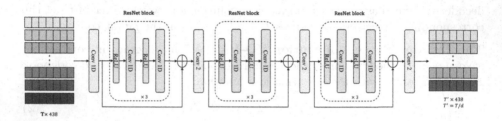

Fig. 4. The structure of Music Feature Extractor. Music features are further extracted through one-dimensional convolutional layers and residual connections.

As shown in Fig. 4, our detailed implementation uses a 1-D convolution and residual structure to downsample the music feature.

3.3 Motion GPT

After working with the above 3D Dance Motion VQ-VAE, we can generate unique dance pose clips with physical meaning according to the trained codebook Z. Now that we can explore Beats-Aligned Motion GPT, which is a Transformer [8,28] based network. The GPT [8,21] model must focus on generating future dance motion codes with corresponding music styles under the given music conditions and current starting quantized pose position codes. The Beats-Aligned Motion GPT achieved as Feature Embedding, Positional Embedding and 12 Transformer Decoders, as shown in Fig. 5.

In order to align the spatial dimensions of the input musical features with the quantized pose position codes. The quantized upper and lower body and music features are subjected to feature embedding operations as $u, l \in \mathbb{R}^{T' \times C_{pose}}$

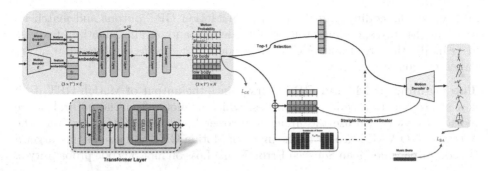

Fig. 5. Motion GPT Structure. The GPT model is designed to apply the encoded upper and lower body pose codes a_u, a_l and music features a_m to generate the target future motion probability p^u, p^l.

and $m \in \mathbb{R}^{T' \times C_{music}}$, respectively. Adding a learned position embedding assigns a unique vector representation for each position, which contains the encoding of positional information. Then, the feature vector is propagated through 12 successive Transformers [28], and mapped to motion probability $p \in R^{((3 \times T') \times C)}$. When generating future motion sequences, the upper and lower body pose codes $p^u = p_{T':2T'-1}$, $p^l = p_{2T':3T'-1}$ extracted from motion probability features.

Cross-Conditional Causal Attention. We adopt "Cross-conditional causal attention" [24] which encodes the features across different time series and only allows the information passing from the past. It is defined as follows:

$$\text{Attention}(Q, K, V) = \text{Softmax}\left(\frac{QK^T + M}{\sqrt{d}}\right) V$$

where $\mathbf{d}$ is the number of channels in the attention layer. $\mathbf{Q}, \mathbf{K}, \mathbf{V}$ denote the query, key, and value from the input feature, and M is the mask matrix used to determine the type of attention sub-layers.

3D Dance Motion Generation. In order to make the model overall differentiable, we use Straight-Through estimator [1] to combine VQ-VAE Decoder and codebooks to generate gradient-derivable sequences of actual future motions $\hat{P}_{1:T}$. As shown in Fig. 5, we simulate the generated result of motion GPT as Quantified pose code $\hat{e}_e$, similar to the output feature of Motion Encoder. The method is to perform matrix multiplication of the action encoding probability $p_{1:T'}$ generated by the motion GPT and the codebook Z. Because matrix multiplication can be considered as a function, when computing matrix multiplication, the gradient can be computed using the automatic differentiation method, which calculates the partial derivative of the output matrix with respect to the input matrix (see Equations below).

$$\hat{e}_e = p_{1:T'} \cdot Z, \quad \hat{e}_d = \arg\min_Z \|p_{1:T'}\|, \quad \hat{e}_d = \hat{e}_e + sg[\hat{e}_d - \hat{e}_e]$$

On the contrary, the forward propagation process in the motion GPT generation stage, we perform top-1 selection (select the index with the highest probability) on the coding probability $p_{1:T'}$ of Motion GPT output and match its index to the corresponding position in Codebook and obtain Quantized features $\hat{e}_k$. Finally, the pre-trained VQ-VAE Decoder decodes $\hat{e}_k$ to obtain the future dance motion sequence.

The training of Motion GPT. First, since the output of Motion GPT itself is a set of N action code probabilities, which cannot reflect the natural dance motions anyway, we use the upper and lower body motion features from the output of VQ-VAE Encoder as the input of Motion GPT and also as the target a for computing the Mean Squared Error(MSE) Loss on motion code probability p:

$$\mathcal{L}_{CE} = \frac{1}{T'} \sum_{t=1}^{T'} \sum_{b=u,l} \left\| p_t^b - a_t^b \right\|_2 \tag{6}$$

where p_t^b is the motion code probability, and a_t^b is the target sequence of VQ-VAE Encoder outputs. Due to the generation mode of the upper and lower half body pose separation, the results of two different loss functions must be combined.

We design Beat Align Loss as Eq. (7) to allow the model to produce more accurate dance motions with music rhythm.

$$\mathcal{L}_{BA_{l2}} = \left\| B_d - B_m \right\|_2 \tag{7}$$

where B_d is the dance motion beats and B_m is the music beats. We identify dance beats by the difference between the physical position of the action in the front and back frames and use a Gaussian filter to calculate the probability of dance beats. The smaller the distance between the front and back frames, the higher the probability that the frame is a dance beat, as follows:

$$B_d = \exp \left(-\frac{\left\| \hat{x}_{(0:T-1)} - \hat{x}_{(1:T)} \right\|_2}{\sigma^2} \right) \tag{8}$$

where $\hat{x} \in \mathbb{R}^{(T \times 24 \times 3)}$ is the prediction of the motion sequence. The probability of motion beats is evaluated by calculating the difference between the motion speed of the two nearly frames.

4   Experimental Result

Our proposed method is evaluated on the AIST++ dataset published by [17], containing 1,408 3D human dance motion sequences represented as joint rotations and root trajectories. All dance motions are paired with their corresponding 60 music clips.

Implementation Details. We train the MIDGET model as a two-stage task with the Adam [11] optimiser. The model framework is trained on an NVIDIA RTX 4090 GPU for 24 h with a batch size of 64.

In our experiments, we separate dance and music into a 240-frame format. The size of the trainable upper and lower body dance memory codebooks in the VQ-VAE model is N=512 dimensions. The downsampling rate in the VQ-VAE Encoder and Music Encoder sections are 8, so the dimensionality of the quantized codebooks also results from downsampling $Z_{up}, Z_{low} \in \mathbb{R}^{30 \times 512}$.

Evaluation Metrics and Baslines. We adopt the *Fréchet Inception Distances* (FID) [5,17,24] to evaluate our method by following existing works. We similarly measure diversity by calculating the average Euclidean distance between different dance motions in the geometric and kinematic feature space. Finally, using the Beat Align score [17,24,26] and Beat Consistency score [15] to measure the relevance between music and dance motions. Precisely, the explanation and equations are as follows:

$$BA = \frac{1}{|B^m|} \sum_{t^m \in B^m} \exp \left\{ \frac{\min_{\forall t^d \in B^d} ||t^d - t^m||_2}{2\sigma^2} \right\}$$

$$BC = \frac{1}{|B^d|} \sum_{t^d \in B^d} \exp \left\{ \frac{\min_{\forall t^m \in B^m} ||t^d - t^m||_2}{2\sigma^2} \right\}$$

(9)

where t_i^m is the music beat in B^m, t_i^d is the dance motion beat in B^d, and σ is the normalized parameter set to 3 simultaneously.

4.1   Results

We compare the performance of our proposed model with the state-of-the-art methods including FACT [17], Bailando [24] and EDGE [26]. Table 1 shows that our proposed model consistently outperforms Bailando [24] in all evaluations under the same underlying model framework and avoids action freezing problem. Specifically, our approach improves by 6.8% and 22.3% in FID_k and FID_g, respectively, for evaluating the physical features of dance.

Moreover, **MIDGET** can generate more diverse dances, as reflected in the improvement of Div_k and Div_g metrics by 10.3% and 1.6%. Finally, the Beats Align Score achieves 10.5% higher, suggesting that generated dance motions are better aligned with music beats. In addition, compared with Li *et al.*[18] and FACT [17], our proposed model has a significant advantage in all evaluations. Although **MIDGET** does not outperform EDGE [26], it has lighter structure.

Table 1. Evaluation of Existing models on AIST++ dataset. Compared to the Ground Truth and three recent state-of-the-art methods,

Method	FID$_k$ ↓	FID$_g$ ↓	DIV$_k$ ↑	DIV$_g$ ↑	BA Score ↑	BC Score ↑
Ground Truth	17.10	10.60	8.19	7.45	0.2374	0.2083
Li *et al.* [18]	86.43	43.46	6.85	7.45	0.201	0.203
FACT [17]	35.35	15.55	5.94	6.18	0.221	0.203
Bailando [24]	30.43	11.42	7.83	6.34	0.233	0.208
EDGE [26]	-	-	10.03	6.67	0.263	0.210
MIDGET(Ours)	28.51	8.87	8.64	6.44	0.254	0.212

However, although our model does not outperform the EDGE model [26], requires to pre-train a model, namely Jukebox [3] to extract features for all music sequences for up to three days, which is extremely time consuming. We achieved similar performance using fewer resources than the EDGE model with a light-weight feature extractor.

4.2   Ablation Studies

We perform the ablation study on 3D Motion VQ-VAE module, the music feature extractor and the Motion GPT estimator. The results are shown in Table 2.

Table 2. Ablation Study on AIST++ testset. The experiments cover Motion VQ-VAE, Music Feature Extractor, Best Match, and generation strategies.

Method	$FID_k \downarrow$	$FID_g \downarrow$	BA Score ↑	BC Score ↑
w/o. VQ-VAE	120.45	41.24	0.259	0.200
w/o. upper/lower	64.07	14.82	0.246	0.203
w/o Extractor	30.43	11.42	0.243	0.203
w/o $\mathcal{L}_{BA}$	29.10	7.25	0.239	0.205
Single Generated	29.34	9.84	0.288	0.209
MIDGET	28.51	8.87	0.254	0.212

Motion VQ-VAE. When we remove the quantized codebook in VQ-VAE [24, 27] to store the dance motion features, the GPT model cannot generate realistic and physically meaningful body poses while the FID_k and FID_g becomes very high. Moreover, we also analyzed whether it is necessary to separate the upper and lower body for independent codebook training. Thus, we just adopt the entire body action sequence as input to train VQ-VAE and Motion GPT. The obtained FID_k and FID_g are 124.7% and 66.9% which is worse than estimating the code for the upper and lower body, respectively.

Music Feature Extractor. We adopt Beat Align Score to evaluate the effectiveness of our proposed Music Feature Extractor. Results in Table 2 show that the BA Score obtained with Music Feature Extractor is higher than the downsampling of music features alone (4.53% improvement). This indicates that our proposed Music Feature Extractor can extract more effective music features.

Beats Match. With our introduced Beat Align Loss, the Motion GPT significantly improves the alignment between actions and music beats, which has increased by 20% and 10% on BA Score and consistency scores (BC Score), respectively. The detailed beat alignment effect is shown in Fig. 6.

Generation Strategy. We evaluate the generation strategy by just predicting 1-dimensional pose code (8 frames) only using Motion GPT during training. The long sequence is generated in an auto-regressive manner. While the generated sequence has a higher BA Score (see Table 2), the results suffer from severe motion freeze issue.

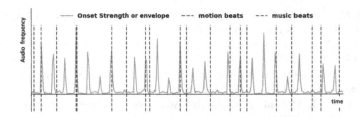

Fig. 6. Beats alignment between Music and Generated Dance. Visualize the alignment of dance beats (purple dashed lines) and music beats (green dashed lines). (Color figure online)

5 Conclusion

We proposed MIDGET, which can generate realistic, and smooth long-sequence dance motions. We have introduced a Beat Align Loss and the Straight-Through Estimator to achieve the end-to-end training, and a simple music feature extractor to improve music-feature learning. The method can largely solve the motion freeze issue for long-sequence generation evidenced by the experiments on AIST++ dataset and achieves superior performance.

References

1. Bengio, Y., Léonard, N., Courville, A.: Estimating or propagating gradients through stochastic neurons for conditional computation. arXiv preprint arXiv:1308.3432 (2013)
2. Chen, X., He, K.: Exploring simple siamese representation learning. In: Proceedings of the IEEE/CVF Conference on Computer Vision and Pattern Recognition, pp. 15750–15758 (2021)
3. Dhariwal, P., Jun, H., Payne, C., Kim, J.W., Radford, A., Sutskever, I.: Jukebox: a generative model for music. arXiv preprint arXiv:2005.00341 (2020)
4. Fachner, J.: Time is the key-music and altered states of consciousness. Altering Conscious.: Multidisc. Perspect. **1**, 355–376 (2011)
5. Heusel, M., Ramsauer, H., Unterthiner, T., Nessler, B., Hochreiter, S.: GANs trained by a two time-scale update rule converge to a local Nash equilibrium. In: Advances in Neural Information Processing Systems, vol. 30 (2017)
6. Ho, J., Jain, A., Abbeel, P.: Denoising diffusion probabilistic models. Adv. Neural. Inf. Process. Syst. **33**, 6840–6851 (2020)
7. Hochreiter, S., Schmidhuber, J.: Long short-term memory. Neural Comput. **9**(8), 1735–1780 (1997)
8. Hu, Z., Dong, Y., Wang, K., Chang, K.W., Sun, Y.: GPT-GNN: generative pre-training of graph neural networks. In: Proceedings of the 26th ACM SIGKDD International Conference on Knowledge Discovery & Data Mining, pp. 1857–1867 (2020)
9. Huang, R., Hu, H., Wu, W., Sawada, K., Zhang, M., Jiang, D.: Dance revolution: Long-term dance generation with music via curriculum learning. arXiv preprint arXiv:2006.06119 (2020)

10. Jain, L.C., Medsker, L.R.: Recurrent Neural Networks: Design and Applications, 1st edn. CRC Press Inc, USA (1999)
11. Kingma, D.P., Ba, J.: Adam: a method for stochastic optimization. arXiv preprint arXiv:1412.6980 (2014)
12. Lawrance, A., Lewis, P.: An exponential moving-average sequence and point process (ema1). J. Appl. Probab. **14**(1), 98–113 (1977)
13. Lee, H.Y., et al.: Dancing to music. In: Advances in Neural Information Processing Systems, vol. 32 (2019)
14. Lee, M., Lee, K., Park, J.: Music similarity-based approach to generating dance motion sequence. Multimedia Tools Appl. **62**, 895–912 (2013)
15. Li, B., Zhao, Y., Zhelun, S., Sheng, L.: Danceformer: music conditioned 3d dance generation with parametric motion transformer. In: Proceedings of the AAAI Conference on Artificial Intelligence, vol. 36, pp. 1272–1279 (2022)
16. Li, R., et al.: Magic: multi art genre intelligent choreography dataset and network for 3d dance generation. arXiv preprint arXiv:2212.03741 (2022)
17. Li, R., Yang, S., Ross, D.A., Kanazawa, A.: AI choreographer: music conditioned 3d dance generation with AIST++. In: Proceedings of the IEEE/CVF International Conference on Computer Vision, pp. 13401–13412 (2021)
18. Li, Z., Zhou, Y., Xiao, S., He, C., Huang, Z., Li, H.: Auto-conditioned recurrent networks for extended complex human motion synthesis. arXiv preprint arXiv:1707.05363 (2017)
19. Ofli, F., Erzin, E., Yemez, Y., Tekalp, A.: Learn2dance: learning statistical music-to-dance mappings for choreography synthesis. IEEE Trans. Multimed. **14**(3), 747–759 (2012)
20. Qi, Y., Liu, Y., Sun, Q.: Music-driven dance generation. IEEE Access **7**, 166540–166550 (2019)
21. Radford, A., Narasimhan, K., Salimans, T., Sutskever, I., et al.: Improving language understanding by generative pre-training (2018)
22. Razavi, A., Van den Oord, A., Vinyals, O.: Generating diverse high-fidelity images with VQ-VAE-2. In: Advances in Neural Information Processing Systems, vol. 32 (2019)
23. Shiratori, T., Nakazawa, A., Ikeuchi, K.: Synthesizing dance performance using musical and motion features. In: Proceedings 2006 IEEE International Conference on Robotics and Automation, 2006. ICRA 2006, pp. 3654–3659 (2006). https://doi.org/10.1109/ROBOT.2006.1642260
24. Siyao, L., et al.: Bailando: 3d dance generation via actor-critic GPT with choreographic memory. In: CVPR (2022)
25. Steinberg, N., et al.: Range of joint movement in female dancers and nondancers aged 8 to 16 years: anatomical and clinical implications. Am. J. Sports Med. **34**(5), 814–823 (2006)
26. Tseng, J., Castellon, R., Liu, C.K.: Edge: editable dance generation from music. arXiv preprint arXiv:2211.10658 (2022)
27. Van Den Oord, A., Vinyals, O., et al.: Neural discrete representation learning. In: Advances in Neural Information Processing Systems, vol. 30 (2017)
28. Vaswani, A., et al.: Attention is all you need. In: Advances in Neural Information Processing Systems, vol. 30 (2017)

Machine Learning and Data Mining

Minimum Message Length Inference of the Weibull Distribution with Complete and Censored Data

Enes Makalic[1]($\boxtimes$)(iD) and Daniel F. Schmidt[2](iD)

[1] Melbourne School of Population and Global Health, The University of Melbourne, Parkville, VIC 3010, Australia
emakalic@unimelb.edu.au
[2] Department of Data Science and AI, Monash University, Clayton, VIC 3800, Australia
daniel.schmidt@monash.edu

Abstract. The Weibull distribution, with shape parameter $k > 0$ and scale parameter $\lambda > 0$, is one of the most popular parametric distributions in survival analysis. It is well established that the maximum likelihood estimate of the Weibull shape parameter is inadequate due to the associated large bias when the sample size is small or the proportion of censored data is large. This manuscript demonstrates how the Bayesian information-theoretic minimum message length principle, coupled with a suitable choice of weakly informative prior distributions, can be used to infer Weibull distribution parameters given either complete data or data with censoring. Empirical experiments show that the proposed minimum message length estimate of the shape parameter is superior to the maximum likelihood estimate and is competitive with other recently proposed modified maximum likelihood estimates in terms of Kullback-Leibler risk.

Keywords: Weibull distribution · parameter estimation · model selection

1 Introduction

The Weibull distribution is one of the most important probability distributions in analysis of lifetime data. The probability density function and cumulative density function of a Weibull random variable T with shape parameter $k > 0$ and scale parameter $\lambda > 0$ are

$$
p_T(t|k, \lambda) = \left(\frac{k}{\lambda^k}\right) t^{k-1} \exp\left(-\left(\frac{t}{\lambda}\right)^k\right), \quad F_T(t|k, \lambda) = 1 - \exp\left(-\left(\frac{t}{\lambda}\right)^k\right),
\tag{1}
$$

respectively. With lifetime data, we often do not observe complete data and instead have joint realisations of the random variables $(Y = y, \Delta = \delta)$ where

$$Y = \min(T, C) \tag{2}$$

$$\Delta = \mathrm{I}(T \le C) = \begin{cases} 1, & \text{if } T \le C \text{ (observed survival)} \\ 0, & \text{if } T > C \text{ (observed censoring)} \end{cases} \tag{3}$$

where the random variables T and C are assumed to be independent and denote the survival time and the censoring time of an item, respectively. In words, we observe the survival time $T = t$ of an item if it is less than the corresponding censoring time $C = c$ (i.e., $t \le c$) ; otherwise, we only know that the item survived past time c (i.e., $t > c$). The censoring time may be a fixed constant (say, c) or a random variable that may depend on other factors. Given n i.i.d. lifetime data points $\mathbf{y} = (y_1, \ldots, y_n)$ and the corresponding censoring indicators, an important problem is to estimate the unknown parameters k and λ and thus learn about the survival distribution of the items.

The most common approach to parameter estimation is the method of maximum likelihood, where the unknown parameters are set to values that maximise the (log-) likelihood of the observed data. Unfortunately, in the case of the Weibull shape parameter the corresponding maximum likelihood estimate is known to have large bias with both complete and censored data (see, for example, [5,8,12]) and this is especially evident for small sample sizes and/or large amounts of censoring. This manuscript introduces the Bayesian minimum message length (MML) approach to inductive inference and demonstrates how MML can be used to estimate Weibull parameters in both the complete and censored data setting. We show that with an appropriate choice of prior distributions the MML estimate of the shape parameter improves on the maximum likelihood estimate, given censored or complete data, and is competitive with alternative proposals that modify the maximum likelihood estimate to reduce bias.

2   Minimum Message Length

The minimum message length (MML) principle [13,14,16] is a Bayesian framework for inductive inference that provides a unified approach to parameter estimation and model selection. Given data, the key step in applying MML is the computation of the length of a message that describes (encodes) the data, with the assumption that the message comprises two components:

1. the *assertion*, encoding of the structure of the model, including all model parameters $\boldsymbol{\theta} \in \boldsymbol{\Theta} \subseteq \mathbb{R}^p$; and
2. the *detail*, encoding the data D using the model $p(D|\boldsymbol{\theta})$ from the assertion.

The length of the assertion measures the complexity of the model, with simpler models having a shorter assertion compared to more complex models. The length of the detail measures how well the model named in the assertion fits the data; more complex models will have shorter detail lengths compared to simpler models. The length of the combined two-part message, $I(D, \boldsymbol{\theta})$, is

$$I(D, \boldsymbol{\theta}) = \underbrace{I(\boldsymbol{\theta})}_{\text{assertion}} + \underbrace{I(D|\boldsymbol{\theta})}_{\text{detail}} \tag{4}$$

i.e., the sum of the length of the assertion, $I(\boldsymbol{\theta})$, and the length of detail, $I(D|\boldsymbol{\theta})$. Inference in the MML framework proceeds by finding the model

$$\hat{\boldsymbol{\theta}}(D) = \arg\min_{\theta \in \Theta} \{I(D, \boldsymbol{\theta})\} \tag{5}$$

that minimises the length of the two-part message. Minimising the MML code-length requires balancing complexity of a model (assertion) with the corresponding fit to the data (detail) with the preferred model being the simplest model that fits the data sufficiently well. A key advantage of MML is that the unit of measurement, the codelength (generally measured in $\log_e$ digits, called nits or nats), is universal in the sense that allows inference and comparison of models with different model structures (e.g., linear regression vs. decision tree) and parameters within a single, unified framework.

Precise computation of codelengths is a NP-hard problem in general. There exist many MML approximations to the codelength (4) [13], with the MML87 approximation [13, 16] being the most widely applied due to its relative computational simplicity. Under suitable regularity conditions, the MML87 codelength approximates (4) by

$$I_{87}(D, \boldsymbol{\theta}) = \underbrace{-\log \pi_\theta(\boldsymbol{\theta}) + \frac{1}{2}\log|J_\theta(\boldsymbol{\theta})| + \frac{p}{2}\log \kappa_p + \frac{p}{2}}_{\text{assertion}} \underbrace{- \log p(D|\boldsymbol{\theta})}_{\text{detail}} \tag{6}$$

where $\pi_\theta(\boldsymbol{\theta})$ is the prior distribution of the parameters $\boldsymbol{\theta}$, $|J_\theta(\boldsymbol{\theta})|$ is the determinant of the expected Fisher information matrix, $p(D|\boldsymbol{\theta})$ is the likelihood function of the model and κ_p is a quantization constant [1]; when the number of parameters p is small, we have

$$\kappa_1 = \frac{1}{12}, \quad \kappa_2 = \frac{5}{36\sqrt{3}}, \quad \kappa_3 = \frac{19}{192 \times 2^{1/3}}, \tag{7}$$

while, for moderate to large p, κ_p is well-approximated by [13]:

$$\frac{p}{2}(\log \kappa_p + 1) \approx -\frac{p}{2}\log 2\pi + \frac{1}{2}\log p\pi - \gamma, \tag{8}$$

where $\gamma \approx 0.5772$ is the Euler–Mascheroni constant. The MML87 approximation is invariant under smooth one-to-one reparametarizations of the likelihood function and is asymptotically equivalent to the Bayesian information criterion (BIC) [10] as $n \to \infty$ with $p > 0$ fixed. There exist many successful applications of the MML principle in statistics and machine learning literature, including factor analysis [17] and time series analysis [9], among others.

3 Complete Data

3.1 Maximum Likelihood Estimates

Consider first the setting of complete data with no censoring. The negative log-likelihood of data $\mathbf{y} = (y_1, \ldots, y_n)$ is

$$- \log p_T(\mathbf{y}|k, \lambda) = n \log \left(\frac{\lambda^k}{k} \right) - (k-1) \left(\sum_{i=1}^{n} \log y_i \right) + \sum_{i=1}^{n} \left(\frac{y_i}{\lambda} \right)^k. \quad (9)$$

The maximum likelihood (ML) estimates of k, λ are

$$\hat{\lambda}^k(\mathbf{y}) = \frac{1}{n} \sum_{i=1}^{n} y_i^{\hat{k}(\mathbf{y})}, \quad (10)$$

where $\hat{k}(\mathbf{y})$ is defined implicitly by

$$\frac{n}{k} + \sum_{i=1}^{n} \log y_i - \frac{n \sum_i y_i^k \log y_i}{\sum_i y_i^k} = 0 \quad (11)$$

and must be solved for numerically. While the ML estimate of λ is reasonable, the ML estimate of k is known to exhibit large bias and perform poorly in terms of squared error risk, especially for small sample sizes [12].

Several attempts have been made to construct a modified ML estimate with improved performance. Ross [8] derives the simple adjustment formula

$$\hat{k}_R(\mathbf{y}) = \left(\frac{n-2}{n-0.68} \right) \hat{k}_{ML}(\mathbf{y}) \quad (12)$$

for the ML estimate that reduces the bias to typically better than about 0.05%, though this adjustment applies to complete data only. Similarly, Hirose [5] derives tables with correction coefficients that can be used to obtain modified ML estimates of both k and λ with reduced bias. Tanaka et al. [12] derive second-order improved ML estimators and show that the ML estimator of the shape parameter is always second-order inadmissible. In a somewhat different approach, Yang and Xie [20] apply the modified profile likelihood proposed by Cox and Reid [2,3] to derive a penalized maximum likelihood estimate of k. Specifically, the Yang and Xie estimate of λ is equivalent to the ML estimate while the new estimate of the shape parameter k is obtained by numerically solving

$$\frac{n-2}{k} + \sum_{i=1}^{n} \log y_i - \frac{n \sum_i y_i^k \log y_i}{\sum_i y_i^k} = 0. \quad (13)$$

which is similar to (11), the only difference being $(n-2)$ in the numerator of the first term. Yang and Xie empirically show that their estimate of k is less biased than the ML estimate and is more statistically efficient than Ross' simple modification (12).

3.2 Minimum Message Length Estimates

To derive the MML87 codelength (6) we require the determinant of the expected Fisher information matrix

$$|J(k, \lambda)| = \frac{n^2 \pi^2}{6\lambda^2}, \tag{14}$$

and prior distributions for both parameters. Assuming that k and λ are *a priori* independent, we opt for the half-Cauchy distributions

$$\pi(k, \lambda) = \pi(k)\pi(\lambda), \quad \pi(k) = \frac{2}{\pi(1+k^2)}, \quad \pi(\lambda) = \frac{2}{\pi(1+\lambda^2)}. \tag{15}$$

As λ is a scale parameter, a heavy tailed distribution like the half-Cauchy is appropriate and recommended in, for example, [7]. Additionally, the half-Cauchy distribution is suitable for the shape parameter k as $k = 1$ denotes a constant failure rate and decreasing ($k < 1$) and increasing ($k > 1$) failure rate are equally likely *a priori*. The MML87 codelength for the Weibull distribution is

$$I_{87}(D, k, \lambda) = -\log\left(\frac{4}{\pi^2(1+k^2)(1+\lambda^2)}\right) + \frac{1}{2}\log\left(\frac{n^2\pi^2}{6\lambda^2}\right) - \log p_T(\mathbf{y}|k,\lambda) + 1 + \log \kappa_2 \tag{16}$$

where the negative log-likelihood function is given in (9) and $\kappa_2 = 5/(36\sqrt{3})$ (see Sect. 2). Unfortunately, with this selection of prior distributions, the MML87 estimates of k and λ must be obtained by numerically minimising (16).

It is straightforward to show that the modified maximum likelihood estimate of Yang and Xie (13) is the MML87 estimate obtained under the prior distribution

$$\pi(k, \lambda) = \pi(k)\pi(\lambda), \quad \pi(k) \propto \frac{1}{k^2}, \quad \pi(\lambda) \propto \frac{1}{\lambda}. \tag{17}$$

which is improper unless lower and upper bound limits are imposed on both the shape and scale parameters. The implied prior distribution for λ is the usual scale invariant distribution often used to model a scale parameter while the prior distribution for the shape parameter k is heavy tailed and Cauchy-like asymptotically. As the aforementioned implied prior distributions are similar to (15) in their behaviour, it is expected that both the Yang and Xie modified maximum likelihood estimate and the MML87 estimate proposed in this manuscript will yield similar parameter estimates with virtually identical properties.

4 Censored Data

Consider first the fixed (type I) censoring setup where observations are censored after some period of time $c > 0$. In particular, we observe the lifetime of an item only if $T_i \leq c$, otherwise we observe the censoring time c. The likelihood function of n observed data points $D = \{(y_1, \delta_1), \ldots, (y_n, \delta_n)\}$ is

$$p(D) = \prod_{i=1}^{n} p_T(y_i)^{\delta_i}(1 - F_T(y_i))^{1-\delta_i} \tag{18}$$

where $\delta_i = 1$ if the survival time is observed, and $\delta_i = 0$ otherwise. In contrast, under random censoring, both the lifetime T_i and the censoring time C_i are assumed to be mutually independent random variables. Here, the likelihood function of n observed data points $D = \{(y_1, \delta_1), \ldots, (y_n, \delta_n)\}$ can be written as

$$p(D) = \left(\prod_{i=1}^{n} p_T(y_i)^{\delta_i} (1 - F_T(y_i))^{1-\delta_i} \right) \left(\prod_{i=1}^{n} p_C(y_i)^{1-\delta_i} (1 - F_C(y_i))^{\delta_i} \right),$$

where $p_T(t|\theta)$ and $F_T(t|\theta)$ denote the density and the cumulative density function of the random variable T, respectively. We also consider the random censoring setup examined in [4], where both T_i and C_i are Weibull random variables

$$T_i \sim \text{Weibull}(\theta, \beta), \quad C_i \sim \text{Weibull}(\theta, \alpha), \quad i = 1, \ldots, n, \qquad (19)$$

where $\alpha, \beta > 0$ are the scale parameters and $\theta > 0$ is the common shape parameter. The joint probability density function of $Y_i = \min(T_i, C_i)$ and $\Delta_i = I(T_i < C_i)$ is

$$p_{Y,\Delta}(y, \delta | \alpha, \beta, \theta) = \left(\frac{\theta}{\alpha^\theta} \right) \left(\frac{\alpha}{\beta} \right)^{\delta_i \theta} y^{\theta-1} \exp\left(-\left(\frac{1}{\alpha^\theta} + \frac{1}{\beta^\theta} \right) y^\theta \right). \qquad (20)$$

4.1   Maximum Likelihood Estimates

Consider the **type I censoring** setup with a fixed censoring point c as described in Sect. 4. The likelihood of n data points $D = \{(y_1, \delta_1), \ldots, (y_n, \delta_n)\}$ is

$$p(D|k, \lambda) = \left(\frac{k}{\lambda^k} \right)^d \exp\left(-\frac{1}{\lambda^k} \sum_{i=1}^{n} y_i^k \right) \prod_{i=1}^{n} y_i^{\delta_i(k-1)}. \qquad (21)$$

The maximum likelihood (ML) estimates of k, λ are

$$\hat{\lambda}^k(\mathbf{y}) = \frac{1}{d} \sum_{i=1}^{n} y_i^{\hat{k}(\mathbf{y})}, \qquad (22)$$

where $d = \sum_{i=1}^{n} \delta_i$ and $\hat{k}(\mathbf{y})$ is given implicitly by

$$\frac{d}{k} + \sum_{i=1}^{n} \delta_i \log y_i - \frac{d \sum_i y_i^k \log y_i}{\sum_i y_i^k} = 0. \qquad (23)$$

The maximum likelihood estimate of k is known to exhibit large bias in small samples and when the proportion of censoring is high. Sirvanci and Yang [11] propose the alternative estimate

$$\hat{k}^{-1}(D) = \frac{1}{dg(d/n)} \sum_{i=1}^{n} \delta_i (\log c - \log y_i), \qquad (24)$$

where the function $g(\cdot)$, given by

$$g(p) = \log\log(1-p)^{-1} - \frac{1}{p}\int_0^p \log\log(1-t)^{-1}\,dt, \tag{25}$$

is a bias correction factor for the bias in estimating $1/k$. Sirvanci and Yang derive finite sample properties of this estimate and show that it has high relative efficiency in estimating $1/k$ over a range of censoring levels ($10\% - 90\%$ censoring) provided $0 < d < n$. Using the same strategy as in the complete data case (see Sect. 3.1), Yang and Xie [20] propose a new modified maximum likelihood estimate of the shape parameter k that is obtained by solving

$$\frac{d-1}{k} + \sum_{i=1}^{n}\delta_i\log y_i - \frac{d\sum_i y_i^k\log y_i}{\sum_i y_i^k} = 0. \tag{26}$$

However, this approach requires $d > 1$ to yield a positive estimate for k.

Next, we examine the **random censoring** setup described in Sect. 4. The likelihood of the data under the random censoring model is

$$p_D(D|\alpha,\beta,\theta) = \left(\frac{\theta}{\alpha^\theta}\right)^n\left(\frac{\alpha}{\beta}\right)^{d\theta}\exp\left(-\left(\frac{1}{\alpha^\theta}+\frac{1}{\beta^\theta}\right)\sum_{i=1}^n y_i^\theta\right)\prod_{i=1}^n y_i^{\theta-1} \tag{27}$$

where, as before, $d = \sum_{i=1}^n \delta_i$. From this, the maximum likelihood estimates of (α,β) are

$$\hat{\alpha}_{\mathrm{ML}} = \left(\frac{\sum_{i=1}^n y_i^\theta}{n-d}\right)^{1/\theta}, \quad \hat{\beta}_{\mathrm{ML}} = \left(\frac{\sum_{i=1}^n y_i^\theta}{d}\right)^{1/\theta} \tag{28}$$

while the maximum likelihood estimate of θ must be obtained by numerical optimisation. Clearly, the maximum likelihood estimates $(\hat{\alpha}_{\mathrm{ML}},\hat{\beta}_{\mathrm{ML}})$ exist only if $d \in (0,n)$. Alternatively, consider the following decomposition.

Lemma 1. *The joint probability density function of (Y_i, Δ_i) can be written as*

$$p_{Y,\Delta}(y,\delta|\alpha,\beta,\theta) = p_\Delta(\delta|\phi)\,p_Y(y|k,\lambda), \tag{29}$$

where $\Delta \sim \mathrm{binom}(n,\phi)$ and $Y \sim \mathrm{Weibull}(k,\lambda)$ and

$$\phi = P(T \le C) = \frac{\alpha^\theta}{\alpha^\theta+\beta^\theta}, \quad k = \theta, \quad \lambda = \frac{\beta}{(1+(\beta/\alpha)^\theta)^{1/\theta}}. \tag{30}$$

The proof is straightforward and is omitted. By Lemma 1 and invariance of the maximum likelihood estimate, the maximum likelihood estimates of α,β and θ can also be obtained from the usual maximum likelihood estimates for the binomial and Weibull distributions

$$\hat{\phi}_{\mathrm{ML}} = \frac{1}{n}\sum_{i=1}^n \delta_i, \quad \hat{\lambda}_{\mathrm{ML}}^{\hat{k}} = \frac{1}{n}\sum_{i=1}^n y_i^{\hat{k}}, \tag{31}$$

where $\hat{k}$ is given implicitly by

$$\frac{1}{n}\sum_{i=1}^{n}\log y_i + \frac{1}{k} - \frac{\sum_i y_i^k \log y_i}{\sum_i y_i^k} = 0 \tag{32}$$

and by noting that

$$\theta = k, \quad \alpha = \lambda(1-\phi)^{-1/\theta}, \quad \beta = \lambda\phi^{-1/\theta}. \tag{33}$$

These estimates exist only if $\phi_{\mathrm{ML}} \in (0,1)$ or, equivalently, $d \in (0,n)$.

4.2   Minimum Message Length Estimates

We consider first MML inference under the **type I censoring** setup described in Sect. 4. Let

$$z_c = \left(\frac{c}{\lambda}\right)^k, \quad p = 1 - \exp(-z_c), \tag{34}$$

where p is the probability that an observation is not censored. As with the complete data setting, we assume independent half-Cauchy prior distributions (see (15)) for both the shape and the scale parameters. The expected Fisher information matrix with type I censoring is [18]

$$J(k,\lambda) = n \begin{pmatrix} \frac{p+2\gamma^{(1)}(1,z_c)+\gamma^{(2)}(1,z_c)}{k^2} & -\frac{p+\gamma^{(1)}(1,z_c)}{\lambda} \\ -\frac{p+\gamma^{(1)}(1,z_c)}{\lambda} & p\left(\frac{k}{\lambda}\right)^2 \end{pmatrix}$$

where $\gamma(\cdot,\cdot)$ is the incomplete gamma function

$$\gamma(z,x) = \int_0^x t^{z-1}\exp(-t)dt \quad \text{and} \quad \gamma^{(j)}(z,x) = \frac{d^j\gamma(z,x)}{dz^j}.$$

The determinant of the expected Fisher information matrix

$$|J(k,\lambda)| = \left(\frac{n}{\lambda}\right)^2 (\gamma^{(2)}(1,z_c)p - \gamma^{(1)}(1,z_c)^2), \tag{35}$$

is a complicated function of the probability of no censoring, p. The MML87 codelength for the Weibull distribution with type I censoring is

$$I_{87}(D,\theta) = -\log\left(\frac{4}{\pi^2(1+k^2)(1+\lambda^2)}\right) + \frac{1}{2}\log|J(k,\lambda)| - \log p(D|k,\lambda) + 1 + \log\kappa_2 \tag{36}$$

where the negative log-likelihood function $-\log p(D|k,\lambda)$ is given in (21) and $\kappa_2 = 5/(36\sqrt{3})$ (see Sect. 2). As with the complete data case the MML87 estimates of k and λ must be obtained by numerically minimising (36).

Consider next the **random censoring** setup described in Sect. 4 where the lifetime T_i and the censoring time C_i are mutually independent Weibull random variables with a common shape parameter. From Lemma 1, the joint density of (Y_i, Δ_i) can be written as a product of a binomial distribution $\Delta \sim \mathrm{binom}(n,\phi)$ and a Weibull distribution $Y|\Delta \sim \mathrm{Weibull}(k,\lambda)$. This implies that an MML code for the data D could comprise two messages with the first message encoding the binary censoring indicators $\boldsymbol{\delta} = (\delta_1, \ldots, \delta_n)$, followed by another message that encodes the lifetimes $\mathbf{y} = (y_1, \ldots, y_n)$ given the censoring data $\boldsymbol{\delta}$. With this encoding, the total MML codelength for the data $D = \{(y_1, \delta_1), \ldots, (y_n, \delta_n)\}$ is

$$I_{87}(D, \alpha, \beta, \theta) = I_{87}(\boldsymbol{\delta}, \phi) + I_{87}(\mathbf{y}, k, \lambda | \boldsymbol{\delta}), \tag{37}$$

where ϕ is the probability of observing an uncensored datum. As with maximum likelihood, MML87 is invariant under one-to-one parameter transformations implying that MML87 estimates of (α, β, θ) can be obtained from MML87 estimates of (ϕ, k, λ) using the relations (33).

The MML87 codelength of the binomial distribution was derived in, for example, [13, 15] and, for a uniform prior distribution on ϕ, is given by

$$I_{87}(\boldsymbol{\delta}, \phi) = -\left(d + \frac{1}{2}\right) \log \phi - \left(n + \frac{1}{2} - d\right) \log(1 - \phi) + \frac{1}{2}(1 + \log(n/12)) \tag{38}$$

where, as before, $d = \sum_i \delta_i$. The minimum of the codelength is at the MML87 estimate

$$\hat{\phi}_{87}(\boldsymbol{\delta}) = \frac{d + 1/2}{n + 1}. \tag{39}$$

The conditional codelength of $\mathbf{y}$ given the censoring indicators $\boldsymbol{\delta}$, $I_{87}(\mathbf{y}, k, \lambda | \boldsymbol{\delta})$ is the MML87 codelength (16) for the Weibull distribution discussed in Sect. 3.2; as before this can be minimised numerically for k and λ. Once we have $(\hat{\phi}_{87}, \hat{k}_{87}, \hat{\lambda}_{87})$ we can transform these to $(\hat{\alpha}_{87}, \hat{\beta}_{87}, \hat{\theta}_{87})$ via (33).

5 Experiments

5.1 Complete Data

The MML87 estimate of the shape parameter k derived in Sect. 3.2 is now compared to the maximum likelihood (MLE) estimate (11) and the modified maximum likelihood (MMLE) estimate (13) using simulated data. In each simulation run, n data points were generated from the model Weibull$(k, \lambda = 1)$ where $n = \{10, 20, 50\}$ and the shape parameter was set to $k \in \{0.5, 1, 5, 10\}$. Given the data, MLE, MMLE and MML87 estimates were computed and compared in terms of bias and mean squared error. For each value of (k, n) 10^5 simulations were performed and the average bias and mean squared error results are shown in Table 1 for each estimate.

It is clear that the MMLE and MML87 estimates improve significantly on the maximum likelihood estimate in terms of both bias and mean squared error for each tested value of (n, k). We further note that the MMLE estimate of k is slightly less biased than the proposed MML87 estimate, though the two estimates are virtually indistinguishable in terms of the average mean squared error. As discussed in Sect. 3.2, the MMLE estimate is a special case of the MML87 estimator for a particular choice of the prior distribution with complete data, and it is therefore expected that the two estimates will have similar behaviour.

Table 1. Bias and mean squared error for maximum likelihood (MLE), modified maximum likelihood (MMLE) and MML87 estimates of k computed over 10^5 simulations runs with $\lambda = 1$.

n	k	Bias			Mean Squared Error		
		MLE	MMLE	MML87	MLE	MMLE	MML87
10	0.5	0.085	**0.008**	0.063	0.038	**0.023**	0.029
	1.0	0.168	**0.015**	0.085	0.152	**0.094**	0.099
	5.0	0.850	**0.085**	0.117	3.836	2.352	**2.336**
	10.0	1.692	**0.164**	0.181	14.973	9.143	**9.124**
20	0.5	0.038	**0.004**	0.030	0.012	**0.009**	0.011
	1.0	0.076	**0.008**	0.040	0.048	**0.037**	0.038
	5.0	0.371	**0.031**	0.045	1.194	0.927	**0.923**
	10.0	0.774	**0.093**	0.100	4.881	3.761	**3.757**
50	0.5	0.015	**0.002**	0.012	0.004	**0.003**	0.003
	1.0	0.029	**0.004**	0.016	0.015	**0.013**	0.014
	5.0	0.143	**0.016**	0.021	0.366	0.329	**0.328**
	10.0	0.279	**0.025**	0.028	1.456	1.311	**1.310**

5.2   Censored Data

We also compared the MML87 estimate (see Sect. 4.2) to the maximum likelihood estimate (MLE) (23) and the modified maximum likelihood estimate (MMLE) (26) under type I censored data. The experimental setup was identical to that for complete data with the following changes: (i) the proportion of uncensored observations was set to $p \in \{0.3, 0.5, 0.7, 0.9\}$, and (ii) $n \in \{20, 30, 40\}$ data points were generated during each simulation run. We restricted the experiments to exclude data sets where the number of uncensored observations $d(= \sum_i \delta_i) <$ 2, as the MLE and MMLE estimates are not defined for small d. In addition to the bias and the mean squared error in estimating the shape parameter, we computed the Kullback–Leibler (KL) divergence [6] between the data generating model and each estimated model. The results averaged over 10^5 simulations runs for each combination of (n, p, k) are shown in Table 2.

We again observe that the MLE estimate of k is strongly biased particularly for small k and p. While the MMLE is less biased than the proposed MML87 estimate, the MML87 estimate achieves smaller mean squared error and smaller KL divergence compared to the MMLE in all experiments. Additionally, we observe that the KL divergence for the MMLE model is similar to the MLE model, despite the significant reduction in bias of estimating the shape parameter k achieved by the MMLE. The proposed MML87 estimate is an improvement over the MLE and highly competitive against estimators that are designed to reduce MLE bias (e.g., Yang and Xie [20]).

6   Discussion

This manuscript demonstrates how the MML framework of inductive inference can be applied to the Weibull distribution with either complete or censored

Table 2. Bias, mean squared error and Kullback–Leibler (KL) divergence for maximum likelihood (MLE), modified maximum likelihood (MMLE) and MML87 estimates of k computed over 10^5 simulations runs with $\lambda = 1$; p denotes the proportion of uncensored observations.

n	p	k	Bias			Mean Squared Error			KL Divergence		
			MLE	MMLE	MML87	MLE	MMLE	MML87	MLE	MMLE	MML87
20	0.3	0.5	0.114	**0.002**	0.070	0.158	0.077	**0.040**	0.069	0.060	**0.042**
		1.0	0.055	**0.005**	0.038	0.042	0.031	**0.026**	0.060	0.056	**0.043**
		5.0	0.037	**0.006**	0.021	0.020	0.017	**0.016**	0.057	0.054	**0.044**
		10.0	0.019	**-0.001**	0.006	0.011	0.010	**0.010**	0.048	0.046	**0.039**
	0.5	0.5	0.114	**0.002**	0.070	0.158	0.077	**0.040**	0.069	0.060	**0.042**
		1.0	0.055	**0.005**	0.038	0.042	0.031	**0.026**	0.060	0.056	**0.043**
		5.0	0.037	**0.006**	0.021	0.020	0.017	**0.016**	0.057	0.054	**0.044**
		10.0	0.019	**-0.001**	0.006	0.011	0.010	**0.010**	0.048	0.046	**0.039**
	0.7	0.5	0.114	**0.002**	0.070	0.158	0.077	**0.040**	0.069	0.060	**0.042**
		1.0	0.055	**0.005**	0.038	0.042	0.031	**0.026**	0.060	0.056	**0.043**
		5.0	0.037	**0.006**	0.021	0.020	0.017	**0.016**	0.057	0.054	**0.044**
		10.0	0.019	**-0.001**	0.006	0.011	0.010	**0.010**	0.048	0.046	**0.039**
	0.9	0.5	0.114	**0.002**	0.070	0.158	0.077	**0.040**	0.069	0.060	**0.042**
		1.0	0.055	**0.005**	0.038	0.042	0.031	**0.026**	0.060	0.056	**0.043**
		5.0	0.037	**0.006**	0.021	0.020	0.017	**0.016**	0.057	0.054	**0.044**
		10.0	0.019	**-0.001**	0.006	0.011	0.010	**0.010**	0.048	0.046	**0.039**
30	0.3	0.5	0.067	**0.002**	0.048	0.059	0.039	**0.026**	0.042	0.039	**0.028**
		1.0	0.035	**0.003**	0.024	0.020	0.017	**0.016**	0.038	0.036	**0.029**
		5.0	0.023	**0.004**	0.013	0.012	0.010	**0.010**	0.036	0.035	**0.030**
		10.0	0.016	**0.003**	0.008	0.007	0.007	**0.007**	0.034	0.033	**0.029**
	0.5	0.5	0.067	**0.002**	0.048	0.059	0.039	**0.026**	0.042	0.039	**0.028**
		1.0	0.035	**0.003**	0.024	0.020	0.017	**0.016**	0.038	0.036	**0.029**
		5.0	0.023	**0.004**	0.013	0.012	0.010	**0.010**	0.036	0.035	**0.030**
		10.0	0.016	**0.003**	0.008	0.007	0.007	**0.007**	0.034	0.033	**0.029**
	0.7	0.5	0.067	**0.002**	0.048	0.059	0.039	**0.026**	0.042	0.039	**0.028**
		1.0	0.035	**0.003**	0.024	0.020	0.017	**0.016**	0.038	0.036	**0.029**
		5.0	0.023	**0.004**	0.013	0.012	0.010	**0.010**	0.036	0.035	**0.030**
		10.0	0.016	**0.003**	0.008	0.007	0.007	**0.007**	0.034	0.033	**0.029**
	0.9	0.5	0.067	**0.002**	0.048	0.059	0.039	**0.026**	0.042	0.039	**0.028**
		1.0	0.035	**0.003**	0.024	0.020	0.017	**0.016**	0.038	0.036	**0.029**
		5.0	0.023	**0.004**	0.013	0.012	0.010	**0.010**	0.036	0.035	**0.030**
		10.0	0.016	**0.003**	0.008	0.007	0.007	**0.007**	0.034	0.033	**0.029**
40	0.3	0.5	0.047	**0.002**	0.035	0.033	0.025	**0.018**	0.029	0.028	**0.021**
		1.0	0.025	**0.002**	0.017	0.014	0.012	**0.011**	0.027	0.026	**0.022**
		5.0	0.017	**0.003**	0.009	0.008	0.008	**0.007**	0.027	0.026	**0.023**
		10.0	0.014	**0.004**	0.008	0.005	0.005	**0.005**	0.026	0.026	**0.023**
	0.5	0.5	0.047	**0.002**	0.035	0.033	0.025	**0.018**	0.029	0.028	**0.021**
		1.0	0.025	**0.002**	0.017	0.014	0.012	**0.011**	0.027	0.026	**0.022**
		5.0	0.017	**0.003**	0.009	0.008	0.008	**0.007**	0.027	0.026	**0.023**
		10.0	0.014	**0.004**	0.008	0.005	0.005	**0.005**	0.026	0.026	**0.023**
	0.7	0.5	0.047	**0.002**	0.035	0.033	0.025	**0.018**	0.029	0.028	**0.021**
		1.0	0.025	**0.002**	0.017	0.014	0.012	**0.011**	0.027	0.026	**0.022**
		5.0	0.017	**0.003**	0.009	0.008	0.008	**0.007**	0.027	0.026	**0.023**
		10.0	0.014	**0.004**	0.008	0.005	0.005	**0.005**	0.026	0.026	**0.023**
	0.9	0.5	0.047	**0.002**	0.035	0.033	0.025	**0.018**	0.029	0.028	**0.021**
		1.0	0.025	**0.002**	0.017	0.014	0.012	**0.011**	0.027	0.026	**0.022**
		5.0	0.017	**0.003**	0.009	0.008	0.008	**0.007**	0.027	0.026	**0.023**
		10.0	0.014	**0.004**	0.008	0.005	0.005	**0.005**	0.026	0.026	**0.023**

data. By minimising a single inferential quantity, the codelength, we obtain new parameter estimates that have advantages (in terms of bias and mean squared estimation error) over the usual maximum likelihood estimates. MML inference can also be applied to data with type II censoring. In type II censoring, the experiment begins with n items under observation and is stopped after the first m failures are observed. The determinant of the expected Fisher information matrix for the Weibull distribution under Type II censoring is known [19]. Assuming the same half-Cauchy prior distribution for the parameters, experiments (not shown) demonstrate that the MML estimates are less biased and have better mean squared error compared to the maximum likelihood estimates in this setting.

References

1. Conway, J.H., Sloane, N.J.A.: Sphere Packing, Lattices and Groups, 3rd edn. Springer-Verlag, Cham (1998). https://doi.org/10.1007/978-1-4757-6568-7
2. Cox, D.R., Reid, N.: Parameter orthogonality and approximate conditional inference. J. Roy. Stat. Soc. (Ser. B) **49**(1), 1–39 (1987)
3. Cox, D.R., Reid, N.: A note on the difference between profile and modified profile likelihood. Biometrika **79**(2), 408–411 (1992)
4. Danish, M.Y., Aslam, M.: Bayesian inference for the randomly censored Weibull distribution. J. Stat. Comput. Simul. **84**(1), 215–230 (2012). https://doi.org/10.1080/00949655.2012.704516
5. Hirose, H.: Bias correction for the maximum likelihood estimates in the two-parameter Weibull distribution. IEEE Trans. Dielectr. Electr. Insul. **6**(1), 66–68 (1999). https://doi.org/10.1109/94.752011
6. Kullback, S., Leibler, R.A.: On information and sufficiency. Ann. Math. Stat. **22**(1), 79–86 (1951)
7. Polson, N.G., Scott, J.G.: On the half-Cauchy prior for a global scale parameter. Bayesian Anal. **7**(4) (2012). https://doi.org/10.1214/12-BA730
8. Ross, R.: Formulas to describe the bias and standard deviation of the ML-estimated Weibull shape parameter. IEEE Trans. Dielectr. Electr. Insul. **1**(2), 247–253 (1994). https://doi.org/10.1109/94.300257
9. Schmidt, D.F., Makalic, E.: Minimum message length analysis of multiple short time series. Statist. Probab. Lett. (2016). https://doi.org/10.1016/j.spl.2015.09.021
10. Schwarz, G.: Estimating the dimension of a model. Ann. Stat. **6**(2), 461–464 (1978)
11. Sirvanci, M., Yang, G.: Estimation of the Weibull parameters under type I censoring. J. Am. Stat. Assoc. **79**(385), 183–187 (1984). https://doi.org/10.2307/2288354
12. Tanaka, H., Pal, N., Lim, W.K.: On improved estimation under Weibull model. J. Stat. Theory Pract. **12**(1), 48–65 (2017). https://doi.org/10.1080/15598608.2017.1305921
13. Wallace, C.S.: Statistical and Inductive Inference by Minimum Message Length. first edn. In: Jordan, M.I. (ed.) Information Science and Statistics. Springer, Cham (2005). https://doi.org/10.1007/0-387-27656-4
14. Wallace, C.S., Boulton, D.M.: An information measure for classification. Comput. J. **11**(2), 185–194 (1968). http://www.allisons.org/ll/MML/Structured/1968-WB-CJ/
15. Wallace, C.S., Dowe, D.L.: MML clustering of multi-state, Poisson, von Mises circular and Gaussian distributions. Stat. Comput. **10**(1), 73–83 (2000)

16. Wallace, C.S., Freeman, P.R.: Estimation and inference by compact coding. J. Roy. Stat. Soc. (Ser. B) **49**(3), 240–252 (1987)
17. Wallace, C.S., Freeman, P.R.: Single-factor analysis by minimum message length estimation. J. Roy. Stat. Soc. (Ser. B) **54**(1), 195–209 (1992)
18. Watkins, A., John, A.: On the expected fisher information for the Weibull distribution with type I censored data. Int. J. Pure Appl. Math. **15**(4), 401–412 (2004)
19. Watkins, A., John, A.: On the expected fisher information for the Weibull distribution with type II censored data. Int. J. Pure Appl. Math. **26**(1), 91–104 (2006)
20. Yang, Z., Xie, M.: Efficient estimation of the Weibull shape parameter based on a modified profile likelihood. J. Stat. Comput. Simul. **73**(2), 115–123 (2003). https://doi.org/10.1080/00949650215729

Multiple Teacher Model for Continual Test-Time Domain Adaptation

Ran Wang[ID], Hua Zuo[✉][ID], Zhen Fang[ID], and Jie Lu

Australian Artificial Intelligence Institute, Faculty of Engineering and IT,
University of Technology Sydney, Sydney, NSW, Australia
ran.wang-2@student.uts.edu.au, {hua.zuo,zhen.fang,jie.lu}@uts.edu.au

Abstract. Test-time adaptation (TTA) without accessing the source data provides a practical means of addressing distribution changes in testing data by adjusting pre-trained models during the testing phase. However, previous TTA methods typically assume a static, independent target domain, which contrasts with the actual scenario of the target domain changing over time. Using previous TTA methods for long-term adaptation often leads to problems of error accumulation or catastrophic forgetting, as it relies on the capability of a single model, leading to performance degradation. To address these challenges, we propose a multiple teacher model approach (MTA) for continual test-time domain adaptation. Firstly, we reduce error accumulation and leverage the robustness of multiple models by implementing a weighted and averaged multiple teacher model that provides pseudo-labels for enhanced prediction accuracy. Then, we mitigate catastrophic forgetting by logging mutation gradients and randomly restoring some parameters to the weights of the pre-trained model. Our comprehensive experiments demonstrate that MTA outperforms other state-of-the-art methods in continual time adaptation.

Keywords: Domain Adaptation · Test-time Adaptation

1 Introduction

Deep neural network models can yield remarkable results when both training and testing data come from the same distribution. However, real-world circumstances, such as weather conditions or device aging, can corrupt testing data or induce domain shifts. Test-time adaptation methodologies deal with this issue by continuously improving the adaptability of the model during the testing phase [2, 15, 24, 26]. Previous approaches of test-time adaptation typically require access to source data and assume the target domain as static and independent. However, in reality, access to source data is often not possible during inference time due to privacy or legal restrictions. Furthermore, testing samples are consistently online and subject to distribution changes over time [19]. For instance, the performance of an autonomous vehicle might considerably decrease during a transition from sunny to rainy weather conditions [1, 6, 11]. Therefore, continual

© The Author(s), under exclusive license to Springer Nature Singapore Pte Ltd. 2024
T. Liu et al. (Eds.): AI 2023, LNAI 14471, pp. 304–314, 2024.
https://doi.org/10.1007/978-981-99-8388-9_25

adaptation during test-time is crucial for effectively managing real-world domain transfer issues.

Current test-time adaptation methods include entropy minimization, image augmentation, and pseudo-labeling. However, existing methods may introduce biases towards extreme samples in continually shifting environments (e.g., noise [20]), and continual adaptation to new samples could lead to catastrophic forgetting of previously learned knowledge [8, 25]. Additionally, most of the current methods rely on a single model. However, the performance of a single model varies across different scenarios, and using a single model disregards the advantages and strengths of other models in handling different types of samples. Consequently, it fails to deliver optimal performance in all circumstances.

To address the challenges of continual adaptation, we propose multiple teacher model for continual test-time domain adaptation (MTA). Our method consists of two components: (1) To tackle model accumulation errors and improve robustness, we utilize multiple models as teacher model. This enhances the quality of pseudo-labels and provides more precise predictions through weighted averaging. (2) To mitigate catastrophic forgetting and retain the knowledge of the source model, we freeze the teacher model to ensure the preservation of a portion of the source model. Additionally, we record mutation gradients to randomly restore some parameters to the pre-trained source model. By leveraging the integrated learning capabilities and diversity of the multiple models, MTA generates more accurate pseudo-labels, reduces error accumulation, and enables adaptation to continuous changes in the test environment over time.

Our approach is easy to implement and can be integrated with any existing pre-trained models without the need for retraining on source data and needing fewer trainable parameters. We assess MTA and state-of-the-art continuous testing time adaptation baselines on multiple datasets, including common TTA benchmarks (CIFAR-10C, CIFAR-100C, and ImageNet-C). Our contributions can be summarized as follows:

· We incorporate multiple models as teacher model, enhancing the accuracy of predictions through weighted-average pseudo-labels and average predictions.
· By freezing multiple teacher model and randomly restoring parameters, we effectively mitigate catastrophic forgetting.

2   Related Work

2.1   Unsupervised Domain Adaptation

The goal of unsupervised domain adaptation (UDA) is to better adapt to domain offsets in the source and unlabelled target domains [7,27,28]. UDA methods usually use convolutional neural network models such as ResNet18, ResNet50 [3], etc., which have powerful feature extraction capabilities, to align the source and target domain distributions through adversarial learning [23] or distance loss [16]. Adversarial learning aims to find a shared representation between the source and target distributions, while distance loss minimizes the distance between them.

However, the models of these methods are fixed during the testing phase and cannot better adapt to the changing test distribution during testing. In order to improve the performance of domain adaptation during testing, the test-time adaptation method is proposed.

2.2 Test-Time Adaptation

Test-time adaptation refers to domain adaptation during the testing phase [24, 26]. Unlike domain adaptation, which needs both source and target domain data, test-time adaptation does not require access to source data. Previous methods include: entropy minimization [19,24] makes model predictions more certain by minimising test entropy, batch normalisation [26] adjusts the normalized data during testing to make the model stable, and pseudo-labeling [18] helps to update the model by assigning predictive labels to unlabelled test data. However, these methods mainly target single-target domain scenarios, and after adapting to one changing test scenario, they reset the model to adapt to the next scenario.

Recently, CoTTA [25] introduced a continual adaptation setting that does not reset the model, thus enabling continuous long-term adaptation across multiple scenes. This setup better reflects real-world conditions but also poses challenges related to error accumulation and catastrophic forgetting. To address these issues, CoTTA proposes to use incremental predictions from weighted prediction pseudo-labels as pseudo-labels to obtain more accurate average predictions through exponentially averaged teacher-student models. NOTE [8] uses instance-aware batch normalization to correct for out-of-distribution samples to improve the performance of test samples as well as the RMT [5] method, which utilizes the average teacher model approach to obtain more robust predictions. However, these methods rely heavily on the robustness of a single model, while the adaptive ability of various models in various scenarios under continuous adaptation varies, which can lead to a decrease in the generalization ability of a single model in multiple scenarios.

2.3 Multi-model Transfer Learning

Multi-model transfer learning strives to transfer pre-trained models with varying algorithms, datasets, architectures, etc., to the target task through adaptive assembly and learning optimization [12,13]. By combining the strengths of different pre-trained models, multi-model can enhance the robustness of the model by compensating for each other's strengths and the ability to deal with more complex tasks by expanding the scope of the model's understanding. Recent approaches include the fusion of teacher-student model representations through fusion representation, so that the student model gets knowledge from multiple teacher models [14]. By fusing all layer parameters for different models with the same structure to build a completely new fused model [22]. Weight selection by activating model-specific layer parameters [21]. However, these methods require

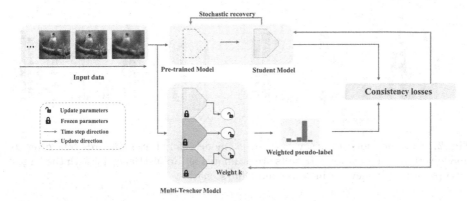

Fig. 1. An overview of multiple teacher model for continual test-time adaptation (MTA). MTA mitigates error accumulation by obtaining weighted average pseudo-labeled predictions from a multiple-teacher model. The prediction weights of the teacher model and the weights of the student model are updated by consistency loss. Student model weights are recovered randomly to mitigate the effects of catastrophic forgetting.

fine-tuning of the model parameters and are computationally intensive, time-consuming, and more suitable for less complex situations. To speed up this process and achieve more robust model predictions, we reduce the computational effort by freezing the model parameters and adjusting only the out-of-model weight parameters, which reduces the computational effort.

3   Methodology

Previous testing time adaptation typically involves adapting to a single scene (e.g., Gaussian noise, blur) at a time. In contrast, our approach to continual test-time adaptation differs from previous methods as it involves continual adaptation to multiple scenes over time [5, 25].

3.1   Problem Definition

Given a pre-trained model f_θ with parameters θ, the aim of full test-time adaptation is to adapt f_θ to a stream of online unlabeled target domain data without accessing the source data or re-training. The target domain data is represented as x^T, which is a batch of corrupted samples from the sample set S^T, provided in sequential order as $x_t^T \rightarrow x_{t+1}^T$, while S^T changes over time as $S_t^T \rightarrow S_{t+1}^T$. At time t, the model f_θ adapts and predicts on x^T until it adapts to the entire stream of data. The key differentiation in this testing methodology is whether the model is reset for each scene. Continual test-time adaptation setting does not reset the model.

Figure 1 shows a brief overview of MTA. Firstly, the test data is fed into both the student model and the teacher model in chronological order. While

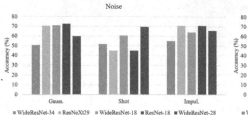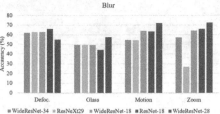

Fig. 2. Classification accuracy (%) of 5 models for the standard ImageNet-to-ImageNetC adaptation task. Results are evaluated on RobustBench [3] with the largest corruption severity level 5 in Noise and Blur scenarios.

the student model relies on a single pre-trained model to make predictions, the multi-teacher model utilizes multiple pre-trained models for predictions. The predictions from the multi-teacher model are then weighted to obtain a weighted average prediction. Subsequently, this weighted average prediction, along with the student model's predictions, calculates a consistency loss. The consistency loss is used to update the parameters of the student model and the prediction weight parameters of the multi-teacher model. Throughout this process, changes in gradients are recorded. When gradient exceeds the threshold, some of the student model's parameters are reset to the source pre-trained model.

3.2   Multiple Teacher Model Weighted Pseudo-labels

Pre-trained Model. To address the burden of re-training and not accessing the source data, the selection of pre-trained models becomes crucial, as a well-chosen pre-trained model can lead to better adaptation results. Previous continual test-time adaptation methods, such as CoTTA [25], are limited to a single pre-trained model, relying on the robustness of one model. However, each model may have different adaptation capabilities for various data [3], as demonstrated in Fig. 2, which shows the accuracy of 5 models on two corruption scenarios with severity level 5 in ImageNet-C. The accuracy of each model differs in different scenarios. To enhance the utilization of the unique features of each model and enhance the resilience of the adaptation phase, a straightforward and intuitive solution is to employ multiple models as teacher models.

Multi-model Weighted Pseudo-label. In the continual adaptation, the accuracy of pseudo-labels is highly sensitive to extreme samples. For example, CoTTA [25] employs image augmentations and weighted average predictions for a single model to improve the accuracy of pseudo-labels. However, such image augmentations add additional noise and require more computational resources. Inspired by the idea of increasing robustness and reducing computation by using multiple models [14,21,22], we adopt k models as teacher models $f_{\theta_{t_1}}, \ldots, f_{\theta_{t_k}}$. To reduce the number of parameters, we freeze the teacher model parameters. At time-step t, we first predict pseudo-labels $y_1, \ldots, y_k$ for each teacher model and apply a weight parameter β_k to each prediction to obtain a multi-model weighted prediction pseudo-label:

$$\langle y_t \rangle_w = \sum_{k=1}^{K} \beta_k f_{\theta_{t_k}} \tag{1}$$

where $\langle y_t \rangle_w$ is the weighted pseudo-label of the teacher model.

In handling the problem of distribution shift, consistency loss can autonomously adjust the model's stability [2]. In MTA, we use multi-teacher model predictions as pseudo-labels and enforce consistency loss between the weighted prediction pseudo-label $\langle y_t \rangle_w$ and student model f_{θ_s}. Given target data x_t^T, the consistency loss can be defined as:

$$\mathcal{L}_c = -\sum_{i=1}^{N} \langle y_{t,i} \rangle_w \log(f_{\theta_s}(x_{t,i}^T)) \tag{2}$$

where N is the classes, for each time step t, compare each category i the weighted pseudo-label $\langle y_{t,i} \rangle_w$ with the student model predicted probability $\log(f_{\theta_s}(x_{t,i}^T))$ The student model parameters will then be updated over time step as $\theta_t \to \theta_{t+1}$. Our final prediction for target data x_t^T comes from weighted pseudo-label $\langle y_t \rangle_w$.

By using weighted pseudo-labeling with multiple models, not only the accumulation of errors is effectively avoided in long-term adaptation, but also the robustness to different test data distributions can be increased on the basis of a single model, and more accurate predictions can be achieved by combining the advantages of multiple models.

3.3 Maximally Gradient-Stochastic Recovery

The problem of catastrophic forgetting is to be faced when more accurate pseudo-labeling can mitigate the accumulation of errors over long periods of continual adaptation. Prolonged test-time adaptation can bias the direction of model updating towards new samples, which can lead to forgetting the original knowledge of the model. In particular, strong domain bias can lead to the model being biased in the direction of the bias and not being able to accurately predict either the new distribution or the original distribution.

In order to mitigate the detrimental impact of long-term continual adaptation, we choose to record the gradient that is most disturbed by the new samples, previous research [9] has revealed that these samples with high gradient exacerbate model bias. To address this, we randomly reset the parameters of the student model to the source pre-trained model after a certain number of iterations. By employing this random resetting, the student model reacquires the original knowledge from the source.

As the time step transitions from $t \to t+1$, the update of the student model's parameters can be defined as from $\theta_t \to \theta_{t+1}$, with the gradient update being $\nabla L(\mathbf{x}_t, y_t)$. The update of the student model's parameters will be as follows:

$$\theta_{t+1} = \theta_t - \lambda \nabla L(\mathbf{x}_t, y_t) \tag{3}$$

We define the gradient value and maximal gradient as

$$G_t = \|\nabla L(\mathbf{x}_t, y_t)\| \tag{4}$$

$$\forall t \le t' \le t + k, \quad G_{t'} > \alpha \to G_{t'} = \max_{t \le \tau \le t+k} G_\tau \tag{5}$$

If the gradient value G_t consistently exceeds the threshold α within a continuous time period, i.e., from time step t to $t + k$, we consider that the gradient value has reached its maximum within this specific time frame. When the gradient reaches its maximum, we implement a random parameter recovery strategy on the student model [25]. Assuming at time t, a convolutional layer parameter of the student model f_{θ_s} is W_t, which updates to W_{t+1} after one-time step, the update rule for random recovery can be expressed by the following equation:

$$W_{t+1} = \beta \odot W_0 + (1 - \beta) \odot W_{t+1} \tag{6}$$

where β is a same shape probability matrix of W_{t+1} between 0 and 1, and $\odot$ denotes element-wise multiplication. This update rule allows the update process by keeping part of the current parameters in W_{t+1} and restoring part of the initial parameters in W_0. Stochastic recovery can be considered as a special form of dropout. With the stochastic recovery of some of the parameters to the source model, the model does not deviate too far from the source model, which can be effective in overcoming a certain degree of catastrophic forgetting and can prevent the model from collapsing.

4   Experiments

We implemented our results with other baselines on three continual test-time adaptation benchmarks via PyTorch framework: CIFAR10-to-CIFAR10-C, CIFAR100-to-CIFAR100-C, and ImageNet-to-ImageNet-C.

4.1   Implementation Details

Dataset and Baselines. CIFAR10-C, CIFAR100-C and ImageNet-C are the most commonly used TTA benchmarks for evaluating the robustness to corruptions [10], where CIFAR10-C and CIFAR100-C have 10,000 test images including 10/100 categories. ImageNet-C contains 50,000 test images with 1000 categories. CIFAR10-C, CIFAR100-C and ImageNet-C provide 15 types of corruptions, which are noise (gaussian, shot, impulse), blur (defocus, glass, motion, zoom) weather (snow, frost, fog) and digital (brightness, contrast, elastic trans, pixelate, jpeg). In order to be consistent with the previous methodology, we tested using the most severe damage level 5. We consider the classical models of TTA and the state-of-the-art test-time adaptation models, which include RMT [5], NOTE [8], CoTTA [25], Source [25], BN [25] and TENT [24].

Table 1. Classification **accuracy (%)** for the standard CIFAR10-to-CIFAR10-C, CIFAR100-to-CIFAR100-C and ImageNet-to-ImageNet-C adaptation task with the highest corruption severity level 5 at time step t. The **bold** number indicates best result.

$t \text{-----------------------------------} \rightarrow$

	Method	Gau.	Shot	Imp.	Defo.	Glass	Mot.	Zoom	Snow	Frost	Fog	Brit.	Cont.	Elas.	Pixel	JPG	Avg
CIFAR10-C	Source	27.7	34.3	27.1	53.1	45.7	65.2	58.0	74.9	58.7	74.0	90.7	53.3	73.4	41.5	69.7	56.5
	BN	71.9	73.9	63.7	87.2	64.7	85.8	87.9	82.7	82.6	84.7	91.6	87.4	76.2	80.3	72.7	79.6
	TENT	75.2	79.4	71.4	85.6	68.9	83.5	85.9	80.9	81.4	81.4	87.8	79.7	74.3	79.2	75.1	79.3
	NOTE	76.2	77.0	68.9	88.2	69.1	88.2	88.1	84.7	86.0	86.7	91.4	92.5	78.8	83.1	77.0	82.4
	CoTTA	75.7	78.7	73.4	88.4	72.4	87.8	89.7	85.2	85.9	87.6	92.5	89.4	81.7	86.6	82.7	83.9
	RMT	78.1	81.4	75.9	89.2	76.4	88.0	89.6	87.0	**87.6**	**88.6**	91.7	**89.9**	84.8	**88.7**	85.4	85.5
	MTA	**85.0**	**82.7**	**82.0**	**91.0**	**79.9**	**88.8**	**91.6**	**87.3**	86.0	84.3	**93.9**	86.1	**87.3**	84.0	**89.4**	**86.6**
CIFAR100-C	Source	27.0	32.0	60.6	70.7	45.9	69.2	71.2	60.5	54.2	49.7	70.5	44.9	62.8	25.3	58.8	53.6
	BN	39.1	40.1	34.3	66.3	42.4	63.5	64.8	53.3	53.1	57.2	67.7	64.4	54.2	56.4	44.5	53.4
	TENT	**62.8**	**64.2**	58.3	62.1	48.8	51.7	51.5	41.6	36.3	28.9	29.6	17.7	12.0	11.5	9.6	39.1
	NOTE	34.4	37.4	28.0	63.2	39.5	65.1	63.3	60.4	58.3	57.7	71.4	67.7	56.2	52.3	49.1	53.6
	CoTTA	58.8	62.1	60.5	72.3	63.0	**71.6**	**74.3**	67.0	67.7	59.3	**75.4**	72.9	66.9	71.5	66.1	67.3
	RMT	60.9	62.0	64.5	**72.8**	66.0	69.8	70.7	**71.0**	**71.9**	67.7	73.3	**74.1**	71.5	**72.8**	67.9	69.1
	MTA	61.7	62.6	**69.0**	70.5	**66.7**	69.9	72.4	68.8	70.9	**69.6**	72.4	73.5	**74.9**	71.0	**70.5**	**69.6**
ImageNet-C	Source	2.2	2.9	1.8	18.3	10.2	14.8	22.0	16.5	22.9	24.1	58.7	5.5	17.5	20.7	31.4	18.0
	BN	10.6	11.5	11.0	9.2	10.0	18.7	30.2	27.4	26.2	37.4	55.7	7.9	35.5	39.7	29.3	24.0
	TENT	18.4	10.3	9.0	6.9	7.8	15.3	27.6	38.4	37.0	48.3	61.8	27.9	49.2	52.6	46.7	30.5
	NOTE	12.4	14.3	12.8	16.7	16.8	26.4	34.6	35.0	31.4	42.1	56.5	24.1	38.8	45.9	37.2	29.7
	CoTTA	15.3	17.9	19.4	18.7	21.0	31.4	42.5	39.7	39.5	44.7	63.4	**33.9**	42.2	**58.8**	**50.0**	35.9
	RMT	22.1	26.9	30.1	26.5	**28.9**	**36.9**	**42.9**	40.6	40.8	49.6	57.1	32.9	**51.0**	34.3	33.1	36.9
	MTA	**43.0**	**43.8**	**45.9**	**27.9**	27.9	29.0	38.8	36.0	36.0	35.7	**64.8**	19.8	29.0	48.7	41.5	**37.9**

Hyperparameters. We follow the implementation of previous methods [5, 24, 25]. The pre-trained models are from RobustBench [3]. We choose three models for the teacher model in each task (k = 3). In more detail, for CIFAR10-C, we use ResNet-18 [6], WideResNet-28 [17] and WideResNet-18 [4]. For CIFAR100-C, we use ResNeXt-29 [1], WideResNet-28 [6] and WideResNet-18 [17]. We use ResNet-50 [3] for ImageNet-C. The batch size of all experiments is 100. The data stream begins with the Gaussian dataset and ends with the JPEG dataset. For optimization, we use a learning rate of 1e−4, we set the initial teacher model weight as 0.1, 0.2, 0.7, restoration probability β as 0.05 and threshold α as 15 for all our experiments. Our results are based on the average of 5 times results.

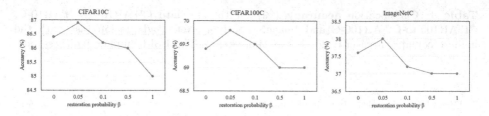

Fig. 3. The average classification accuracy of different restoration probability parameter β in the three benchmarks for MTA.

4.2   Experiment Results

Table 1 presents the performance of several models under the CIFAR10-to-CIFAR10-C scenario. When the source model is directly applied without adaptation, the accuracy is lower at 56.5%. However, simple adaptation improves the accuracy to 79.3%, indicating that directly using the source model is not effective in managing severe corruption. On the other hand, test-time adaptation single-model methods show notable improvements in accuracy. Notably, NOTE, CoTTA, and RMT have achieved an accuracy rate of 85.5%. However, the adaptability of a single model remains viable, and MTA achieves superior performance on CIFAR10-C, with accuracy improving to 86.6%. This suggests that weight prediction of multi-models is more precise than that of a single model without additional computational overhead.

Compared to CIFAR10-C, CIFAR100-C is a more challenging categorization task, with categories expanding from 10 to 100. The results for the Source and simple adaptation with BN remain stable, but the accuracy of TENT drops to 39.1%, attributable to the accumulation of severe errors leading to a significant dip in the final damage type prediction accuracy. Other methods such as NOTE, CoTTA, and RMT, which can combat error accumulation, outperform others, but MTA still yields the highest accuracy. ImageNet-C is the most complex dataset, which contains 1000 categories. For more difficult classification problems, the Source model delivers very low accuracy, while simple adaptation brings about considerable improvement and MTA achieve the best performance.

In Fig. 3, we discuss the choice of the restoration probability parameter β, and a comparison of the three benchmarks gives the highest average accuracy when $\beta = 0.05$.

5   Conclusion

In this paper, we present an effective and easy-to-implement method for continual test-time adaptation. The approach mitigates the problems of error accumulation and catastrophic forgetting during test-time adaptation by combining multiple models and introducing maximum gradient stochastic recovery. This technique significantly improves the performance and stability of the model. The experiments demonstrate the superiority of MTA as evaluated under multiple successive damage scenarios. Compared to the previous single-model TTA approach, our method achieves a 1–3% accuracy improvement across all scenarios. Additionally, our approach is highly flexible and scalable, allowing us to combine a variety of excellent pre-trained models based on actual needs. This work represents a meaningful step towards further advancement in testing time-adaptive algorithms.

Acknowledgements. This work is supported by the Australian Research Council under Discovery Early Career Researcher Award DE220101075.

References

1. Addepalli, S., Jain, S., Babu R., V.: Efficient and effective augmentation strategy for adversarial training. In: NeurIPS (2022)
2. Chen, L., Zhang, Y., Song, Y., Shan, Y., Liu, L.: Improved test-time adaptation for domain generalization. CoRR abs/2304.04494 (2023)
3. Croce, F., et al.: RobustBench: a standardized adversarial robustness benchmark. In: NeurIPS (2021)
4. Diffenderfer, J., Bartoldson, B.R., Chaganti, S., Zhang, J., Kailkhura, B.: A winning hand: compressing deep networks can improve out-of-distribution robustness. In: Advances in Neural Information Processing Systems 34: Annual Conference on Neural Information Processing Systems 2021, NeurIPS 2021, 6–14 December 2021, virtual, pp. 664–676 (2021)
5. Döbler, M., Marsden, R.A., Yang, B.: Robust mean teacher for continual and gradual test-time adaptation. CoRR abs/2211.13081 (2022)
6. Erichson, N.B., Lim, S.H., Xu, W., Utrera, F., Cao, Z., Mahoney, M.W.: NoisyMix: boosting model robustness to common corruptions (2022)
7. Fang, Z., Lu, J., Liu, F., Zhang, G.: Semi-supervised heterogeneous domain adaptation: theory and algorithms. IEEE Trans. Pattern Anal. Mach. Intell. **45**(1), 1087–1105 (2023). https://doi.org/10.1109/TPAMI.2022.3146234
8. Gong, T., Jeong, J., Kim, T., Kim, Y., Shin, J., Lee, S.: NOTE: robust continual test-time adaptation against temporal correlation. In: NeurIPS (2022)
9. Gu, Y., Yang, X., Wei, K., Deng, C.: Not just selection, but exploration: online class-incremental continual learning via dual view consistency. In: IEEE/CVF Conference on Computer Vision and Pattern Recognition, CVPR 2022, New Orleans, LA, USA, pp. 7432–7441 (2022)
10. Hendrycks, D., Dietterich, T.: Benchmarking neural network robustness to common corruptions and perturbations (2019)
11. Hendrycks, D., Mu, N., Cubuk, E.D., Zoph, B., Gilmer, J., Lakshminarayanan, B.: AugMix: a simple data processing method to improve robustness and uncertainty. In: 8th International Conference on Learning Representations, ICLR 2020, Addis Ababa, Ethiopia, 26–30 April 2020 (2020)
12. Li, K., Lu, J., Zuo, H., Zhang, G.: Dynamic classifier alignment for unsupervised multi-source domain adaptation. IEEE Trans. Knowl. Data Eng. **35**(5), 4727–4740 (2023). https://doi.org/10.1109/TKDE.2022.3144423
13. Li, K., Lu, J., Zuo, H., Zhang, G.: Multidomain adaptation with sample and source distillation. IEEE Trans. Cybern. 1–13 (2023). https://doi.org/10.1109/TCYB.2023.3236008

14. Liu, I., Peng, J., Schwing, A.G.: Knowledge flow: improve upon your teachers. In: 7th International Conference on Learning Representations, ICLR 2019, New Orleans, LA, USA, 6–9 May 2019. OpenReview.net (2019)
15. Liu, Y., Kothari, P., van Delft, B., Bellot-Gurlet, B., Mordan, T., Alahi, A.: TTT++: when does self-supervised test-time training fail or thrive? In: NeurIPS (2021)
16. Long, M., Cao, Y., Wang, J., Jordan, M.I.: Learning transferable features with deep adaptation networks. In: Proceedings of the 32nd International Conference on Machine Learning, ICML 2015, Lille, France, 6–11 July 2015, pp. 97–105. JMLR.org (2015)
17. Modas, A., Rade, R., Ortiz-Jiménez, G., Moosavi-Dezfooli, S., Frossard, P.: PRIME: a few primitives can boost robustness to common corruptions. In: Avidan, S., Brostow, G., Cissé, M., Farinella, G.M., Hassner, T. (eds.) ECCV 2022. LNCS, vol. 13685, pp. 623–640. Springer, Cham (2022). https://doi.org/10.1007/978-3-031-19806-9_36
18. Morerio, P., Volpi, R., Ragonesi, R., Murino, V.: Generative pseudo-label refinement for unsupervised domain adaptation. In: IEEE Winter Conference on Applications of Computer Vision, WACV 2020, Snowmass Village, CO, USA, pp. 3119–3128 (2020)
19. Niu, S., et al.: Towards stable test-time adaptation in dynamic wild world. In: The Eleventh International Conference on Learning Representations, ICLR 2023, Kigali, Rwanda, 1–5 May 2023. OpenReview.net (2023)
20. Recht, B., Roelofs, R., Schmidt, L., Shankar, V.: Do CIFAR-10 classifiers generalize to CIFAR-10? CoRR abs/1806.00451 (2018)
21. Shu, Y., Cao, Z., Zhang, Z., Wang, J., Long, M.: Hub-pathway: transfer learning from A hub of pre-trained models. In: NeurIPS (2022)
22. Shu, Y., Kou, Z., Cao, Z., Wang, J., Long, M.: Zoo-tuning: adaptive transfer from A zoo of models. In: Proceedings of the 38th International Conference on Machine Learning, ICML 2021, 18–24 July 2021, Virtual Event (2021)
23. Tzeng, E., Hoffman, J., Saenko, K., Darrell, T.: Adversarial discriminative domain adaptation. In: 2017 IEEE Conference on Computer Vision and Pattern Recognition, CVPR 2017, Honolulu, HI, USA, 21–26 July 2017, pp. 2962–2971. IEEE Computer Society (2017)
24. Wang, D., Shelhamer, E., Liu, S., Olshausen, B.A., Darrell, T.: Tent: fully test-time adaptation by entropy minimization. In: ICLR (2021)
25. Wang, Q., Fink, O., Gool, L.V., Dai, D.: Continual test-time domain adaptation. In: IEEE/CVF Conference on Computer Vision and Pattern Recognition, CVPR 2022, New Orleans, LA, USA, 18–24 June 2022, pp. 7191–7201. IEEE (2022)
26. Zhang, M., Levine, S., Finn, C.: MEMO: test time robustness via adaptation and augmentation. In: NeurIPS (2022)
27. Zhang, Y., Liu, F., Fang, Z., Yuan, B., Zhang, G., Lu, J.: Learning from a complementary-label source domain: theory and algorithms. IEEE Trans. Neural Netw. Learn. Syst. 33(12), 7667–7681 (2022). https://doi.org/10.1109/TNNLS.2021.3086093
28. Zhong, L., Fang, Z., Liu, F., Yuan, B., Zhang, G., Lu, J.: Bridging the theoretical bound and deep algorithms for open set domain adaptation. CoRR (2020). https://arxiv.org/abs/2006.13022

Causal Disentanglement for Adversarial Defense

Ji-Young Park$^{(\boxtimes)}$, Lin Liu, Jixue Liu, and Jiuyong Li

University of South Australia, Adelaide, Australia
ji-young.park@mymail.unisa.edu.au,
{lin.liu,jixue.liu,jiuyong.li}@unisa.edu.au

Abstract. Representation learning that seeks the high accuracy of a classifier is a key contribute to the success of state-of-the-art DNNs. However, DNNs face the threat of adversarial attacks and their robustness is in peril. While the adversarial defense has been widely studied, much of the research is based on a statistical association and causality based defense approach is a relatively open area. We present CDAD (_C_ausal _D_isentanglement for _A_dversarial _D_efense), a novel defense method that learns and utilizes causal representations for robust prediction. We take inspiration from a recent study that takes a causal perspective on the adversarial problem and considers the susceptibility of DNNs to adversarial examples come from their reliance on spurious associations between non-causal features and labels, such that an adversary exploits the associations to succeed in the attack. Causal representations are robust as the causal relationship between a cause of the label and the label is invariant under different environments. However, discovering causal representations is a challenging task, especially in the context of image data. Harnessing the recent advancement in representation learning with VAE (_V_ariational _A_uto_E_ncoder), we design CDAD as a VAE based causal disentanglement representation learning method to decouple causal and non-causal representations. CADA uses the invariance property of causal features as a constraint in the disentanglement of causal features and non-causal features. Experimental results show CDAD's highly competitive performance compared to the state-of-the-art defense methods, while possessing a causal foundation.

Keywords: Causality · Adversarial machine learning · Representations

1 Introduction

Obtaining good representations is a key factor in the current success of DNNs (Deep Neural Networks). However, pursuing high accuracy has led to an undesired consequence of representation learning such that DNNs tend to learn any representations which can enhance prediction accuracy. The features learned by

T. Liu et al. (Eds.): AI 2023, LNAI 14471, pp. 315–327, 2024.
https://doi.org/10.1007/978-981-99-8388-9_26

these DNNs often heavily depend on spurious correlations rather than causality [18]. Importantly, the learning, which disregards causality is problematic, especially, under the adversarial setting.

An adversarial example [19] is a sample that contains carefully designed malicious perturbations aiming for the misclassification of a classifier. Let x denote the input and y the true label of x. An adversary or attack method aims to compromize a classifier k with $x' = x + \delta$, where x' is the adversarial example, with δ being the adversarial perturbation. The attacker's goal is to achieve $k(x') = y'$ by maximizing $\mathcal{L}(k(x + \delta), y)$, where $\mathcal{L}$ denotes the loss function and y' denotes a wrong label. Meanwhile, a defender or defense method aims to safeguard machine learning models from adversarial examples.

Due to the surprising vulnerability of machine learning models to adversarial examples, adversarial machine learning has drawn huge attention and a large body of works has been proposed in both adversarial attack methods [1,3,5,6,21] and defense methods [4,10,12,17]. However, there has been an ongoing competition between attack methods and defense methods such that a new robust defense method is often broken shortly after its release by a recently developed stronger attack method. Thus far, PGD-Adversarial Training (PGD-AT) [11] is considered to be the most robust defense method and many other defense methods have been proposed to improve PGD-AT.

The armed race between attack methods and defense methods supports an argument in [15] that the approach based on statistical associations between features and labels may not be robust to generalize well beyond the i.i.d setting (e.g. adversarial setting), thus, understanding the underlying causal relations which can help with tackling adversarial vulnerability is necessary.

A recent work [25] sought representations that are strongly relevant to labels under the adversarial setting, but as the learning still relies on correlations, the relationship between representations learned may still be spurious and thus the model built with them is not robust. Although there is not a single definitive reason that can fully explain the vulnerability of DNNs, a recent study [28] proposed a view for understanding the generation process of adversarial examples and showed that the adversary exploits the spurious correlation between non-causal features and labels. However, the proposed defense method in [28] aims to achieve the alignment of the natural distribution and adversarial distribution to increase adversarial robustness, not the learning for causal representations.

In this paper, we propose CDAD(*C*ausal *D*isentanglement for *A*dversarial *D*efense), a defense method that aims to learn and use causal representations to perform predictions. Specifically, we learn and disentangle the representations into causal and non-causal representations respectively, and only use the causal representations in the classification to improve adversarial robustness. However, it is challenging to identify causal features and distinguish them from non-causal ones purely from an observational dataset. Inspired by the causal invariance property studied in the causal inference area [14], our fundamental idea is that causal features are capable of producing stable predictions across different environments with different experimental or interventional settings (i.e. causal invariant) while non-causal features are not. We consider the natural setting and the adversarial setting as two different environments and use the invariance property of causal features across the environments as a constraint for the disentanglement of representations, specifically for identifying causal representations. Meanwhile, based on the argument that adversaries exploit the spurious relationships between non-causal features and the label to achieve misclassification, we use this as another constraint to further help with distinguishing non-causal representations from causal representations. Accordingly, the objective function of CDAD incorporates the above described constraints with a VAE's objective.

We summarize our contributions as follows:

- We propose a new adversarial defense method, CDAD that learns and utilizes causal representations to achieve adversarial robustness. To disentangle causal and non-causal representations, we adopt the invariant property of causal features across different environments and the exploitability of spurious associations between non-causal features and labels, and use them in the regularization of a VAE's overall objective. The novel application of causal invariance, the integration of VAE for causal disentanglement for adversarial defense opens up a promising direction for the enhancement of adversarial robustness.
- Experimental results show that using causal features for building predictive models is a highly competitive and effective way to defend different adversarial attacks under different adversarial attack scenarios. Specifically, CDAD's performance in black-box scenario indicates that using causal features for prediction appears to be a promising way to safeguard machine learning models from adversarial examples generated by the information of unseen models.

2   The Proposed CDAD Method

Overview of the Causal Disentanglement Approach. Following the formalization in [28] of the generation process of an image and the labeling of an image as shown in Fig. 1, we assume that an image (denoted as X) comprises two components, contents (C) and style (S), and the image label (Y) is directly determined by its contents only. Using the terminology of causal graphical modeling, C and S are parents (direct causes) of X, and Y only has C as its parent. From the causal graph, we have C and S independent of each other, i.e. $C \perp\!\!\!\perp S$, but as X is a collider node (a child of multiple nodes, i.e. C and S here), for a given image $X = x$, S and C will be correlated, i.e. $C \not\perp\!\!\!\perp S \mid X$. This means that conditioning on X, in data, there is a spurious correlation between S and C, and thus, there

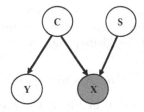

Fig. 1. A causal graph illustrating the generation mechanism of an image and its label, where X, Y, C, and S denote an image, label of the image, image contents and style respectively.

is a spurious correlation between the style information S and label Y too since C is a direct cause of Y. Previous study has shown that DNNs have a high reliance on the spurious correlation between S and Y to make accurate predictions and such spurious correlations have been exploited by adversarial attacks [28]. In [28], an approach was proposed to align the conditional distribution $P(Y|S,X)$ in the natural setting and adversarial setting to eliminate adversarial perturbations for robust predictions.

In this paper, we propose to utilize the invariance property of causal features [15] to build more robust models. With the causal invariance property, and the assumed causal structure in Fig. 1, let $e1$ and $e2$ denote two different environments (e.g. with different interventions on variables other than Y) then we have $P(Y^{e1}|C^{e1}) = P(Y^{e2}|C^{e2})$, i.e. the conditional distributions of Y given all its direct causes in different environments are the same. However, non-causal features do not have such invariance property. This indicates that if we build a classifier using content features only, then the performance of the classifier is expected to be stable in different environments, including adversarial settings.

Our idea thus is to utilize the causal features for Y, specifically the content features for building a DNN for robustness to adversarial examples. However, it is difficult to identify content features directly from an image dataset. Therefore we propose to (1) leverage the disentanglement ability of the VAE model [8] to learn and disentangle the representation of X into content representation, C and style representation, S; (2) consider the settings where the natural samples and the adversarial samples are obtained respectively as two different environments, and use the causal invariance property that is expected to be held by C as a constraint in the disentanglement. Additionally, inspired by the analysis in [28], which suggests that for adversarial examples, there can be a strong correlation between their style and the targeted class given an adversarial image, we use the correlation as a constraint to help better distinguish S from C. The causal invariance property has been used in literature to construct stable prediction models or select causal features [9,26], but as far as we know, most of the work requires the use of datasets from different sources, and the property has not been fully explored for adversarial defense. Moreover, no work in adversarial defense has used the causal invariance property together with a VAE based disentanglement approach.

2.1   The Architecture of CDAD

As shown in Fig. 2, the architecture of CDAD is an extension of the standard VAE structure. In addition to the encoder E and decoder D for learning z and z' (latent representations of the nature and adversarial examples x and x' respectively), we incorporate three networks f, g, and h into CDAD to implement the constraints introduced in the previous section for disentangling the content and style representations. Specifically, f is a classifier that uses the causal/content representation learned (i.e. c or c' in Fig. 2) only as the features and aims to predict the true label (i.e. y) for either natural samples or adversarial samples. The encoder E and network f, when put together, is the robust classifier we aim

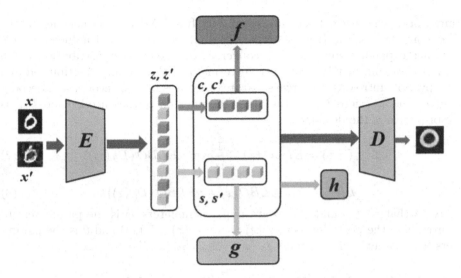

Fig. 2. An overview of CDAD architecture. As an extension of standard VAE, in addition to an encoder E and decoder D, CDAD is built with three incorporated networks f, g, and h. With given natural samples x and adversarial examples x' as inputs, a VAE learns representations of natural samples z and adversarial examples z'. The network f uses content information of x or x' (i.e. c and c') only to predict the label. The network g uses style information of adversarial examples (s') to predict the wrong label that an attack would draw. The network h uses the entire information of adversarial examples ($c' + s'$) and support the VAE's representation learning. The decoder D uses ($c' + s'$) to generate images which are similar to adversarial examples x.

to obtain. Network g implements the constraint that for adversarial examples, their style information (s', the non-causal representation learned) should have a strong correlation with the targeted class (y'). Moreover, we use another network, h to fully use the information encoded in adversarial examples to assist the representation learning by the VAE.

Corresponding to the designed components of the architecture, the objective function of CDAD is a composition of four sub-objectives, $\mathcal{L}_z$, the objective of the VAE for learning the latent representation of nature and adversarial samples, $\mathcal{L}_f$, the objective for network f for learning causal representation, $\mathcal{L}_g$, the objective for network g for better identifying the non-causal representation, and $\mathcal{L}_h$, the objective for disentanglement and supporting g.

Thus, high-level objective of CDAD is as follows.

$$\mathcal{L}_{CDAD} = \mathcal{L}_z + \beta_1 \mathcal{L}_f + \beta_2 \mathcal{L}_g + \beta_3 \mathcal{L}_h \tag{1}$$

2.2 $\mathcal{L}_z$: Learning Latent Representation

To learn the latent representation z (or z', but for simplicity of presentation, we only show the formulas of z below), we use Soft-IntroVAE [20] as the baseline framework. The distinct advantages of Soft-IntroVAE are; its encoder in the VAE converges to the true posterior distribution $p_\theta(z|x)$ and induces reliable inference capability, and the VAE converges to a generative distribution which minimizes a sum of KL divergence in terms of both the data distribution and entropy. Soft-IntroVAE uses two separate objectives for the encoder and decoder as given below where $\mathcal{L}_{E_\phi}$ is the objective function of the encoder and $\mathcal{L}_{D_\theta}$ is the objective of the decoder.

$$\mathcal{L}_{E_\phi}(x,z) = ELBO(x) - \frac{1}{\alpha}\exp(\alpha ELBO(D_\theta(z))) \tag{2}$$

$$\mathcal{L}_{D_\theta}(x,z) = ELBO(x) + \gamma ELBO(D_\theta(z)) \tag{3}$$

Note that $\alpha \geq 0$ and $\gamma \geq 0$ are hyper-parameters, ϕ is the parameters to approximate the posterior as $q_\phi(z|x) = \mathcal{N}(\mu_\phi(x), \Sigma_\phi(x))$ and θ is the parameters for a latent variable model $p_\theta(x) = \mathbb{E}_{p(z)}[p_\theta(x|z)]$.

2.3 $\mathcal{L}_f$, $\mathcal{L}_g$, and $\mathcal{L}_h$: Disentangling Causal and Non-causal Representations

While the representations learned with the objective functions in Eq. 2 and Eq. 3 are good enough for generating high quality images, there is no causal objective involved in the representation learning, hence, the representations learned may not be causal features. Thus, we posit the invariance property of causal features as a constraint that identifies causal features and use the spurious correlations exploited by adversarial examples to distinguish non-causal features from causal features. Specifically, we disentangle z (or z'), the representation learned from x (or x') into two parts c (or c') and s (or s'), by defining the two functions f and g on representations and let f learn c (or c') and g learn s' we have

$$f(c) = y, \ f(c') = y \tag{4}$$

$$g(s') = y' \tag{5}$$

To support g and the disentanglement, we use another classifier h such that:

$$h(c' + s') = y' \tag{6}$$

We learn the parameters of f by minimizing the loss between the predicted label using c and the ground truth label, and also minimizing the loss between the predicted label using c' and the ground truth label. For g, we aim to learn parameters that minimize the loss between the predicted label using s' and the wrong label that the attacker would obtain. We learn the parameters of h by minimizing the loss between the predicted label using $c' + s'$ and the wrong label. Thus, $\mathcal{L}_f$, $\mathcal{L}_g$, and $\mathcal{L}_h$ are defined as follows.

$$\mathcal{L}_f = CE(f(c'), y) + CE(f(c), y) \tag{7}$$

$$\mathcal{L}_g = CE(g(s'), y') \tag{8}$$

$$\mathcal{L}_h = CE(h(c' + s'), y') \tag{9}$$

3   Experiments

3.1   Experiment Setup

Baselines. In this section, we evaluate the efficacy of causal prediction on standard samples and adversarial examples using disentangled invariant features. We compare CDAD's performance with a CNN in [27] (ST for short) as a standard baseline model and the state-of-the-art defense methods, including PGD-Adversarial Training (PGD-AT) [11] and FGSM-Adversarial Training (FGSM-AT) [24]. We choose the ST since it demonstrated the efficacy of prediction in [27]. We use PGD-AT as a baseline defense method in the comparison as it is considered to be the most robust defense method so far and it is widely used to validate the robustness of newly proposed defense methods. However, despite the robustness provided by PGD-AT, it has high computational complexity and a new defense method FGSM-AT [24] was recently proposed to significantly reduce training time while it maintaining competitive robustness by combining FGSM and random initialization. Hence, we also include FGSM-AT as a baseline defense method for the evaluation. For all of the baselines, we use publically available source codes published by authors.

Evaluation. Following the suggestions in [2] for reliable verification, We use different attack settings under two types of attack scenarios. We will evaluate CDAD's standard accuracy and robustness against different attacks in black-box and white-box scenarios and will provide the motivation of conducting each type of the evaluations and the evaluation details in the corresponding subsections below.

Datasets. We use the MNIST and Fashion-MNIST (FMNIST) datasets in the evaluation. MNIST is a dataset of handwritten digits, containing 60K of training samples and 10K of test samples. FMNIST is a fashion images dataset, with 60K training samples and 10K of test samples. Each image in the two datasets is a 28×28 pixel greyscale image.

Implementation. CDAD comprises a VAE and three simple auxiliary classifiers f, g, and h. We use Soft-IntroVAE [20] as a baseline architecture. Each of the auxiliary classifiers has two fully connected layers. For both datasets MNIST and FMNIST, we pre-train CDAD on the combination of standard training data and their adversarial examples as described in Sect. 2.3. The adversarial examples used in the training are generated by PGD attack method and fed to CDAD based on PGD-AT method in [11] for training. Specifically, we use $\epsilon = 0.3$ for MNIST, and $\epsilon = 0.1$ for FMNIST to generate the adversarial examples. Using the training samples, we simultaneously train an encoder, three classifiers, and a decoder as an entity of CDAD.

As illustrated in Fig. 2, we give batches of standard samples and their adversarial examples to E the encoder to discover representations which help D the decoder to generate quality images. However, the discovered representations by minimizing the sum of KL divergence appear to have spurious correlations

Table 1. Results on standard accuracy.

	CDAD	ST	PGD-AT	FGSM-AT
MNIST	99.18	99.07	99.47	98.50
FMNIST	86.09	88.82	86.07	68.38

Table 2. Results (Robustness in %) of **black-box attack scenario** regarding **MNIST**.

Attacks	Attack strength	CDAD	PGD-AT	FGSM-AT
PGD	epsilon = 0.1	98.90	99.17	97.90
PGD	epsilon = 0.2	98.31	98.74	97.26
MIFGSM	epsilon = 0.1	98.79	99.20	97.91
MIFGSM	epsilon = 0.2	98.14	98.62	97.09
AutoAttack	epsilon = 0.1	98.89	99.26	98.09
AutoAttack	epsilon = 0.2	98.56	98.90	97.30
Jitter	epsilon =0.1	98.93	99.34	98.14
RFGSM	epsilon = 0.1	98.87	99.23	97.98
Onepixel	pixel =1	98.89	99.20	97.85
CW	iteration = 20	98.94	99.32	97.98

between the input and its label. Thus, instead of giving whole representations to D the decoder, we disentangle the representations. For the disentanglement, we split the latent representation z (or z') as same size for to c and s (or c' and s'). We set the dimension for z the representations of standard samples and z' the representations of adversarial examples to 128 and 64 for MNIST and FMNIST respectively. Then, the disentangled causal and non-causal features are given to the three classifiers as described in Sect. 2.3. We use Adam optimizer and set the same learning rate to 2e-4 for the encoder, decoder, f, g, and h. For the VAE's reconstruction error, we use $\|x - \hat{x}\|$. We set both β_1 and β_2 the hyperparameters in Eq. 1 to 0.06, and β_3 to 0.01.

3.2 Standard Accuracy

It is known that robust defense methods tend to present lower standard accuracy than normally trained models [23]. However, as a defense method needs to work under both natural and adversarial settings, it is important for it to have competent performance with natural examples, hence we first evaluate the standard accuracy of CDAD. The results are shown in Table 1, where we see that CDAD shows highly competitive performance compared to other models on MNIST and FMNIST. This indicates that prediction only using causal features is an effective way to achieve state-of-the-art standard accuracy.

3.3 Adversarial Robustness

We verify CDAD's robustness to different attack methods under both black-box attack scenario and white-box scenario. Following the suggestion in [22], we use the adaptive attack approach for CDAD and use various types of attack methods. We use the most strong adversarial attack methods including Autoattack [3], PGD [11], RFGSM [21], Jitter [16], and C&W [1]. In the black-box scenario, we additionally use the MIFGSM [5] attack, which uses a momentum term that increases the transferability under the black-box setting. For the attack implementations, we use Cleverhans [13] for C&W and Torchattacks [7] for the other attack methods.

Table 3. Results (Robustness in %) of **black-box attack scenario** regarding **FMNIST**. "-" indicates a result is not applicable due to the performance instability.

Attacks	Attack strength	CDAD	PGD-AT	FGSM-AT
PGD	epsilon = 0.05	84.95	85.41	76.47
PGD	epsilon = 0.15	79.98	67.58	61.17
MIFGSM	epsilon = 0.1	83.46	84.24	51.92
MIFGSM	epsilon = 0.2	39.31	20.16	-
AutoAttack	epsilon = 0.1	84.31	84.94	52.54
AutoAttack	epsilon = 0.2	57.56	34.64	-
Jitter	epsilon = 0.1	85.41	85.62	62.49
RFGSM	epsilon = 0.1	84.13	84.75	66.82
Onepixel	pixel =1	85.19	85.70	67.11
CW	iteration = 20	85.58	85.99	67.66

Black-Box Attack Scenario. As CDAD has been designed to learn causal representations which are causally related to the labels and use the causal representations learned for label prediction, we expect CDAD generalizes well to adversarial examples generated with the information of unseen source models, i.e. to adversarial examples under the black-box attack scenario, which is a main advantage of CDAD over non-causal defense methods. Thus, the focus of our evaluation of robustness is on the black-box attack scenario, to demonstrate the advantage of CDAD. In the black-box attack scenario, the adversary does not know about any information of a target model. Thus, in this setting, the adversary uses the surrogate model to use its information for the generation of adversarial examples, then feed the generated adversarial examples to the target model as inputs. We use ST (i.e. the CNN in [27]) as a surrogate model to generate adversarial examples. We train the surrogate model of black-box attack scenario for MNIST and FMNIST with 40 epochs and 50 epochs respectively.

Table 4. Results (Robustness in %) of **white-box attack** scenario regarding **MNIST**.

Attacks	Attack strength	CDAD	ST	PGD-AT	FGSM-AT
PGD	epsilon = 0.1	97.87	68.63	98.88	96.94
PGD	epsilon = 0.2	95.73	5.29	97.77	94.67
Jitter	epsilon = 0.1	98.02	79.28	98.94	96.97
Jitter	epsilon = 0.2	96.91	58.12	98.1	95.11
RFGSM	epsilon = 0.1	97.89	68.19	98.9	96.93
RFGSM	epsilon = 0.2	95.49	6.73	97.58	94.45
AutoAttack	epsilon = 0.1	98.66	67.97	98.83	96.67
AutoAttack	epsilon = 0.2	97.13	0.10	97.09	93.25
CW	iteration = 100	94.92	1.55	94.79	83.07
CW	iteration = 400	76.47	1.19	89.31	70.04

With MNIST in black-box scenario, to generate adversarial examples, we set iteration = 20 for C&W, iteration = 40 for PGD and MIFGSM, and iteration = 10 for Jitter and RFGSM. In addition, for a comprehensive evaluation, we use different values of ϵ for the diverse strength of adversarial perturbations. We use $\epsilon = 0.1$ for Jitter and RFGSM, and use $\epsilon = (0.1, 0.2)$ for PGD, MIFGSM, and Autoattack. We report the evaluation result of the black-box scenario on MNIST in Table 2, where we see CDAD outperforms FGSM-AT to all different attack methods on MNIST. When compared to PGD-AT, CDAD shows significantly competitive robustness to different types of attack methods and a range of perturbation strengths on MNIST.

With FMNIST in black-box scenario, we use the same iterations that used on MNIST for the attack methods. We use $\epsilon = 0.1$ for Jitter and RFGSM, and $\epsilon = (0.1, 0.2)$ for MIFGSM and Autoattack. For PGD, we use $\epsilon = (0.05, 0.15)$. We report the evaluation result on FMNIST in Table 3. CDAD outperforms FGSM-AT to all different attack methods on FMNIST. When compared to PGD-AT, CDAD shows significantly competitive robustness to different types of attack methods with a range of perturbation strengths on FMNIST. Specifically, CDAD outperforms FGSM-AT to all types of attack methods and outperforms PGD-AT to PGD with $\epsilon = 0.15$, MIFGSM and Autoattack with $\epsilon = 0.2$.

White-Box Attack Scenario. For further investigation, we continue to evaluate CDAD's robustness in white-box attack scenario. In the white-box scenario, the adversary has complete knowledge of a target model. In practice, once CDAD is trained, it will be deployed with the encoder E and the network f as a predictive model, without g, h and D, to perform the prediction. Thus, we give full information of the encoder and f to the adversary for the white-box scenario. The adversary takes input samples and extracts causal features, then obtains corresponding logits to compute adversarial perturbations.

Table 5. Results (Robustness in %) of **white-box attack** scenario regarding **FMNIST**.

Attacks	Attack strength	CDAD	ST	PGD-AT	FGSM-AT
PGD	epsilon = 0.05	81.07	12.07	82.79	3.81
PGD	epsilon = 0.15	53.16	0.03	11.30	-
Jitter	epsilon = 0.05	81.06	17.85	82.69	2.27
Jitter	epsilon = 0.1	75.00	8.01	78.10	0.24
RFGSM	epsilon = 0.1	74.08	0.24	78.08	0.92
RFGSM	epsilon = 0.2	27.47	0.01	13.34	0.79
AutoAttack	epsilon = 0.1	80.60	0.06	76.56	0.02
AutoAttack	epsilon = 0.2	8.51	0.00	0.70	0.01
CW	iteration = 50	85.96	34.29	84.12	52.71
CW	iteration =200	85.17	13.40	81.59	18.52

For the white-box attack evaluation on MNIST, we use iteration = 100 and 400 for C&W, iteration = 40 for PGD, Jitter, and RFGSM. We present evaluation results under white-box scenario on MNIST in Table 4. Under the white-box scenario, CDAD outperforms FGSM-AT to all of the attack methods on MNIST. When compared to PGD-AT, CDAD demonstrate comparable results to most attack methods. Moreover, CDAD outperforms PGD-AT to Autoattack using ϵ = 0.2, and C&W with iteration = 100.

Table 5 presents the evaluation result of white-box attacks on FMNIST. We use the same iterations for attack methods used for MNIST with the exception of C&W which uses iteration = (50, 200). As shown in Table 5, CDAD outperforms FGSM-AT to all attack methods. CDAD also demonstrates highly competitive results when compared to PGD-AT. Specifically, CDAD outperforms PGD-AT against C&W with iteration = (50, 200), PGD with ϵ = 0.15, RFGSM with ϵ = 0.2 and Autoattack with ϵ = (0.1, 0.2).

4   Conclusion

In this study, we have proposed CDAD, a novel VAE-based causal disentanglement representation learning method for adversarial defense. CDAD decouples representations into causal and non-causal representations, then uses causal representations only in the prediction to improve adversarial robustness. For the disentanglement of causal and non-causal representations, CDAD uses the invariance property of causal features as a constraint. We evaluate CDAD's robustness to different types of attack methods under white-box and black-box scenarios. Experimental results show that CDAD is highly competitive compared to other state-of-the-art defense methods with respect to standard accuracy and robustness on MNIST and FMNIST. Especially, the experimental result in black-box scenario is promising as the result may indicate using causal invariant features in the prediction of adversarial examples is a more effective way than using undisentangled features under the setting that requires a certain commonality or invariance over different environments.

Acknowledgements. We acknowledge the Australian Government Research Training Program Scholarship. This work has been partially supported by funding from Saab Australia Pty. Ltd. and the Australian Research Council (grant number: DP230101122).

References

1. Carlini, N., Wagner, D.: Towards evaluating the robustness of neural networks. In: 2017 IEEE Symposium on S&P, pp. 39–57 (2017)
2. Carlini, N., et al.: On evaluating adversarial robustness. arXiv:1902.06705 (2019)
3. Croce, F., Hein, M.: Reliable evaluation of adversarial robustness with an ensemble of diverse parameter-free attacks. In: ICML, pp. 2206–2216. PMLR (2020)
4. Dong, Y., et al.: Adversarial distributional training for robust deep learning. In: Advances in NeurIPS, vol. 33, pp. 8270–8283 (2020)
5. Dong, Y., et al.: Boosting adversarial attacks with momentum. In: CVPR, pp. 9185–9193 (2018)
6. Goodfellow, I.J., et al.: Explaining and harnessing adversarial examples. In: ICLR (2015)
7. Kim, H.: Torchattacks: a PyTorch repository for adversarial attacks. arXiv:2010.01950 (2020)
8. Kingma, D.P., Welling, M.: Auto-encoding variational bayes. In: ICLR (2014)
9. Liu, J., et al.: Stable adversarial learning under distributional shifts. In: Proceedings of the AAAI, pp. 8662–8670 (2021)
10. Ma, X., et al.: Characterizing adversarial subspaces using local intrinsic dimensionality. In: ICLR (2018)
11. Madry, A., et al.: Towards deep learning models resistant to adversarial attacks. In: ICLR (2018)
12. Pang, T., et al.: Mixup inference: better exploiting mixup to defend adversarial attacks. In: ICLR (2020)
13. Papernot, N., et al.: Technical report on the CleverHans v2.1.0 adversarial examples library. arXiv preprint arXiv:1610.00768 (2018)
14. Peters, J., et al.: Causal inference by using invariant prediction: identification and confidence intervals. J. Roy. Stat. Soc. **78**, 947–1012 (2016)
15. Peters, J., et al.: Elements of Causal Inference: Foundations and Learning Algorithms. The MIT Press, Cambridge (2017)
16. Schwinn, L., et al.: Exploring misclassifications of robust neural networks to enhance adversarial attacks. Appl. Intell. **53**, 19843–19859 (2023). https://doi.org/10.1007/s10489-023-04532-5
17. Sen, S., et al.: EMPIR: ensembles of mixed precision deep networks for increased robustness against adversarial attacks. In: ICLR (2020)
18. Sun, X., et al.: Recovering latent causal factor for generalization to distributional shifts. In: Advances in NeurIPS, vol. 34, pp. 16846–16859 (2021)
19. Szegedy, C., et al.: Intriguing properties of neural networks. In: ICLR (2014)
20. Daniel, T., Tamar, A.: Soft-IntroVAE: analyzing and improving the introspective variational autoencoder. In: Proceedings of the IEEE/CVF CVPR, pp. 4391–4400 (2021)

21. Tramer, F., et al.: Ensemble adversarial training. In: ICLR (2018)
22. Tramer, F., et al.: On adaptive attacks to adversarial example defenses. In: NeurIPS (2020)
23. Tsipras, D., et al.: Robustness may be at odds with accuracy. In: ICLR (2019)
24. Wong, E., et al.: Fast is better than free. In: ICLR (2020)
25. Yang, K., et al.: Class-disentanglement and applications in adversarial detection and defense. In: Advances in NeurIPS (2021)
26. Yu, K., et al.: Multi-source causal feature selection. PAMI **42**, 2240–2256 (2020)
27. Zhang, H., et al.: Theoretically principled trade-off between robustness and accuracy. In: ICML, pp. 7472–7482 (2019)
28. Zhang, Y., et al.: Adversarial robustness through the lens of causality. In: ICLR (2022)

Gemini: A Dual-Task Co-training Model for Partial Label Learning

Beibei Li[1], Senlin Shu[1], Beihong Jin[2], Tao Xiang[1(✉)], and Yiyuan Zheng[2]

[1] College of Computer Science, Chongqing University, Chongqing, China
txiang@cqu.edu.cn
[2] State Key Laboratory of Computer Science, Institute of Software,
Chinese Academy of Sciences, Beijing, China

Abstract. Partial-Label Learning (PLL) is an important weakly supervised learning task that assumes each training instance is annotated with a set of candidate labels. In recent years, self-training PLL models, which learn label confidence vectors and train the models with them iteratively, have attracted increasing attention and achieve state-of-the-art performance. However, they suffer from error accumulation problem due to that mistakenly disambiguated instances could mislead the model with false positive labels and causes performance degradation. To address this issue, we propose a novel dual-task co-training PLL model named Gemini that trains two structurally identical networks by optimizing two distinct tasks. Specifically, one is trained by optimizing PLL task with the original partially labeled data, and the other is trained by optimizing the noisy label learning (NLL) task with noisy pseudo labels generated according to the disambiguated label confidence. Moreover, we interrupt the error accumulation via a carefully-designed cooperation mechanism between the two networks, which comprises prediction alignment and correctly-disambiguated instance selection. Extensive experiments on benchmark datasets demonstrate the effectiveness and robustness of Gemini on both uniform and instance-dependent partially labeled data. The code is available at https://github.com/libeibeics/Gemini.

Keywords: Machine Learning · Partial Label Learning · Co-training

1 Introduction

Training Deep Neural Networks (DNN) via supervised learning requires massive accurately-annotated data, which are, however, expensive to be collected. To overcome this problem, weakly supervised learning [34], which trains models with weak supervision, including incomplete, inaccurate, and inexact supervision, has been widely studied recently. Partial-label learning (PLL) [4] is a typical task of weakly supervised learning with inaccurate supervision, which assumes that

Supported by Chongqing Overseas Chinese Entrepreneurship and Innovation Support Program, Chongqing Artificial Intelligence Innovation Center, China.

each training instance is annotated with a candidate label set that contains the ground-truth labels. Since label ambiguity is pervasive in data annotations, partial-label learning has been widely applied in various real-world applications, such as automatic image annotation [3] and multimedia content analysis [31].

Recent research interests on PLL [23, 26] mainly concentrate on identification-based methods, which regard the ground-truth label as a latent variable and try to recognize the ground-truth label by conducting label disambiguation. They are built based on various techniques, such as maximum margin [29], graph models [18, 22, 27, 33], clustering [16], contrastive learning [23], and manifold consistency regularization [26]. Besides, several self-training deep models [9, 17, 25], which learn label confidence vectors and train the models with them iteratively, have arisen in recent years. However, they suffer from error accumulation problem due to that complicated instances are difficult to classify and easy to be mistakenly disambiguated, further could mislead the model with false positive labels and causes performance degradation.

An intuitive way to address the drawbacks of self-training models is co-training [2, 19]: train two networks simultaneously and let them interact with each other. Since the two networks are initialized differently, they have different learning abilities and can help correct each other. The idea of training two networks simultaneously has been extensively explored in noisy label learning (NLL) [10, 20, 24], but are still not well studied in PLL. In recent years, Yao et al. [28] propose a novel PLL approach that trains two networks collaboratively by letting them interact with each other. However, it actually transforms PLL into NLL via data duplication before training the model with a representative co-training NLL method [11], which results in training datasets with a high noise rate and high time and spatial overhead. Besides, the two networks are trained with the same input data and loss function, thus, they can easily reach a consensus and reduce to a single network in function.

In this paper, we explore a dual-task co-training strategy in PLL, which force the two networks to obtain different classification abilities by optimizing them with different tasks explicitly. Firstly, we build a self-training progressive disambiguation-based PLL model to learn confidence of each candidate label, which consists of a DNN-based feature encoder that learns instance representation from raw features and a classifier. Then, we find that, though complicated instances are hard to disambiguate, the label confidence learned by the PLL model can identify the ground-truth labels of the majority of training instances, which motivates us to train another network with this clarified information. Therefore, we generate pseudo labels for instances by annotating each training instance with the most confident label, so that we transform the partially labeled data into noisy labeled data and train another network with them. We interrupt error accumulation from two aspects. One is to ensure that the two networks are trained collaboratively and can learn from each other, the other is to avoid the influence of mistakenly-disambiguated instances on model training. Therefore, we design a network cooperation mechanism that comprises of prediction alignment and instance selection. Finally, the problem can be alleviated effectively.

Our contributions can be summarized as follows:

(1) We propose a dual-task co-training deep PLL model Gemini, which simultaneously trains two networks by optimizing NLL and PLL tasks with the original partially labeled data and the generated noisy pseudo labels, respectively.
(2) We interrupt error accumulation by aligning the prediction of the two networks and force Gemini pay more attention to correctly-disambiguated training instances via small-loss instance selection.
(3) Extensive experimental results on benchmark datasets demonstrate the effectiveness of Gemini on both uniform and instance-dependent partially labeled data.

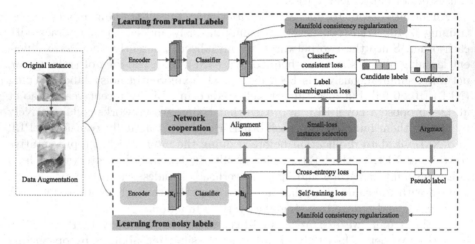

Fig. 1. Architecture of Gemini. Gemini consists of four components, i.e., two structurally-identified networks that mainly learn from original partial and generated noisy pseudo labels respectively, network cooperation mechanism, and manifold consistency regularization.

2 Methodology

In this section, we describe the proposed PLL model Gemini. As shown in Fig. 1, Gemini consists of two structurally-identical networks that are trained collaboratively, where one is trained using the original partially labeled data, and the other is trained with generated noisy pseudo labels. The two networks interact with each other via a network cooperation mechanism. Further, in order to exploit more self-supervised information, we construct manifold consistency regularization between the feature space and the label space.

Notations. We denote $\mathcal{X} \subset \mathbb{R}^d$ as the features of instances, $\mathcal{Y} \subset \{1, 2, \ldots, m\}$ as the labels, and $\mathcal{D} = \{(\boldsymbol{x}_i, y_i)\}_{i=1}^n$ as the training dataset. Each training instance $\mathbf{x}_i \in \mathcal{X}$ is labeled with a candidate label set $\mathcal{Y}_i \subset \mathcal{Y}$ that contains the ground-truth label y_i. Our goal is to train a multi-classifier $f(\cdot)$, from the feature space to label space with the partially labeled data, i.e., $f : \mathcal{X} \to \mathcal{Y}$.

2.1 Learning from Partial Labels

In order to train a classifier from partially labeled data, for each instance $x_i \in \mathcal{X}$, we firstly obtain its encoded feature $\mathbf{x}_i$ from its raw feature with a deep feature encoder. As for image instances, the feature encoder can be built on deep neural networks, such as LeNet [15], ResNet [12] and WideResNet [30]. Then, we construct a classifier to compute the classification logits, and obtain the classification probability $\mathbf{p}_i \in \mathbb{R}^m$ via the softmax function. The k-th element of $\mathbf{p}_i$ denotes the probability that the i-th instance is classified into k-th category, and is calculated as follows:

$$\mathbf{p}_{ik} = \frac{\exp\left(f_k\left(\mathbf{x}_i\right)/\tau\right)}{\sum_j \exp\left(f_j\left(\mathbf{x}_i\right)/\tau\right)}, \tag{1}$$

where τ is a temperature parameter, the larger τ is, the smoother the classification probabilities are.

Data augmentation, which produces richer and harder instances by making small changes to the original instances, can benefit classification performance. Following the state-of-the-art PLL model DPLL [26], we apply two augmentation methods, i.e., Autoaugment [5] and Cutout [6], to generate two augmentation views for each instance, which are $x_i' = Aug_1\left(x_i\right)$ and $x_i'' = Aug_2\left(x_i\right)$. We denote the representations and classification probabilities of the augmented instances as $\mathbf{x}_i', \mathbf{x}_i''$ and $\mathbf{p}_i', \mathbf{p}_i''$, respectively.

We adopt the data augmentation-enhanced Classification-Consistent (CC) loss function [9] as the basic loss to train one of the two networks, which assumes each candidate label set is uniformly sampled from the $2^{(m-1)} - 1$ possible candidate label sets that contains the ground-truth label. The empirical risk estimator is formulated as follows:

$$\mathcal{L}_{\text{basic}} = -\frac{1}{3n} \sum_{i=1}^n \sum_{x_i^* \in \left\{x_i, x_i', x_i''\right\}} \log\left(\sum_{k \in \mathcal{Y}_i} \mathbf{p}_{ik}^*\right). \tag{2}$$

The basic loss above assigns the same weight to each candidate label, which makes the ground-truth label easily overwhelmed by other false positive labels. Therefore, we distinguish the importance of each candidate label and learn a label confidence vector $\mathbf{c}_i \in \mathbb{R}^m$ for each instance. Specifically, during training, we adopt the classification probabilities of the original instance and the augmented instances to calculate the confidence vector. For example, the confidence of instance x_i classified as the k-th category in the t-th epoch $\mathbf{c}_{ik}^{(t)}$ is calculated as follows:

$$\mathbf{a}_{ik}^{(t)} = \sqrt[3]{\prod_{x_i^* \in \{x_i, x_i', x_i''\}} \mathbf{p}_{ik}^{*(t-1)}}, \tag{3}$$

$$\mathbf{c}_{ik}^{(t)} = \begin{cases} \dfrac{\mathbf{a}_{ik}^{(t)}}{\sum_{j \in \mathcal{Y}_i} \mathbf{a}_{ij}^{(t)}} & \text{if } k \in \mathcal{Y}_i \\ 0 & \text{otherwise.} \end{cases} \tag{4}$$

Here, we set confidence of non-candidate labels as 0, larger confidence indicates higher probability to be the ground-truth label. Confidence of each candidate label is initialized uniformly. Finally, we construct a label disambiguation loss as following:

$$\mathcal{L}_{\text{ld}} = -\frac{1}{3n} \sum_{i=1}^{n} \sum_{y_k \in \mathcal{Y}_i} \sum_{x_i^* \in x_i, x_i', x_i''} \mathbf{c}_{ik} \log \mathbf{p}_{ik}^*. \tag{5}$$

Since minimizing the loss drives the classification probabilities of both the original and augmented instances close to the confidence vector simultaneously, it encourages the network's output to be invariant to small changes applied to the feature space.

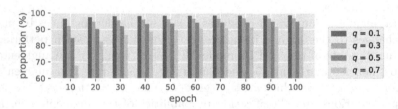

Fig. 2. The proportion of instances with ground-truth labels correctly identified by confidence vectors (on CIFAR-10), where q is the probability that incorrect labels turned into candidate labels.

2.2 Learning from Noisy Labels

As the label disambiguation proceeds, the majority of instances get largest confidence on their ground-truth labels. As shown in Fig. 2, after training the network with partially labeled data for 10 epochs, in the manually constructed partially labeled CIFAR-10 dataset (details are introduced in Sect. 3.1), more than 95% training instances get the largest confidences on their ground-truth labels and more instances are disambiguated correctly as the training progresses. In order to make full use of the identified labels, we annotate each training instance with the label corresponding to the maximum confidence. Assuming instance x_i gets the largest confidence on the k'-th label, i.e., $k' = \text{argmax}_k \{\mathbf{c}_{ik} | 1 \le k \le |\mathcal{Y}_i|\}$, we generate an one-hot label vector $\tilde{\mathbf{y}}_i$ for it, where $\tilde{y}_{ik'} = 1$. Then, we obtain a new dataset, each of which is annotated with an exact label. Since some instances may be mistakenly disambiguated and annotated, there are still some noisy data in the generated dataset, which results in a noisy label learning task. We take the generated noisy dataset to train the second network.

Firstly, we construct a supervised multi-class cross-entropy loss function as Eq. 6, where $\mathbf{h}_i$ denotes the classification probability of instance x_i predicted by the second network.

$$\mathcal{L}_{\text{sup}}(x_i) = -\tilde{\mathbf{y}}_i \log \mathbf{h}_i. \tag{6}$$

Then, we construct a self-supervised loss to keep the consistency between the classification probabilities of original instances and augmented instances, in order to make the model invariant to small changes in instances. Specifically, we normalize the classification probabilities of original instances by Eq. (7), where we set the probability of non-candidate labels to 0. Then, we construct self-supervised loss based on cross-entropy shown in Eq. (8):

$$\tilde{\mathbf{h}}_{ik} = \begin{cases} \frac{\mathbf{h}_{ik}}{\sum_{j \in \mathcal{Y}_i} \mathbf{h}_{ij}} & \text{if} \quad k \in \mathcal{Y}_i, \\ 0 & \text{otherwise.} \end{cases} \tag{7}$$

$$\mathcal{L}_{\text{ssl}}(x_i) = -\sum_{x_i^* \in \{x_i, x_i', x_i''\}} \tilde{\mathbf{h}}_i \log(\mathbf{h}_i^*). \tag{8}$$

Overall, the loss function to train the first network with partially labeled data is as follows:

$$\mathcal{L}_{\text{pll}} = \mathcal{L}_{\text{basic}} + \mathcal{L}_{\text{ld}} + \lambda \mathcal{L}_{\text{sim\_pll}}, \tag{9}$$

where λ is the weight of manifold regularization, a dynamic hyperparameter that gradually increases. λ in t-th epoch is calculated by $\lambda = \min(\lambda_{\max}, \lambda_{\max}\frac{t}{T})$, where $\lambda_{\max}$ is the maximum λ and T is the total number of training epochs.

2.3 Network Cooperation

We interrupt the error accumulation by designing a network cooperation mechanism between the two networks, which comprises two parts, i.e., prediction alignment and correctly-disambiguated instance selection.

Since the two networks are trained with different data and loss functions, they have different modeling abilities. We construct a alignment loss shown in Eq. (10) to let them interact with each other.

$$\mathcal{L}_{\text{align}}(x_i) = KL(\mathbf{h}_i \| \mathbf{p}_i) + KL(\mathbf{p}_i \| \mathbf{h}_i). \tag{10}$$

Complicated instances are easy to be mistakenly disambiguated at the beginning of training. Gradients from them will mislead the model and cause error accumulation. Thus, stopping mistaken gradients benefits model training.

Due to the memorization effect of DNNs [1], i.e., DNN would first memorize training data with correct labels and then those with noisy labels, instances with correct pseudo labels have higher probabilities to get smaller losses at the beginning of training. Thus, instances with small supervised losses $\mathcal{L}_{\text{sup}}$ are more likely to be correctly annotated. Besides, if an instance gets a small align loss $\mathcal{L}_{\text{align}}$, the predictions of the two networks reach a consensus and are more confident. Therefore, we pick up small-loss instances based on the sum of the supervised loss and alignment loss, and treat them as clean data. When minimizing the label disambiguation loss $\mathcal{L}_{\text{ld}}$ and supervised loss $\mathcal{L}_{\text{sup}}$, only the gradients from clean data are back-propagated, which reduces the impact of mistakenly-disambiguated instances on model training and alleviates error accumulation. For each mini-batch, the small-loss instance rate is set to μ, where $0 \leq \mu \leq 1$. As the training process proceeds, more instances are identified correctly, and μ can be increased gradually.

2.4 Manifold Consistency Regularization

Due to the manifold relationship between the feature space and label space, instances with similar confidence vectors should have similar representations. With confidence getting accurate gradually, confidence similarities among instances can be employed to guide the learning of feature encoder. Thus, we construct a manifold consistency regularization.

Specifically, for each partially labeled instance pair $\langle x_i, x_j \rangle$ in a mini-batch $\mathcal{B}$, where $x_i \in \mathcal{B}$ and $x_j \in \mathcal{B}$, we calculate their cosine similarity scores, and normalize them via softmax. For example, the similarity between instance i and instance j is calculated as follows:

$$s_{ij}^c = \frac{\exp\left(\text{sim}\left(\mathbf{c}_i, \mathbf{c}_j\right)/\epsilon\right)}{\sum_{x_k \in \mathcal{B}} \exp\left(\text{sim}\left(\mathbf{c}_i, \mathbf{c}_k\right)/\epsilon\right)}, \tag{11}$$

where $\text{sim}(\cdot)$ is cosine similarity function, ϵ is a temperature parameter, smaller ϵ makes the similarities more differentiated.

Similarly, we calculate the similarity of encoded features between each instance pair s_{ij}^f with the cosine similarity between features, i.e., $\text{sim}\left(\mathbf{x}_i, \mathbf{x}_j\right)$.

Since that the similarity distribution in the feature space should be as heterogeneous as possible with that of the confidence space, we construct the following KL divergence-based manifold consistency loss function:

$$\mathcal{L}_{\text{sim\_pll}} = \frac{1}{n} \sum_{i=1}^{n} \frac{1}{n} \sum_{j=1}^{n} KL(\text{stop-grad}(s_{ij}^c)) \| s_{ij}^f), \tag{12}$$

where $\text{stop-grad}(s_{ij}^c)$ denotes stop the gradient from s_{ij}^c.

On the other hand, for the generated noisy labeled dataset, I_{ij} denotes whether instance x_i and x_j are annotated with the same label or not, if yes, then $I_{ij} = 1$, otherwise, $I_{ij} = 0$. If the identified labels of two instances are the same, it means that they have a high probability to belong to the same class, thus, we constrain their encoded features to be similar using the loss as Eq. 13, where $\sigma(\cdot)$ is sigmoid function.

$$\mathcal{L}_{\text{sim\_nll}} = -\sum_i^{|\mathcal{X}|} \sum_j^{|\mathcal{X}|} I_{ij} \log \sigma \left(\text{sim} \left(\mathbf{x}_i, \mathbf{x}_j \right) / \epsilon \right). \tag{13}$$

The loss function to train the second network with the generated noisy labeled data is as follows:

$$\mathcal{L}_{\text{nll}} = \mathcal{L}_{\text{sup}} + \mathcal{L}_{\text{ssl}} + \lambda \mathcal{L}_{\text{sim\_nll}}. \tag{14}$$

2.5 Loss Function

The final loss function is as follows:

$$\mathcal{L} = \mathcal{L}_{\text{pll}} + \mathcal{L}_{\text{nll}} + \mathcal{L}_{\text{align}}. \tag{15}$$

As the training progresses, the predictive performance of the two networks gradually converges. In the prediction phase, we take the average scores of the two networks as the final prediction.

Table 1. Accuracy (mean±std) comparison on CIFAR-10 and SVHN with uniform partial labels on different ambiguity levels. (The best performance in each column is highlighted in bold. The second best performance in each column is underlined.)

Datasets	Models	$q = 0.1$	$q = 0.3$	$q = 0.5$	$q = 0.7$
SVHN	CC	97.348 ± 0.100%	97.139 ± 0.048%	96.978 ± 0.020%	96.377 ± 0.020%
	RC	97.292 ± 0.085%	97.243 ± 0.128%	97.050 ± 0.049%	95.898 ± 0.108%
	PRODEN	97.081 ± 0.077%	96.445 ± 0.290%	96.183 ± 0.325%	94.573 ± 0.492%
	PiCO	95.680 ± 0.080%	95.585 ± 0.015%	95.630 ± 0.020%	95.150 ± 0.024%
	DPLL	97.261 ± 0.029%	97.062 ± 0.013%	96.797 ± 0.033%	94.972 ± 0.106%
	NCPD	97.469 ± 0.011%	97.431 ± 0.045%	97.325 ± 0.041%	7.048 ± 0.479%
	Gemini	**97.715 ± 0.038%**	**97.760 ± 0.011%**	**97.711 ± 0.018%**	**97.420 ± 0.016%**
	Improv.	0.246%	0.329%	0.386%	1.043%
CIFAR-10	CC	94.129 ± 0.181%	93.226 ± 0.261%	92.102 ± 0.155%	88.846 ± 0.031%
	RC	94.950 ± 0.100%	94.610 ± 0.054%	94.139 ± 0.059%	92.423 ± 0.051%
	PRODEN	94.443 ± 0.213%	93.845 ± 0.326%	93.466 ± 0.243%	91.259 ± 0.780%
	PiCO	94.357 ± 0.109%	94.183 ± 0.179%	93.697 ± 0.238%	92.157 ± 0.209%
	DPLL	95.905 ± 0.052%	95.654 ± 0.208%	95.365 ± 0.140%	93.856 ± 0.366%
	NCPD	96.284 ± 0.050%	95.280 ± 0.110%	95.280 ± 0.110%	76.583 ± 0.522%
	Gemini	**96.923 ± 0.061%**	**96.524 ± 0.055%**	**96.328 ± 0.040%**	**95.455 ± 0.113%**
	Improv.	0.639%	0.87%	0.963%	1.599%

3   Experiments

3.1   Experimental Setup

Datasets. We conduct experiments on three widely used benchmark datasets, including CIFAR-10, CIFAR-100 [14], and SVHN [21]. We generate both uniform and instance-dependent partially labeled datasets following [26]. Check the supplementary materials for more details.

Compared Methods. To evaluate the performance of Gemini, we choose the following deep partial label learning methods as competitors: (1) CC [9], which is a classifier-consistent method based on the assumption that candidate label sets are generated uniformly. (2) RC [9], a risk-consistent method based on the importance of re-weighting strategy. (3) PRODEN [17], which is a progressive identification method accomplishing classifier learning and label identification simultaneously. (4) PiCO [23], which achieves partial label learning via a contrastive learning module along with a novel class prototype-based label disambiguation algorithm. (5) DPLL [26], which utilizes consistency regularization for deep partial label learning.

Implementation Details. We implement the model with PyTorch, employ 18-layer ResNet as the backbone network, and optimize the model with SGD optimizer. In order to estimate the selected small-loss instance ratio, we sample some training instances and label them manually as correctly as possible to get their ground-truth label, then utilize their identification accuracy of their ground-truth label to determine the ratio.

For a fair comparison, we adopt the same backbone networks, learning rate, and optimizer for all compared methods. For methods that did not adopt data augmentation in the original paper, such as RC, CC, and PRODEN, we enhance the models with the same data augmentation methods in Gemini.

3.2   Performance Comparison

The results of the performance comparison experiments are shown in Table 1, 2 and 3.

Table 2. Accuracy (mean±std) comparison on CIFAR-100 with uniform partial labels on different ambiguity levels.

Datasets	Models	$q = 0.01$	$q = 0.05$	$q = 0.1$	$q = 0.2$
CIFAR-100	CC	75.560 ± 0.537%	75.138 ± 0.154%	73.224 ± 1.017%	69.035 ± 0.339%
	RC	76.252 ± 0.168%	75.689 ± 0.129%	74.737 ± 0.282%	72.708 ± 0.358%
	PRODEN	76.147 ± 0.291%	75.682 ± 0.097%	74.604 ± 0.285%	72.512 ± 0.212%
	PiCO	73.145 ± 0.035%	72.585 ± 0.145%	59.365 ± 0.445%	25.545 ± 0.715%
	DPLL	79.300 ± 0.262%	78.855 ± 0.165%	78.064 ± 0.050%	76.316 ± 0.232%
	NCPD	78.190 ± 0.080%	76.990 ± 0.041%	71.923 ± 0.042%	42.701 ± 0.832%
	Gemini	**82.186 ± 0.119%**	**81.778 ± 0.081%**	**81.284 ± 0.046%**	**80.110 ± 0.097%**
	Improv.	2.886%	2.923%	3.22%	3.794%

Table 3. Accuracy (mean±std) comparison on CIFAR-10 and SVHN with instance-dependent partial labels.

Models	CIFAR-10	SVHN
CC	93.701 ± 0.006%	96.072 ± 0.041%
RC	93.270 ± 0.013%	96.899 ± 0.087%
PRODEN	92.409 ± 0.041%	95.626 ± 0.084%
PiCO	92.715 ± 0.055%	95.615 ± 0.045%
DPLL	93.657 ± 0.104%	95.796 ± 0.015%
NCPD	94.011 ± 0.011%	96.633 ± 0.056%
Gemini	**95.292 ± 0.079%**	**97.451 ± 0.025%**
Improv.	1.281%	0.552%

Firstly, the proposed model Gemini significantly outperforms all the competitors in three datasets under different q, for example, on CIFAR-100, when the p is 0.01, 0.05, 0.1, and 0.2, the performance exceeds the state-of-the-art model DPLL 2.886%, 2.923%, 3.22%, and 3.794% respectively. This fully demonstrates the effectiveness of the proposed model.

Secondly, as q increases, the accuracy rates of most compared methods drops obviously. For example, on SVHN, when $q = 0.1$, the accuracy rates of most competitors can reach 97%, but when $q = 0.7$, they drop significantly. However, the accuracy of our proposed model ranges from 97.715% to 97.420% when q varies from 0.1 to 0.7, where the performance drop is much more slight and is only 0.3%. Moreover, compared to the best competitor, the accuracy improvement of the proposed model becomes more obvious as q increases. For example, on CIFAR-10, when $q = 0.1$ and $q = 0.7$, the performance improvements are 0.639% and 1.599% respectively, and on CIFAR-100, when $q = 0.01$ and $q = 0.2$, the accuracy improvements are 2.886% and 3.794% respectively. The above observations show that the proposed model has strong robustness.

Furthermore, compared methods show different performances on different datasets. Generally, on CIFAR-10 and CIFAR-100, DPLL achieves the best results, but on SVHN, RC and CC enhanced by data augmentation perform extraordinarily. This may result from the fact that SVHN is for a 0–9 digital classification task, which is much easier than the image classification task that the CIFAR dataset serves for. Thus it can be well-solved by simple models.

Last but not least, in addition to the outstanding performance on the uniformly generated partially label datasets, as shown in Table 3, the proposed model also surpasses other compared models on instance-dependent partially labeled datasets. It achieves a performance improvement of 1.281% on CIFAR-10 and achieves an accuracy of 97.451% on SVHN, which demonstrates the effectiveness and robustness of the proposed model.

Table 4. Ablation study.

Model Variants	CIFAR-10, $q = 0.7$	SVHN, $q = 0.7$	CIFAR-100, $q = 0.2$
Gemini-L.N.L.	94.48 ± 0.09%	96.25 ± 0.08%	77.91 ± 0.24%
Gemini-Man.Reg	95.42 ± 0.03%	97.25 ± 0.02%	79.95 ± 0.01%
Gemini	**95.57 ± 0.06%**	**97.42 ± 0.02%**	**80.11 ± 0.10%**

3.3 Ablation Study

In this section, an ablation study is conducted to examine the contributions of the two key components of Gemini. Two model variants are created by eliminating the second network that learns from noisy labels and the manifold learning module, referred to as 'Gemini-LNL' and 'Gemini-Man.Reg.', respectively. The performance is presented in Table 4, from which we can conclude that each component contributes to the improvement in accuracy. Furthermore, it is observed that the impact of learning from noisy labels outweighs that of manifold constraints on performance. This finding substantiates the rationality and effectiveness of the model design.

4   Related Work

Traditional PLL methods can be divided into two categories, average-based methods, and identification-based methods. The average-based methods treat each label in a candidate label set equally [4,13,32]. But, the ground truth label of each instance is easily overwhelmed, especially when the number of candidate labels is large. To alleviate the problem, identification-based methods try to disambiguate ground-truth label from candidate label sets [29,32]. Besides, manifold consistency regularization, which assumes that similar instances are supposed to have similar label distributions, has been widely employed in PLL to estimate the label confidence and learn the classifier simultaneously [7,8,29,32,33]. However, these traditional methods are usually linear or kernel-based models, which are hard to deal with large-scale datasets.

With the powerful modeling capability of deep learning, deep PLL methods can handle high-dimensional features and outperform traditional methods [9,23,26]. Among these models, self-training models, which learn label confidence vectors and train the models with them iteratively, have attracted increasing attention [17,25], even achieve state-of-the-art performance. However, they suffer from error accumulation. NCPD [28] converts PLL task to NLL task, then trains two networks collaboratively. Nevertheless, the task conversion in NCPD results in a high noise rate and calculation overhead, and the two networks are trained with the same input data and loss functions, making them easy to reach a consensus and cannot correct for each other effectively, which naturally motivates us to improve them in our research.

5   Conclusion

In this paper, we explore the error accumulation problem of self-training PLL models and establish a PLL model Gemini based on co-training. By explicitly training two networks with different data and loss functions, Gemini enhances the probability of two neural networks acquiring distinct capabilities. By aligning prediction and conducting small-loss instance selection, Gemini allows the two networks to communicate with each other and alleviate the influence of mistakenly-disambiguated instances on model training, thus, interrupts the error accumulation. Extensive experiments on benchmark datasets fully demonstrate the effectiveness and robustness of Gemini on both uniform and instance-dependent PLL datasets. We will further conduct research on different co-training architectures and network cooperation mechanisms to tap the potential of dual-task co-training models for partial label learning.

References

1. Arpit, D., et al.: A closer look at memorization in deep networks. In: ICML, pp. 233–242 (2017)
2. Blum, A., Mitchell, T.: Combining labeled and unlabeled data with co-training. In: COLT, pp. 92–100 (1998)
3. Chen, C.H., Patel, V.M., Chellappa, R.: Learning from ambiguously labeled face images. TPAMI **40**(7), 1653–1667 (2018)
4. Cour, T., Sapp, B., Taskar, B.: Learning from partial labels. JMLR **12**, 1501–1536 (2011)
5. Cubuk, E.D., Zoph, B., Mané, D., Vasudevan, V., Le, Q.V.: AutoAugment: learning augmentation strategies from data. In: CVPR, pp. 113–123 (2019)
6. DeVries, T., Taylor, G.W.: Improved regularization of convolutional neural networks with cutout (2017). arXiv:1708.04552 [cs]
7. Feng, L., An, B.: Leveraging latent label distributions for partial label learning. In: Proceedings of the 27th International Joint Conference on Artificial Intelligence, IJCAI 2018, pp. 2107–2113. AAAI Press, Stockholm (2018)
8. Feng, L., An, B.: Partial label learning with self-guided retraining. In: AAAI, vol. 33, no. 01, pp. 3542–3549 (2019)
9. Feng, L., et al.: Provably consistent partial-label learning. In: NeurIPS, vol. 33, pp. 10948–10960 (2020)
10. Han, B., et al.: Co-teaching: robust training of deep neural networks with extremely noisy labels. In: NeurIPS, vol. 31 (2018)
11. Han, B., et al.: Co-teaching: robust training of deep neural networks with extremely noisy labels. In: NeuIPS, NIPS 2018, pp. 8536–8546. Curran Associates Inc., Red Hook (2018)
12. He, K., Zhang, X., Ren, S., Sun, J.: Deep residual learning for image recognition. In: CVPR, pp. 770–778 (2016)
13. Hüllermeier, E., Beringer, J.: Learning from ambiguously labeled examples. Intell. Data Anal. **10**(5), 419–439 (2006)
14. Krizhevsky, A.: Learning multiple layers of features from tiny images. In: Computer Science (2009)

15. LeCun, Y., Bottou, L., Bengio, Y., Haffner, P.: Gradient-based learning applied to document recognition. Proc. IEEE **86**(11), 2278–2324 (1998)
16. Liu, L.P., Dietterich, T.G.: A conditional multinomial mixture model for superset label learning. In: NIPS, pp. 548–556 (2012)
17. Lv, J., Xu, M., Feng, L., Niu, G., Geng, X., Sugiyama, M.: Progressive identification of true labels for partial-label learning. In: ICML, pp. 6500–6510 (2020)
18. Lyu, G., Feng, S., Wang, T., Lang, C., Li, Y.: GM-PLL: graph matching based partial label learning. TKDE **33**(2), 521–535 (2021)
19. Ma, F., Meng, D., Xie, Q., Li, Z., Dong, X.: Self-paced co-training. In: ICML, pp. 2275–2284 (2017)
20. Malach, E., Shalev-Shwartz, S.: Decoupling "when to update" from "how to update". In: NIPS, pp. 961–971 (2017)
21. Netzer, Y., Wang, T., Coates, A., Bissacco, A., Wu, B., Ng, A.Y.: Reading digits in natural images with unsupervised feature learning. In: NIPS Workshop (2011)
22. Wang, D.B., Li, L., Zhang, M.L.: Adaptive graph guided disambiguation for partial label learning. In: SIGKDD, pp. 83–91 (2019)
23. Wang, H., et al.: Contrastive label disambiguation for partial label learning. In: ICLR (2022)
24. Wei, H., Feng, L., Chen, X., An, B.: Combating noisy labels by agreement: a joint training method with co-regularization, Seattle, WA, USA. In: CVPR, pp. 13723–13732 (2020)
25. Wen, H., Cui, J., Hang, H., Liu, J., Wang, Y., Lin, Z.: Leveraged weighted loss for partial label learning. In: ICML, vol. 139, pp. 11091–11100 (2021)
26. Wu, D.D., Wang, D.B., Zhang, M.L.: Revisiting consistency regularization for deep partial label learning. In: ICML, vol. 162, pp. 24212–24225 (2022)
27. Xu, N., Lv, J., Geng, X.: Partial label learning via label enhancement. In: AAAI, pp. 5557–5564 (2019)
28. Yao, Y., Gong, C., Deng, J., Yang, J.: Network cooperation with progressive disambiguation for partial label learning. In: Hutter, F., Kersting, K., Lijffijt, J., Valera, I. (eds.) ECML PKDD 2020. LNCS (LNAI), vol. 12458, pp. 471–488. Springer, Cham (2021). https://doi.org/10.1007/978-3-030-67661-2_28
29. Yu, F., Zhang, M.L.: Maximum margin partial label learning. In: ACML, vol. 45, pp. 96–111 (2016)
30. Zagoruyko, S., Komodakis, N.: Wide residual networks. arXiv (2016)
31. Zeng, Z., et al.: Learning by associating ambiguously labeled images. In: CVPR, pp. 708–715 (2013)
32. Zhang, M.L., Yu, F.: Solving the partial label learning problem: an instance-based approach. In: IJCAI, pp. 4048–4054 (2015)
33. Zhang, M.L., Zhou, B.B., Liu, X.Y.: Partial label learning via feature-aware disambiguation. In: SIGKDD, pp. 1335–1344 (2016)
34. Zhou, Z.H.: A brief introduction to weakly supervised learning. Natl. Sci. Rev. **5**(1), 44–53 (2018)

Detecting Stress from Multivariate Time Series Data Using Topological Data Analysis

Hieu Vu Tran$^{(\boxtimes)}$, Carolyn McGregor , and Paul J. Kennedy

Australian Artificial Intelligence Institute and UTS Ontario Tech University Joint Research Centre in AI for Health and Wellness, School of Computer Science, University of Technology Sydney, Sydney, Australia
HieuVu.Tran@student.uts.edu.au, c.mcgregor@ieee.org, Paul.Kennedy@uts.edu.au

Abstract. Stress can have dangerous effects to human mental and physical health. Statistics, machine learning and novel data analytics approaches have been used to detect stress from physiological time series data. However such data is noisy which can limit the effectiveness of algorithms. Topological Data Analysis (TDA) is a novel approach that can handle noisy data and may be promising for physiological time series data analysis. However, TDA is currently in the early stages of development, with researchers still grappling with the problem of feature extraction from TDA signatures for machine learning. Current state-of-the-art in TDA handles only small computer vision or univariate time series data analysis due to its computational expense. We present a TDA method for stress detection and validate it on the public Wearable Stress and Affect Detection (WESAD) dataset. We contribute a complete TDA method to classify long multivariate physiological time series data that overcomes some of the computational expensive and demonstrate its effectiveness compared to state of the art methods.

Keywords: Topological Data Analysis · stress detection · multivariate time series

1 Introduction

Stress currently poses a critical danger to human beings. There are currently many researchers trying to detect stress, mainly from data collected from wearable devices. This is because wearable devices can measure changes in physiological features caused by the stressors. Most often machine learning or statistics is used to detect stress by classifying stress and non-stress states from the data. However, physiological time series data collected from wearable devices is noisy. Researchers currently preprocess data to successfully detect stress. This requires a deep understanding of both computer science and clinical aspects. Topological data analysis (TDA) is a novel approach with the potential to help with noise. TDA is a combination of computational, statistical and topological methods

T. Liu et al. (Eds.): AI 2023, LNAI 14471, pp. 341–353, 2024.
https://doi.org/10.1007/978-981-99-8388-9_28

which find the "shape" of the data [11] and is robust to noise. By "shape" we mean the topological features present in the data.

This paper presents a TDA approach for classifying long multivariate physiological time series data. However, TDA is considered to be computationally expensive and many researchers consider it is only effective to handle small amounts of data [26,29]. Consequently TDA research with long and multivariate time series data is underdeveloped. Our proposed method uses a time window and suitable parameters to transform data from time series to TDA, and the use of a 1D CNN model to achieve good results. We evaluate our method using the Wearable Stress and Affect Detection (WESAD) benchmark dataset [27]. This is the most recent benchmark stress dataset containing about 60 million records from 15 participants and recording Galvanic Skin Response (GRS), Electrocardiogram (ECG), Electromyogram (EMG), respiration data, skin temperature, and 3-axis accelerometer data. To the best of our knowledge, our paper is one of the first approaches using TDA for long multivariate physiological time series data. With such a huge data set, TDA generates copious data, taking a prolonged implementation time and requiring tremendous amounts of memory when building the persistence diagram and extracting TDA features.

The contribution of this paper is a TDA approach to classify long multivariate physiological time series data, one of the first with TDA on such data. It overcomes the problem of long processing times and high memory requirements. Extension from the univariate to multivariate case is non-trivial. Compared with state-of-the-art multivariate time series classification methods and stress detection methods, we achieve strong results on the WESAD benchmark data.

The rest of this paper is organized as follows. Section 2 presents TDA theory. Section 3 presents related work. Section 4 introduces the WESAD dataset. Section 5 presents experiments and results. Finally, Sect. 6 concludes the paper.

2   Topological Data Analysis

2.1   Overview of Topological Data Analysis

Topological Data Analysis (TDA) is a geometric method with the ability to capture the shape and pattern of high-dimensional data [19]. It uses persistent homology to create persistence diagrams and extract topological characteristics of the data [4,5]. It can generate a 1-dimensional representation of data [22]. Edelsbrunner et al. [8] described the concept of persistent homology and built a graphical representation for persistent homology known as the persistence diagram. Carlsson et al. [3], redefining persistent homology, introduced the persistence bar code, an equivalent tool to the persistence diagram.

TDA brings four distinct advantages over other methods. First, TDA takes a coordinate-free point of view, relying only on the distance between objects [19,23]. Secondly, results from TDA remain invariant with small changes in input, which means that TDA is robust to noise [13,19,23]. Thirdly, TDA can convert data to a compressed form to more effectively analyze the data [19].

Lastly, input data to TDA can come from many forms such as images, time series, and graphs [13].

2.2 Core Concepts of Topological Data Analysis

Persistent Homology. The primary tool in TDA is persistent homology: data is analysed by observing homology changes under the effect of a filtration parameter ϵ, that represents the radius of a ball around each data point. Persistent homology represents the topological changes when the data is surveyed at many filtration parameter values [13,36]. When the filtration parameter changes, then topological features of the data appear and disappear. Topological signatures are a summary representation of topological features [13]. Topological features of the data are the connected components in the case of 0-dimension, loops in the case of 1-dimension, and voids in the case of 2-dimensions. The events associated with the appearance and disappearance of the topological features are birth and death, respectively. The whole collection of sequences of births and deaths can be represented by a persistence diagram or a persistence bar code [13].

One explanation of persistent homology is to imagine a ball attached to every data point as shown in Fig. 1. When the radius of the ball (i.e., the filtration parameter ϵ) is zero, then a "birth" event will be associated with each ball (Fig. 1a). As the radius gradually grows, then balls will start to overlap other balls. Overlaps create 0-dimensional features (connected components) and "death" events. The persistence bar code now has some complete sequences from starting points (birth) and the ending points (death) (Fig. 1b). As the radius continues to grow, it may reach a value when 1-dimensional features appear. The example in Fig. 1c, shows one 1-dimensional feature (loop) with a corresponding sequence appearing in the persistence bar code. As the radius continues to grow, the circle grows more prominent and the 1-dimensional feature is filled, with the corresponding line ending in the persistence bar code (Fig. 1d). The barcode can also be presented as a persistence diagram with birth and death values corresponding to x-, and y-coordinates (Fig. 1f). One approach to describe a topological signature is the use of Betti numbers (β_k), that measure the number of k-dimensional surfaces in a space: β_0 is the number of connected components, β_1 is the number of loops and β_2 is the number of voids or cavities.

3 Related Work

TDA has gained success in several fields such as material science, cellular data and drug design [4,35]. It shows promise for enhancing machine learning [17] with feature extraction or reducing the need for preprocessing.

Use of TDA in time series is at an early stage with significant limitations. Current researches on TDA and time series classification are based on subsampling, but their effectiveness on large data sets is limited. Two groups generated TDA features from time series [7,9]. Subsampling methods of both groups were based on K-nearest neighbour density and random and max–min algorithms. However, Dirafzoon et al. [7] omitted to mention the size of the data and Emrani

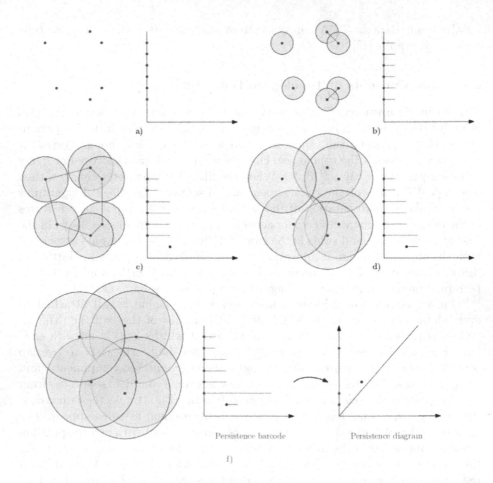

Fig. 1. Persistent homology and filtration parameter ϵ builds persistence bar codes and diagrams. Subfigures described in the text. Graphs a-f x-axis is ϵ, y = data point, For persistence diagram, x = births and y = deaths. Figure from [5].

et al. [9] used short breathing sound signals containing only several thousand point clouds. Other methods did not have effective methods to classify time series using TDA. Mostly distance-based methods such as k-nearest neighbour (kNN) are used [21,28]. Chung et al. [6] proposed a classification method based on persistent curve, which gained good results on UCR time series data with a few hundred to a few thousand data points. Whilst these are appropriate with small data sets, the effort of calculating distances between persistence diagrams to apply these methods is too expensive for larger datasets [15]. Recent methods make modifications using time windows but these methods are still somewhat limited. Majumdaret al. [20] showed a TDA method to classify univariate time series data. Tymochko et al. [33] used non-overlapping windows to detect sleep states and extract persistent images from the windows but their model overfitted.

In summary, current TDA approaches for time series require modification to be suitable for longer or multivariate time series. Their computational cost hinders the use with time series, even data with a few thousand data points [25].

4 WESAD Dataset

The Wearable Stress and Affect Detection dataset [27] contains data from 15 graduate student participants in a lab environment, none of whom were pregnant, heavy smokers or had recorded mental or cardiovascular issues. Their ages were 27.5 ± 2.4 years and there were twelve male and three female participants. Whilst the number of subjects is small, the time series lengths are long. Physiological data collected includes galvanic skin response (GRS) measured in microsiemens (μS), electrocardiogram (ECG) measured in millivolts (mV), electromyogram (EMG) measured in millivolts, respiration data measured in %, skin temperature measured in °C, and a 3–axis accelerometer measured in g (1 g = 9.80665 m/s^2) [27]. All data was sampled at 700 Hz. Each participant had an index from 2 to 17 (data from participants 1 and 12 were not saved due to errors in the recording progress). Data was classified to eight types; however, label zero is considered as a not defined or transient state, and labels five, six and seven are considered as should be ignored. Four labels, then, are usable: label one as the baseline, two as the stress state, three as the amusement state, and label four as the meditation state. However, in this paper, to facilitate comparison to state of the art, we omit meditation.

5 Experiment and Results

This experiment classifies stress based on Betti numbers using a convolutional neural network. Our methodology applies four steps as shown in Fig. 2. Step 1 transforms a windowed time series into point clouds using an adapted Takens embedding. We do this because a point cloud is a standard input to TDA. The existing Takens embedding converts from a univariate time series to a point cloud. We have adapted this to the multivariate case. Step 2 transforms the point cloud into a persistence diagram. Step 3 extracts Betti numbers combining them into one sequence. Step 4 inputs the sequence into a 1D convolutional neural network (CNN) model to detect stress for three labels: "baseline", "stress", and "amusement". For this paper, only records with labels corresponding to these three classes are kept. In summary 37,601,196 of the total 60,807,600 records were removed so that our experiment is comparable to other WESAD experiments. The giotto-tda library [31] was used for TDA operations in steps 2 and 3.

5.1 Step 1: Convert Multivariate Time Series to Point Cloud

First, the multivariate time series data is divided into equal length time windows with the length is 1.1 s, each of which is converted into a point cloud. Note that

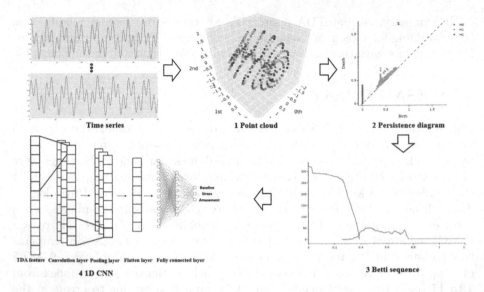

Fig. 2. Schematic of our experimental methodology consisting of four steps (shown as arrows). Step 1 transforms a time series to a point clouds. Step 2 builds the persistence diagram. Step 3 extracts the Betti sequence. Step 4 classifies data into three classes with a 1D CNN model.

a label is associated to each time window, not the entire participant. For this we adapt the Takens embedding [30] to the multivariate case. Starting from a multivariate time series $f(t_1, t_2, ..., t_k)$ with k the number of variables and t_i relating to univariate time series i, each of length l. Given an embedding dimension d and time delay parameter τ, the sequence of vectors is defined as $(f(t_{1j}), f(t_{1j} + \tau), f(t_{1j} + 2\tau), ..., f(t_{1j} + (d-1)\tau), ..., f(t_{kj}), f(t_{kj} + \tau), f(t_{kj} + 2\tau), ..., f(t_{kj} + (d-1)\tau))$. The index j ranges from 1 to $l - d + 1$ and $f(t_{ij})$ is value of time lag j, univariate time series i. For our experiments, we choose $d = 3$ and $\tau = 1$. The Takens embedding is applied to this resulting sequence resulting in a single point cloud.

5.2   Step 2: Generate the Persistence Diagram

In the next step, the point cloud data is used to build a persistence diagram. With increasing value of filtration parameters, this can be a slow computational process. Consequently we apply a multi-threaded approach. The TDA edge removal technique [2] is also applied to keep only critical connections between points when applying persistent homology whilst maintaining consistent topological characteristics. Only topological features from level 0 to 2 are generated. These are connected components, loops and voids respectively.

Determining the persistent homology requires generating topological features with increasing value of the filtration parameter [24,29]. For a specific point cloud, the filtration parameter can range from 0 to the maximum distance between any two points in the point cloud. Calculation of the this maximum distance is computationally expensive, so we make the assumption that there is one point in the cloud that has all maximum values in all dimensions, and there is another point has all minimum values in all dimensions. The distance between these two points is the maximum possible distance between any two points and defines the maximum possible value of the filtration parameter. When the filtration parameter increases beyond this value no further topological features are generated, so it can be used as a maximum possible value of the filtration parameter, with a consequent reduction in computational expense.

Based on our proposed method for transforming a multivariate time series from the WESAD dataset to a point cloud, using $d = 3$ and $\tau = 1$, then one point in the Takens embedding space has $8 \times 3 = 24$ dimensions: 3 dimensions for each of the 8 features: Galvanic skin response (GRS), Electrocardiogram (ECG), Electromyogram (EMG), respiration data, skin temperature, and a 3-axis accelerometer (x, y, z). Let $\boldsymbol{p}_i$ be defined as the vector $(min_{xi}, min_{yi}, min_{zi}, min_{GRSi}, min_{ECGi}, min_{EMGi}, min_{rdi}, min_{sti})$ and $\boldsymbol{q}_i$ be defined as $(max_{xi}, max_{yi}, max_{zi}, max_{GRSi}, max_{ECGi}, max_{EMGi}, max_{rdi}, max_{sti})$, which are the maximum and minimum values of the respective features of participant i. The resulting function to calculate the maximum possible filtration parameter $F_{\max}$ of participant i is defined as

$$F_{\max}(i) = \sqrt{\tau(\boldsymbol{q}_i - \boldsymbol{p}_i)^2} \tag{1}$$

Given the WESAD dataset, the maximum possible filtration parameter of all participants is for participant 9 and takes the value $F_{\max} = 80$.

5.3   Step 3: Extract the Betti Numbers and Combine into One Sequence

The next step calculates $(\beta_0, \beta_1, \beta_2)$, the vector of number of 0-dimension, 1-dimension, and 2-dimension topological features. We use a vector $(\beta_0, \beta_1, \beta_2)$ of Betti numbers to partly describe a dataset as input to a deep neural network. Given the maximum filtration parameter value of 80 each with three β values, we result in $80 \times 3 = 240$ input values for each time window input into the deep learning model.

5.4   Step 4: 1D CNN Model

Following some experimentation (not presented here), we determined the most effective parameter settings and ended up training a deep neural model (shown in Fig. 3) containing two 1D convolutional, two dropout, one max pooling, one flatten, and three dense layers. The first two dense layers utilized L2 regularization. Dropout and L2 regularization are applied to reduce overfitting. The pooling layer reduces number of built features, allowing the model to concentrate on essential features. Constructed features are transformed to 1-dimensional data using flattening. The model is optimised using Adam with a categorical cross-entropy loss function. Model performance is evaluated with balanced accuracy and F1 as the data set is unbalanced with respect to the label. Experiments ran on four machines with the configuration 2× AMD EPYC 7532 2.40 GHz 32 cores 256M L3 Cache, 1024 GB 3200 MHz ECC DDR4-RAM, 2× 1.92 TB SSD SAS Drives and 2× 3.84 TB SSD SAS Drives, NVIDIA 16 GB Graphics Card.

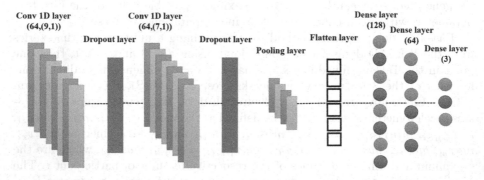

Fig. 3. Schematic of our 1D CNN model giving sizes of each layer.

5.5 Results and Comparisons

In a first experiment, we train a model on each participant separately using a 60/20/20 split for the training, validation, and test respectively. Results including F1 score for each label, balanced accuracy, and elapsed time are presented in Table 1. Results are averaged over three runs with random initial weights, but other parameter values kept the same, including dropout rate 0.5, learning rate 5×10^{-4}, and L2 0.01. The last row of the table show averages over all participants. The size of the time window is 1.1 s (which equates to 770 time lags) and the stride is $1/700$ s (which is a one time lag).

In a second experiment, we explored different window sizes. Table 2 presents results for 140 (equivalent to 0.2 s), 280 (0.4 s), 420 (0.6 s), 560 (0.8 s), 700 (1 s) and 770 (1.1 s) and shows that a window size of 770 gives the best results for both F1 score of stress and also balanced accuracy compared to other window sizes. Values in this table are the averaged results for participants.

In a third experiment we compare our approach to state of the art methods for multivariate time series classification. Methods compared are Time Convolutional Neural Network (Time CNN) and Fully Connected Neural Network (FCN) [38], Multi Layer Perceptron Network (MLP) and ResNet [34], LSTM-FCN [16], Time series attentional prototype network (TapNet) [37], and InceptionTime [10]. Default parameter settings from `sktime` were used. Average results over three runs are shown in Table 3. Compared to the state of the art multivariate time series classification methods, our proposed method has the

Table 1. Test results when applying deep learning model on each participant.

Participants	F1 baseline	F1 stress	F1 amusement	Balanced accuracy	Time
2	1	0.993	0.987	0.993	9 h 25 m
3	0.993	0.997	0.987	0.99	10 h 23 m
4	1	1	0.997	1	10 h 46 m
5	0.993	0.993	0.977	0.99	12 h 55 m
6	1	1	1	1	10 h 29 m
7	0.987	0.993	0.977	0.987	10 h 42 m
8	0.99	0.987	0.977	0.987	11 h 14 m
9	1	1	1	1	10 h 56 m
10	0.92	0.92	0.75	0.893	11 h 44 m
11	1	1	1	1	10 h 40 m
13	1	0.997	0.99	1	11 h 22 m
14	1	1	0.993	1	10 h 31 m
15	0.993	0.993	0.99	0.993	10 h 50 m
16	1	1	0.997	1	10 h 52 m
17	0.99	0.997	0.977	0.99	11 h 53 m
Average	0.991	0.991	0.973	0.988	11 h 58 m 48 s

Table 2. Average test results when training with different window sizes.

Window size	F1 baseline	F1 stress	F1 amusement	Balanced accuracy
140	0.85	0.839	0.723	0.823
280	0.931	0.905	0.861	0.913
420	0.955	0.941	0.875	0.937
560	0.951	0.944	0.859	0.935
700	0.951	0.952	0.841	0.933
770	0.991	0.991	0.973	0.988

best F1 score in term for stress detection and third best balanced accuracy. Tap-Net, a recent deep learning method, generally performs slightly better than our approach.

Finally, in Table 4 we compare our results against published results for competitor methods on the WESAD dataset. Average F1 score for stress and average balanced accuracy score of all participants is presented. Lisowska et al. used a 1D CNN to classify stress from the raw WESAD data [18]. Huynh et al. applied a customized deep neural network model [14]. Two teams utilized CNNs [1,12]. Tiwari et al. used Random Forest [32]. Our method has the highest F1 and accuracy.

Table 3. Comparisons with the state of the art time series classification methods

Methods	F1 baseline	F1 stress	F1 amusement	Balanced accuracy
Time CNN	0.657	0.174	0.093	0.418
Full CNN	**0.998**	0.99	0.971	0.991
MLP	0.693	0	0	0.379
LSTMFCN	0.968	0.969	0.923	0.961
InceptionTime	0.942	0.926	0.893	0.929
TapNet	**0.998**	0.983	**0.982**	**0.992**
ResNet	0.996	0.987	0.944	0.984
Proposed method	0.991	**0.991**	0.9673	0.988

Table 4. Comparison with other methods, some other methods only provide accuracy or F1 score. '-' in the table represent values that were not provided.

Methods	Average F1 score	Average accuracy
Lisowska et al. 2021 [18]	0.705	-
Huynh et al. 2021 [14]	-	0.835
Alshamrani 2021 [1]	0.84	0.85
Gupta et al. 2021 [12]	-	0.854
Tiwari et al. 2023 [32]	0.98	0.98
Proposed method	**0.991**	**0.988**

6   Conclusion

This paper presented a TDA-based multivariate time series classification method for long time series. We implemented a Takens embedding for multivariate time series to convert the time series to point clouds and then extract TDA features (Betti numbers) at various values of a filtration parameter to represent the persistent homology of these data. These extracted features were used to predict stress labels using a 1D CNN model. Our experiments confirmed that our method performs well against state-of-the-art multivariate time series classification methods and competing state-of-the-art methods on the WESAD data.

Acknowledgments. H.V.T thanks Vingroup Science and Technology Scholarship Program for their generous support and stipend.

References

1. Alshamrani, M.: An advanced stress detection approach based on processing data from wearable wrist devices. Int. J. Adv. Comput. Sci. Appl. **12**, 399–405 (2021)

2. Boissonnat, J.D., Pritam, S.: Edge collapse and persistence of flag complexes. In: SoCG 2020–36th International Symposium on Computational Geometry (2020)
3. Carlsson, G., Zomorodian, A., Collins, A., Guibas, L.J.: Persistence barcodes for shapes. Int. J. Shape Model. **11**(02), 149–187 (2005)
4. Carrière, M., Chazal, F., Ike, Y., Lacombe, T., Royer, M., Umeda, Y.: PersLay: a neural network layer for persistence diagrams and new graph topological signatures. In: International Conference on Artificial Intelligence & Statistics, pp. 2786–2796. PMLR (2020)
5. Chazal, F., Michel, B.: An introduction to topological data analysis: fundamental and practical aspects for data scientists. Front. Artif. Intell. **4**, 1–28 (2021)
6. Chung, Y.M., Cruse, W., Lawson, A.: A persistent homology approach to time series classification. arXiv e-prints arXiv: 2003.06462 (2020)
7. Dirafzoon, A., Lokare, N., Lobaton, E.: Action classification from motion capture data using topological data analysis. In: 2016 IEEE Global Conference on Signal and Information Processing (GlobalSIP), pp. 1260–1264. IEEE (2016)
8. Edelsbrunner, H., Letscher, D., Zomorodian, A.: Topological persistence and simplification. In: Proceedings 41st Annual Symposium on Foundations of Computer Science, pp. 454–463. IEEE (2000)
9. Emrani, S., Gentimis, T., Krim, H.: Persistent homology of delay embeddings and its application to wheeze detection. IEEE Sig. Process. Lett. **21**(4), 459–463 (2014)
10. Fawaz, H.I., et al.: InceptionTime: finding AlexNet for time series classification. Data Min. Knowl. Disc. **34**(6), 1936–1962 (2020). https://doi.org/10.1007/s10618-020-00710-y
11. Gidea, M., Katz, Y.: Topological data analysis of financial time series: landscapes of crashes. Phys. A **491**, 820–834 (2018)
12. Gupta, D., Bhatia, M., Kumar, A.: Resolving data overload and latency issues in multivariate time-series IoMT data for mental health monitoring. IEEE Sens. J. **21**(22), 25421–25428 (2021)
13. Hofer, C., Kwitt, R., Niethammer, M., Uhl, A.: Deep learning with topological signatures. In: Advances in Neural Information Processing Systems, vol. 30 (2017)
14. Huynh, L., Nguyen, T., Nguyen, T., Pirttikangas, S., Siirtola, P.: StressNAS: affect state and stress detection using neural architecture search. In: Adjunct Proceedings of the 2021 ACM International Joint Conference on Pervasive and Ubiquitous Computing and Proceedings of the 2021 ACM International Symposium on Wearable Computers, pp. 121–125 (2021)
15. Karan, A., Kaygun, A.: Time series classification via topological data analysis. Expert Syst. Appl. **183**, 115326 (2021)
16. Karim, F., Majumdar, S., Darabi, H., Harford, S.: Multivariate LSTM-FCNs for time series classification. Neural Netw. **116**, 237–245 (2019)
17. Kim, K., Kim, J., Zaheer, M., Kim, J., Chazal, F., Wasserman, L.: PLLay: efficient topological layer based on persistence landscapes. In: NeurIPS 2020–34th Conference on Neural Information Processing Systems (2020)
18. Lisowska, A., Wilk, S., Peleg, M.: Catching patient's attention at the right time to help them undergo behavioural change: stress classification experiment from blood volume pulse. In: Tucker, A., Henriques Abreu, P., Cardoso, J., Pereira Rodrigues, P., Riaño, D. (eds.) AIME 2021. LNCS (LNAI), vol. 12721, pp. 72–82. Springer, Cham (2021). https://doi.org/10.1007/978-3-030-77211-6_8
19. Lum, P.Y., et al.: Extracting insights from the shape of complex data using topology. Nat. Sci. Rep. **3**(1), 1–8 (2013)

20. Majumdar, S., Laha, A.K.: Clustering and classification of time series using topological data analysis with applications to finance. Expert Syst. Appl. **162**, 113868 (2020)
21. Marchese, A., Maroulas, V.: Signal classification with a point process distance on the space of persistence diagrams. Adv. Data Anal. Classif. **12**, 657–682 (2018). https://doi.org/10.1007/s11634-017-0294-x
22. Munch, E.: A user's guide to topological data analysis. J. Learn. Anal. **4**(2), 47–61 (2017)
23. Musa, S.M.S.S., Noorani, M.S.M., Razak, F.A., Ismail, M., Alias, M.A., Hussain, S.I.: Using persistent homology as preprocessing of early warning signals for critical transition in flood. Nat. Sci. Rep. **11**(1), 1–14 (2021)
24. Pereira, C.M., de Mello, R.F.: Persistent homology for time series and spatial data clustering. Expert Syst. Appl. **42**(15–16), 6026–6038 (2015)
25. Sanderson, N., Shugerman, E., Molnar, S., Meiss, J.D., Bradley, E.: Computational topology techniques for characterizing time-series data. In: Adams, N., Tucker, A., Weston, D. (eds.) IDA 2017. LNCS, vol. 10584, pp. 284–296. Springer, Cham (2017). https://doi.org/10.1007/978-3-319-68765-0_24
26. Savle, K., Zadrozny, W., Lee, M.: Topological data analysis for discourse semantics? In: Proceedings of the 13th International Conference on Computational Semantics-Student Papers, pp. 34–43 (2019)
27. Schmidt, P., Reiss, A., Duerichen, R., Marberger, C., Laerhoven, K.V.: Introducing WESAD, a multimodal dataset for wearable stress and affect detection. In: Proceedings of the 20th ACM International Conference on Multimodal Interaction, pp. 400–408 (2018)
28. Seversky, L.M., Davis, S., Berger, M.: On time-series topological data analysis: new data and opportunities. In: Proceedings of the IEEE Conference on Computer Vision and Pattern Recognition Workshops, pp. 59–67 (2016)
29. Skaf, Y., Laubenbacher, R.: Topological data analysis in biomedicine: a review. J. Biomed. Inform. **130**, 104082 (2022)
30. Takens, F.: Detecting strange attractors in turbulence. In: Rand, D., Young, L.-S. (eds.) Dynamical Systems and Turbulence, Warwick 1980. LNM, vol. 898, pp. 366–381. Springer, Heidelberg (1981). https://doi.org/10.1007/BFb0091924
31. Tauzin, G., et al.: giotto-tda: a topological data analysis toolkit for machine learning and data exploration. J. Mach. Learn. Res. **22**(1), 1834–1839 (2021)
32. Tiwari, S., Chandra, R., Agarwal, S.: An optimized hybrid solution for IoT based lifestyle disease classification using stress data. In: Tanveer, M., Agarwal, S., Ozawa, S., Ekbal, A., Jatowt, A. (eds.) ICONIP 2022. CCIS, vol. 1794, pp. 433–445. Springer, Cham (2023). https://doi.org/10.1007/978-981-99-1648-1_36
33. Tymochko, S., Singhal, K., Heo, G.: Classifying sleep states using persistent homology and Markov chains: a pilot study. In: Demir, I., Lou, Y., Wang, X., Welker, K. (eds.) Advances in Data Science. AWMS, vol. 26, pp. 253–289. Springer, Cham (2021). https://doi.org/10.1007/978-3-030-79891-8_11
34. Wang, Z., Yan, W., Oates, T.: Time series classification from scratch with deep neural networks: a strong baseline. In: 2017 International Joint Conference on Neural Networks (IJCNN), pp. 1578–1585. IEEE (2017)
35. Wu, C., Hargreaves, C.A.: Topological machine learning for multivariate time series. J. Exp. Theor. Artif. Intell. **34**(2), 311–326 (2022)
36. Yen, P.T.W., Cheong, S.A.: Using topological data analysis (TDA) and persistent homology to analyze the stock markets in Singapore and Taiwan. Front. Phys. **9**, 1–19 (2021)

37. Zhang, X., Gao, Y., Lin, J., Lu, C.T.: TapNet: multivariate time series classification with attentional prototypical network. In: Proceedings of the AAAI Conference on Artificial Intelligence, vol. 34, pp. 6845–6852 (2020)
38. Zhao, B., Lu, H., Chen, S., Liu, J., Wu, D.: Convolutional neural networks for time series classification. J. Syst. Eng. Electron. **28**(1), 162–169 (2017)

Mining Label Distribution Drift in Unsupervised Domain Adaptation

Peizhao Li[1](✉), Zhengming Ding[2], and Hongfu Liu[1]

[1] Brandeis University, Waltham, USA
peizhaoli@brandeis.edu
[2] Tulane University, New Orleans, USA

Abstract. Unsupervised domain adaptation targets to transfer task-related knowledge from labeled source domain to unlabeled target domain. Although tremendous efforts have been made to minimize domain divergence, most existing methods only partially manage by aligning feature representations from diverse domains. Beyond the discrepancy in data distribution, the gap between source and target label distribution, recognized as label distribution drift, is another crucial factor raising domain divergence, and has been under insufficient exploration. From this perspective, we first reveal how label distribution drift brings negative influence. Next, we propose Label distribution Matching Domain Adversarial Network (LMDAN) to handle data distribution shift and label distribution drift jointly. In LMDAN, label distribution drift is addressed by a source sample weighting strategy, which selects samples that contribute to positive adaptation and avoid adverse effects brought by the mismatched samples. Experiments show that LMDAN delivers superior performance under considerable label distribution drift.

Keywords: Unsupervised Domain Adaptation · Label Distribution Drift · Transfer Learning · Deep Learning

1 Introduction

Domain adaptation is a fundamental research topic in the machine learning and computer vision field [2,16]. It aims to build models on labeled source data and related target data, then make models adapt and generalize on target domain. Different settings for domain adaptation are applicable for complicated real-world problems [9,10,13,24,25,27,30–32]. Unsupervised domain adaptation, containing no label in target domain, is a challenging but practical setting, owing to that actual scenes are actually suffering from the lack of specific annotations.

The mitigation of domain shift, which aims to reduce the domain divergence between source and target, is the primary solution for unsupervised domain

Supplementary Information The online version contains supplementary material available at https://doi.org/10.1007/978-981-99-8388-9_29.

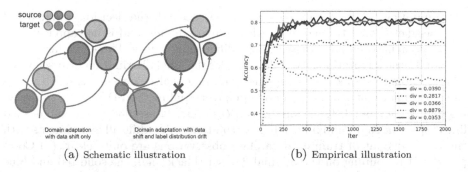

(a) Schematic illustration   (b) Empirical illustration

Fig. 1. (a) Schematic illustration on domain adaptation with and w/o label distribution drift. Source and target domain differ in the color of the borders of the circles. Circles with different colors inside denote different categories, and the size indicates the number of samples within that category. Straight lines denote the decision boundary learned by the classifier. Adaptation under label distribution drift makes features misaligned at the categorical level and decision boundary not applicable to target domain. (b) Empirical illustration on the performance of DANN under varying degrees of label distribution drift on *Office-31* dataset from source domain *Amazon* to target domain *Webcam*. The black line indicates training with the original label distribution, while the blue and red lines denote sample drop rates at 50% and 75%, respectively. Solid lines indicate that dropped samples come from the first 15 classes for both source and target domain, while dashed lines indicate drops come from the first 15 and last 16 classes in source and target domain. The legend provides KL-divergence between source and target label distribution.

adaptation problems. Existing methods [12,15,18,20,33–35,42,44] mainly focus on alleviating negative influence brought by domain shift in feature representations. They reduce the discrepancy by pushing feature distribution from two separate domains close to each other. Consequently, models are expected to be generalized favorably to a related target data distribution. Adversarial learning is recently introduced into domain adaptation with promising performance [18,34,40,41]. By executing generated features to confuse the discriminator, meanwhile forming the discriminator to distinguish from source to target, domain adversarial training aligns features from separate domains through a min-max optimization and delivers domain-invariant representations.

Existing deep domain adaptation methods mainly focus on feature-level alignment. Unfortunately, such ill-judged consideration is not enough to guarantee the success of a beneficial adaptation. As another component of domain shift, the disparity in distributions of labels between source and target domain, *i.e.*, the number of samples in each category differs from source to target, is named as label distribution drift in corresponding to the shift in the distribution of samples across domains. As presented in Fig. 1a, label distribution drift is pernicious according to two aspects. First, along with the adapted processing, features belonging to a large-scale category in target domain are inevitably approaching features in mismatched categories in source domain due to the imbalanced adaptation toward label distribution. As a result, the alignment corrupts feature representations of those misaligned samples. Second, the decision boundary of the classifier is only trained on labeled source samples, and is not applicable to target domain when label distributions differ significantly. These two inside reasons make the adaptation power down under the label distribution drift scenario. The foregoing perspective is supported by practical evidence. Figure 1b shows the performance of the classical domain adversarial method DANN [18] on varying degrees of label distribution drift on *Office-31* dataset [38]. Solid and dashed lines represent slight and huge label distribution drift, while all experiments with the same amount of training data. Two observations are quite clear: (1) Compared to the training on the original dataset (black line), the solid red and blue lines deliver similar results, indicating that even if the size of classes in the same domain is imbalanced, high performance can still be achieved by DANN under the scenario of accordant source and target label distribution. Although training sets in solid lines contain different samples, dropped samples do not bring negative effects on the performance; (2) The disparity between solid and dashed lines indicates that when there is a significant label distribution drift between two domains, the performance drops dramatically. The adaptation performance of DANN becomes much worse with a larger divergence between source and target label distribution. More challenging, different from sample distribution shift between domains, label distribution cannot be aligned directly by existing methods because of the unknown target label distribution, and it becomes more difficult to address this problem when a considerable label distribution drift exists.

In this paper, we consider label distribution drift in visual unsupervised domain adaptation, and present a solution whereby managing data shift together with label distribution drift in a unified framework. As mentioned above, domain adaptation with only feature space alignment is only the partial picture. Therefore, we attempt to align two domains on the premise that corresponding label distributions are roughly matched, and continually alleviate data shift and label distribution drift simultaneously during training. To this end, we propose the **L**abel distribution **M**atching **D**omain **A**dversarial **N**etwork (LMDAN). We propose a novel weighting strategy for source sample re-weighting, which explores training samples that benefit advantageous adaptation while mitigating negative influences from aligning irrelevant categories across domains. The proposed re-weighting function is capable of contributing to both the adversarial feature alignment and classified boundary learning, hence alleviating the two-fold negative impacts brought by considerable label distribution drift simultaneously.

2   Related Work

Feature alignment aims to reduce the domain divergence in feature space. Traditional methods construct projections for two domains, mapping two feature distributions into the manifold space or subspace to address the domain shift problem [11,14,17,20]. Recently, Long *et al.* [33,35] use deep models to reduce the discrepancy between feature spaces in multiple layer levels. Further, by the success of Generative Adversarial Network [21], adversarial learning in deep models for unsupervised domain adaptation delivers favorable performance [5,34,44].

Ganin *et al.* [18] is the first to employ an adversarial learning-based domain adaptation model and pave to many following works. Tzeng *et al.* [40] uses two separate encoders for adaptation while decomposing the transfer process in an end-to-end fashion. Besides, the incorporation of conditional distribution [7,34] in adaptation is also a promising way to reach domain-invariant representations. Optimal transport for domain adaptation [1,3,8,37] is another interesting line of research, where source samples are mapped into target domain with minimal cost transportation. Apart from domain adaptation, feature alignment between different groups of samples also has been used in techniques addressing machine learning fairness [6,26,28,29,39]. Although great efforts have been made to seek better feature alignment, only handling the feature divergence is not sufficient to guarantee a good adaptation without negative transfer.

Besides the aforementioned challenge and the spring-up of solutions, label distribution drift is another inherent barrier in domain adaptation problems but with less exploration compared to the divergence in feature space. The barrier derives from the divergence between known source label and unknown target label distribution. Previously, some works do indeed formulate label distribution drift problems or so-called target shift [43], however, their emphasis is not on deep visual domain adaptation. Liang *et al.* [19] focus on the negative transfer and imbalanced distributions in multi-source transfer learning, while Ming *et al.* [36] exploit label and structural information within and across domains based on the maximum mean discrepancy.

Holding a different emphasis from existing studies, we tackle the deep visual domain adaptation problem under considerable label distribution drift situations and conduct a comprehensive cognition on label distribution drift from both experimental and methodological perspectives. Based on these, we introduce a novel label-matching strategy by continually seeking samples that benefit positive adaptation and simultaneously prevent the negative transfer.

3   Label Distribution Drift

Notation. We start from the basic notations. Consider taking n_s labeled samples $\{(x_i^s, y_i^s)\}_{i=1}^{n_s} \in (\mathcal{X} \times \mathcal{Y})^{n_s}$ under domain distribution $\mathcal{D}^s$, where $\mathcal{X}$ and $\mathcal{Y}$ denotes the corresponding feature and label space, and n_t unlabeled samples with the same label set as source samples $\{x_j^t\}_{j=1}^{n_t} \in \mathcal{X}^{n_t}$ taken from target domain distribution $\mathcal{D}^t$. The goal of unsupervised domain adaptation is to utilize labeled source data for the predictions on unlabeled target samples. Suppose an encoder $\mathcal{F}$ that is designed for projecting samples drawn i.i.d. from the input space $\mathcal{D}^s$ and $\mathcal{D}^t$ to a shared feature space.

Label Distribution Drift. Tremendous efforts have been made to explore solutions for unsupervised domain adaptation. Unfortunately, most existing studies only focus on feature space divergence by minimizing $Div(\mathbb{E}_{x \sim \mathcal{D}^s}[\mathcal{F}(x)], \mathbb{E}_{x \sim \mathcal{D}^t}[\mathcal{F}(x)])$, while ignoring the negative effects brought by

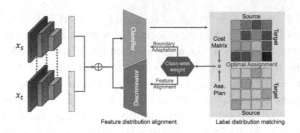

Fig. 2. Label distribution Matching Domain Adversarial Network.

label distribution drift. Here, $Div(,)$ can be understood as the distribution divergence in terms of any distribution divergence measurement, *e.g.*, KL divergence, or Wasserstein distance.

From a theoretical perspective, a generalization bound for domain adaptation problem towards the expected error on target samples [2] is given as follows:

$$\epsilon^t(h) \leq \epsilon^s(h) + \frac{1}{2}d_{\mathcal{H}\Delta\mathcal{H}}(\mathcal{D}^s, \mathcal{D}^t) + \lambda^* + \Omega, \tag{1}$$

where $\epsilon^t(h)$ and $\epsilon^s(h)$ are expected errors on target and source domain, respectively; $\mathcal{H}$ is hypothesis space, $\lambda^* := \epsilon^s(h^*) + \epsilon^t(h^*)$ is the optimal joint risk among source and target samples, and Ω is a constant related to the numbers of samples, dimensions, confidence level, and VC-dimension of $\mathcal{H}$. With the assumption that source and target label distribution are close enough, methods with only feature alignment could achieve small target error $\epsilon^t(h)$ by reducing domain distance term $\frac{1}{2}d_{\mathcal{H}\Delta\mathcal{H}}(\mathcal{D}^s, \mathcal{D}^t)$. However, as pointed out by Zhao *et al.* [45], when the above assumption does not hold, the huge label distribution gap between two domains leads the joint error term λ^* increase oppositely during the optimization towards the domain distance term, and might counteract with the reduction in domain distance term, which increases the upper bound.

Figure 1b further supports the illustrated statement with practical evidence. Based on the above theoretical analysis and empirical illustration, merely realizing the alignment of feature distribution is still far away from the success of the adaptation. The above exploration motivates us to provide a unified problem formulation of unsupervised domain adaptation on both data distribution shift and label distribution drift.

Problem Formulation. Two challenges for unsupervised domain adaptation problems brought by domain divergence are as follows.

Data Distribution Shift. Usually, samples and extracted feature representations from source to target domain are different, *i.e.*, $\mathcal{P}(x^s) \neq \mathcal{P}(x^t)$, and prohibits models from learning a classifier with labeled source samples that can be directly applied for target sample predictions. For this reason, feature alignment can be achieved by minimizing the divergence:

$$Div(\mathbb{E}_{x^s \sim \mathcal{D}_s}[\mathcal{F}(x^s)], \mathbb{E}_{x^t \sim \mathcal{D}_t}[\mathcal{F}(x^t)]), \tag{2}$$

or by minimizing the distribution divergence that is conditional on the categorical belonging of samples to narrow the data distribution shift for domain adaptation:

$$Div(\mathbb{E}_{(x^s,y^s)\sim\mathcal{D}^s}[\mathcal{F}(x^s)|y^s], \mathbb{E}_{(x^t,y^t)\sim\mathcal{D}^t}[\mathcal{F}(x^t)|y^t]), \ \forall y^s = y^t. \qquad (3)$$

Label Distribution Drift. Beyond the inconsistency in feature space, domain divergence also occurs in label space, where $\mathcal{P}(y^s) \neq \mathcal{P}(y^t)$. It is more challenging to handle label distribution drift than data shift due to $\mathcal{P}(y^t)$ being an agnostic distribution in unsupervised domain adaptation.

When considerable label distribution divergence exists, since we always equally sample from two domains and feed the sampling into adversarial alignment, the excessive training towards feature distribution alignment will mislead to the unaligned minimization:

$$Div(\mathbb{E}_{(x^s,y^s)\sim\mathcal{D}^s}[\mathcal{F}(x^s)|y^s], \mathbb{E}_{(x^t,y^t)\sim\mathcal{D}^t}[\mathcal{F}(x^t)|y^t]), \ \forall y^s \neq y^t, \qquad (4)$$

which aligns target feature representations to irrelevant categories in source domain during training. These can induce corruption in the level of categorical feature representations and rise increasing predicted error when predicting y^t in target domain, and further turn out negative effects on the adaptation.

Most deep visual unsupervised domain adaptation methods consider domain divergence merely from data shift, and take no drift in label space for granted. We need to admit that addressing the partial picture is enough for experimental datasets, since current datasets do not include a considerable label distribution drift. We lament that such success is not only far away from real scenarios but also suffers from degraded performance due to label distribution drift (See Fig. 1b). Consider adverse effects can be easily accessed when the inconsistent label distribution and the alignment between irrelevant classes across domains exist, not all samples can be fully used during the training process for the reason that positive adaptation only comes from the part of a correctly matched pair of source-target samples. Consequently, we try to exploit and emphasize the part of correctly matched samples in two domains, while mitigating the alignment on class-mismatched samples. Thus we further increase the ratio of positive adaptation and avoid the negative transfer brought by the alignment between irrelevant categories across source and target domains. This can be viewed as an unsupervised sample selection in that we are continually seeking samples that benefit from adaptation and avoid negative transfer concurrently.

4   Label Distribution Matching

Overview. Figure 2 shows the framework of the proposed LMDAN. It minimizes the domain divergence embedded in feature space on the premise of close source-target label distribution. Specifically, to align the source and target domain under label distribution drift, LMDAN contains two interactive parts: adversarial training for domain-invariant features generation, and the class-wise re-weighting strategy through the optimal assignment for source sample selection.

In adversarial feature alignment, the encoder $\mathcal{F}$ tries to extract feature $\mathcal{F}(x^s)$ and $\mathcal{F}(x^t)$ from two domains and confuse the discriminator $\mathcal{D}$, while $\mathcal{D}$ tries to distinguish $\mathcal{F}(x^s)$ and $\mathcal{F}(x^t)$ from each other. Finally, $\mathcal{F}$ is trained to map data distribution from two domains close enough. In the source samples weighting part, by adding class-wise weights on both adversarial training and supervision on the classifier, we manipulate feature alignment in adversarial training and decision boundary of the classifier to tackle the label drift scenario simultaneously. The dual weighting strategy makes the network adapt to target domain by two sides: (1) The weighting for the min-max game emphasizes features in the same category to get closer across domains, at the same time mitigate the misalignment, and (2) The weighting for classifier makes decision boundary adapt to the target label distribution. In the following, we emphatically illustrate the source sample weighting strategy, and then provide details for model training.

4.1 Label Distribution Matching

Label distribution matching is one of the crucial components in the LMDAN framework. It disposes of label distribution drift towards source-target sample matching. Here, we expect to exploit samples in the parts of classes matched across source and target domain by the optimal assignment, then enlarge matched classes and shrink the size of less relevant classes in source domain. As a result, samples in source domain engaging in the adversarial feature alignment are able to approach to target domain in terms of label distribution, and further contribute to increasing positive and mitigating negative transfer.

To achieve this, we employ the classified probability g with $||g||_1 = 1$ of every sample to measure the degree of matching. Based on the measurement of distance and optimal matching, mismatched pairs result in a larger distance, while matched pairs perform inversely. Consider a cost function $c : \mathcal{C} \times \mathcal{C} \rightarrow \mathbb{R}^+$ and g_i^s and g_j^t the classified probabilities obtained by the classifier $\mathcal{G}$ for source sample x_i^s and target sample x_j^t, respectively, and the output space $\mathcal{C} : g \in \mathcal{C}$. Based on optimal assignment [23], we seek for a joint probability distribution γ according to g^s and g^t:

$$\gamma^* = \underset{\gamma \in \prod(\mathcal{C} \times \mathcal{C})}{\arg\min} \int_{\mathcal{C} \times \mathcal{C}} c(g^s, g^t) d\gamma(g^s, g^t). \tag{5}$$

This indicates the optimal assignment based on classified probabilities from source to target with the minimum cost.

As for the discrete version for implementation, we employ Euclidean distance to build the cost matrix $M = \{m_{ij}\} \in \mathbb{R}^{n_s \times n_t}$ between source and target domain,

$$m_{ij} = c(g_i^s, g_j^t) = ||g_i^s - g_j^t||_2, \tag{6}$$

and other distance functions can be used as well. Based on the cost matrix M, the optimal assignment is written as:

$$\gamma^* = \underset{\gamma \in \mathbb{R}^{n_s \times n_t}}{\arg\min} \langle \gamma, M \rangle_F, \text{ s.t.} : \gamma \mathbf{1}_{n_t} = \sum_i^{n_s} g_i^s, \ \gamma \mathbf{1}_{n_s} = \sum_j^{n_t} g_j^t, \tag{7}$$

where $\langle \cdot, \cdot \rangle_F$ indicates Frobenius inner product, and $\mathbf{1}_n$ is an all-one n-dimension vector.

We then incorporate the distance within classified probabilities into the optimal assignment plan and make the conjunct term guide class-wise weights for each class. We obtain the weight guiding matrix $\mathbf{T} = \{t_{ij}\} \in \mathbb{R}^{n_s \times n_t}$ by

$$\mathbf{T} = \gamma^* \circ \mathbf{M}, \tag{8}$$

where $\circ$ denotes the Hadamard product. By matching classification probabilities with the minimal cost, the weight guiding matrix provides guidance for misaligned samples. Moreover, following the above step, we compute the class-wise weight w_k for the class with index k in source domain by:

$$w_k = \left((\sum_{i=1}^{n_s} \mathbb{1}_{y_i^s = k})^\alpha \cdot \sum_{i=1}^{n_s} t_{ij} \mathbb{1}_{y_i^s = k} \right)^{-1}, \tag{9}$$

where $\mathbb{1}$ is the indicator function. Note that w_k consists of two parts, where the first term manages the imbalanced class size within source domain itself, and the second awards or punishes the matched or mismatched pairs between source and target accordingly. α is the parameter to control the influence of source class imbalanced scale.

Using weights in terms of categories according to the optimal matching toward classified probabilities, we are able to distinguish classes that are misaligned and less relevant to positive transfer from well-aligned ones during training. By re-weighting samples in source domain by class-wise weights, the sizes or corresponding categories are enlarged or shrunk accordingly, then further push the source label distribution to the unknown target one dynamically.

4.2   Objective and Solution

Finally, we provide the objective functions and the corresponding optimizing solution for LMDAN. We first calculate class-wise weights on each mini-batch sample and then optimize toward the min-max game in adversarial learning integrated with subsequent classification by a dual weighting strategy. Loss functions for LMDAN can be written as:

$$\min_{\mathcal{F},\mathcal{G}} \mathcal{L}_1(\mathcal{F},\mathcal{G},\mathcal{D}) + \lambda \mathcal{L}_2(\mathcal{F},\mathcal{D}) \text{ and } \max_{\mathcal{D}} \mathcal{L}_2(\mathcal{F},\mathcal{D}), \text{ with} \tag{10}$$

$$\mathcal{L}_1 = \mathop{\mathbb{E}}_{(x^s, y^s) \sim \mathcal{D}^s} w_i \mathcal{L} \left(\mathcal{G} \left(\mathcal{F}(x^s) \right), y^s \right),$$

$$\mathcal{L}_2 = \mathop{\mathbb{E}}_{x^s \sim \mathcal{D}^s} w_i \log \left[\mathcal{D} \left(\mathcal{F} \left(x^s \right) \right) \right] + \mathop{\mathbb{E}}_{x^t \sim \mathcal{D}^t} \log \left[1 - \mathcal{D} \left(\mathcal{F} \left(x^t \right) \right) \right],$$

where w_i is the corresponding weight of the class where $x^s \sim \mathcal{D}^s$ belongs to, and λ is the trade-off hyperparameter for classification loss and adversarial loss. In our objective functions, the weighting strategy is conducted in two places. The weighted classifier $\mathcal{G}$ captures label distribution drift for better decision

Table 1. Results for unsupervised domain adaptation with label distribution drift on *Office-31*[0.75;0.75] dataset.

Method	A → W	A → D	W → A	W → D	D → A	D → W	Average
ResNet50 [22]	66.1 ± 4.3	65.8 ± 1.5	53.3 ± 3.1	**87.8 ± 2.9**	53.0 ± 3.7	79.4 ± 2.6	67.6 ± 1.5
DANN [18]	50.7 ± 2.6	54.0 ± 2.7	35.4 ± 3.4	62.6 ± 4.2	34.6 ± 3.8	56.3 ± 2.9	49.0 ± 0.8
JAN [35]	51.2 ± 3.2	49.5 ± 2.4	46.1 ± 3.9	72.9 ± 4.1	40.9 ± 5.1	71.8 ± 2.6	55.4 ± 1.6
WMMD [42]	39.1 ± 5.2	43.3 ± 4.1	38.4 ± 2.7	67.8 ± 4.8	34.1 ± 3.2	68.1 ± 7.1	48.5 ± 3.4
CDAN [34]	65.7 ± 3.2	62.8 ± 4.8	52.5 ± 2.7	78.1 ± 4.7	39.8 ± 4.5	73.5 ± 4.4	62.1 ± 1.7
RAAN [4]	59.4 ± 3.8	65.7 ± 2.9	48.5 ± 5.0	76.4 ± 3.5	45.8 ± 6.9	77.4 ± 3.6	62.2 ± 3.2
SymNets [44]	57.1 ± 4.0	54.6 ± 2.7	41.9 ± 6.3	67.0 ± 5.1	32.4 ± 4.8	57.2 ± 6.7	51.7 ± 2.7
BSP [5]	61.5 ± 2.1	58.9 ± 2.6	47.5 ± 3.2	85.0 ± 3.6	40.4 ± 2.9	84.1 ± 3.0	62.9 ± 2.2
LMDAN	**73.1 ± 1.7**	**71.0 ± 2.5**	**56.5 ± 2.4**	84.4 ± 2.6	**57.8 ± 4.9**	**88.8 ± 3.5**	**71.9 ± 2.1**

Table 2. Results for unsupervised domain adaptation with label distribution drift on *ImageCLEF-DA*[0.75;0.75] dataset.

Method	C → I	C → P	I → C	I → P	P → C	P → I	Average
ResNet50 [22]	76.9 ± 3.2	63.8 ± 1.7	87.1 ± 1.8	71.3 ± 0.7	81.7 ± 3.7	73.4 ± 4.2	75.7 ± 2.1
DANN [18]	47.4 ± 2.8	40.8 ± 2.7	55.0 ± 1.2	50.4 ± 2.3	55.0 ± 3.1	51.2 ± 3.6	50.0 ± 1.4
JAN [35]	34.2 ± 2.8	27.9 ± 1.0	38.8 ± 3.8	49.0 ± 3.4	36.7 ± 4.4	44.1 ± 3.0	38.5 ± 0.7
WMMD [42]	42.4 ± 1.1	30.4 ± 3.5	65.2 ± 3.9	70.8 ± 3.0	47.2 ± 3.5	56.4 ± 1.9	52.0 ± 2.9
CDAN [34]	58.1 ± 3.8	52.2 ± 3.2	76.3 ± 4.0	62.7 ± 1.8	66.2 ± 9.5	59.2 ± 1.2	63.1 ± 2.2
RAAN [4]	62.9 ± 1.3	54.6 ± 3.3	78.3 ± 1.7	63.6 ± 3.6	71.0 ± 6.6	65.4 ± 2.4	66.0 ± 2.3
SymNets [44]	59.2 ± 5.0	53.8 ± 3.2	70.5 ± 3.9	57.2 ± 3.8	63.4 ± 7.6	54.3 ± 1.5	59.7 ± 1.4
BSP [5]	52.6 ± 1.7	43.4 ± 2.5	70.5 ± 2.9	58.6 ± 4.6	67.0 ± 4.3	62.9 ± 1.8	59.2 ± 1.8
LMDAN	**79.1 ± 2.8**	**67.7 ± 2.7**	**89.8 ± 2.3**	**71.6 ± 2.8**	**88.1 ± 2.4**	**80.5 ± 1.0**	**79.5 ± 0.8**

boundary adaptation on target domain, and the weighted discriminator $\mathcal{D}$ and encoder $\mathcal{F}$ further adjust feature alignment to fit label distribution drift as well.

In our implementation, we utilize cross-entropy loss as the loss function for $\mathcal{L}$, and set the trade-off parameter λ default to 1 for all experiments. Since the complexity of the optimal assignment is not scalable to the whole dataset, the mini-batch label matching is applied as well. Two benefits are clear. Mini-batch training makes the complexity of the optimal matching affordable in big data adaptation. Besides, equivalent numbers of data points from source and target domain can be sampled, rendering the matching and feature alignment balanced. We use pre-trained ResNet-50 [22] as the feature extractor. Following by [18], we set the initial learning rate $lr = 0.01$ for SGD optimizer, then gradually adjust the learning rate for the classifier by $lr_c = lr(1 + 10p)^{-0.75}$, where p is the training process changed from 0 to 1 linearly. The learning rate for discriminator is $lr_d = \frac{1-\exp(-10p)}{1+\exp(-10p)} lr$.

5 Experimental Analysis

Due to the space limitation, we defer some experimental settings and most of the experimental results (more benchmarking results, ablation studies, and visualization) to supplementary materials.

Table 1 reports quantitative results for unsupervised domain adaptation on *Office-31* [0.75;0.75] dataset. The performance of all competitive methods significantly drops under the huge label distribution divergence and even becomes worse than non-adapted ResNet-50. This indicates that only aligning the feature divergence is not enough for a positive adaptation, for label distribution drift is also a crucial component of domain shift, and has not been paid sufficient attention in domain adaptation area. To dispose of this problem, LMDAN considers source-target sample pairs with different weights, enlarges and shrinks weights for matched and mismatched samples on classified probabilities. By this means, LMDAN outperforms other competitive methods by a large margin. To be noticed, WMMD and RAAN also embed source sample re-weighting strategies into training. However, their weights highly rely on predictions of the target label distribution and make them struggle to handle huge label distribution drift (Table 2).

6 Conclusion

We proposed the Label distribution Matching Domain Adversarial Network (LMDAN) framework for unsupervised domain adaptation. We designed the label distribution matching and weighting strategy for source samples re-weighting, and matched the known source label distribution with the agnostic target one. Experimental results demonstrated the superior performance of LMDAN over other state-of-the-art methods.

References

1. Balaji, Y., Chellappa, R., Feizi, S.: Normalized Wasserstein for mixture distributions with applications in adversarial learning and domain adaptation. In: Proceedings of the IEEE International Conference on Computer Vision (2019)
2. Ben-David, S., Blitzer, J., Crammer, K., Kulesza, A., Pereira, F., Vaughan, J.W.: A theory of learning from different domains. Machine Learning (2010)
3. Damodaran, B.B., Kellenberger, B., Flamary, R., Tuia, D., Courty, N.: DeepJDOT: deep joint distribution optimal transport for unsupervised domain adaptation. In: Ferrari, V., Hebert, M., Sminchisescu, C., Weiss, Y. (eds.) ECCV 2018. LNCS, vol. 11208, pp. 467–483. Springer, Cham (2018). https://doi.org/10.1007/978-3-030-01225-0_28
4. Chen, Q., Liu, Y., Wang, Z., Wassell, I., Chetty, K.: Re-weighted adversarial adaptation network for unsupervised domain adaptation. In: Proceedings of the IEEE Conference on Computer Vision and Pattern Recognition (2018)

5. Chen, X., Wang, S., Long, M., Wang, J.: Transferability vs. discriminability: batch spectral penalization for adversarial domain adaptation. In: Proceedings of the 36th International Conference on Machine Learning. Proceedings of Machine Learning Research (2019). http://proceedings.mlr.press/v97/chen19i.html

6. Chhabra, A., Li, P., Mohapatra, P., Liu, H.: Robust fair clustering: a novel fairness attack and defense framework. In: The Eleventh International Conference on Learning Representations (2022)

7. Cicek, S., Soatto, S.: Unsupervised domain adaptation via regularized conditional alignment. In: Proceedings of the IEEE International Conference on Computer Vision, October 2019

8. Courty, N., Flamary, R., Tuia, D., Rakotomamonjy, A.: Optimal transport for domain adaptation. IEEE Trans. Pattern Anal. Mach. Intell. (2017). https://doi.org/10.1109/TPAMI.2016.2615921

9. Ding, Z., Nasrabadi, N.M., Fu, Y.: Semi-supervised deep domain adaptation via coupled neural networks. IEEE Trans. Image Process. **27**(11), 5214–5224 (2018)

10. Ding, Z., Fu, Y.: Deep domain generalization with structured low-rank constraint. IEEE Trans. Image Process. **27**(1), 304–313 (2017)

11. Ding, Z., Fu, Y.: Deep transfer low-rank coding for cross-domain learning. IEEE Trans. Neural Netw. Learn. Syst. **30**(6), 1768–1779 (2019)

12. Ding, Z., Li, S., Shao, M., Fu, Y.: Graph adaptive knowledge transfer for unsupervised domain adaptation. In: Ferrari, V., Hebert, M., Sminchisescu, C., Weiss, Y. (eds.) ECCV 2018. LNCS, vol. 11206, pp. 36–52. Springer, Cham (2018). https://doi.org/10.1007/978-3-030-01216-8_3

13. Ding, Z., Liu, H.: Marginalized latent semantic encoder for zero-shot learning. In: Proceedings of the IEEE Conference on Computer Vision and Pattern Recognition (2019)

14. Ding, Z., Shao, M., Fu, Y.: Deep low-rank coding for transfer learning. In: Twenty-Fourth International Joint Conference on Artificial Intelligence (2015)

15. Ding, Z., Shao, M., Fu, Y.: Robust multi-view representation: a unified perspective from multi-view learning to domain adaption. In: IJCAI, pp. 5434–5440 (2018)

16. Ding, Z., Zhao, H., Fu, Y.: Deep domain adaptation. In: Jain, L.C., Wu, X. (eds.) Learning Representation for Multi-View Data Analysis. AIKP, pp. 203–249. Springer, Cham (2019). https://doi.org/10.1007/978-3-030-00734-8_9

17. Fernando, B., Habrard, A., Sebban, M., Tuytelaars, T.: Unsupervised visual domain adaptation using subspace alignment. In: Proceedings of the IEEE International Conference on Computer Vision (2013)

18. Ganin, Y., et al.: Domain-adversarial training of neural networks. J. Mach. Learn. Res. **17**(1), 2096–2030 (2016)

19. Ge, L., Gao, J., Ngo, H., Li, K., Zhang, A.: On handling negative transfer and imbalanced distributions in multiple source transfer learning. ASA Data Sci. J. Stat. Anal. Data Mining **7**(4), 254–271 (2014)

20. Gong, B., Shi, Y., Sha, F., Grauman, K.: Geodesic flow kernel for unsupervised domain adaptation. In: Proceedings of the IEEE Conference on Computer Vision and Pattern Recognition (2012). https://doi.org/10.1109/CVPR.2012.6247911

21. Goodfellow, I., et al.: Generative adversarial nets. In: Advances in Neural Information Processing Systems (2014)

22. He, K., Zhang, X., Ren, S., Sun, J.: Deep residual learning for image recognition. In: Proceedings of the IEEE Conference on Computer Vision and Pattern Recognition (2016)

23. Kantorovich, L.V.: On the translocation of masses. J. Math. Sci. **133**(4), 1381–1382 (2006)

24. Kodirov, E., Xiang, T., Fu, Z., Gong, S.: Unsupervised domain adaptation for zero-shot learning. In: Proceedings of the IEEE International Conference on Computer Vision (2015)
25. Li, P., et al.: SelfDoc: self-supervised document representation learning. In: Proceedings of the IEEE/CVF Conference on Computer Vision and Pattern Recognition, pp. 5652–5660 (2021)
26. Li, P., Liu, H.: Achieving fairness at no utility cost via data reweighing with influence. In: International Conference on Machine Learning, pp. 12917–12930. PMLR (2022)
27. Li, P., Wang, P., Berntorp, K., Liu, H.: Exploiting temporal relations on radar perception for autonomous driving. In: Proceedings of the IEEE/CVF Conference on Computer Vision and Pattern Recognition, pp. 17071–17080 (2022)
28. Li, P., Wang, Y., Zhao, H., Hong, P., Liu, H.: On dyadic fairness: exploring and mitigating bias in graph connections. In: International Conference on Learning Representations (2020)
29. Li, P., Zhao, H., Liu, H.: Deep fair clustering for visual learning. In: Proceedings of the IEEE/CVF Conference on Computer Vision and Pattern Recognition, pp. 9070–9079 (2020)
30. Li, S., Song, S., Huang, G., Ding, Z., Wu, C.: Domain invariant and class discriminative feature learning for visual domain adaptation. IEEE Trans. Image Process. **27**(9), 4260–4273 (2018)
31. Li, S., et al.: Deep residual correction network for partial domain adaptation. IEEE Trans. Pattern Anal. Mach. Intell. **43**(7), 2329–2344 (2020)
32. Liu, H., Shao, M., Fu, Y.: Structure-preserved multi-source domain adaptation. In: Proceedings of the IEEE 16th International Conference on Data Mining (2016)
33. Long, M., Cao, Y., Wang, J., Jordan, M.I.: Learning transferable features with deep adaptation networks. In: Proceedings of the 32nd International Conference on Machine Learning (2015). http://dl.acm.org/citation.cfm?id=3045118.3045130
34. Long, M., Cao, Z., Wang, J., Jordan, M.I.: Conditional adversarial domain adaptation. In: Advances in Neural Information Processing Systems (2018)
35. Long, M., Zhu, H., Wang, J., Jordan, M.I.: Deep transfer learning with joint adaptation networks. In: Proceedings of the 34th International Conference on Machine Learning (2017)
36. Ming Harry Hsu, T., Yu Chen, W., Hou, C.A., Hubert Tsai, Y.H., Yeh, Y.R., Frank Wang, Y.C.: Unsupervised domain adaptation with imbalanced cross-domain data. In: Proceedings of the IEEE International Conference on Computer Vision (2015)
37. Perrot, M., Courty, N., Flamary, R., Habrard, A.: Mapping estimation for discrete optimal transport. In: Advances in Neural Information Processing Systems (2016)
38. Saenko, K., Kulis, B., Fritz, M., Darrell, T.: Adapting visual category models to new domains. In: Daniilidis, K., Maragos, P., Paragios, N. (eds.) ECCV 2010. LNCS, vol. 6314, pp. 213–226. Springer, Heidelberg (2010). https://doi.org/10.1007/978-3-642-15561-1_16
39. Song, H., Li, P., Liu, H.: Deep clustering based fair outlier detection. In: Proceedings of the 27th ACM SIGKDD Conference on Knowledge Discovery & Data Mining, pp. 1481–1489 (2021)
40. Tzeng, E., Hoffman, J., Saenko, K., Darrell, T.: Adversarial discriminative domain adaptation. In: Proceedings of the IEEE International Conference on Computer Vision (2017)
41. Xia, H., Ding, Z.: Structure preserving generative cross-domain learning. In: Proceedings of the IEEE/CVF Conference on Computer Vision and Pattern Recognition, pp. 4364–4373 (2020)

42. Yan, H., Ding, Y., Li, P., Wang, Q., Xu, Y., Zuo, W.: Mind the class weight bias: weighted maximum mean discrepancy for unsupervised domain adaptation. In: Proceedings of the IEEE Conference on Computer Vision and Pattern Recognition (2017)
43. Zhang, K., Schölkopf, B., Muandet, K., Wang, Z.: Domain adaptation under target and conditional shift. In: Dasgupta, S., McAllester, D. (eds.) Proceedings of the 30th International Conference on Machine Learning (2013)
44. Zhang, Y., Tang, H., Jia, K., Tan, M.: Domain-symmetric networks for adversarial domain adaptation. In: Proceedings of the IEEE Conference on Computer Vision and Pattern Recognition (2019)
45. Zhao, H., Combes, R.T.D., Zhang, K., Gordon, G.: On learning invariant representations for domain adaptation. In: Proceedings of the 36th International Conference on Machine Learning (2019)

Automatic Classification of Sensors in Buildings: Learning from Time Series Data

Mashud Rana[1][✉], Ashfaqur Rahman[1], Mahathir Almashor[2],
John McCulloch[1], and Subbu Sethuvenkatraman[2]

[1] Data61, CSIRO, Eveleigh, Australia
{mdmashud.rana,ashfaqur.rahman,john.mcculloch}@data61.csiro.au
[2] Energy, CSIRO, Eveleigh, Australia
{mahathir.almashor,subbu.sethuvenkatraman}@csiro.au

Abstract. Smart buildings are generally equipped with thousands of heterogeneous sensors and control devices that impact the operation of their electrical systems. Analytical tools that aim to optimise the energy efficiency within such complex systems requires prior mapping or (classification) of diverse set of sensors according to a standard. Prior research primarily focuses on exploiting the similarities in sensor names (text metadata) to categorise them into identical classes (or groups). However, the sensors within and across buildings often follow distinct naming conventions by different vendors. In addition the definition of the classes or groups also varies significantly amongst researchers. This limits the usability and portability of prior techniques when applied across buildings. There are standard ontologies (Brick, Haystack etc.) that provide a set of standardized classes for the sensors in the buildings. The work herein follows a new avenue to address this challenging classification problem by (i) utilizing only time-series data of sensors and not text metadata, (ii) developing a simple, effective and hitherto unexplored Machine Learning (ML) model to classify the sensors into a set of standard Brick classes, and (iii) evaluating the model on a large proprietary dataset comprising of 129 buildings. Experimental results demonstrate promising performance of the presented data driven model, with average classification accuracy in terms of weighted F-score at 0.78 ($\pm$0.14), and statistically significant improvements over prior methods.

Keywords: Sensor Metadata · Building Automation · Time Series

1 Introduction

Building Automation Systems (BAS) integrate various applications that monitor, analyse, and optimise the energy usage of modern buildings. Their increasing adoption reduces building operational costs, overall emissions, and enable us to achieve urgent sustainability goals [7]. Successful integration of diverse energy analytics with BAS requires access to both the semantic information of a large

T. Liu et al. (Eds.): AI 2023, LNAI 14471, pp. 367–378, 2024.
https://doi.org/10.1007/978-981-99-8388-9_30

number of entities (e.g., sensors, control points), and their associated operational data. The semantic information (aka *metadata*) is required to identify the equipment, locations, physical phenomenon being sensed, and the relationships between those entities. However, such information is often unavailable, unstructured or inconsistent [15]. Reasons for this include: changes in the physical configuration of the buildings over time, and heterogeneous entities that are installed, managed, and named by different vendors. As a result, the usability and portability of energy analytics across buildings are severely limited.

Consequently, several standard schemata like Brick [2] and Haystack [1] have been developed. These schemata describe the heterogeneous sensors and control devices in buildings and the complex relationships among them in a structured and consistent format using a predefined set of *classes* (or *tags*). This allows the use of machine readable representations of the different subsystems of buildings for diverse analytical applications. However, the mapping of entities in buildings by following a standard schema is a time and labour intensive, requiring significant effort from highly specialised domain experts and yet ultimately, is an exercise that is susceptible to errors. Our goal here is to investigate the automated and data-driven mapping of building sensors based on their time series data. In this paper, we use *sensors* and *entities* interchangeably.

Existing approaches predominantly address this classification problem from a purely text processing perspective, using information retrieved from entity names as inputs to the classifiers. The main reason for this is that important properties of the entities are embedded in their names in most cases. The embedded information can help to infer their types, locations, and relationships [10]. However, the success of these approaches are heavily dependent on intrinsic similarities of entity names in both source and target buildings. Any variations of entity naming conventions between source and target buildings negatively affect the portability and usability of the models across buildings. In such case, it is required to integrate domain expertise (e.g., [3]) or use knowledge from target buildings (e.g., [8]) which is not feasible. In contrast, the Time-Series (TS) data associated with similar types of sensors is expected to be consistent within and across buildings. Such data contains patterns that can be utilised as signatures for accurately classifying different types of sensors, regardless of their deployed buildings. Therefore, we've focused on the classification of sensor types based on TS data. Our contributions can be summarized as follows:

- We present a TS-data-driven approach for automatically classifying the sensors in buildings by utilizing XGBoost [4]. A set of statistical features representing the patterns in TS data is explored for classification. Since the approach only requires TS data, it provides better portability and usability across buildings compared to the approaches utilising entity names.
- In contrast to the existing approaches that use user-defined classes or Haystack tags, our approach aims to classify sensors according to the popular Brick schema and at a more granular level (as opposed to shallower levels in current literature) in the class hierarchy to facilitate more widespread applications.
- We evaluate the approach against 129 buildings contained within a proprietary dataset from Australian Data Cleaning House (DCH). The operational

patterns across its diverse building range varies greatly, leading to sensor data with different characteristics and statistical distributions. Using this dataset imposes a stricter challenge, as opposed to datasets that display more homogeneous properties.

- We provide a systematic comparison of XGboost against: (i) other ML classifiers such as Random Forest (RF), Neural Networks (NNs), and Support Vector Machines (SVM) used in prior data-driven approaches; and (ii) the Building Adapter model [11] which utilises both TS and text name space.

2   Related Work

Prior approaches for classification of sensors in buildings can be broadly categorised based on the type of data used [7]: (i) **name space**; and (ii) **time series**. Koh et al. [15] reviewed and implemented several state-of-the-art approaches from both groups. The first group of studies focus on utilising encoded information in text metadata (entity names, vendor specified descriptions, specification of target buildings) as inputs for classification. Balaji et al. [3] proposed "Zodiac", a semi-automatic model that grouped entities by processing their names thru a bag-of-words method and utilising the resultant count values. The entities in each cluster were classified by repeatedly training a Random Forest (RF) model and incorporating feedback from domain experts. He and Wang [8] utilized information extraction principles to merge differing text corpuses gathered from buildings. Text from source buildings were combined with additional knowledge (known as "specification files") that were synthesised from target buildings, which were then fed into a Bi-LSTM model. Scrabble [14] used the combination of Conditional Random Fields and Neural Networks (NNs) on text metadata to apply Brick classes to the sensors. Other prominent studies that adopted similar approaches include [12,18].

In contrast to the abundance of studies utilising text metadata, there are very few studies that rely on TS data to train ML classifiers. Gao et al. [7] utilised several statistical features (e.g., mean, mode, quantiles and deciles) from TS as input to train a set of ML models that including RF, k-Nearest Neighbour (kNN), and SVM. They trained these models for classifying both composite tags and individual tags from Haystack ontology. Hong et al. [9] studied clustering of TS data based on a similarity metric (*cross-predictability*) that was applied to group four types of sensors. The labels used in their study were defined manually instead of following any standard ontology. TS data was also used in Koc et al. [13] but only for inferring spatial relationships between sensors.

There exists studies that incorporate both TS and name space data for classification. The Building Adapter (BA) model in [11] stands as a prominent example. A group of classifiers (SVM, RF, and Logistic Regression (LR)) were trained using 44 TS-based features as inputs. It then transferred the knowledge from source to target building based on clustering of entities using text data from target buildings. Mishra et al. [17] also adopted a similar approach based on RF and SVM classifiers trained on TS data, and clustering on text data.

3   Datasets and Problem Statement

3.1   Datasets

We consider a dataset from DCH, which contains energy and operational data for more than 150 buildings from 30 sites across Australia. After excluding a small subset due to data quality issues, 129 buildings encompassing corporate offices, libraries and research labs remained. Some buildings recorded comprehensive sensor data, whilst others were limited to specific subsystems such as electrical systems. These buildings were modelled manually using the Brick schema by domain experts. The entities in different buildings are named using different naming conventions and by different vendors. Table 1 presents a sample of metadata in our dataset (with anonymised site and building names).

Table 1. Sample metadata for different entities.

Bld id	Entity name	Brick class
Bld_2(Site_5)	dsapi-tacit-silent-start-Site_5_Bld_2.Bld2LandP.VoltageBN	Voltage Sensor
Bld_1(Site_2)	dsapi-thick-gusty-rest-Building_1-Site_2.E5_07_OfficeL2Lighting.CurrentPhaseA	Current Sensor
Bld_1(Site_7)	dsapi-synonymous-tender-campaign-Site_7.B_1...EMeter_3_DB_LG_P2_kVAr	Electrical Power Sensor

Our classification approach relies on TS data attached to the entities in each building, which were typically recorded over several years. Nonetheless, we only considered data from 1st Jan 2022 onwards due to quality issues and to avoid pandemic related anomalies. Moreover, the TS data for buildings in DCH were recorded at different resolutions varying from 5 to 45 min. When modeling the data, we extracted several features representing their statistical properties and patterns, and this is elaborated upon in Sect. 4.1.

3.2   Problem Statement

Given the following:

1. the TS data $TS_B^N = \{ts_1, ts_2, ..., ts_N\}$ of a set of N entities $E_B = \{e_1, e_2, ..., e_N\}$ in a building B where $ts_i = [o_i^1, o_i^2, ..., o_i^L]$ is vector of L time ordered numerical observations of a phenomenon sensed by an entity e_i.
2. the class label y_i of each entity $e_{i \in \{1\ to\ N\}}$. $Y_B = \{y_1, y_2, ..., y_N\}$ is the set of class labels for all entities in the same building.

The Goal is to develop a model, M, to classify the class labels for the entities in the set of E using information from TS_B^N. In other words, the model M intends to learn the mapping function $F(X) \rightarrow Y_B$ where X is the input feature set computed from TS_B^N.

In this study, the entity represents the set of sensors measuring different phenomena (e.g., current, voltage, energy usages, etc.) of electrical systems as well as the outside air temperatures of buildings. The class labels of the entities or sensors are assigned from Brick version 1.2, and we are specifically focused on sensors belonging to five main classes {*Electrical Power Sensor, Voltage Sensor, Current Sensor, Energy Sensor, Outside Air Temperature Sensor*} and their associated sub-classes in the Brick ontology.

4   Proposed Data-Driven Approach

Figure 1 presents a schematic diagram of the proposed approach for sensor types classification. It consists of two main steps: i) *model development*, and ii) *testing*.

4.1   ML Model Development

Time Series Feature Extraction. Feature selection is a crucial step for ML model development. It is the process of identifying a set of informative inputs that can represent the statistical distribution and the patterns in the data [16]. An appropriate feature set helps the ML model to learn both linear and nonlinear relationships between inputs and target, and reduce the chances of overfitting that consequently lead to better performance.

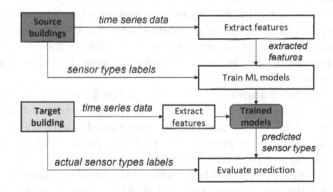

Fig. 1. Schematic diagram of the proposed ML approach for sensor type classification.

The TS data belonging to different groups of sensors shows different statistical distribution over times. Therefore, we aim to extract a set of features from each TS based on a windowing technique as described in [11]. We first segment each TS data into a set of fixed length window where each segment has 50% overlaps with previous one. The length of the window varies depending on the gaps between the consecutive samples in the TS since our dataset has TS with varying sampling rate. For TS with gaps between consecutive samples $\geq$15 min, we set the window length to be 1 h considering possible hourly pattern in TS. Otherwise, it is set dynamically such that the window contains a minimum number of samples (10 in our case). Following [11], for each window we then computed 11 statistical features representing 4 different statistical properties: i) extreme: *min and max*, ii) variability: *median, root mean square, 1st quartile, 3rd quartile, inter quartile range*, iii) moments: *variance, skewness, and kurtosis*, and iv) shape: *slope*. For each of the 11 statistical features, we then compute 4 summary statistics *min, max, variance, and standard deviation* from the series of values for all window. This will results in 44 final features (*11 features × 4 summary statistics* for each feature) for each TS.

Model Training. As the classifier, we adapt the Extreme Gradient Boosting (XGBoost) model. XGBoost is a scalable and distributed gradient-boosted decision tree model that can be applied for supervised classification and regression. The main motivations of choosing XGBoost over other classical ML models like NNs include: i) it is more robust to the noisy data, ii) it is highly parallelizable and hence faster to train on large datasets, iii) it requires less computational resources, and iv) it has less parameters and is easier to tune.

XGBoost trains a set of base learners (shallow Decision Trees (DT)) iteratively such that each base learner focuses on the examples that were difficult to classify by the previous one. In other words, in each iteration XGBoost trains a new base learner which aims to minimise the error of the previous learner. The final prediction is computed by combining the predictions from all base learners based on their weights that are determined based on their performance on training data. In contrast to the RF algorithm which also applies a set of DTs each trained on separate subset of training data chosen based on bootstrap sampling and minimizes the variance and over-fitting, the XGBoost focuses on minimizing the bias and under-fitting during the training process. For more details on the theory of XGBoost we refer to [6].

The 44 statistical features extracted from all the TS streams and their respective Brick class labels, from all the source buildings are then fed into XGBoost model. The model learns the mapping between input features and targets (Brick classes) through a training process. We tune the parameters of the XGBoost model by applying a grid search strategy based on 10-folds cross validation of training data from all the source buildings combined. The searching space the of different parameters are presented in Table 2. After finding the best combination of parameters, the model is trained on the entire training data.

4.2 Testing

The evaluation of the trained model using the data from the target buildings begins with feature extraction. For each entity in a target building, we first compute the 44 features explained in Sect. 4.1. These features are the provided to the trained model as inputs and the model predicts the sensors types (Brick classes) for the entities in target buildings. The predicted sensor type labels are then compared with ground-truth Brick class labels to compute the performance of the model.

Table 2. Parameters of the XGBoost model used for grid searching.

Parameters	Description	Values used for grid search
n_estimators	number of base learners (gradient boosted trees) to train	[50, 100, 200]
max_depth	maximum depth each base learner can be expanded	[5, 10, 20, None (fully expanded)]
learning_rate	regularisation to control the contribution of base learners	[1, 0.5, 0.1]
max_leaves	maximum umber of leave nodes each base leaner can have	[0 (no limit), 10, 20, 50]

5   Results and Discussion

5.1   Evaluation Process

The performance of the proposed approach for sensor type classification are evaluated using two metrics: *accuracy* and *F-score*. *Accuracy* represents the percentage of total number of sensors that are correctly classified by the model. For an imbalanced classification problem like ours, *accuracy* alone may not provide sufficient insight of model's performance since it doesn't consider the ratio of observations in different classes. F-score is another assessment metric which evaluates the predictive skill of a model by considering its class-wise performance. For a binary classification task, it can be defined as the harmonic mean of *precision* and *recall* as in (1) where *precision* is the proportion of correctly predicted positive class relative to the all positive predictions and *recall* indicates the fraction of correctly predicted positive class with respect to the total number of observations belonging to actual positive class. For our multi-class classification task, we consider the weighted F-score which is computed as the average of the F-score of each class where the weights are determined by the number of observations in each class.

$$F score = 2 \times \frac{(precision \times recall)}{(precision + recall)} \tag{1}$$

Moreover, we evaluate the performance of the model on each building separately. Specifically, out of N sites in our DCH dataset, we consider one site $Site_{i \in \{1 \, to \, N\}}$ as the target site and remaining $N-1$ sites $\{Site_{j \, = \, 1 \, to \, N \, \& \, j \neq i}\}$ as the source sites. The model was trained on the data from all buildings in source sites and tested on each building from the target sites. This process is repeated N times, each time we have a different set of source sites and a target site.

5.2   Model's Performance

Table 3 presents the performance of our model. Although our dataset consists of 129 buildings, for brevity we included the results for 11 buildings each having at least 100 sensors to be classified.

Classification results shows that performance of the proposed approach varies for different site/building pairs. The accuracy is the range of 0.53 to 0.94 and the F-score is between 0.55 to 0.95. The model shows the best performance on the buildings $Bld\_1(Site\_5)$, $Bld\_1(Site\_6)$, and $Bld\_1(Site\_8)$ with both accuracy and F-score over 0.90. On the other hand, the classification accuracy of the sensors belonging to the three buildings $Bld\_1(Site\_7)$, $Bld\_1(Site\_3)$, and $Bld\_1(Site\_9)$ are the lowest (0.53, 0.56, 0.61, respectively). The same trend is also observed if F-score is considered as the assessment metric. The main reason for the comparatively lower accuracy of the model for these three buildings is relatively poor quality of TS data. Although the TS data from most of the buildings was recorded for at least 1 year (Jan–Dec 2022), the sensors in buildings $Bld\_1(Site\_7)$, $Bld\_1(Site\_3)$, and $Bld\_1(Site\_9)$ has only few short bursts of data possibly due to outages. This consequently made the extracted features atypical due to the lack of sufficient samples and representative patterns associated to different sensor types. The accuracy or F-score of the model on the data from other buildings is ≥ 0.70.

Table 3. Performance of the proposed approach.

Bld id	Samples	Accuracy	F-score
$Bld\_1(Site\_1)$	1872	0.76	0.72
$Bld\_2(Site\_1)$	252	0.83	0.83
$Bld\_3(Site\_1)$	179	0.83	0.72
$Bld\_1(Site\_2)$	1003	0.70	0.71
$Bld\_1(Site\_3)$	312	0.56	0.55
$Bld\_1(Site\_4)$	276	0.88	0.87
$Bld\_1(Site\_5)$	218	0.94	0.95
$Bld\_1(Site\_6)$	216	0.92	0.91
$Bld\_2(Site\_6)$	165	0.87	0.88
$Bld\_1(Site\_7)$	131	0.53	0.61
$Bld\_1(Site\_8)$	124	0.91	0.91
$Bld\_1(Site\_9)$	118	0.61	0.59

Moreover, the distribution of the number of sensors in different buildings shows that buildings $Bld\_1(Site\_1)$ and $Bld\_1(Site\_2)$ have the highest number of sensors (1872 and 1003, respectively). The accuracy (or F-score) of the model computed using the sensors from these two buildings is 0.76 (0.72) and 0.70 (0.71), respectively. In addition, the overall accuracy/F-score (averaged over all the buildings) is 0.78 with a standard deviation of 0.14. All these results are obtained using the features extracted from TS data only. This highlights that the TS data contains signature information or patterns that can be utilised in conjunction with ML algorithms to classify the sensors in buildings.

5.3   Comparison

We assess and compare the performance of the proposed approach from different perspectives. Firstly, we assess the advantage and generalisation ability the XGBoost classifier for sensor type classification task by integrating several other ML algorithms as the classifiers in our approach. Secondly, we compare the performance of the proposed approach with a state-of-the-art model (BA [11]).

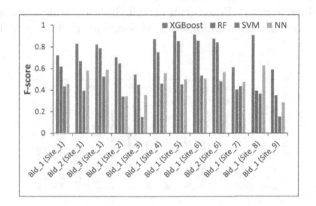

Fig. 2. Comparison of XGBoost with different ML models.

To study the effectiveness of using XGBoost classifier in our proposed approach, we evaluate the performance of proposed approach with three most widely used classical ML algorithms in the literature that include NNs, SVM, and RF. For a fair comparison, the evaluation is conducted using the same feature sets and following the same process applied with XGBoost.

Figure 2 presents the performance of the proposed approach with different ML models used as classifiers, evaluated using F-score. The graph using accuracy is similar and hence not included here. It shows that proposed approach achieves the best performance using the XGBoost as the classifier for all the buildings. The main reason for better performance of XGBoost is its robustness to noisy data as ours TS dataset. The pairwise differences of accuracy/F-score for XGBoost and any classifiers used for comparison is statistically significant (measured by Wilcoxon rank-sum test) at $p \leq 0.05$ for all the buildings except $Bld\_2(Site\_6)$. Among the three classifiers used for comparison, RF which uses ensemble of decision trees provides the highest classification accuracy. This highlights the better generalisation ability of ensemble based models in mapping input-output relationship for sensor types classification.

Moreover, the BA [11] model implemented for comparison utilises both TS and text names space of the entities as inputs. BA first trains a group of ML models (SVM, RF, and LR) using the 44 feature (discussed in Sect. 4.1) as inputs, extracted from the TS data of the sensors in the source buildings. The classification of sensors from a target building is done in two steps. Firstly, it

groups the sensors in the target building into different clusters by using the a set
of text features formed from the entity names by applying k-mers [5] method.
Basically, these text features are the count values of each sub-string of consecutive characters of length k from the entity names. Secondly, the prediction from
the ML models are weighted per instance basis based on the similarity between
the neighbouring graphs produced from both clustering and prediction of base
models for each entity in target buildings. Since our approach used the same
TS feature set as in the BA, comparison with BA allows us to investigate if the
different classifier (e.g., XGBoost) can provide better classification accuracy and
to investigate whether the combination of text and TS feature is beneficial for
our dataset.

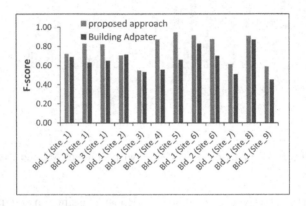

Fig. 3. Comparison with the Building Adapter model [11].

Figure 3 present the performance (F-score) of our proposed approach and the
BA model implemented for comparison. The proposed approach using XGBoost
as the classifier provides better classification accuracy compared to the BA which
utilises SVM, LR and RF to model TS data. The proposed approach provides
better classification accuracy for 9 (out of 11) buildings and for remaining 2
buildings ($Bld\_1(Site\_2)$, $Bld\_1(Site\_3)$) both show similar performance. Overall, the average F-score over all the buildings is 0.78 ± 0.14 for the proposed
approach vs 0.65 ± 0.12 for BA. The improvement of classification over the BA
model is also statistically significant at $p \leq 0.05$. This indicate that utilisation
XGBoost instead of other classifiers used in BA to model TS data leads to better
classification accuracy for our dataset. Moreover, the better performance of the
proposed approach is obtained by using TS data only as opposed to both TS
and text name space data in BA.

Although it is expected that name space data can provide useful additional
information to the model for classification, in our case the BA model which
utilises text data in addition to TS did not show better performance. The relatively lower performance of BA can be explained from two perspectives. Firstly,

similar to our approach, BA trains the ML models on TS data from source buildings and utilises the text data from target buildings to provide weights to ML models. However, the among three ML models in BA, RF provided the most accurate prediction with high confidence. Hence the weighting based clustering utilising text data doesn't make big difference in the contributions of the base classifiers to compute final prediction. Secondly, to obtain features from text data, our implementation of k-mers used $k = 3$ following the actual BA model. However, sub-strings of lengths 3 obtained from entity names possibly cannot distinguish the sensors accurately and different values of k set based on empirical evaluation could be a better choice for our DCH dataset.

6   Conclusions

We presented a straight-forward and effective data-driven approach for classifying varying types of electrical and temperature sensors within buildings. The classification was performed in accordance with widely used Brick ontology, and at a more granular hierarchical level than in prior art. This approach was evaluated using a large Australian dataset comprising 129 buildings, with experimentation showing performance at up to 95% accuracy and an average F-score of 0.78 across all buildings. We also found that the classification accuracy for few buildings were not as high as others, with F-scores below 0.6. However, in contrast to the approaches based on text metadata, the main advantages of the proposed approach is its portability and usability. It can be applied across buildings without worrying about the variation of naming conventions of entities in source and target buildings, and it does not require any knowledge from target buildings at all. Building on these promising preliminary results, future work will focus on: (i) augmenting the feature set by using advanced signal processing methods; and (ii) adapting deep learning based classifiers to further improve classification accuracy.

References

1. Project haystack (2001). https://project-haystack.org/
2. Balaji, B., et al.: Brick: towards a unified metadata schema for buildings. In: Proceedings of the 3rd ACM International Conference on Systems for Energy-Efficient Built Environments, pp. 41–50 (2016)
3. Balaji, B., Verma, C., Narayanaswamy, B., Agarwal, Y.: Zodiac: organizing large deployment of sensors to create reusable applications for buildings. In: Proceedings of the 2nd ACM International Conference on Embedded Systems for Energy-Efficient Built Environments, pp. 13–22 (2015)
4. Chen, T., Guestrin, C.: XGBoost: a scalable tree boosting system. In: Proceedings of the 22nd ACM SIGKDD International Conference on Knowledge Discovery and Data Mining, pp. 785–794 (2016)
5. Eskin, E., Weston, J., Noble, W., Leslie, C.: Mismatch string kernels for SVM protein classification. In: Advances in Neural Information Processing Systems 15 (2002)

6. Friedman, J.H.: Greedy function approximation: a gradient boosting machine. Ann. Stat. **29**, 1189–1232 (2001)
7. Gao, J., Ploennigs, J., Berges, M.: A data-driven meta-data inference framework for building automation systems. In: Proceedings of the 2nd ACM International Conference on Embedded Systems for Energy-Efficient Built Environments, pp. 23–32 (2015)
8. He, F., Wang, D.: Cloze: a building metadata model generation system based on information extraction. In: Proceedings of the 9th ACM International Conference on Systems for Energy-Efficient Buildings, Cities, and Transportation, pp. 109–118 (2022)
9. Hong, D., Gu, Q., Whitehouse, K.: High-dimensional time series clustering via cross-predictability. In: Proceedings of the 20th International Conference on Artificial Intelligence and Statistics, pp. 642–651 (2017). ISSN 2640-3498
10. Hong, D., Ortiz, J., Whitehouse, K., Culler, D.: Towards automatic spatial verification of sensor placement in buildings. In: Proceedings of the 5th ACM Workshop on Embedded Systems For Energy-Efficient Buildings, pp. 1–8 (2013)
11. Hong, D., Wang, H., Ortiz, J., Whitehouse, K.: The building adapter: towards quickly applying building analytics at scale. In: Proceedings of the 2nd ACM International Conference on Embedded Systems for Energy-Efficient Built Environments, pp. 123–132 (2015)
12. Jiao, Y., Li, J., Wu, J., Hong, D., Gupta, R., Shang, J.: SeNsER: learning cross-building sensor metadata tagger. In: Findings of the Association for Computational Linguistics: EMNLP 2020, pp. 950–960 (2020)
13. Koc, M., Akinci, B., Bergés, M.: Comparison of linear correlation and a statistical dependency measure for inferring spatial relation of temperature sensors in buildings. In: Proceedings of the 1st ACM Conference on Embedded Systems for Energy-Efficient Buildings, pp. 152–155 (2014)
14. Koh, J., Balaji, B., Sengupta, D., McAuley, J., Gupta, R., Agarwal, Y.: Scrabble: transferrable semi-automated semantic metadata normalization using intermediate representation. In: Proceedings of the 5th Conference on Systems for Built Environments, pp. 11–20 (2018)
15. Koh, J., Hong, D., Gupta, R., Whitehouse, K., Wang, H., Agarwal, Y.: Plaster: an integration, benchmark, and development framework for metadata normalization methods. In: Proceedings of the 5th Conference on Systems for Built Environments, pp. 1–10 (2018)
16. Koprinska, I., Rana, M., Agelidis, V.G.: Correlation and instance based feature selection for electricity load forecasting. Knowl.-Based Syst. **82**, 29–40 (2015)
17. Mishra, S., et al.: Unified architecture for data-driven metadata tagging of building automation systems. Autom. Constr. **120**, 103411 (2020)
18. Waterworth, D., Sethuvenkatraman, S., Sheng, Q.Z.: Advancing smart building readiness: automated metadata extraction using neural language processing methods. Adv. Appl. Energy **3**, 100041 (2022)

An Integrated Federated Learning and Meta-Learning Approach for Mining Operations

Venkat Munagala[1]($\boxtimes$)(iD), Sankhya Singh[1](iD), Srikanth Thudumu[1]($\boxtimes$)(iD),
Irini Logothetis[1](iD), Sushil Bhandari[2](iD), Amit Bhandari[2], Kon Mouzakis[1](iD),
and Rajesh Vasa[1](iD)

[1] Applied Artificial Intelligence Institute (A2I2), Deakin University,
Geelong, VIC 3216, Australia
{T.Munagala,Sankhya.Singh,Srikanth.Thudumu,
Rena.Logothetis,Kon.Mouzakis,Rajesh.Vasa}@deakin.edu.au
[2] MineExcellence, Bundoora, VIC 3083, Australia
{Sushil,Amit}@mineexcellence.com

Abstract. Mining operations are increasingly adopting advanced technologies to improve operations. These technologies include sensors, drones, robotic drills, and autonomous haul trucks and generate large quantities of data. Analyzing these large datasets can significantly improve mining practices by optimizing outcomes. For example, in drill and blast designs, we can optimize through learning from distributed data processing across multiple sites. However, mines are not equipped to process and handle large amounts of data. The advent of Machine Learning (ML) has enabled the effective handling of large volumes of complex data and the extraction of valuable insights. To accelerate the development of machine learning models for drill and blast design development, we propose an architecture based on two emerging fields of machine learning: Federated Learning (FL) and meta-learning. The proposed architecture facilitates collaboration and encourages sharing best practices among mines while ensuring the data remains local to each mine. In this research, we analyze the proposed architecture, detailing its components and functionalities. Specifically, we aim to determine the optimal learning rate and boosting rounds for unseen mine data using meta-learning. Our evaluation demonstrates that our learning rate and boosting rounds significantly decrease the Root Mean Squared Error (RMSE) value over a standard learning rate by 28.6%, showing the efficacy of the proposed architecture. We examine the potential advantages and inherent challenges of developing a robust system with our architecture, setting the groundwork for future research.

Keywords: Data Privacy · Drill and Blast Design · Federated Learning · Machine Learning · Meta-Learning · Mining

T. Liu et al. (Eds.): AI 2023, LNAI 14471, pp. 379–390, 2024.
https://doi.org/10.1007/978-981-99-8388-9_31

1 Introduction

Rock blasting is a conventional method employed in mining to break rock into small fragments for ore extraction. Drilling and blasting is a fundamental yet cost-intensive process in rock blasting, accounting for 15–60% of total mining costs. It involves various controllable and uncontrollable factors that can have far-reaching operational and environmental implications if miscalculated. These factors include (i) technical considerations, such as design parameters and geological conditions, (ii) non-technical aspects, such as production requirements and geographical factors, (iii) compliance with local environmental laws and mine infrastructure, and (iv) resource constraints. One critical challenge in drilling and blasting operations is controlling flyrock resulting from explosions. Flyrock is the unwanted displacement of rock fragments that poses severe safety hazards and risk of equipment damage [30]. Empirical methods are commonly used to develop optimal drill and blast designs for controlling flyrock [27]. These are resource-intensive, time-consuming, and expensive as they warrant several experimental blasts to collect data to derive site-specific constants on-site.

The recent IoT revolution has simplified data collection and transmission from remote locations, such as drone-based geological mapping and sensor data [25, 28]. The surge in data volume and variety can improve optimizing drill and blast designs. Yet, the challenge is in effectively handling and processing this data to extract valuable insights, as manual processing is not feasible for a mine engineer. Machine learning (ML) has been leveraged to extract insights from complex, varied data [22, 35]. It has proven effective in mining applications by uncovering hidden patterns within datasets, specifically in mineral exploration, drilling and blasting operations, and production optimization [13, 17, 20]. However, tuning hyperparameters for ML models often requires manual effort and trial and error. Meta-learning techniques have been proven to enhance ML model performance and address this issue by automating the tuning process [34]. However challenges arising from operational differences and data privacy concerns necessitate local development and storage of these models, resulting in high costs and time investments.

Particularly, meta-learning techniques have been shown to enhance model performance by tuning ML model hyperparameters, reducing trial-and-error [34]. However, challenges arising from operational differences, such as data privacy and the complexity of tuning ML models, necessitate local development and storage of these models resulting in high costs and time investments.

Our study proposes a novel approach combining Federated Learning (FL) and meta-learning to develop efficient ML models for flyrock distance prediction and control. Our architecture incorporates a meta-learner operating within an FL framework, facilitating shared learning while ensuring data privacy. This meta-learner predicts optimal model hyperparameters for new mine drill and blast data by utilizing FL's decentralized training across sites [1, 18].

We contribute to the creation and evaluation of a hybrid architecture that merges FL and meta-learning in a simulated environment using synthetic and real-world drilling and blasting datasets. This hybridization accelerates ML model development in local frameworks by learning from other mines via a centralized framework without direct communication. It ensures data stays local, maintaining privacy and security. Additionally, this approach saves time and resources that would otherwise be required for numerous experimental blasts when using empirical methods to achieve optimal drill and blast designs. This paper includes a background and related work section (Sect. 2), research methodology (Sect. 3), results and discussion (Sect. 4), and conclusion (Sect. 5).

2   Background and Related Work

2.1   Federated Learning

A conventional federated learning architecture contains a central server with a global model and multiple clients, where the global model is trained on the local data of each client. While the model training is local to the client, the final configuration is shared with the central server allowing it to update the global model through aggregation [18]. This update involves a method known as FederatedAveraging, where the server computes a weighted average of the model parameters received from the clients in Eq. 1

$$\theta = \frac{\sum_{i=1}^{K} n_i \cdot \theta_i}{\sum_{i=1}^{K} n_i} \tag{1}$$

where θ_i and n_i represent the model parameters and the number of data samples at client i, respectively, and K is the total number of clients.

This architecture allows clients to have control over their data while benefiting from machine learning model training that occurs at other sites. Despite the benefits of FL in various domains, we are unaware of any studies applying this method to optimize drilling and blasting operations in the mining industry. Previous research presents the adaption of FL in a production environment and its ability to be extended to other industries [9, 19]. However, FL faces challenges when the data across clients is unbalanced and non-independent and identically distributed (non-i.i.d) [15].

2.2   Meta-learning

Meta-learning is where a model learns from prior knowledge of task characteristics, model parameters, and model evaluations [34]. It is the concept of 'learning to learn,' facilitating adaption to new tasks with minimal change in data or fine-tuning. Meta-learning has demonstrated effectiveness in domain generalization [14, 16] and enhancing ML model development [11] for multi-tasks and single-tasks where the problem is solved iteratively [10]. In an FL framework, using a meta-learner as an aggregator algorithm can address data heterogeneity while preserving data privacy and security. However, the potential of meta-learning to enhance the accuracy of ML models in mining remains unexplored. Taking inspiration from applications in networking [7, 36], image processing [24], and recommendation systems [2], our study aims to explore this potential.

3   Research Methodology

We present an integrated approach that combines federated learning and meta-learning to address mining-related challenges. Our methodology is a two-stage learning process designed to expedite model development, reduce resources spent on hyperparameter tuning, and improve privacy by leveraging local data storage. In the first stage, we simulate different mining conditions by creating synthetic datasets that correspond to distinct mine types. We then develop and implement a FL framework with local and central components. The second stage introduces a meta-learning approach that provides optimal configurations for an ML model based on the specific characteristics of the data at each mining site. To demonstrate our method, we apply it to the prediction and control of flyrock distance. Flyrock distance is a key safety measure in mining operations, and controlling it improves cost, time, and process efficiency in drilling and blasting. Figure 1 presents a visual representation of the overall system architecture followed by sections covering the data preparation, learning framework design, and experimental setup.

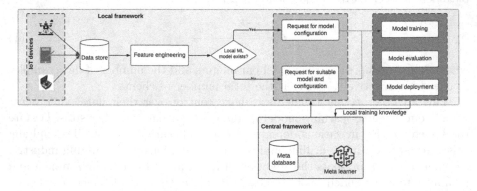

Fig. 1. An overview of the proposed architecture

3.1   Design and Setup of Federated Learning Framework

We start by implementing an FL framework using synthetic datasets to simulate distinct mining conditions as a foundation for local learning.

Dataset Construction. For a realistic representation of diverse mining conditions, we generated synthetic datasets corresponding to three different types of mines - granite, limestone, and copper. For the remainder of this paper, where we demonstrate the concept, these mines are referred to as clients. These datasets were constructed from samples in literature [3,21,32,33]. These datasets incorporate various factors essential to the drill and blast design process and flyrock distances. We used Synthetic Data Vault (SDV) library to create our datasets [23], maintaining the statistical properties observed in literature-sourced samples. We validated these new datasets with domain experts to reinforce their practical applicability. Additionally, we used a real-world dataset as a benchmark to assess our approach. The comparative results demonstrated the applicability of our methodology.

Local Learning: Implementation of XGBoost Model. For local learning at each mine site, we employed the XGBoost model. Its renowned performance in various mining-related tasks and adept handling of different data types and distributions make it ideal for our FL architecture [6, 26].

To optimize our XGBoost model, we focused on tuning two key hyperparameters - learning rate and number of boosting rounds. Initially, we deployed a manual random search approach to identify the most promising range for the XGBoost hyperparameters at each mine site. Afterward, we conducted a more detailed search within these ranges to locate the optimal configurations, as summarized in Table 1. We then trained an XGBoost model for each mine, testing multiple hyperparameter combinations within the defined optimal ranges. We evaluated each unique combination of the learning rate and boosting rounds using the standard measure for regression tasks, Root Mean Squared Error (RMSE). To prevent overfitting and improve the generalizability of the model, we employed a five-fold cross-validation strategy. We divided the data into five subsets, using 80% of the data for model training and the remaining 20% for validation in each round. After five independent rounds of training and validation, we reported the final performance as the average RMSE across the testing sets. Each hyperparameter configuration and its corresponding RMSE were stored as a meta-feature set, linking the local learning outcomes to the subsequent meta-learning phase.

Table 1. Hyperparameter Ranges for Each Mine

Mine	Learning Rate	Step Size	Boosting Rounds
Mine 1 (Granite)	[0.01, 1.0]	0.01	[50, 1000]
Mine 2 (Limestone)	[0.01, 0.5]	0.01	[50, 400]
Mine 3 (Copper)	[0.01, 0.6]	0.01	[50, 300]

3.2 Incorporation of Meta-learning into the Framework

Our hybrid architecture has the advantages of federated learning and meta-learning, where the accuracy and performance of the system are enhanced by improving locally-trained models by integrating meta-learning. This approach captures broad trends across various mining operations and translates this knowledge into actionable recommendations for hyperparameter configurations. This reduces the need for extensive manual effort in selecting and fine-tuning the model hyperparameters.

Creating Meta-Database from Mine Data. To use meta-learning, a meta-database needs to be created that links individual mines to the meta-learner. This meta-database captures the relationships between each mine's data characteristics and the corresponding performance (RMSE) of different hyperparameter configurations. This forms the meta-features that guide the meta-learner.

The meta-database consolidates two types of information: data characteristics and ML model details. The data characteristics include statistical measures of key drill and blast design factors such as the burden-to-spacing ratio, stemming length, hole depth, and flyrock, as listed in Table 2. ML model details are the specific hyperparameter configuration (learning rate and number of boosting rounds) and associated RMSE. Populating the meta-database was a two-step process. First, we extracted the characteristics of local datasets using the statistical functions $describe()$, $kurtosis()$, and $skew()$. Then, we trained the local ML model and recorded its performance over a predetermined range of hyperparameters (refer to Sect. 3.1). We merged the information returned from these processes into a meta-feature set and transmitted it to the central framework forming an entry into the meta-database. We repeated this process for the three clients of the central server.

Table 2. Summary of dataset characteristics

Mine Type	Factor	Min	Max	Std. Deviation	Skewness	Kurtosis
Mine 1 (Granite)	Burden/Spacing	0.40	0.95	0.14	−0.87	−0.18
	Stemming Length	1.40	4.50	0.94	0.81	−0.70
	Hole Depth	10.0	28.0	5.57	1.06	−0.42
	Flyrock	67.0	350.0	60.44	1.52	1.74
Mine 2 (Limestone)	Burden/Spacing	0.48	1.0	0.07	−0.42	1.18
	Stemming Length	2.90	3.70	0.14	−0.20	0.34
	Hole Depth	5.0	10.30	0.74	−2.69	11.22
	Flyrock	22.0	54.0	7.94	−0.44	−0.78
Mine 3 (Copper)	Burden/Spacing	0.67	0.95	0.05	0.02	−0.11
	Stemming Length	2.87	4.07	0.38	0.50	−1.12
	Hole Depth	11.76	13.99	0.54	−0.21	−0.71
	Flyrock	25.32	48.60	5.81	1.49	1.10

Global Learning: Integration of Meta-Learner. The global learning phase involves leveraging collective knowledge stored in the meta-database to enhance the learning process. It learns from the past experiences of individual local learning models. To achieve this, we incorporated a central meta-learner model that facilitates global learning by utilizing the meta-database to predict the optimal ML model hyperparameters for a new dataset based on its unique data characteristics. We have chosen to use DecisionTreeRegressor as our meta-learner model because of its interpretability and simplicity in providing a concise understanding of the relationship between data characteristics and optimal hyperparameters [29]. We used the consolidated meta-database to train the DecisionTreeRegressor meta-learner model to learn the associations between the meta-features and their corresponding RMSE. This enabled the meta-learner to predict suitable hyperparameters for new datasets, streamlining the local model optimization process at new mines and reducing the time and resources required for hyperparameter tuning [8].

3.3   Experiment Setup

We implemented this approach using Python in two stages: (i) the training and (ii) the testing of the meta-learner, as depicted in Fig. 2.

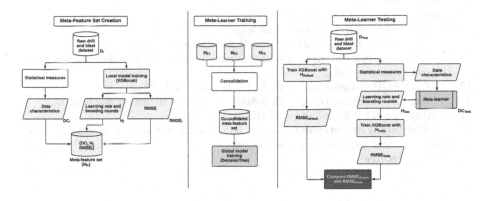

Fig. 2. An Overview of Experiment Setup

Meta-Learner Training. The training stage involves a local XGBRegressor and a global DecisionTreeRegressor as a meta-learner. We used the three mine datasets corresponding to the granite, limestone, and copper clients. At each client, we trained a local XGBoost model and identified the optimal ranges for learning_rate and n_estimators through 5-fold cross-validation, using 'squared-error' as the objective function. We collected the associated RMSE for all combinations of these hyperparameters and in conjunction with the data characteristics of each mine, we formed the meta-feature set M_{F_i}. The meta-feature set for each mine was then passed to a central framework via an API call, consolidating the data into the meta-database D_{meta}. This was then used to train the meta-learner, which predicts the optimal hyperparameters based on the given drill and blast data characteristics. Algorithm 1 outlines the process for training and testing the meta-learner.

Algorithm 1 Meta-Learner Training and Testing

1: **Input:** Training data for each mine → D_C, D_G, D_L, Test data D_{Test}
2: **Output:** Trained meta-learner, RMSE comparison between meta-learner and default XGBoost parameters
3: **procedure** METALEARNERTRAINING
4:    **for each** $D_i \in \{D_C, D_G, D_L\}$ **do**
5:        Derive data characteristics → DC_i
6:        Determine hyperparameter configuration → H_i
7:        Train XGBoost on D_i with various H_i configurations
8:        Evaluate RMSE for each configuration
9:        Calculate $RMSE_i$ for each configuration
10:       Construct meta-feature set $M_{F_i} = \{DC_i, H_i, RMSE_i\}$
11:   **end for**
12:   Combine M_{F_i} to form $D_{meta} = \sum_{i \in \{C,G,L\}} M_{F_i}$
13:   Train DecisionTreeRegressor on D_{meta} to predict learning rate and number of boosting rounds
14: **end procedure**
15: **procedure** METALEARNERTESTING
16:   Derive data characteristics → DC_{Test}
17:   Concatenate DC_{Test} with an RMSE value of zero → M_{TD}
18:   Predict optimal learning rate and number of boosting rounds for M_{TD} using the meta-learner → H_{meta}
19:   Train XGBoost on D_{Test} with H_{meta} to obtain $RMSE_{meta}$
20:   Train XGBoost on D_{Test} with default parameters to obtain $RMSE_{default}$
21:   Compare $RMSE_{meta}$ and $RMSE_{default}$
22:   **Additional Experiment:** Include '50%' from describe function in DC_{Test} and repeat above steps to obtain $RMSE_{median}$
23: **end procedure**

Meta-Learner Testing. In the testing stage, we evaluated our trained meta-learner using unseen real-world data from a new mine D_{Test}. We analyzed its data characteristics (DC_{Test}) using the *describe*(), *kurtosis*(), and *skew*() functions, analogous to the meta-features. We appended an ideal RMSE value of zero to these characteristics, forming a meta-feature set M_{TD} as input to the meta-learner. The meta-learner predicted the optimal learning rate and number of boosting rounds for the given M_{TD}. These predicted values were used to train the XGBoost model on the unseen test data and we collected the resulting RMSE ($RMSE_{meta}$). We also trained another XGBoost model using default parameters and calculated its corresponding RMSE ($RMSE_{default}$) for comparative analysis. Additionally, we experimented by including the median of the data (represented as the '50%' value from the *describe*() function) in DC_{Test} to understand its impact on the meta-learner's predictions and the performance of the local model.

4 Results and Discussion

Evaluating our hybrid framework across three distinct mine sites, we noticed promising results that support our proposed methodology. By consolidating metadata shared independently by each site, our meta-learner improved precision in predicting optimal hyperparameters. When applied to an XGBoost model, the predicted learning rate (0.55) and boosting rounds (175) outperformed the default setup, with a decrease in RMSE from 16.14 to 11.52 (Table 3).

Table 3. Comparison of XGBoost Regressor Performance on Test Set: Default, Predictive without Median, and Predictive with Median

Model	Type	Learning-rate	Boosting-rounds	RMSE
XGBoost	Default	0.03	10	16.14
	Predicted (w/o Median)	0.55	175	11.52
	Predicted (w/ Median)	0.48	225	13.47

This decrease is an improvement in predicting flyrock distance. Despite the small difference in value, it is substantial in real-world settings. Enhanced prediction accuracy can amplify safety during drilling and blasting operations by facilitating accurate demarcation of safety zones, thereby minimizing potential damage to equipment and any risk of injuries. Moreover, improved prediction precision can optimize explosive usage, contributing to cost savings. Such improvements in blasting operations can result in substantial annual savings [3,4,31].

Incorporating the median value (from the *describe()* function) into the metadata yielded a different result. Although the resulting RMSE (13.47) was slightly higher than the predictive model without the median, it was still lower than the default configuration. This discrepancy underscores the nuances of data characteristics in model performance and the careful selection of meta-features.

The advantages of our framework include safeguarding data privacy, promoting model customization, and enabling generalization capabilities for out-of-distribution data. However, it is important to acknowledge the unique challenges associated with our architecture with respect to our research and the nature of mining operations:

Challenges

- Meta-learning can be computationally demanding, particularly with larger meta-databases and complex models [14].
- Interactions between local and central frameworks may escalate communication costs and network security concerns [5].
- Variable mining practices may instigate data imbalance across features, affecting data quality and, consequently, model performance [12].

– Factors such as data distribution, meta-feature selection, and the choice of ML model for local and global training could influence the effectiveness of the meta-learner and the overall system performance.

Thus, despite our promising results, they are limited to this research as a proof-of-concept evaluated with a synthetic dataset. Our objective was to focus on the experimental nature of our study and provide context for the performance of the proposed framework. A more comprehensive assessment using real-world scenarios and diverse datasets will enhance our understanding and validate the efficacy of our approach.

5 Conclusion

In this research, we have presented a novel hybrid framework that integrates federated learning and meta-learning to optimize drilling and blasting operations in mining. We applied our framework across multiple mock mine sites, demonstrating significant improvements in model accuracy, and ensuring safer and more efficient mining operations. By effectively predicting optimal hyperparameters, this approach uniquely caters to specific requirements at each mining site, allowing optimal customization of ML models and protecting data privacy. Moreover, its capability to generalize out-of-distribution data has shown promising results and opens up a multitude of possibilities for its future applications in diverse mining scenarios.

Our research can be improved and refined to minimize computational burdens, communication costs, and data imbalances. We hope that this research will serve as a stepping stone for future endeavours aimed at harnessing the synergy of federated learning and meta-learning in varied industrial scenarios.

References

1. AbdulRahman, S., Tout, H., Ould-Slimane, H., Mourad, A., Talhi, C., Guizani, M.: A survey on federated learning: the journey from centralized to distributed on-site learning and beyond. IEEE Internet Things J. **8**(7), 5476–5497 (2020)
2. Arambakam, M., Beel, J.: Federated meta-learning: democratizing algorithm selection across disciplines and software libraries. In: 7th ICML Workshop on Automated Machine Learning (AutoML) (2020)
3. Armaghani, D.J., Koopialipoor, M., Bahri, M., Hasanipanah, M., Tahir, M.: A SVR-GWO technique to minimize flyrock distance resulting from blasting. Bull. Eng. Geol. Env. **79**, 4369–4385 (2020)
4. Bilim, N., Çelik, A., Kekeç, B.: A study in cost analysis of aggregate production as depending on drilling and blasting design. J. Afr. Earth Sci. **134**, 564–572 (2017)
5. Bonawitz, K., et al.: Towards federated learning at scale: system design. Proc. Mach. Learn. Syst. **1**, 374–388 (2019)
6. Chandrahas, N.S., Choudhary, B.S., Teja, M.V., Venkataramayya, M., Prasad, N.K.: XG boost algorithm to simultaneous prediction of rock fragmentation and induced ground vibration using unique blast data. Appl. Sci. **12**(10), 5269 (2022)

7. Dong, F., et al.: PADP-FedMeta: a personalized and adaptive differentially private federated meta-learning mechanism for AIoT. J. Syst. Archit. **134**, 102754 (2023)
8. Feurer, M., Hutter, F.: Hyperparameter optimization. Autom. Mach. Learn. Methods Syst. Challenges 3–33 (2019)
9. Hard, A., et al.: Federated learning for mobile keyboard prediction. arXiv preprint arXiv:1811.03604 (2018)
10. Hospedales, T., Antoniou, A., Micaelli, P., Storkey, A.: Meta-learning in neural networks: a survey. IEEE Trans. Pattern Anal. Mach. Intell. **44**(9), 5149–5169 (2021)
11. Hutter, F., Kotthoff, L., Vanschoren, J.: Automated Machine Learning: Methods, Systems, Challenges. Springer Nature, Cham (2019). https://doi.org/10.1007/978-3-030-05318-5
12. Lawal, A.I.: A new modification to the Kuz-Ram model using the fragment size predicted by image analysis. Int. J. Rock Mech. Min. Sci. **138**, 104595 (2021)
13. Lawal, A.I., Kwon, S.: Application of artificial intelligence to rock mechanics: an overview. J. Rock Mech. Geotech. Eng. **13**(1), 248–266 (2021)
14. Li, D., Yang, Y., Song, Y.Z., Hospedales, T.: Learning to generalize: meta-learning for domain generalization. In: Proceedings of the AAAI Conference on Artificial Intelligence, vol. 32 (2018)
15. Li, T., Sahu, A.K., Talwalkar, A., Smith, V.: Federated learning: challenges, methods, and future directions. IEEE Signal Process. Mag. **37**(3), 50–60 (2020). https://doi.org/10.1109/MSP.2020.2975749
16. Li, W., Wang, S.: Federated meta-learning for spatial-temporal prediction. Neural Comput. Appl. **34**(13), 10355–10374 (2022). https://doi.org/10.1007/s00521-021-06861-3
17. McCoy, J.T., Auret, L.: Machine learning applications in minerals processing: a review. Miner. Eng. **132**, 95–109 (2019)
18. McMahan, B., Moore, E., Ramage, D., Hampson, S., y Arcas, B.A.: Communication-efficient learning of deep networks from decentralized data. In: Artificial Intelligence and Statistics, pp. 1273–1282. PMLR (2017)
19. McMahan, H.B., Ramage, D., Talwar, K., Zhang, L.: Learning differentially private recurrent language models. arXiv preprint arXiv:1710.06963 (2017)
20. Mitchell, T.M., et al.: Machine Learning, vol. 1. McGraw-Hill, New York (2007)
21. Monjezi, M., Khoshalan, H.A., Varjani, A.Y.: Optimization of open pit blast parameters using genetic algorithm. Int. J. Rock Mech. Min. Sci. **48**(5), 864–869 (2011)
22. Obermeyer, Z., Emanuel, E.J.: Predicting the future-big data, machine learning, and clinical medicine. N. Engl. J. Med. **375**(13), 1216 (2016)
23. Patki, N., Wedge, R., Veeramachaneni, K.: The synthetic data vault. In: 2016 IEEE International Conference on Data Science and Advanced Analytics (DSAA), pp. 399–410. IEEE (2016)
24. Połap, D., Woźniak, M.: Meta-heuristic as manager in federated learning approaches for image processing purposes. Appl. Soft Comput. **113**, 107872 (2021)
25. Qi, C.C.: Big data management in the mining industry. Int. J. Miner. Metall. Mater. **27**(2), 131–139 (2020)
26. Qiu, Y., Zhou, J., Khandelwal, M., et al.: Performance evaluation of hybrid WOA-XGBoost, GWO-XGBoost, and BO-XGBoost models to predict blast-induced ground vibration. Eng. Comput. **38**(5), 4145–4162 (2022). https://doi.org/10.1007/s00366-021-01393-9
27. Raina, A.K., Murthy, V., Soni, A.K.: Flyrock in surface mine blasting: understanding the basics to develop a predictive regime. Current Sci. 660–665 (2015)

28. Rogers, W.P., et al.: Automation in the mining industry: review of technology, systems, human factors, and political risk. Min. Metall. Explor. **36**(4), 607–631 (2019)

29. Safavian, S.R., Landgrebe, D.: A survey of decision tree classifier methodology. IEEE Trans. Syst. Man Cybern. **21**(3), 660–674 (1991)

30. Sawmliana, C., Hembram, P., Singh, R.K., Banerjee, S., Singh, P., Roy, P.P.: An investigation to assess the cause of accident due to flyrock in an opencast coal mine: a case study. J. Inst. Eng. (India) Ser. D **101**, 15–26 (2020)

31. Sevelka, T.: Preventing the potentially deadly consequences of flyrock: mandatory minimum setbacks and separation distances required. J. Nat. Resour. **5**(4), 66–98 (2022)

32. Trivedi, R., Singh, T., Gupta, N.: Prediction of blast-induced flyrock in opencast mines using ANN and ANFIS. Geotech. Geol. Eng. **33**, 875–891 (2015)

33. Trivedi, R., Singh, T., Raina, A.: Prediction of blast-induced flyrock in Indian limestone mines using neural networks. J. Rock Mech. Geotech. Eng. **6**(5), 447–454 (2014)

34. Vanschoren, J.: Meta-learning: a survey. arXiv preprint arXiv:1810.03548 (2018)

35. Wang, L., Alexander, C.A.: Machine learning in big data. Int. J. Math. Eng. Manag. Sci. **1**(2), 52–61 (2016)

36. Yue, S., Ren, J., Xin, J., Zhang, D., Zhang, Y., Zhuang, W.: Efficient federated meta-learning over multi-access wireless networks. IEEE J. Sel. Areas Commun. **40**(5), 1556–1570 (2022)

An Augmented Learning Approach for Multiple Data Streams Under Concept Drift

Kun Wang[ID], Jie Lu[✉][ID], Anjin Liu[ID], and Guangquan Zhang[ID]

Australian Artificial Intelligence Institute, University of Technology Sydney, Sydney, Australia
{Kun.Wang,Jie.Lu,Anjin.Liu,Guangquan.Zhang}@uts.edu.au

Abstract. Multiple data streams learning attracts more and more attention recently. Different from learning a single data stream, the uncertain and complex occurrence of concept drift in multiple data streams, bring challenges in real-time learning task. To address this issue, this paper proposed a method called time-warping-based concept drift learning method (TW-CDM) for dealing with multiple data streams. First, a time-warping-based drift identification process is given to recognize the drift region. Second, an augmented learning process is developed by crossly using the located region data. Finally, a selectively augmented learning process is given to reduce the influence of different drift severity. The proposed method is evaluated on both synthetic and real-world datasets, and compared with benchmark methods. The experiment results show the efficiency of the proposed method.

Keywords: Concept drift · Data stream · Multiple streams · Ensemble learning

1 Introduction

Data streams commonly exist in real-world life and have attracted high research attention in recent years [9,17]. Scenarios like weather changes, price fluctuates, and user interest drifts are representative of streaming data. Related research about handling concept drift for data stream learning has made successful progress. However, most of the previous work aims to address the concept drift learning in a single stream, and few of them consider the situation that deals with multiple data streams [25,27]. On the one hand, multiple data streams learning is relatively more complex, since the concept drift situations in each of them are different. Maintaining the learning performance on such complex tasks is a challenge. On the other hand, with the higher requirement of real-world prediction, learning tasks on multiple data streams need to be enhanced for better application. Therefore, it is necessary to consider an efficient method to handle concept drift for multiple data streams learning.

Recently, there are related works that focus on multiple data streams learning and develop several markable outputs. We found that there are three main task settings of multiple data streams learning: Supervised [25,28] and unsupervised [22,24,26] tasks. These previously proposed approaches provide optimization and performance improvement of learning strategies for processing multiple data streams. However, few of them

ⓒ The Author(s), under exclusive license to Springer Nature Singapore Pte Ltd. 2024
T. Liu et al. (Eds.): AI 2023, LNAI 14471, pp. 391–402, 2024.
https://doi.org/10.1007/978-981-99-8388-9_32

consider identifying and adapting different situations of concept drift in such a complex scenario.

Learning concept drift in multiple data streams is more complex than handling a single data stream. First, in multiple data streams, concept drift may occur asynchronously and will lead to model learning decay [5, 27]. Second, the drift severity in multiple data streams may also change differently, which will interfere with model learning. Third, there may be correlation and interaction between multiple data streams, which brings difficulties to model learning. Besides, learning efficiency by selecting appropriate knowledge to deal with multiple streams should also be addressed. Therefore, learning multiple data streams under concept drift is a challenge that should be highly addressed.

Motivated by this, we aim to give a clear definition of concept drift in multiple data streams, and try to find an appropriate learning solution. To address these issues, this paper proposes a time-warping-based concept drift learning method (TW-CDM) to deal with multiple data streams adaptively. The contribution of this paper is shown below:

- A time warping-based strategy has been given to help identify the possible drifted data in multiple data streams. This process can help not only recognize the changeable data, but also reflect the drift severity in time.
- An augmented learning approach has been developed to help adapt to the concept drift that occurs on multiple data streams. The update process is triggered and augmented by the results of the drift-identified process.
- The experiment on multiple data streams with different drift situations shows the effectiveness of the proposed method. And the source code is available online[1].

The rest of this paper is organized as follows. Section 2 summarizes the recent works. The research problem statement and the proposed approaches are introduced in Sect. 3. Section 4 describes the experiment setting, datasets, and benchmark methods. The experiment results are discussed in Sect. 5. Finally, a summary of the conclusion and our future study is given in Sect. 6.

2   Related Work

In this section, we give an overview of research works about multiple data streams learning and concept drift learning.

2.1   Multiple Data Streams Learning

Recently, many research works focusing on multiple data streams learning based on different task settings. To summarize, current studies of multiple data streams learning are mainly based on the settings of supervised and unsupervised tasks. For supervised tasks, research [27] addresses the issue of concept drift in multiple data streams, and a group concept drift detection method has been proposed to support model learning. In

---

addition, a method called fuzzy drift variance (FDV) method has been given to measure the correlated drift patterns among multiple data streams [25]. Aiming to handle multiple relevant data streams, a method called MuNet has been developed to improve learning efficiency by leveraging the dependency among multiple data streams [28]. To address the issue of sequential pattern mining, research [11] proposes a method called PSP-AMS method to help better identify the patterns in multiple data streams.

For unsupervised tasks, to deal with the concept evolving nature of data streams, research [24] proposes a fuzzy hierarchical clustering method for clustering multiple nominal data streams. Besides, a dynamic ensemble clustering algorithm based on evidence accumulation has been generated to address missing data and delayed data in multiple data streams [23]. Also, the data stream clustering method based on multiple mcdoids and medoid voting has been proposed to improve the efficiency of sequence clustering [22]. A histogram-based clustering method has been proposed to deal with online multiple data streams [3]. Besides, focusing on the problem of unlabelled drifting streams, a framework called Learn-to-Adapt (L2A) has been developed to handle concept drift adaptation tasks in multiple data streams [26]. Furthermore, research [5, 10] handle multiple data streams according to the idea of transfer learning [20] and domain adaptation [7].

In addition, it is also needed to consider real-world decision-making [18]. Application tasks like multimodal learning [21] and multi-source learning [19] may also face the impact of uncertainty.

2.2 Concept Drift Learning Under Uncertainty

Concept drift learning is an interesting topic that recently gets highly focused deal with the impact of the uncertain environment [9, 13, 17]. With the issue that the changeable distribution of streaming data may cause learning decay and performance degradation, many research works aim to detect, understand, and adapt to concept drift.

For concept drift detection, commonly used methods include DDM [8], EDDM [2], ADWIN [4], and so on. Most drift detection methods identify changeable data based on drift threshold, error rate, and statistic results. For concept drift understanding, research works [15, 16] introduce the strategy to locate the region of concept drift to support better understanding. This idea not only can identify where the drift occurs, but also can help update the learning model based on how drift occurs. For concept drift, the main target is to figure out the appropriate strategy to help the learning model adapt to the changeable data stream. For example, Learn++.NSE [6] as a classical drift learning method, main the performance based on ensemble learning and dynamically weighted majority voting. Also, the method of dynamic weighted majority [12] has been proposed to help handle data stream with concept drift. Recently, novelty methods have also been proposed to enhance concept drift learning, like adaptative minimax risk classifiers (AMRCs) [1], and data distribution generation for predictable concept drift adaptation (DDG-DA) [14].

However, most of the current works aim to deal with concept drift occurring on a single data stream, few of them consider the scenario when drift occurs on multiple data streams asynchronously, which is a relatively more complex learning task. The same as learning a single data stream, drift identification and learning also plays an important role. Therefore, the target of this paper is to figure out the strategy that can ensure the machine learning model can deal with multiple data streams occurs concept drift.

3   Methodology

This section first gives the problem description of concept drift in multiple data streams, and then develops a time warping-based model for multi-stream learning to help detect and adapt to concept drift.

3.1   Problem Statement

In our problem setting, we initially focus on dealing with two data streams with the same size and homogeneous structure, but different concept drift situations. Given two data streams $S_1 = \{x_i^{S_1}, y_i^{S_1}\}_{i=1}^n$, $S_2 = \{x_i^{S_2}, y_i^{S_2}\}_{i=1}^n$, follow the distribution P, Q. As time goes on, the data distribution of these two streams may change from time t to time $t + 1$, causing different concept drift situations, leading to learning model degradation. Here, we give a definition of concept drift in multiple data streams as below.

Definition 1. *(Concept Drift in Multiple Data Streams) For data streams S_1, S_2, concept drift occurs when the data distribution of data streams changes, denote as*

$$P_t \left(y^{S_1} | x^{S_1} \right) \neq P_{t+1} \left(y^{S_1} | x^{S_1} \right) \vee Q_t \left(y^{S_2} | x^{S_2} \right) \neq Q_{t+1} \left(y^{S_2} | x^{S_2} \right), \tag{1}$$

where P and Q are different distributions followed by streams S_1 and S_2.

When concept drift occurs in multiple data streams, the machine learning model should identify and adapt this uncertain change timely in both of them. To deal with this complex scenario, it is needed to find an appropriate method $\mathcal{F}$ to help reduce the learning loss L when dealing with multiple data streams with concept drift. Therefore, our research target is

$$\mathcal{F} = \arg\min_{\mathcal{F}} \sum_{t=0}^{T} L_t \left(\mathcal{F} \left(\{x^{S_1}, x^{S_2}\} \right), \{y^{S_1}, y^{S_2}\} \right). \tag{2}$$

Aiming to address this issue, the following paper develops a drift identification and adaptation process for multiple data streams learning.

3.2   Time Warping-Based Drift Identification Process

In this section, we first consider real-time drift identification in multiple data streams. We train a model $f(x)$ on chunks $D_t^{S_1}$, $D_t^{S_2}$, and then test it on chunks $D_{t+1}^{S_1}$, $D_{t+1}^{S_2}$. Get the prediction results and calculate the prequential accuracy $\varepsilon_{t+1}^{S_1}$, $\varepsilon_{t+1}^{S_2}$ of two streams.

$$\varepsilon_{t+1}^{S_1} = \left\{ \varepsilon_{t+1,i}^{S_1} \right\}_{i=1}^n, \ \varepsilon_{t+1}^{S_2} = \left\{ \varepsilon_{t+1,i}^{S_2} \right\}_{i=1}^n. \tag{3}$$

Motivated the by dynamic time-warping method, which can measure the similarity of two series data. It can help identify the change and differences between two series of

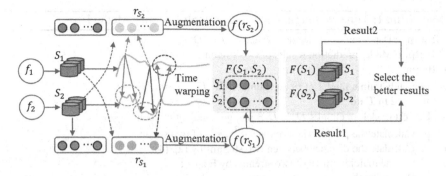

Fig. 1. An illustration of the framework of the time-warping-based concept drift learning method for multiple data stream.

data. Motivated by this, we apply this method to calculate the distance between these two series, denote as

$$d\left(\varepsilon_{t+1}^{S_1}, \varepsilon_{t+1}^{S_2}\right) = \left|\varepsilon_{t+1}^{S_1} - \varepsilon_{t+1}^{S_2}\right|. \tag{4}$$

Thus, the matching path of these two series can be obtained. The matching path means that the parts of the two sequences are similar.

$$\varepsilon_{t+1,i}^{S_1} \leftarrow \varepsilon_{t+1,j}^{S_2} \text{ or } \varepsilon_{t+1,i}^{S_1} \leftarrow \left\{\varepsilon_{t+1,j}^{S_2}, \cdots, \varepsilon_{t+1,j+\eta}^{S_2}\right\}. \tag{5}$$

And we choose the case of $\varepsilon_{t+1,i}^{S_1} \leftarrow \left\{\varepsilon_{t+1,j}^{S_2}, \cdots, \varepsilon_{t+1,j+\eta}^{S_2}\right\}$, and then find the degraded accuracy, and identify drift that occurs in S_1, locate the regions of possible drift occurrence in S_2 (Fig. 1).

3.3   Learning Augmentation for Drift Adaptation

In this section, we try to use the identified regions of data to help support the model learning. The results of time warping reflect that there is a similarity between S_1 and S_2 at some time steps. So, we select the data of two streams crossly to augment the learning process. First, merging the chunks at time t as a new train set, denote as

$$D_t = D_t^{S_1} \cup D_t^{S_2}. \tag{6}$$

Then, train an ensemble model as a base model $F(x)$ at time t and test it on $D_{t+1}^{S_1}$, $D_{t+1}^{S_2}$ at time $t + 1$, and identified changeable region of two streams, r_{S_1}, r_{S_2}. Then update the train set at the current time point for the base model learning as

$$D_{t+1} = \left\{D_{t+1}^{S_1}\right\} \cup \left\{D_{t+1}^{S_2}\right\}. \tag{7}$$

Finally, retrain the base model on the updated train set, denote as $F(D_{t+1})$. Then separately train a single tree model on r_{S_1}, r_{S_2}, denote as $f(r_{S_1}), f(r_{S_2})$. Let λ be a parameter, the updated model for predicting two data streams can be expressed as

$$\hat{y}^{S_1} = F(D_{t+1}) + \lambda f(r_{S_2}), \ \hat{y}^{S_2} = F(D_{t+1}) + \lambda f(r_{S_1}). \tag{8}$$

Algorithm 1: Time-Waping-based Concept Drift Learning Method

   Input : Data chunks of two data streams D^{S_1}, D^{S_2}.
   Output: Model prediction results of two streams.

1 **Begin**
2 Separately train a model $f(x)$ on chunks $D_1^{S_1}$, $D_1^{S_2}$ at time $t = 1$;
3 **for** $t = 2$ **to** T **do**
4 | Test model on chunks $D_t^{S_1}$, $D_t^{S_2}$ and get prediction results;
5 | Calculate the prequential accuracy $\varepsilon_t^{S_1}$, $\varepsilon_t^{S_2}$ of two streams;
6 | Calculate the distance between two streams by Eq. (4);
7 | Get the matching path of two streams by Eq. (5);
8 | Recognize the changeable regions r_{S_1}, r_{S_2} by Eq. (7);
9 | **if** *No changeable regions* **then**
10 | | Update the train set by Eq. (6);
11 | | Retrain base model $F(x)$;
12 | **end**
13 | **if** *Identify changeable regions* **then**
14 | | Update the train set by Eq. (7);
15 | | Retrain base model $F(x)$, train tree models for two streams;
16 | | Get the prediction result by Eq. (8);
17 | | Retrain base model $F(x)$;
18 | | Get the prediction result by Eq. (9);
19 | **end**
20 | Choose the better prediction results by Eq. (10).
21 **end**
22 **End**

3.4 Selectively Augmented Learning Process

In this section, we consider different drift situations and aim to maintain the learning performance. We selectively use the augmented learning process and initial retraining process to reduce the influence of the difference between the two streams. So, we initially retrain the base model on the updated training set and output results, denote as

$$\hat{y}'^{S_1} = F(D_{t+1}), \quad \hat{y}'^{S_2} = F(D_{t+1}). \tag{9}$$

Considering the uncertainty, we selectively choose the better results after all the training processes have been finished, and the results have been output, denote as

$$\min \left(L_{Aug} \left(\{ y^{S_1}, y^{S_2} \}, \{ \hat{y}^{S_1}, \hat{y}^{S_2} \} \right), L_{Ini} \left(\{ y^{S_1}, y^{S_2} \}, \{ \hat{y}'^{S_1}, \hat{y}'^{S_2} \} \right) \right). \tag{10}$$

Moreover, this proposed method currently deals with two parallel streams, it can also be extended to handle a larger number of data streams, which would be our future research target.

4 Experiment

This section introduces the experiment settings, synthetic and real-world datasets description, and experiment results discussion.

4.1 Experiment Setting

In this experiment, we test and evaluate the proposed method on the datasets, then output and discuss the results. First, for the model usage, a gradient boosting decision tree model has been used as the base model, and a decision tree model has been used for learning the augmented data. Both of them are from the Sklearn package with default parameter settings. To reduce the influence of the randomness, the random state of the learning model is set as 0. Both of them are from the sklearn package with default parameter settings. Second, the learning process is data chunk-based and the prequential train and test principle is applied. Moreover, the accuracy and F1-score metrics are chosen for the evaluation of the results. There are three main steps of the experiment, and the detailed procedures are listed as follows:

- **Step 1:** To help recognize the drift that occurs on multiple data streams, we test the proposed time warping-based drift identification process on data chunks and then record the drift frequency.
- **Step 2:** To reduce the influence of drifted data, test the proposed drift adaptation method which embedded the data augmentation process on multiple data streams, and then discuss the experiment results.
- **Step 3:** To maintain the learning performance, we selectively choose the traditional retrain and augmented retrain processes to help adapt to multiple data streams with various drift situations.

4.2 Datasets

In the experiment, we choose both synthetic and real-world datasets to evaluate the proposed method. Since this is our first attempt to deal with multiple data streams, the experiment setting is learning and testing the proposed method on two data streams of the same size at the same time. And the chunk size is set as 100 default. Therefore, we design different scenarios to simulate multiple data streams learning. For the synthetic datasets, we generate six data streams with different drift situations. Each of them contains 10,000 data instances. We simulate the drift occurrence by rotating the decision boundary, the degree of the rotating process is θ. Thus, different drift types have been simulated, such as sudden drift, incremental drift, and a mixture of both of them. drift types have been simulated, such as sudden drift, incremental drift, and a mixture of both of them. Then, we use the first data stream as the base stream, and handle this stream with one of the other streams. For the real-world datasets, we first segment the dataset into two data streams, then learn and test the proposed method on them.

- **Stream 1:** is the base data stream which is generated with sudden drift from the 3,001-st to the 6,000-th time points by changing the decision boundary.
- **Stream 2:** also is generated with sudden drift. The drift severity is the same as Stream 1, but the drift time point is late, and the drift period is relatively longer, which is from the 4,001-st to the 8,000-th time point.
- **Stream 3:** simulates a higher frequency of sudden drift, the decision boundary changes every 2,000 time points with the same drift severity as Stream 1.

Table 1. Datasets Description

Datasets	Instances	Attributes	Classes	Chunk size	Drift type	θ value
Stream 1	10,000	3	2	100	Sudden(short)	0,0,180,180,180,180,0,0,0,0
Stream 2	10,000	3	2	100	Sudden(long)	0,0,0,180,180,180,180,180,0,0
Stream 3	10,000	3	2	100	Sudden(high)	0,0,180,180,0,0,180,180,0,0
Stream 4	10,000	3	2	100	Sudden(low)	0,0,90,90,180,180,-90,-90,0,0
Stream 5	10,000	3	2	100	Incremental	0,45,90,135,180,-135,-90,-45,0,0
Stream 6	10,000	3	2	100	Both	0,45,90,135,180,180,0,0,0,0
Weather	25,626	8	2	365	–	–
Electricity	45,312	8	2	100	–	–

- **Stream 4:** simulates the same drift frequency as Stream 3, that is, the decision boundary changes every 2,000 time points with a relatively lower drift severity.
- **Stream 5:** generates the data occurs with incremental drift, the decision boundary changes every 1,000 time points with a lower drift severity.
- **Stream 6:** contains data with incremental and sudden drift, incremental drift occurs from 1-st to the 5,000-th time point while sudden drift at the 7,001-st time point.
- **Weather:** is the record data collected from the NOAA datasets. This data contains 25,626 data instances, 8 attributes, and 2 classes. The chunk size of this data is set as 365. We segment this data into two sets with the same size for the experiment.
- **Electricity:** contains 8 attributes and 2 classes, and 45,312 data instances. We segment this data into two sets with the same size for the experiment.

The prediction results will be evaluated by accuracy and f1-score. The detailed information of each dataset is shown in Table 1.

4.3   Benchmark Methods

To evaluate the performance of the proposed method on multiple data streams, several benchmark methods have been chosen for comparison. The detailed information on each benchmark method is listed below:

- **Baseline:** combines the initial chunks of multiple data streams in the first time step, and initially trains a base model. Then, testing the model in the data chunk of followed time steps without any update.
- **Retrain:** combines the initial chunks of multiple data streams and trains a base model, then tests and retrains it on the new incoming chunks.
- **Augmented Learning (AL) process:** is embedded with the data augmentation process to help learn multiple data streams. Data selected by the time-warping process have been used in the learning process for augmented learning.
- **Time Warping-based Concept Drift learning Method (TW-CDM):** considers the situation that there is a significant distribution difference in multiple data streams. This method embeds the selectively augmented learning process to maintain the learning performance.

Table 2. Results of time-warping-based change identification process

Tasks	$S_1 \rightarrow S_2$	$S_1 \rightarrow S_3$	$S_1 \rightarrow S_4$	$S_1 \rightarrow S_5$	$S_1 \rightarrow S_6$
Change frequency	19	26	29	33	21

Table 3. Friedman test results with benchmark methods

Methods	Baseline	Retrain	AL	TW-CDM$_{\lambda=0.05}$
p value	$0.00815 < 0.01$	$0.00815 < 0.01$	$0.00815 < 0.01$	$0.05878 > 0.01$

Python 3.7 was used to implement the proposed method. The computational environment is a Red Hat Enterprise Linux Workstation release 7.9 (Maipo), with Intel(R) Xeon(R) Gold 6238R CPU @ 2.20 GHz.

5 Experiment Discussion

This section gives a detailed discussion of the experiment results.

5.1 Evaluate the Change Identification Process

In this experiment, we test the proposed time warping-based drift identification process on both synthetic and real-world datasets, and then record the drift frequency. As shown in Table 2, this process can not only help recognize the drift occurrence in time, but also reflect the drift severity according to the recorded frequency.

Different from some traditional drift detection methods which are based on a manually set drift threshold, the proposed method tries to find the similar part of two data streams based on the prequential accuracy of two streams' data chunks as time series. Thus, not only the change can be found easily, but also the regions that have the same similarity and possibly occur change can also be located (Fig. 2).

5.2 Measure the Augmented Learning Process

In this experiment, we use the data in the located region to help model learning. And we evaluate the proposed augmented training process by testing it on both synthetic and real-world datasets, then calculate the accuracy and f1-score, the results are shown in Tables 4, 5. By comparing with the baseline, the proposed augmented training process got a relatively higher evaluation score. And the data streams that occur sudden/incremental drift can be handled with relatively higher performance. Furthermore, the located changeable regions help the model learning and adaptation.

There are both advantages and disadvantages to this method. According to the experiment results, the data in the located region from other data stream can help increase the learning ability of the base model. That is to say, the knowledge in multiple data streams can be crossly used to maintain the learning performance. But this effectiveness may disappear when the data distribution of streams has significant differences. This is the reason why the results of the augmented training process are relatively lower.

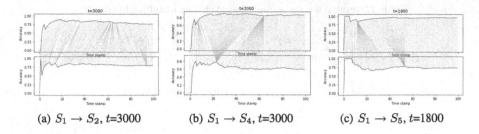

(a) $S_1 \to S_2$, t=3000   (b) $S_1 \to S_4$, t=3000   (c) $S_1 \to S_5$, t=1800

Fig. 2. A plot of the time-warping results of different multiple data streams scenarios.

Table 4. Average chunk accuracy of time-warping-based concept drift learning method (%)

Tasks	$S_1 \to S_2$	$S_1 \to S_3$	$S_1 \to S_4$	$S_1 \to S_5$	$S_1 \to S_6$	Weather	Electricity
Baseline	49.52	59.59	59.12	59.79	64.55	68.51	67.06
Retrain	84.18	77.31	87.61	86.40	91.92	77.85	76.00
AL	83.15	74.79	86.72	85.58	91.50	77.56	75.56
TW-CDM$_{\lambda=0.05}$	84.87	77.40	87.66	86.49	91.97	78.14	76.42
TW-CDM$_{\lambda=0.1}$	84.98	77.51	87.69	86.54	91.99	**78.31**	76.55
TW-CDM$_{\lambda=0.5}$	85.38	**77.97**	**87.95**	**86.66**	92.03	77.94	77.73
TW-CDM$_{\lambda=0.8}$	**85.39**	77.95	87.92	86.64	**92.03**	77.97	**77.76**

5.3   Evaluate the Selectively Augmented Learning Process

In this experiment, we selectively use the proposed augmented training process and traditional retraining process to reduce the influence of different drift severity of multiple data streams. The same as the previous experiment, synthetic and real-world datasets are chosen for model evaluation, and the results are summarized in Tables 3, 4, 5. The results indicate that the proposed TW-CDM method got higher scores on both synthetic and real-world datasets. And the Friedman test results of TW-CDM$_{\lambda=0.5}$ with other benchmark methods show the efficiency. This process can help reduce the influence of different drift severity in multiple data streams. It got a higher evaluation score compared with other methods, but the performance still needs further improvement.

Table 5. Average chunk f1-score of time-warping-based concept drift learning method (%)

Tasks	$S_1 \to S_2$	$S_1 \to S_3$	$S_1 \to S_4$	$S_1 \to S_5$	$S_1 \to S_6$	Weather	Electricity
Baseline	49.45	59.63	59.12	59.52	64.53	72.16	60.84
Retrain	84.95	76.75	87.78	86.12	91.67	82.63	76.42
AL	84.40	75.84	87.20	85.60	91.40	82.97	76.78
TW-CDM$_{\lambda=0.05}$	85.09	76.89	87.85	86.24	91.73	83.04	77.07
TW-CDM$_{\lambda=0.1}$	85.28	77.05	87.89	86.30	91.77	**83.29**	77.32
TW-CDM$_{\lambda=0.5}$	85.79	**77.59**	**88.16**	86.47	**91.80**	82.92	**79.78**
TW-CDM$_{\lambda=0.8}$	**85.80**	77.57	88.14	**86.49**	91.79	82.95	79.76

6 Conclusion

This paper focuses on multiple data streams learning, a time-warping-based concept drift learning method has been proposed to help identify and adapt the drift that occurs in multiple data streams. A time warping-based drift identification process has been proposed. Then, augment the training process based on the recognized data region. Finally, a selectively augmented training process is given to enhance the learning ability. The performance has been measured by testing it on synthetic and real-world datasets.

There are also limitations of the proposed method, the issue of asynchronous drift in multiple data streams and the learning strategy in a larger number of data should be further considered. In our future work, we will continue to focus on handling multiple data streams with complex concept drift scenarios. Trying to give a more available strategy to improve the efficiency of multiple data streams learning.

Acknowledgments. This work was supported by the Australian Research Council under Grants DP200100700 and FL190100149.

References

1. Álvarez, V., Mazuelas, S., Lozano, J.A.: Minimax classification under concept drift with multidimensional adaptation and performance guarantees. In: International Conference on Machine Learning, pp. 486–499. PMLR (2022)
2. Baena-Garcıa, M., del Campo-Ávila, J., Fidalgo, R., Bifet, A.: Early drift detection method. In: In 4th International Workshop on Knowledge Discovery from Data Streams. Citeseer (2006)
3. Balzanella, A., Verde, R.: Histogram-based clustering of multiple data streams. Knowl. Inf. Syst. **62**(1), 203–238 (2020)
4. Bifet, A., Gavalda, R.: Learning from time-changing data with adaptive windowing. In: Proceedings of the 2007 SIAM International Conference on Data Mining, pp. 443–448. SIAM (2007)
5. Chandra, S., Haque, A., Khan, L., Aggarwal, C.: An adaptive framework for multistream classification. In: Proceedings of the 25th ACM International Conference on Information and Knowledge Management, pp. 1181–1190. Indianapolis, IN, USA (2016)
6. Elwell, R., Polikar, R.: Incremental learning of concept drift in nonstationary environments. IEEE Trans. Neural Netw. **22**(10), 1517–1531 (2011)
7. Fang, Z., Lu, J., Liu, F., Xuan, J., Zhang, G.: Open set domain adaptation: theoretical bound and algorithm. IEEE Trans. Neural Netw. Learn. Syst. **32**(10), 4309–4322 (2020)
8. Gama, J., Medas, P., Castillo, G., Rodrigues, P.: Learning with drift detection. In: Bazzan, A.L.C., Labidi, S. (eds.) Advances in Artificial Intelligence – SBIA 2004. SBIA 2004. LNCS, vol. 3171, pp. 286–295. Springer, Berlin, Heidelberg (2004). https://doi.org/10.1007/978-3-540-28645-5_29
9. Gama, J., Žliobaitė, I., Bifet, A., Pechenizkiy, M., Bouchachia, A.: A survey on concept drift adaptation. ACM Comput. Surv. **46**(4), 44 (2014)
10. Haque, A., Wang, Z., Chandra, S., Dong, B., Khan, L., Hamlen, K.W.: Fusion: an online method for multistream classification. In: Proceedings of the 2017 ACM on Conference on Information and Knowledge Management, pp. 919–928. Singapore (2017)
11. Jaysawal, B.P., Huang, J.W.: PSP-AMS: progressive mining of sequential patterns across multiple streams. ACM Trans. Knowl. Discov. Data **13**(1), 1–23 (2018)

12. Kolter, J.Z., Maloof, M.A.: Dynamic weighted majority: an ensemble method for drifting concepts. J. Mach. Learn. Res. **8**, 2755–2790 (2007)
13. Krawczyk, B., Minku, L.L., Gama, J., Stefanowski, J., Woźniak, M.: Ensemble learning for data stream analysis: a survey. Inf. Fusion **37**, 132–156 (2017)
14. Li, W., Yang, X., Liu, W., Xia, Y., Bian, J.: DDG-DA: data distribution generation for predictable concept drift adaptation. In: Proceedings of the AAAI Conference on Artificial Intelligence, pp. 4092–4100 (2022)
15. Liu, A., Lu, J., Liu, F., Zhang, G.: Accumulating regional density dissimilarity for concept drift detection in data streams. Pattern Recogn. **76**, 256–272 (2018)
16. Liu, A., Song, Y., Zhang, G., Lu, J.: Regional concept drift detection and density synchronized drift adaptation. In: IJCAI International Joint Conference on Artificial Intelligence (2017)
17. Lu, J., Liu, A., Dong, F., Gu, F., Gama, J., Zhang, G.: Learning under concept drift: a review. IEEE Trans. Knowl. Data Eng. **31**(12), 2346–2363 (2018)
18. Lu, J., Niu, L., Zhang, G.: A situation retrieval model for cognitive decision support in digital business ecosystems. IEEE Trans. Ind. Electron. **60**(3), 1059–1069 (2012)
19. Lu, J., Yang, X., Zhang, G.: Support vector machine-based multi-source multi-attribute information integration for situation assessment. Expert Syst. Appl. **34**(2), 1333–1340 (2008)
20. Lu, J., Zuo, H., Zhang, G.: Fuzzy multiple-source transfer learning. IEEE Trans. Fuzzy Syst. **28**(12), 3418–3431 (2019)
21. Mihai, N., Alexandru, M., Bala-Constantin, Z.: Multimodal emotion detection from multiple data streams for improved decision making. Procedia Comput. Sci. **214**, 1082–1089 (2022)
22. Nadeem, A., Verwer, S.: SECLEDS: sequence clustering in evolving data streams via multiple medoids and medoid voting. In: Amini, M.R., Canu, S., Fischer, A., Guns, T., Kralj Novak, P., Tsoumakas, G. (eds.) Machine Learning and Knowledge Discovery in Databases. ECML PKDD 2022. LNCS, vol. 13713, pp. 157–173. Springer, Cham (2023). https://doi.org/10.1007/978-3-031-26387-3_10
23. Otero, A., Félix, P., Márquez, D.G., García, C.A., Caffarena, G.: A fault-tolerant clustering algorithm for processing data from multiple streams. Inf. Sci. **584**, 649–664 (2022)
24. Sangma, J.W., Pal, V., Kumar, N., Kushwaha, R., et al.: FHC-NDS: fuzzy hierarchical clustering of multiple nominal data streams. IEEE Trans. Fuzzy Syst. **31**(3), 786–798 (2022)
25. Song, Y., Zhang, G., Lu, H., Lu, J.: A fuzzy drift correlation matrix for multiple data stream regression. In: 2020 IEEE International Conference on Fuzzy Systems, pp. 1–6. IEEE (2020)
26. Yu, E., Song, Y., Zhang, G., Lu, J.: Learn-to-adapt: concept drift adaptation for hybrid multiple streams. Neurocomputing **496**, 121–130 (2022)
27. Yu, H., Liu, W., Lu, J., Wen, Y., Luo, X., Zhang, G.: Detecting group concept drift from multiple data streams. Pattern Recogn. **134**, 109113 (2023)
28. Zhou, M., Song, Y., Zhang, G., Zhang, B., Lu, J.: An efficient Bayesian neural network for multiple data streams. In: 2021 International Joint Conference on Neural Networks (IJCNN), pp. 1–8. IEEE (2021)

Sequence Unlearning for Sequential Recommender Systems

Shanshan Ye[ID] and Jie Lu[✉][ID]

Australian Artificial Intelligence Institute, Faculty of Engineering and IT, University of Technology Sydney, Sydney, NSW, Australia
shanshan.ye@student.uts.edu.au, jie.lu@uts.edu.au

Abstract. Sequential recommender systems, leveraging clients' sequential product browsing history, have become an essential tool in delivering personalized product recommendations. As data protection regulations come into focus, certain clients may demand the removal of their data from the training sets used by these systems. In this paper, we focus on the problem of how specific client information can be efficiently removed from a pre-trained sequential recommender system without the need for retraining, particularly when the change to the data set is not substantial. We propose a novel sequence unlearning method for sequential recommender systems by leveraging label noise injection. Intuitively, our method promotes data unlearning by encouraging the system to produce random predictions for the sequences aiming to unlearn. To further prevent the model from overfitting an incorrect label, which could lead to substantial changes in its parameters, our method incorporates a dynamic process wherein the incorrect label is continually altered during the learning phase. This effectively encourages the model to lose confidence in the original label, while also discouraging it from fitting to a specific incorrect label. To the best of our knowledge, this is the first work to tackle the unlearning problem in sequential recommender systems without accessing the remaining data. Our approach is general and can work with any sequential recommender system. Empirically, we demonstrate that our method effectively helps different recommender systems unlearn specific sequential data while maintaining strong generalization performance on the remaining data across different datasets.

Keywords: Sequential Recommender Systems · Machine Unlearning · Noise Injection

1 Introduction

In the rapidly advancing world of machine learning, sequential recommender systems-leveraging clients' sequential product browsing history-have become indispensable for product recommendations [21,23]. As data protection regulations gain prominence, an evolving challenge emerges: clients may demand the removal of their data from the training sets used by these systems, often due to concerns over privacy and regulatory compliance [1].

T. Liu et al. (Eds.): AI 2023, LNAI 14471, pp. 403–415, 2024.
https://doi.org/10.1007/978-981-99-8388-9_33

Existing traditional methods for recommender systems involve retraining the model with the remaining data after excluding the specific client's information [1, 7]. However, this retraining process is often resource-intensive and can be unfeasible when the original training data is inaccessible [2]. This leads to the central question explored in this paper: How can specific client information be efficiently removed from a pre-trained sequential recommender system without the need for retraining, especially when the change to the dataset is not substantial?

To solve this problem, this paper introduces a novel unlearning method through label noise injection based on the learning behavior of learning models, where learning models usually demonstrate a smaller loss on trained data compared to untrained data. Specifically, consider two trained models: one trained on a complete dataset and another trained on a pruned dataset excluding some sequences that are aimed to be forgotten. The latter model usually has a higher loss on the to be forgotten sequences compared with the one trained on the complete dataset. This happens because the latter model does not have an opportunity to fit the excluded sequences, making it less confident in predicting them. Motivated by this observation, our proposed method introduces label noise into the to be forgotten sequences. By subsequently fine-tuning a pre-trained model only with these to be forgotten sequences, we amplify the fine-tuned model's loss on the original to be forgotten sequences, making these sequences to be unlearned. This technique effectively emulates the results one might expect from a network retrained on the remaining data without undergoing a full retraining process.

However, the intuition brings in some challenges. Introducing incorrect labels via label-noise injection and fine-tuning a pre-trained model on the noisy examples can lead a model to overfit these incorrect labels, dramatically changing its parameters. This change can dramatically affect the model's performance on the remaining data. This issue is particularly critical in sequential recommender systems, where sequences in training data usually correlate with each other. Then if sequences targeted for removal strongly correlate with some remaining sequences, overfitting an incorrect label could greatly degenerate performance on the remaining data. To mitigate the adverse impacts of label-noise injection, we propose a dynamic process wherein the incorrect label is continually modified during the learning phase. This inventive strategy helps the model predict less confidently on the forgotten sequence, while also protecting it from fitting to a specific incorrect label. To our knowledge, this is the first paper to delve into the unlearning issue in sequential recommender systems. Our approach can work with any sequential recommender systems, and presents a comprehensive solution for their unlearning.

2   Related Work

2.1   Sequential Recommender Systems

In the early stages of recommender systems development, techniques primarily focused on mapping users and items into latent factor space, with matrix factorization methods being widely adopted [15,20]. As the field evolved, implicit feedback mechanisms, such as purchases or video views, emerged as valuable sources of user information, inspiring new methods designed to harness these insights [4,11]. However, these advancements overlooked one critical aspect: the sequential patterns in consecutively interacted items. Traditional methods, though effective in certain applications, failed to capture this essential aspect of user behavior. Recognizing that sequential patterns in user-item interactions reveal valuable insights into user preferences and evolving interests marked a turning point in the field [27], highlighting the necessity to transition from static models to systems capable of utilizing rich sequential information within user interactions.

In response to this need, sequential recommender systems emerged, prioritizing recommendations based on a user's historical interaction sequence. Pioneering methods like Markov Chain laid the groundwork for predicting the next item, mainly based on previous interactions, using matrix factorization [24]. Deep learning further enriched sequential recommender systems [10]. For instance, Caser [29] applied convolutional neural networks to sequence modeling, treating the item embedding matrix as an image and using convolution operators to identify local transitions. The field's continued evolution integrated Recurrent Neural Networks [23] and Self-Attention mechanisms [14]. GRU4Rec [12] leveraged Gated Recurrent Units in session-based recommendations, and the success of methods like Transformer [31] and BERT [5] further promoted self-attention in sequential recommendations. Unlike Markov chain and RNN methods, self-attention considers attention scores from all item-item pairs within a sequence. Both SASRec [14] and BERT4Rec [28] highlight the effectiveness of self-attention, achieving leading performance in a next-item recommendation.

2.2 Machine Unlearning

Machine unlearning techniques span two primary domains: exact and approximate unlearning. Exact unlearning strategies ensure absolute data forgetting or complete removal [1], often exploiting a divide-aggregate structure to boost retraining efficiency through model [7,25] and data partitioning [34].

Approximate unlearning, conversely, trades off some level of completeness to enhance efficiency and model utility, thus achieving only statistical forgetting [13]. These methods often work by directly reducing the target data's influence on the model through inverse gradient-based update strategies, enabling efficient forgetting without retraining. Influence functions, originating from robust statistics [9], measure the impact of target data on a trained model, but their application in machine unlearning usually requires the computation of the Hessian matrix, necessitating efficiency optimization [22,32].

2.3 Unlearning Method for Recommender Systems

Several unlearning strategies have emerged within the field of recommender systems. RecEraser [3] implements novel data partition algorithms to divide the training data into balanced groups, preserving collaborative information vital for collaborative filtering. LASER [18], concurrent with RecEraser, follows a similar process but with distinct specifics. Various recommendation unlearning works have adopted influence functions from classification or regression tasks [26], using second-order optimization methods like Newton and quasi-Newton to accelerate re-training [30]. AltEraser [19] enhances optimization efficiency by breaking down a large problem into smaller, independently solvable sub-problems using alternating optimization. SCIF [17] selectively updates user embeddings to reduce parameter updates while preserving collaborative information, enhancing model utility. Other notable methods include IFRU [36], which extends the influence function, and efforts by Yuan et al. [35] in the federated recommendation. Lastly, Unlearn-ALS [33] modifies fine-tuning for bi-linear recommendation models under Alternating Least Squares optimization. To the best of our knowledge, all existing unlearning methods for recommender systems require remaining data and are not designed specifically for sequential unlearning.

3   Sequence Unlearning via Random Label Noise Injection

Problem Setup. Consider a dataset D of user-item interaction sequences with m users, defined as $D = \{\mathcal{S}^{u_1}, \mathcal{S}^{u_2}, \ldots, \mathcal{S}^{u_m}\}$. Within D, each sequence $\mathcal{S}^{u_i} = \{S_1^{u_i}, S_2^{u_i}, \ldots, S_n^{u_i}\}$ is the item interactions of user u_i over n sequential time steps. The primary objective is to predict the coming item in a user's sequence. Specifically, the model, at any time step t, leverages the previous t items from a user's interaction sequence to forecast the next item.

Contrastingly, sequence unlearning seeks to let the model forget a specific training sequence $\mathcal{S}^{u*}$. Specifically, for any time step t, we want the model to generate random predictions for the next item. At the same time, the unlearning should not greatly impact the model's performance on other interaction sequences, ensuring that its general predictive capabilities remain largely unaffected.

3.1   Sequence Unlearning with Dynamic Label Noise Injection

In sequential recommender systems, efficiently unlearning specific sequences without retraining is an emerging critical requirement. To address this challenge, we present a method that dynamically injects label noise into the sequences requested to be removed. Our method centers on noise-generation functions that are crafted to introduce a controlled degree of randomness into the output sequence associated with the sequences the system aims to unlearn. By injecting random label noise into the targeted sequences, the sequential recommender system can be guided to produce less accurate predictions for those sequences.

It's noteworthy that in a sequential recommendation dataset, the sequences in the training data typically exhibit correlations with each other. The label noise may mislead the model's performance on other similar data to the to be forgotten data. To mitigate potential overfitting to these incorrect labels, a situation that could dramatically alter the model's parameters after fine-tuning and negatively impact performance on the remaining data, we propose a dynamic

---

Algorithm 1. Unlearning for Sequential Recommender Systems

Require: Trained recommender systems, User sequence set $\mathcal{S}$, Item set $\mathcal{I}$, To be forgotten set D_f, Noise injection probability ρ

1: **for** each epoch **do**   ▷ Iterate through training epochs
2:     **for** each $\mathcal{S}^{u*}$ in D_f **do**   ▷ Iterate through sequences to be forgotten
3:        $O^{u*} \leftarrow$ Corresponding output sequence of $\mathcal{S}^{u*}$
4:        Initialize an empty perturbed label sequence $O^{u*'}$
5:        **for** each time step t in O^{u*} **do**   ▷ Iterate through each time step
6:           Generate a uniform random number x in $[0, 1]$
7:           **if** $x \leq \rho$ **then**
8:              $s_t^{u*'} \leftarrow$ Random label from $\mathcal{I} \setminus \{s_t^{u*}\}$ ▷ To have incorrect labels
9:           **else**
10:             $s_t^{u*'} \leftarrow s_t^{u*}$ ▷ Retain the original label
11:           **end if**
12:           $O^{u*'} \leftarrow O^{u*'} \cup [s_t^{u*'}]$ ▷ Update the perturbed label sequence
13:        **end for**
14:        Update the recommender system using D_f and $O^{u*'}$ ▷ Fine-tune the system with noisy labels
15:     **end for**
16: **end for**

---

noise injection process. In this process, the incorrect label is continually modified during the learning phase. This innovative approach not only encourages the model to not predict the original label of the to be forgotten data but also ensures that the model does not overfit the label noise. In doing so, we effectively reduce the overfitting problem, maintaining a balance between unlearning specific information and preserving the generalization ability of the model on remaining data. The pseudocode is summarised in Algorithm 1.

Label Noise Injection. During the training process, at the time step t, the model predicts the next item based on the previous t items. Consider that the model's input for a user-item sequence $\mathcal{S}^u$ is $\{s_1^u, s_2^u, \ldots, s_{n-1}^u\}$ and the label sequence (or desired output) is a "shifted" version of the same sequence: $O^u = \{s_2^u, s_3^u, \ldots, s_n^u\}$. For a sequence $\mathcal{S}^{u*}$ targeted for unlearning, our method generates a perturbed label sequence $O^{u*'}$ by injecting random label noise into the original label sequence O^{u*}. By fine-tuning the model using the modified label sequences, the model can effectively unlearn these sequences.

Specifically, let $\mathcal{I}$ denote the set of all items. We define an incorrect-label-generation function ϕ that, with a probability ρ, replaces an item label with a random one from the set. The perturbed label sequence is generated as:

1. For each time step t in the sequence, generate a random number x uniformly between 0 and 1.
2. If $x \leq \rho$, apply function ϕ to randomly select an item label from $\mathcal{I} \setminus \{s_t^{u*}\}$. If $x > \rho$, keep the original label.

This approach introduces label noise into certain portions of the sequence based on ρ, while the rest remain unchanged. The perturbed label sequence for training are represented as $O^{u*'} = \{s_2^{u*'}, s_3^{u*'}, \ldots, s_n^{u*'}\}$. Each label $s_t^{u*'}$ is defined by:

$$s_t^{u*'} = \begin{cases} \phi(s_t^{u*}), & \text{if } x \le \rho; \\ s_t^{u*}, & \text{if } x > \rho, \end{cases}$$

where $\phi(s_t^{u*}) \sim \mathcal{U}(\mathcal{I} \setminus \{s_t^{u*}\})$. Here, $\mathcal{U}(\mathcal{I} \setminus \{s_t^{u*}\})$ signifies a uniform distribution over item set $\mathcal{I}$ excluding s_t^{u*}. By adjusting ρ, we can control the degree of noise infusion and the extent of unlearning, allowing us to find a balance between fulfilling unlearning requests and preserving the system's performance on the remaining data. In our experiment, we set $\rho = 1/p$, where p denotes the total number of items. This enables the model to randomly guess the next item given a sequence aimed to be forgotten.

Unlearning Objective. The unlearning objective is central to our method, serving to guide the fine-tuning of a pre-trained sequential recommender system for the effective unlearning of particular sequences, while also mitigating potential adverse effects of this unlearning. A distinctive feature of our approach is the dynamic label noise injection. Specifically, we randomly adjust labels continuously throughout each training epoch. This dynamic modification ensures the model doesn't overfit to any specific incorrect label, preserving the model's generation ability on remaining data.

Let $f_{\hat{\theta}}$ be a pre-trained sequential recommender system. This system outputs probabilities of different items being the next item given a historical user-item interaction sequence as input. To be precise, given a user's item interaction sequence $\{s_1, s_2, \ldots, s_{t-1}\}$ from the first time step up to $t-1$, the i-th output of the system is $f_{\hat{\theta}}(\{s_1, s_2, \ldots, s_{t-1}\})_i = p_{(s_i,t)}$. Here, $p_{(s_i,t)}$ represents the probability that item s_i will be the next item given the first $t-1$ items in a user-item sequence. We employ the binary cross entropy loss function to design the objective for fine-tuning:

$$L' = - \sum_{\mathcal{S}^{u*} \in D_f} \sum_{t \in \{1,2,\ldots,n\}} \left[\log(p_{(s_t^{u*'},t)}) + \sum_{s_i \ne s_t^{u*'}} \log(1 - p_{(s_i,t)}) \right],$$

where $s_t^{u*'}$ is the randomly picked item via noise injection at time step t given the first $t-1$ items in the sequence $\mathcal{S}^{u*}$ targeted for unlearning. The term $p_{(s_t^{u*'},t)}$ quantifies the probability that the randomly picked item $s_t^{u*'}$ will be the next chosen item as predicted by the pre-trained sequential recommender system. Similarly, $p_{(s_i,t)}$ quantifies the probability that the system predicts an item s_i other than $s_t^{u*'}$ as the next item.

Intuitively, minimizing the loss function above makes the recommender system more likely to predict randomly picked items by adding label noise and less likely to predict other items. By integrating dynamic label noise with the unlearning objective, our method ensures a small influence on the pre-trained sequential recommender system while achieving the unlearning of specified sequences.

Table 1. HIT@10, NDCG@10 and MRR by applying different unlearning methods to SASRec on Beauty dataset.

	Method	Metric	$|D_f|=5$	$|D_f|=10$	$|D_f|=15$	$|D_f|=20$	$|D_f|=25$	$|D_f|=30$	$|D_f|=35$	$|D_f|=40$
D_f ($\downarrow$)	BEFORE	HIT@10	40.00	30.00	33.33	40.00	40.00	43.33	45.71	45.00
		NDCG@10	18.93	14.46	14.23	15.84	14.98	17.26	18.15	18.10
		MRR	16.66	14.33	12.81	12.37	11.36	13.06	13.31	13.34
	All Others	HIT@10	0.00	0.00	0.00	0.00	0.00	0.00	0.00	0.00
		NDCG@10	0.00	0.00	0.00	0.00	0.00	0.00	0.00	0.00
		MRR	0.00	0.00	0.00	0.00	0.00	0.00	0.00	0.00
D_r ($\uparrow$)	BEFORE	HIT@10	56.28	56.29	56.30	56.30	56.30	56.30	56.30	56.30
		NDCG@10	27.24	27.24	27.24	27.24	27.25	27.25	27.25	27.25
		MRR	20.79	20.79	20.79	20.79	20.80	20.80	20.80	20.80
	NegGrad	HIT@10	45.03	43.00	32.72	43.67	19.66	41.12	35.66	32.20
		NDCG@10	22.39	21.30	16.25	21.38	10.02	20.35	17.86	16.15
		MRR	17.66	16.85	13.15	16.85	8.37	16.12	14.36	13.04
	FixRand	HIT@10	50.77	51.42	50.91	51.00	49.17	49.65	45.29	47.71
		NDCG@10	24.75	25.51	25.14	25.25	**25.12**	25.28	**25.04**	24.52
		MRR	19.20	19.91	19.61	19.55	19.54	19.74	19.21	19.38
	DyRand	HIT@10	**51.16**	**51.94**	**51.45**	**51.14**	**50.91**	**50.69**	**50.10**	**47.96**
		NDCG@10	**25.23**	**25.72**	**25.45**	**25.42**	25.04	**25.32**	24.99	**24.77**
		MRR	**19.65**	**19.99**	**19.82**	**19.61**	**19.75**	20.09	**19.65**	**19.71**
D_t ($\uparrow$)	BEFORE	HIT@10	5.68	5.68	5.68	5.68	5.68	5.68	5.68	5.68
		NDCG@10	3.03	3.03	3.03	3.03	3.03	3.03	3.03	3.03
		MRR	2.50	2.50	2.50	2.50	2.50	2.50	2.50	2.50
	NegGrad	HIT@10	4.94	5.09	4.34	4.90	2.44	4.97	4.22	3.76
		NDCG@10	2.64	2.71	2.26	2.60	1.30	2.66	2.30	2.03
		MRR	2.20	2.23	1.85	2.16	1.11	2.21	1.94	1.68
	FixRand	HIT@10	5.54	5.50	5.37	5.42	5.25	5.23	4.99	**5.04**
		NDCG@10	**2.94**	2.94	2.90	**2.91**	2.86	2.80	2.65	**2.71**
		MRR	**2.40**	2.44	2.43	2.42	2.40	2.33	2.19	**2.27**
	DyRand	HIT@10	**5.59**	**5.52**	**5.42**	**5.42**	**5.35**	**5.29**	**5.25**	4.97
		NDCG@10	**2.94**	**2.97**	**2.91**	2.90	**2.90**	**2.89**	**2.83**	2.69
		MRR	**2.40**	**2.47**	**2.43**	**2.41**	**2.42**	**2.43**	**2.36**	2.25

4 Experiments

Evaluation Metrics. Our evaluation framework relies on a selection of widely accepted metrics that comprehensively measure the quality of recommendations. The top-k Hit Ratio (HR@k) represents the average number of positively rated items found in the top-k recommendations for each user. The top-k Normalized Discounted Cumulative Gain (NDCG@k) extends the concept of HR@k by also considering the positions of the positively-rated items within the top-k list. Unlike HR@k and NDCG@k, which focus on the top-k items, Mean reciprocal rank (MRR) assesses the ranking across the entire list. We have chosen to rank all items, an approach that diverges from biased sampling [16], and our analysis reports the averaged metrics over all users, where k is set to 10.

Table 2. HIT@10, NDCG@10 and MRR by applying different unlearning methods to SASRec on Sports and Outdoors dataset.

	Method	Metric	$|D_f|=5$	$|D_f|=10$	$|D_f|=15$	$|D_f|=20$	$|D_f|=25$	$|D_f|=30$	$|D_f|=35$	$|D_f|=40$
D_f (↓)	BEFORE	HIT@10	0.00	0.00	6.67	5.00	4.00	3.33	2.86	5.00
		NDCG@10	0.00	0.00	4.21	3.15	2.52	2.10	1.80	2.30
		MRR	0.00	0.00	4.42	3.70	2.96	2.72	2.33	2.55
	All Others	HIT@10	0.00	0.00	0.00	0.00	0.00	0.00	0.00	0.00
		NDCG@10	0.00	0.00	0.00	0.00	0.00	0.00	0.00	0.00
		MRR	0.00	0.00	0.00	0.00	0.00	0.00	0.00	0.00
D_r (↑)	BEFORE	HIT@10	6.73	6.73	6.73	6.73	6.73	6.73	6.73	6.73
		NDCG@10	3.33	3.33	3.33	3.33	3.33	3.33	3.33	3.33
		MRR	3.10	3.10	3.10	3.10	3.10	3.10	3.10	3.10
	NegGrad	HIT@10	6.43	6.21	4.79	4.06	4.26	2.96	3.26	2.74
		NDCG@10	3.28	3.24	2.35	2.36	2.06	1.51	1.67	1.42
		MRR	2.57	2.53	2.13	2.03	1.83	1.33	1.48	1.26
	FixRand	HIT@10	6.68	6.34	6.31	5.82	5.88	5.77	5.26	5.14
		NDCG@10	3.20	3.10	2.96	2.62	2.54	2.53	2.18	2.05
		MRR	3.02	2.93	2.82	2.47	2.39	2.22	2.08	1.92
	DyRand	HIT@10	**6.71**	**6.60**	**6.52**	**6.42**	**6.33**	**5.86**	**5.45**	**5.17**
		NDCG@10	**3.22**	**3.20**	**3.15**	**2.97**	**2.70**	**2.61**	**2.41**	**2.30**
		MRR	**3.10**	**3.03**	**2.98**	**2.86**	**2.61**	**2.43**	**2.22**	**2.18**
D_t (↑)	BEFORE	HIT@10	2.59	2.59	2.59	2.59	2.59	2.59	2.59	2.59
		NDCG@10	1.37	1.37	1.37	1.37	1.37	1.37	1.37	1.37
		MRR	1.19	1.19	1.19	1.19	1.19	1.19	1.19	1.19
	NegGrad	HIT@10	2.42	2.24	1.72	2.45	1.60	1.46	1.59	1.11
		NDCG@10	1.27	1.19	0.88	1.26	0.80	0.71	0.82	0.55
		MRR	1.11	1.05	0.78	1.08	0.69	0.60	0.71	0.47
	FixRand	HIT@10	**2.57**	2.57	2.38	2.32	2.26	2.16	2.03	1.84
		NDCG@10	1.37	**1.37**	1.26	1.22	1.19	1.14	1.06	0.97
		MRR	1.19	1.19	1.08	1.04	1.00	0.97	0.89	0.82
	DyRand	HIT@10	2.56	**2.58**	**2.39**	**2.54**	**2.28**	**2.31**	**2.16**	**2.01**
		NDCG@10	**1.37**	**1.37**	**1.27**	**1.30**	**1.19**	**1.17**	**1.11**	**1.02**
		MRR	**1.20**	**1.20**	**1.10**	**1.10**	**1.02**	**0.97**	**0.92**	**0.85**

Datasets and Experiment Setup. Our experimental evaluation utilizes four renowned benchmark datasets from Amazon reviews, encompassing a wide range of domains. These datasets collectively include more than 1.2 million users and a collection of over 63,000 items. Noted for their significant sparsity, the Amazon datasets furnish a rich compilation of rating reviews spread across various fields. To create sequential patterns, interactions for each user are organized in chronological order based on rating timestamps. Following the methodology in [6], we designate the most recent interaction as the testing sample, represented by D_t, and the penultimate interaction as the validation sample. We filter to include only users with at least five interactions, aligning with the 5-core setting employed by prominent research in this domain [14,28].

In the unlearning experiment, we randomly select a subset of images from the entire training sample of each dataset to form the unlearning sample D_f. The remaining samples are labeled as D_r. The size of this subset is experimentally

Table 3. HIT@10, NDCG@10 and MRR by applying different unlearning methods to SASRec on Toys and Games dataset.

	Method	Metric	$|D_f|=5$	$|D_f|=10$	$|D_f|=15$	$|D_f|=20$	$|D_f|=25$	$|D_f|=30$	$|D_f|=35$	$|D_f|=40$
D_f (↓)	BEFORE	HIT@10	20.00	30.00	33.33	35.00	40.00	36.67	34.29	32.50
		NDCG@10	12.62	14.48	18.55	17.84	20.68	18.67	17.81	16.83
		MRR	10.87	11.05	14.99	13.78	16.33	14.92	14.51	13.62
	All Others	HIT@10	0.00	0.00	0.00	0.00	0.00	0.00	0.00	0.00
		NDCG@10	0.00	0.00	0.00	0.00	0.00	0.00	0.00	0.00
		MRR	0.00	0.00	0.00	0.00	0.00	0.00	0.00	0.00
D_r (↑)	BEFORE	HIT@10	36.24	36.24	36.24	36.24	36.23	36.23	36.24	36.24
		NDCG@10	17.46	17.46	17.46	17.46	17.46	17.46	17.46	17.46
		MRR	13.89	13.89	13.88	13.89	13.88	13.88	13.88	13.89
	NegGrad	HIT@10	34.29	29.77	24.46	2.79	5.83	13.37	19.14	3.45
		NDCG@10	16.47	14.48	12.13	1.34	2.76	6.49	9.41	1.62
		MRR	12.80	11.77	9.83	1.17	2.38	5.57	7.81	1.35
	DyRand	HIT@10	**35.49**	**34.39**	34.13	33.29	29.89	24.53	22.83	17.28
		NDCG@10	**16.90**	**16.77**	16.27	16.40	16.34	13.52	12.63	9.57
		MRR	**13.84**	**13.59**	13.95	13.03	13.30	11.11	10.42	7.95
	FixRand	HIT@10	34.89	34.37	**34.22**	**34.02**	**34.13**	**33.39**	**30.03**	**28.58**
		NDCG@10	16.62	16.63	**16.77**	**16.68**	**16.68**	**16.50**	**16.00**	**14.98**
		MRR	13.67	13.38	**13.41**	**13.10**	**13.04**	**13.09**	**13.04**	**12.16**
D_t (↑)	BEFORE	HIT@10	6.71	6.71	6.71	6.71	6.71	6.71	6.71	6.71
		NDCG@10	3.74	3.74	3.74	3.74	3.74	3.74	3.74	3.74
		MRR	3.12	3.12	3.12	3.12	3.12	3.12	3.12	3.12
	NegGrad	HIT@10	6.07	5.91	5.08	1.14	2.28	4.11	4.71	1.46
		NDCG@10	3.33	3.19	2.66	0.58	1.12	2.14	2.44	0.78
		MRR	2.77	2.64	2.20	0.52	0.95	1.80	2.00	0.70
	FixRand	HIT@10	**6.72**	**6.71**	6.14	5.38	5.34	4.84	4.62	3.89
		NDCG@10	3.72	**3.76**	3.44	2.97	2.99	2.66	2.56	2.11
		MRR	3.10	**3.14**	2.89	2.46	2.50	2.20	2.13	1.75
	DyRand	HIT@10	6.70	6.59	**6.33**	**5.83**	**5.85**	**5.77**	**5.26**	**5.22**
		NDCG@10	**3.73**	3.64	**3.46**	**3.20**	**3.26**	**3.14**	**2.95**	**2.90**
		MRR	**3.12**	3.02	**2.86**	**2.66**	**2.73**	**2.58**	**2.49**	**2.43**

determined, with set cardinalities ranging from 5 to 40. The AdaGrad optimizer is engaged to conduct the unlearning process, which is promptly halted if the Mean Reciprocal Rank (MMR) on unlearning sample D_f reaches 0.

Baselines. In our study, we compare our proposed unlearning methods with several benchmark techniques. Among these baselines, "BEFORE" refers to the model's state before any unlearning process. The technique "NegGrad" [8] involves fine-tuning the model using negative gradients, while "fixRand" describes a method that introduces random label noise but keeps the labels consistent during the unlearning phase. In contrast, "DyRand," our final approach, injects random label noise and dynamically alters the labels as the unlearning progresses.

We apply these unlearning methods to three distinct sequential recommender systems. Specifically, SASRec [14] is a self-attention-based model that uses a multi-head attention mechanism to recommend the next item. S^3-Rec leverages

Table 4. HIT@10, NDCG@10 and MRR by applying different unlearning methods to SASRec on Yelp dataset.

	Method	Metric	$|D_f|=5$	$|D_f|=10$	$|D_f|=15$	$|D_f|=20$	$|D_f|=25$	$|D_f|=30$	$|D_f|=35$	$|D_f|=40$
D_f ($\downarrow$)	BEFORE	HIT@10	0.00	10.00	6.67	5.00	4.00	3.33	5.71	5.00
		NDCG@10	0.00	3.56	2.37	1.78	1.42	1.19	3.87	3.39
		MRR	0.00	1.92	1.28	0.96	0.77	0.72	3.56	3.12
	All Others	HIT@10	0.00	0.00	0.00	0.00	0.00	0.00	0.00	0.00
		NDCG@10	0.00	0.00	0.00	0.00	0.00	0.00	0.00	0.00
		MRR	0.00	0.00	0.00	0.00	0.00	0.00	0.00	0.00
D_r ($\uparrow$)	BEFORE	HIT@10	4.26	4.26	4.26	4.26	4.26	4.26	4.26	4.26
		NDCG@10	2.20	2.20	2.20	2.20	2.20	2.20	2.20	2.20
		MRR	1.96	1.96	1.96	1.96	1.96	1.96	1.96	1.96
	NegGrad	HIT@10	**4.50**	4.02	5.00	4.14	4.82	4.27	3.57	0.38
		NDCG@10	**2.29**	2.07	2.51	2.15	2.49	2.17	1.98	0.21
		MRR	**1.99**	1.86	2.17	1.91	2.21	1.92	1.73	0.20
	FixRand	HIT@10	4.31	5.42	5.03	5.59	5.29	5.86	3.05	3.23
		NDCG@10	2.22	**2.80**	2.57	2.88	2.70	3.06	1.60	1.65
		MRR	1.98	**2.49**	2.32	2.52	2.39	2.71	1.39	1.42
	DyRand	HIT@10	4.31	**5.43**	**5.76**	**5.44**	**5.65**	**6.11**	**4.24**	**4.20**
		NDCG@10	2.22	2.79	**2.98**	**2.81**	**2.93**	**3.20**	**2.18**	**2.15**
		MRR	1.98	**2.49**	2.63	2.50	2.59	2.81	1.87	1.85
D_t ($\uparrow$)	BEFORE	HIT@10	2.75	2.75	2.75	2.75	2.75	2.75	2.75	2.75
		NDCG@10	1.40	1.40	1.40	1.40	1.40	1.40	1.40	1.40
		MRR	1.20	1.20	1.20	1.20	1.20	1.20	1.20	1.20
	NegGrad	HIT@10	2.26	2.31	2.15	1.98	2.10	2.30	1.25	0.16
		NDCG@10	1.14	1.16	1.09	1.00	1.06	1.14	0.62	0.08
		MRR	0.98	1.01	0.94	0.85	0.91	0.97	0.52	0.08
	FixRand	HIT@10	2.74	2.49	2.50	2.40	2.43	1.91	1.30	1.28
		NDCG@10	**1.39**	1.26	1.27	1.21	1.25	0.82	0.66	**0.64**
		MRR	**1.19**	1.07	1.08	1.03	1.07	0.69	0.57	**0.55**
	DyRand	HIT@10	**2.75**	**2.52**	**2.60**	**2.44**	**2.47**	**2.50**	**1.42**	**1.29**
		NDCG@10	**1.39**	**1.27**	**1.32**	**1.26**	**1.27**	**1.27**	**0.71**	0.64
		MRR	**1.19**	**1.09**	**1.13**	**1.09**	**1.10**	**1.08**	**0.61**	**0.55**

the intrinsic correlations within the data, enhancing representations through pre-training to improve sequential recommendations. Lastly, STOSA employs a Wasserstein self-attention module, enabling it to characterize item-to-item positional relationships within sequences effectively. These diverse methods and systems offer a rich ground for evaluating the efficacy and adaptability of our proposed unlearning techniques.

4.1 Unlearning Performance on Different Datasets

We evaluate the unlearning performance of various methods applied to different datasets and using different sequential recommender systems. We present results for four datasets (Toys and Games, Yelp, Beauty, Sport and Outdoors) employing the SASRec sequential recommender system within the main text. Due to space limitations, we have included the additional results in Appendix A.

Tables 1, 3, 2, and 4 all exhibit similar findings. All baseline methods successfully unlearn the targeted sample D_f, reflected in a score of 0 across different evaluation metrics. However, a clear distinction emerges when we consider the remaining sample D_r and the test sample D_t. Both FixRand and DyRand, methods that employ label noise injection to unlearn samples, consistently surpass the NegGrad baseline across all tests.

Notably, the performance of the baseline NegGrad on both D_r and D_t diminishes sharply with increasing unlearning sample size. On the other hand, our newly proposed unlearning methods that use label noise injection remain resilient in performance. Our final approach, dynamic label noise injection (DyRand), stands out by achieving the best performance across various metrics and unlearning sample sizes in the majority of experiments. These results firmly validate our method's capacity to unlearn specific sequential data while retaining the model's predictive competence over the remaining information.

5 Conclusion

In this paper, we tackled the challenge of sequence unlearning in sequential recommender systems, a vital area in the age of data protection and privacy. We introduced a new unlearning method leveraging label noise injection. Our approach ingeniously encourages the system to make random predictions for sequences to be unlearned, using a dynamic noise-injection approach to prevent overfitting to incorrect labels and dramatic changing of the model parameter. Empirical results demonstrated effectiveness in unlearning specific data while preserving generalization performance.

Acknowledgements. This work is supported by ARC Discovery Projects - (DP220102635) "Robust Meta Learning for Risk-aware Recommender Systems".

References

1. Bourtoule, L., et al.: Machine unlearning. In: SP, pp. 141–159. IEEE (2021)
2. Cha, S., Cho, S., Hwang, D., Lee, H., Moon, T., Lee, M.: Learning to unlearn: instance-wise unlearning for pre-trained classifiers. arXiv:2301.11578 (2023)
3. Chen, C., Sun, F., Zhang, M., Ding, B.: Recommendation unlearning. In: WWW, pp. 2768–2777 (2022)
4. Chen, C., Zhang, M., Zhang, Y., Liu, Y., Ma, S.: Efficient neural matrix factorization without sampling for recommendation. TOIS **38**(2), 1–28 (2020)
5. Devlin, J., Chang, M.W., Lee, K., Toutanova, K.: Bert: pre-training of deep bidirectional transformers for language understanding. arXiv:1810.04805 (2018)
6. Fan, Z., et al.: Sequential recommendation via stochastic self-attention. In: WWW, pp. 2036–2047 (2022)
7. Ginart, A., Guan, M., Valiant, G., Zou, J.Y.: Making AI forget you: Data deletion in machine learning. NeurIPS **32** (2019)
8. Golatkar, A., Achille, A., Soatto, S.: Eternal sunshine of the spotless net: selective forgetting in deep networks. In: Proceedings of the IEEE/CVF Conference on Computer Vision and Pattern Recognition, pp. 9304–9312 (2020)

9. Hampel, F.R.: The influence curve and its role in robust estimation. J. Am. Stat. Assoc. **69**(346), 383–393 (1974)
10. He, R., McAuley, J.: Fusing similarity models with Markov chains for sparse sequential recommendation. In: ICDM, pp. 191–200. IEEE (2016)
11. He, X., Liao, L., Zhang, H., Nie, L., Hu, X., Chua, T.S.: Neural collaborative filtering. In: WWW, pp. 173–182 (2017)
12. Hidasi, B., Karatzoglou, A., Baltrunas, L., Tikk, D.: Session-based recommendations with recurrent neural networks. arXiv:1511.06939 (2015)
13. Izzo, Z., Smart, M.A., Chaudhuri, K., Zou, J.: Approximate data deletion from machine learning models. In: AISTATS, pp. 2008–2016. PMLR (2021)
14. Kang, W.C., McAuley, J.: Self-attentive sequential recommendation. In: ICDM, pp. 197–206. IEEE (2018)
15. Koren, Y., Bell, R., Volinsky, C.: Matrix factorization techniques for recommender systems. Computer **42**(8), 30–37 (2009)
16. Krichene, W., Rendle, S.: On sampled metrics for item recommendation. In: SIGKDD, pp. 1748–1757 (2020)
17. Li, Y., Chen, C., Zheng, X., Zhang, Y., Gong, B., Wang, J.: Selective and collaborative influence function for efficient recommendation unlearning. arXiv:2304.10199 (2023)
18. Li, Y., Zheng, X., Chen, C., Liu, J.: Making recommender systems forget: learning and unlearning for erasable recommendation. arXiv:2203.11491 (2022)
19. Liu, W., Wan, J., Wang, X., Zhang, W., Zhang, D., Li, H.: Forgetting fast in recommender systems. arXiv:2208.06875 (2022)
20. Lu, J., Zhang, Q., Zhang, G.: Recommender systems: advanced developments. World Scientific (2020)
21. Ma, C., Kang, P., Liu, X.: Hierarchical gating networks for sequential recommendation. In: SIGKDD, pp. 825–833 (2019)
22. Mehta, R., Pal, S., Singh, V., Ravi, S.N.: Deep unlearning via randomized conditionally independent hessians. In: CVPR, pp. 10422–10431 (2022)
23. Quadrana, M., Karatzoglou, A., Hidasi, B., Cremonesi, P.: Personalizing session-based recommendations with hierarchical recurrent neural networks. In: RecSys, pp. 130–137 (2017)
24. Rendle, S., Freudenthaler, C., Schmidt-Thieme, L.: Factorizing personalized Markov chains for next-basket recommendation. In: WWW, pp. 811–820 (2010)
25. Schelter, S., Grafberger, S., Dunning, T.: Hedgecut: maintaining randomised trees for low-latency machine unlearning. In: SIGMOD, pp. 1545–1557 (2021)
26. Sekhari, A., Acharya, J., Kamath, G., Suresh, A.T.: Remember what you want to forget: algorithms for machine unlearning. NeurIPS **34**, 18075–18086 (2021)
27. Shambour, Q., Lu, J.: An effective recommender system by unifying user and item trust information for B2B applications. J. Comput. Syst. Sci. **81**(7), 1110–1126 (2015)
28. Sun, F., et al.: BERT4Rec: Sequential recommendation with bidirectional encoder representations from transformer. In: CIKM, pp. 1441–1450 (2019)
29. Tang, J., Wang, K.: Personalized top-n sequential recommendation via convolutional sequence embedding. In: WSDM, pp. 565–573 (2018)
30. Tsai, C.H., Lin, C.Y., Lin, C.J.: Incremental and decremental training for linear classification. In: SIGKDD, pp. 343–352 (2014)
31. Vaswani, A., et al.: Attention is all you need. In: NeurIPS, pp. 5998–6008 (2017)
32. Wu, G., Hashemi, M., Srinivasa, C.: Puma: performance unchanged model augmentation for training data removal. In: AAAI, vol. 36, pp. 8675–8682 (2022)

33. Xu, M., Sun, J., Yang, X., Yao, K., Wang, C.: Netflix and forget: efficient and exact machine unlearning from bi-linear recommendations. arXiv:2302.06676 (2023)
34. Yan, H., Li, X., Guo, Z., Li, H., Li, F., Lin, X.: Arcane: an efficient architecture for exact machine unlearning. In: IJCAI, pp. 4006–4013 (2022)
35. Yuan, W., Yin, H., Wu, F., Zhang, S., He, T., Wang, H.: Federated unlearning for on-device recommendation. In: WSDM, pp. 393–401 (2023)
36. Zhang, Y., et al.: Recommendation unlearning via influence function. arXiv:2307.02147 (2023)

MPANet: Multi-scale Pyramid Attention Network for Collaborative Modeling Spatio-Temporal Patterns of Default Mode Network

Hang Yuan[1,2], Xiang Li[1,2], and Benzheng Wei[1,2]([✉])

[1] Center for Medical Artificial Intelligence, Shandong University of Traditional Chinese Medicine, Qingdao 266112, Shandong, China
2021111399@sdutcm.edu.cn, lixiang.vision@foxmail.com, wbz99@sina.com
[2] Qingdao Academy of Chinese Medical Sciences, Shandong University of Traditional Chinese Medicine, Qingdao 266112, Shandong, China

Abstract. The functional activity of the default mode network (DMN) in the resting state is complex and spontaneous. Modeling spatio-temporal patterns of DMN based on four-dimensional Resting-state functional Magnetic Resonance Imaging (Rs-fMRI) provides a basis for exploring spontaneous brain functional activities. However, how to utilize spatio-temporal features to complete the multi-level description of 4D Rs-fMRI with diverse characteristics in the shallow stage of the model and accurately characterize the DMN holistic spatio-temporal patterns remains challenging in the current DMN spatio-temporal patterns modeling. To this end, we propose a Multi-scale Pyramid Attention Network (MPANet) to focus on shallow features and model the spatio-temporal patterns of resting-state personalized DMN. Specifically, in the spatial stage, we design a multi-scale pyramid block in the shallow layer to expand the receptive field and extract granular information at different levels, which realize feature enhancement and guides the model to characterize the DMN spatial pattern. In the temporal stage, parallel guidance from spatial to the temporal pattern is achieved through the fast downsampling operation and introduction of multi-head attention blocks for a more effective fusion of spatio-temporal features. The results based on a publicly available dataset demonstrate that MPANet outperforms other state-of-the-art methods. This network presents a robust tool for modeling the spatio-temporal patterns of individuals with DMN, and its exceptional performance suggests promising potential for clinical applications.

Keywords: Shallow feature characterization · Holistic modeling · Spatio-temporal patterns · Default mode network

1 Introduction

As the core functional network of the most basic state of the brain, the default mode network (DMN) plays a role in regulating of human cognition and emo-

T. Liu et al. (Eds.): AI 2023, LNAI 14471, pp. 416–425, 2024.
https://doi.org/10.1007/978-981-99-8388-9_34

tion. Even the pathological process of some mental diseases is closely related to DMN. Therefore, using effective modeling methods based on four-dimensional Resting-state functional Magnetic Resonance Imaging (Rs-fMRI) to accurately model DMN spatio-temporal patterns to model DMN spatio-temporal patterns accurately is of great significance for exploring the mechanisms of functional brain networks and identifying potential disease factors [1]. However, due to the spontaneity and cooperativity of DMN, which is in a long-term activation state regardless of resting state or task state and interacts with other functional brain networks [2]. These all pose challenges for DMN modeling. How to design applicative methods to accurately locate and characterize the spatio-temporal patterns of DMN has become an urgent problem to be addressed in the research field of DMN modeling [3]. How to design applicative methods to locate and accurately characterize the spatio-temporal patterns of DMN has become an urgent problem to address in the research field of DMN modeling [3].

4D Rs-fMRI data contains 1D temporal series and 3D spatial information, which was difficult to process in early studies due to computational limitations. In order to simplify the data, researchers converted the 4D spatio-temporal data into a two-dimensional pattern of 'time × space', and then modeled the DMN through the selected prior seed points through a model-driven approach. The representative methods include cross-correlation analysis (CCA) [4] and general linear model (GLM) [5]. However, these methods require prior brain regions as seed points, and the accuracy of the characterized individual brain is insufficient. Later, the practice of mathematical analysis was applied to the modeling of DMN. Examples of these methods include principal component analysis (PCA) [6], independent component analysis (ICA) [7] and sparse representation (SR) [8]. They decompose or sparse the transformed spatio-temporal 2D matrix, and finally decompose the corresponding DMN spatio-temporal patterns. It should be noted that firstly, such methods still simplify the four-dimensional spatio-temporal data at the data level, which may ignore the holistic information of the spatio-temporal characteristics of DMN. Secondly, at the method level, such methods are often based on the hypothesis that each brain region is independent from each other to separate the functional brain regions, which may cause the wrong decomposition of DMN, so that the accuracy and interpretability of the results are insufficient.

With the continuous development of deep learning technology, it is possible to construct the spatio-temporal holistic characteristics of 4D fMRI [9]. Initially, researchers still followed the previous research approach by transforming spatio-temporal data into temporal and spatial patterns [10]. Then, deep learning techniques such as convolutional autoencoder (CAE) [11] and restricted Boltzmann machine (RBM) [12] are used to model the dynamic temporal pattern, and the corresponding spatial pattern is obtained under the guidance of the temporal pattern. Although such methods have improved the results, they still ignore the overall characteristics of the spatio-temporal patterns of fMRI, and fail to achieve the mutual guidance and holistic effect of spatio-temporal information. Subsequently, ST-CNN [13] and Multi-Head GAGNN [14] are developed to construct the spatio-temporal patterns of fMRI, which solves the problem of comprehensive utilization of spatio-temporal characteristics of fMRI, and has a high improvement in performance. However, these methods still model spatio-temporal patterns in different stages, failing to realize the collaborative optimization and holistic modeling of spatio-temporal patterns representation process, and the extraction strategy of spatio-temporal features can also be improved. More importantly, some existing methods are designed to focus more on the deep feature extraction of the network. While one this can make the model lightweight, the other it generally makes the result look better. Nevertheless, we need to consider that 4D fMRI as spatio-temporal data contains rich multivariate features, and the features extracted in the deep layer may have biases and errors learned by the model in the shallow stage. So is the perfect result really as perfect as it seems? We can't help but ask: is there a way for the model to learn the rich features of 4D fMRI at a shallow stage, and at the same time, the model can better distinguish the overlapping functional brain network features, so that the results are more similar to the real world?

To address the above limitations and problems, we propose a novel Multi-scale Pyramid Attention Network (MPANet) to achieve precise characterization of DMN spatio-temporal patterns and spatio-temporal collaborative optimization modeling. In the spatial phase, in order to better learn the global features and accurately distinguish the resting-state overlapping brain network features, we introduce multi-scale pyramid convolution to expand the shallow stage feature receptive field. Meanwhile, the coarse-grained features of different levels are extracted to make up for the feature loss in the down-sampling process, so as to realize the effective modeling of personalized DMN. In the temporal phase, we design a multi-head attention block to achieve parallel guidance of spatial pattern information without affecting the holistic spatio-temporal features, thereby characterizing the temporal patterns of personalized DMN. Based on the selected Rs-fMRI data and compared with the most advanced methods, the proposed MPANet shows superior ability in simultaneously optimizing the construction of individual DMN spatio-temporal patterns.

2 Method

The MPANet is divided into two stages: time and space, and the specific structure of the network is illustrated in Fig. 1. 4D Rs-fMRI data is the overall input to the network, and the DMN spatio-temporal patterns are the final output. The network firstly performs the spatial stage (see Fig. 1.(a)) training, and then guides the construction of the temporal stage (see Fig. 1.(b)) after obtaining the spatial pattern with high credibility. Finally, the holistic spatio-temporal patterns of the individual DMN is the output of the MPANet.

2.1 Spatial Stage of Multi-scale Pyramid Attention Network

The feature is miscellaneous in the shallow coding stage, and it is laborious to extract ponderable information. However, this also means that if the personalized

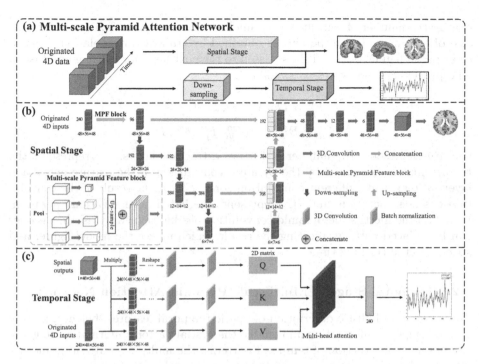

Fig. 1. The (a) is an overview of MPANet framework. The detailed model architecture of MPANet consists of (b) spatial stage and (c) temporal stage.

detailed features of DMN can be identified in the shallow stage, the overlapping brain network can be better distinguished and the DMN can be characterized. Therefore, Therefore, in order to focus on the appropriate features from the messy shallow information, and realize the multi-level description to strengthen the correctness of the extracted features. We design a multi-scale pyramid convolutional block to expand the receptive field as much as possible in the shallow stage of encoding, and perceive the global information at different levels. The coarse-grained and fine-grained feature information of functional brain regions was found in the resting state, and then the features of overlapping brain regions were distinguished to construct refined individual DMN.

Specifically, Multi-scale pyramid convolutional block fuses four different pyramid scale features, with each layer scale corresponding to the subsequent downsampling scale. On one hand, by designing four different scales, information of varying dimensions can be aggregated and strengthened, allowing for subsequent convolution to quickly detect DMN features. On the other hand, the concept of a multi-scaled pyramid design can reduce feature loss during the downsampling process and improve the accuracy of the model. The convolution kernel size is set to $1 \times 1 \times 1$, and the outputs size is set as $6 \times 7 \times 6$, $12 \times 14 \times 12$, $24 \times 28 \times 24$, $48 \times 56 \times 48$ to extract different levels of coarse and fine grained features respectively. Each level of the pyramid is followed by a $1 \times 1 \times 1$ convolution to change the number of convolution channels at different levels to 96, and then, the size of the image before pooling is obtained by trilinear interpolation. Finally, after batch regularization, it was passed to the next layer for splicing and fusion. The process can be expressed as follows.

$$F_{p_i}^n = ConV_{p_i}^n(x), i = 1, 2, 3, 4 \tag{1}$$

$$F_{out}^n = BN(Cat(F_{p_1}^n, F_{p_2}^n, F_{p_3}^n, F_{p_4}^n)) \tag{2}$$

where n represents the number of feature extraction layers, $F_{p_i}^n$ represents the feature matrix generated by the pyramid convolution of the feature extraction layer, p_i represents different convolution cores, F_{out}^n is the result of the convolution cascade operation, and BN represents the batch normalization operation. Based on the multi-scale pyramid convolution block designed by us, the decoder can be reduced without affecting the result, which makes the model lighter and speeds up the training time.

2.2   Temporal Stage of Multi-scale Pyramid Attention Network

To address the holistic description of spatio-temporal features, better integrate spatio-temporal information and enable the spatial pattern information to guide the parallel construction of temporal patterns, the multi-head attention mechanism is selected as the main module in the temporal part.

The module still follows the basic definition of an attention mechanism:

$$Attention = Soft\max(\frac{QK^T}{\sqrt{M}}) \times V \tag{3}$$

where Q is the temporal feature information matrix under the guidance of spatial pattern, K and V represent the feature matrix extracted from Rs-fMRI data respectively. T is the transpose operation and M is the characteristic number of the K matrix. Before obtaining the QKV of the temporal phase, we first perform the downsampling operation on the input, in order to adjust the weight ratio of spatial and temporal information to avoid the wrong attention problem of temporal attention, and remove redundant spatial feature redundancy to speed up the training. This method is defined as follow:

$$M_{reshape} = ConV_{down}(x) \tag{4}$$

where $M_{reshape}$ is the reshape operation on the output matrix, and $ConV_{down}$ and is down-sampling 3D convolution. After three times of down-sampling, the 4D matrix of $d \times h \times w \times T$ is sampled into a 2D matrix of $\left[\frac{d}{8} \times \frac{h}{8} \times \frac{w}{8}\right] \times T$. After down-sampling, the output of the spatial phase is used to guide the generation of Q, and KV is obtained based on the input data of the temporal phase. We apply attention mechanism on the extracted temporal feature information to obtain a two-dimensional matrix of size $[252] \times T$, and then perform average operation to output the DMN temporal pattern of size $1 \times T$.

3   Experimental

3.1   Data Preparation and Preprocessing

We conduct experiments on healthy Rs-fMRI data from the 1000 Functional Connectomes Project in the NeuroImaging Tools & Resources Collaboratory (NITRC) to verify the effectiveness of MPANet. After screening, 176 cases of data finally met the experimental standards. The images were obtained using an echo-planar imaging sequence with the following parameters: 33 axial slices, thickness/gap = 3/0.6 mm, in-plane resolution = 64 × 64, TR = 2000 ms, TE = 30 ms, flip angle = 90°, FOV = 200 × 200 mm. All data are preprocessed using DPABI, including time layer correction, head motion correction, EPI template space standardization and strength normalization. To reduce excessive spatial redundancy before input to the model, the temporal layer remains unchanged and the spatial layer will be further downsampled to a spatial size of 48 × 56 × 48. We selected the most representative DMN resting-state network [15] as the brain template, and trained DMN labels separately for each subject by dictionary learning and sparse coding.

3.2   Experimental Design of Multi-scale Pyramid Attention Network

We selected 'overlap rate' and 'Pearson Correlation' (PC) [16] as the evaluation criteria of the model. Among them, the 'overlap rate' is used to evaluate the spatial pattern similarity between the extracted spatial DMN feature patterns and the labels, and the PC is used to calculate the temporal similarity between the temporal outputs and the labels. The formula for 'overlap rate' and PC are given below.

$$overlap\ rate = \frac{sum(\min(S_a, S_b))}{(sum(S_a) + (sum(S_b))/2} \tag{5}$$

$$PC = \frac{t\sum_{i=1}^{t} m_i l_i - \sum_{i=1}^{t} m_i \sum_{i=1}^{t} l_i}{\sqrt{\left(t\sum_{i=1}^{t} m_i^2 - \left(\sum_{i=1}^{t} m_i\right)^2\right)\left(t\sum_{i=1}^{t} l_i^2 - \left(\sum_{i=1}^{t} l_i\right)^2\right)}} \tag{6}$$

Among them, S_a represents the spatial output pattern of MPANet, while S_b represents the training label for spatial patterns. m and l are temporal patterns of the modeled pattern and the training label, respectively, t is the length of the time series. The learning rate is set to 0.0001, and Adam is employed as the optimizer of the model. The total number of training iterations is set to 200, with the temporal stage commencing at round 150 and incorporating spatial information for guidance. The loss functions for those two stages are defined as '1 − overlap rate' and '1 − Pearson Correlation'.

3.3   Verification of Multi-scale Pyramid Attention Network

To verify the effectiveness of MPANet, firstly, we calculate the spatio-temporal similarity between the modeled DMN spatio-temporal patterns and the training label patterns. Second, we also calculate the spatial similarity between the constructed pattern and the standard RSN spatial template. Third, we compare MPANet with SR and Multi-Head GAGNN. In addition, we compare the proposed model architecture with three other different model designs. The first is to discard the multi-scale pyramid blocks. The second is to restore the reduced layer part of the decoding stage of the U-net structure. The last is to transfer the Multi-scale pyramid block to the deep layers of the encoding stage.

4   Result

4.1   Results of the Modeled DMN Spatio-Temporal Patterns

Table 1. Comparison of DMN spatio-temporal patterns results of different methods

Case	overlap rate with label		overlap rate with DMN template			PC
	GAGNN	MPA Net	SR	GAGNN	MPANet	
mean	0.178±0.063	0.192±0.071	0.462±0.073	0.425±0.011	0.546±0.133	0.762±0.121
sub1	0.200	0.253	0.580	0.425	0.596	0.387
sub2	0.125	0.180	0.413	0.425	0.402	0.846
sub3	0.248	0.293	0.241	0.425	0.644	0.693

We applied the trained model to the test data, and compared the results with SR and Multi-Head GAGNN. Table 1 reports the specific results and we can find that, compared with Multi-Head GAGNN, the spatial overlap rate between Multi-scale pyramid attention Network and labels is better, and the spatial pattern is more similar. After calculating and comparing the overlap rate with the standard RSN template, the designed method also reveals better performance compared to SR and Multi-Head GAGNN. In terms of temporal pattern, the MPANet also achieves satisfactory results. Furthermore, we randomly selected a test subject for visualization to more intuitively demonstrate the superiority of our method. From the results illustrated in Fig. 2, it can be observed that the personalized DMN spatio-temporal patterns modeled has holistic excellence. Firstly, it has less noise points, which indicates that it better separates the overlapping brain networks in the resting state and has better integrity of functional network areas. Secondly, with the mutual guidance of spatio-temporal patterns, the DMN activity characterized by MPANet is satisfactory.

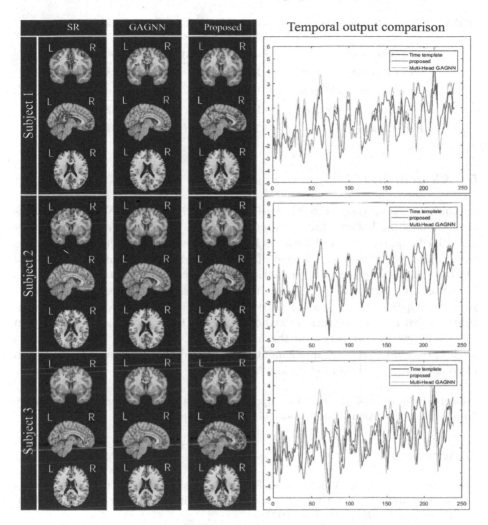

Fig. 2. Spatio-temporal modeling and comparison results of DMN in Rs-fMRI of random three subjects.

4.2 Results of Different Model Architectures

In order to illustrate the effectiveness of our designed framework architecture, we show the spatio-temporal patterns similarity modeled under different model architectures. As shown in Table 2, our designed architecture achieves the best DMN modeling capability, this demonstrates the effectiveness of the multi-scale pyramid block for DMN modeling and the feasibility of the model to characterize 4D fMRI as a whole in the shallow stage.

Table 2. Spatio-temporal patterns results of DMN for different model architectures.

Ablation studies	Spatial similarity	Temporal similarity
Shallow pyramid-removing	0.175±0.069	0.446±0.628
Deep pyramid-addition	0.174±0.067	0.493±0.160
Decoding layer restore	0.189±0.070	**0.768±0.135**
Proposed	**0.192±0.071**	0.762±0.121

5   Conclusion

In this paper, we propose a Multi-scale Pyramid Attention Network to accurately model the holistic spatio-temporal patterns of DMN by extracting multi-granularity spatio-temporal features in the shallow layers of the model. To the best of our knowledge, this is one of the first works to implement the mutual guidance and collaborative optimization of spatio-temporal patterns. The results based on 176 cases of Rs-fMRI show that the designed model has higher performance compared with other advanced methods, which also indicates the feasibility of feature representation extraction for 4D fMRI in the shallow stage. In the future, we intend to extend this work to the DMN identification of patients with early mild cognitive impairment, and expand the modeling objects to ten classical functional brain networks.

Acknowledgements. This work is supported by the Qingdao Science and Technology Huimin Demonstration Project under Grant No.23-2-8-smjk-2-nsh.

References

1. Chang, Z., Wang, X., Wu, Y., et al.: Segregation, integration and balance in resting-state brain functional networks associated with bipolar disorder symptoms. Hum. Brain Mapp. **44**(2), 599–611 (2023)
2. Malotaux, V., Dricot, L., Quenon, L., et al.: Default-mode network connectivity changes during the progression toward Alzheimer's dementia: a longitudinal functional magnetic resonance imaging study. Brain Connect. **13**(5), 287–296 (2023)
3. Jiang, X., Yan, J., Zhao, Y., et al.: Characterizing functional brain networks via spatio-temporal attention 4D convolutional neural networks (STA-4DCNNs). Neural Netw. **158**, 99–110 (2023)
4. Li, K., Guo, L., Nie, J., et al.: Review of methods for functional brain connectivity detection using fMRI. Comput. Med. Imaging Graph. **33**(2), 131–139 (2009)
5. Xu, J., Potenza, M.N., Calhoun, V.D., et al.: Large-scale functional network overlap is a general property of brain functional organization: reconciling inconsistent fMRI findings from general-linear-model-based analyses. Neurosci. Biobehav. Rev. **71**, 83–100 (2016)
6. Zhou, Z., Chen, Y., Ding, M., et al.: Analyzing brain networks with PCA and conditional Granger causality. Hum. Brain Mapp. **30**(7), 2197–2206 (2009)
7. Du, Y., Fan, Y.: Group information guided ICA for fMRI data analysis. Neuroimage **69**, 157–197 (2013)

8. Zhang, W., Lv, J., Li, X., et al.: Experimental comparisons of sparse dictionary learning and independent component analysis for brain network inference from fMRI data. IEEE Trans. Biomed. Eng. **66**(1), 289–299 (2018)
9. Li, H., Srinivasan, D., Zhuo, C., et al.: Computing personalized brain functional networks from fMRI using self-supervised deep learning. Med. Image Anal. **85**, 102756 (2023)
10. Lv, J., Jiang, X., Li, X., et al.: Sparse representation of whole-brain fMRI signals for identification of functional networks. Med. Image Anal. **20**(1), 112–134 (2015)
11. Huang, H., Hu, X., Zhao, Y., et al.: Modeling task fMRI data via deep convolutional autoencoder. IEEE Trans. Med. Imaging **37**(7), 1551–1561 (2017)
12. Zhang, W., Zhao, S., Hu, X., et al.: Hierarchical organization of functional brain networks revealed by hybrid spatiotemporal deep learning. Brain Connect. **10**(2), 72–82 (2020)
13. Zhao, Y., Li, X., Huang, H., et al.: Four-dimensional modeling of fMRI data via spatio-temporal convolutional neural networks (ST-CNNs). IEEE Trans. Cogn. Dev. Syst. **12**(3), 451–460 (2019)
14. Yan, J., Chen, Y., Xiao, Z., et al.: Modeling spatio-temporal patterns of holistic functional brain networks via multi-head guided attention graph neural networks (Multi-Head GAGNNs). Med. Image Anal. **80**, 102518 (2022)
15. Smith, S.M., Fox, P.T., Miller, K.L., et al.: Correspondence of the brain's functional architecture during activation and rest. Proc. Natl. Acad. Sci. **106**(31), 13040–13045 (2009)
16. Adler, J., Parmryd, I.: Quantifying colocalization by correlation: the Pearson correlation coefficient is superior to the Mander's overlap coefficient. Cytometry A **77**(8), 733–742 (2010)

Optimization

Dynamic Landscape Analysis for Constrained Multiobjective Optimization Problems

Hanan Alsouly[1,2,3]($\boxtimes$), Michael Kirley[1,2], and Mario Andrés Muñoz[1,2]

[1] School of Computing and Information Systems, The University of Melbourne, Melbourne, Australia
halsouly@student.unimelb.edu.au, {mkirley,munoz.m}@unimelb.edu.au
[2] ARC Centre in Optimisation Technologies, Integrated Methodologies and Applications (OPTIMA), Melbourne, Australia
[3] College of Computing and Information Sciences, Imam Mohammad Ibn Saud Islamic University, Riyadh, Saudi Arabia

Abstract. Landscape analysis is a data-driven approach that involves sampling the search space of an optimization problem to generate a range of statistical features. These features serve to characterize the 'problem difficulty'. However, the computational costs associated with offline independent sampling can be excessive, and this approach often overlooks valuable information accumulated by the optimization algorithm. This paper aims to expand our understanding of landscape analysis in the domain of black-box constrained multiobjective optimization problems. We demonstrate the potential of leveraging optimization algorithm trajectories to measure landscape features. Our findings underscore the significance of utilizing landscape features as a means to approximate algorithm performance, particularly in cases involving new instances lacking a known reference set. Ultimately, our goal is to employ landscape analysis to dynamically adapt algorithm constraint handling techniques.

Keywords: Constrained multiobjective optimization · Adaptive landscape analysis · Instance space analysis · Evolutionary algorithms · Constraint handling technique

1 Introduction

Many landscape analysis features have been proposed to quantify the characteristics of a problem landscape [7]. They are often used to understand complex problems, predict algorithm performance, and automate algorithm selection. In addition, they can be incorporated into the search procedure of algorithms to enhance their effectiveness, leading to what is known as *landscape-aware algorithms*. In

This research was partially funded by the Australian Government through the Australian Research Council Industrial Transformation Training Centre in Optimisation Technologies, Integrated Methodologies, and Applications (OPTIMA), Project ID IC200100009.

the constrained multiobjective optimization problems (CMOPs) domain, limited research has explored the landscape characteristics [1,11,15]. Moreover, the proposed feature extraction methods in CMOPs suffer from the following limitations: (1) the sampling strategy is independent of the optimization process; (2) they do not consider the dynamics of the search process, *i.e.*, how the local landscape information looks like during the optimization; and (3) some features are computationally expensive.

This paper takes initial exploratory steps into understanding the dynamics of the search process in landscape analysis within the domain of CMOPs. Specifically, we focus on how a constrained multiobjective evolutionary algorithm (CMOEA) explores the local landscape information of CMOPs and how the features subsequently evolve—a concept we refer to as *dynamic landscape analysis*. Building upon recent work by Alsouly *et al.*, [1], which examined the relationship between the performance of CMOEAs and the characteristics of CMOPs using Instance Space Analysis, we utilize the instance space to investigate the trajectory of features over time as observed by the optimization algorithm. Preliminary results demonstrate that monitoring changes in landscape features over time yields valuable insights into the performance of the optimization algorithm. This study provides the foundation for the design of landscape-aware algorithms, enabling the development of more effective algorithms in the context of CMOPs.

The paper is organized as follows: Sect. 2 provides a comprehensive review of related work. Section 3 outlines our methodology. Specifically, we describe the instance space, the sampling methodology and the experimental design. The results are presented and discussed in Sect. 4. Finally, Sect. 5 concludes the paper.

2   Related Work

In landscape analysis, a set of statistical features is used to quantify landscape characteristics of a problem such as ruggedness, evolvability, and variable scaling [8]. The feature values can be approximated from a sample of evaluated solutions. In the context of CMOPs, Picard and Schiffmann [11] extended features from the single constrained optimization problems domain [6] in order to measure the 'disjointedness' of the feasible region, and proposed two features to quantify the relationship between the objectives and constraints. Those features rely on independent uniform samplings and progressive random walks. Vodopija *et al.*, [15] proposed a set of features to quantify the violation landscape's multimodality and local structure, as well as the correlation between the objectives and constraints. To produce samples they used Latin hypercube sampling, simple random walks and adaptive walks. Alsouly *et al.*, [1] proposed a set of features to quantify several characteristics of the violation landscapes and multiobjective-violation landscape as well as the relationship between the true, constrained, Pareto Front (PF) and the unconstrained PF (UPF). Those features values are obtained from uniform samplings and simple random walks. Each of the landscape analysis studies in CMOPs [1,11,15] approximates the feature values by using *offline sampling*, which means that the samples are generated independently of the optimization process.

In a step towards *dynamic landscape analysis*, a method to remove bias from a sample obtained during an algorithm run was proposed by Muñoz *et al.*, [9]. This sampling method yielded similar features to offline sampling. Nevertheless, the tested sample size was too expensive to be used for algorithm selection. Janković and Doerr [5] analyzed how the local features of a single-objective optimization problem change as CMA-ES algorithm's population changes, and compared the local features with the global features. Once a target objective value had been reached, 2000 sample points were generated around the current reached population to compute the features at that moment. They found that the fitness landscape as seen by the algorithm differs significantly from that seen by offline sampling. Moreover, there was no clear trend in how features evolve. Meanwhile, Wang *et al.*, [16] designed a dynamic measure of population evolvability. This feature considers a combination of population-based algorithms and problem properties without using extra sample points. The paper demonstrated how the proposed feature can be used to select an algorithm. However, the algorithm selection model had a drawback: it required each algorithm to consume 20% of the allocated budget to measure the proposed dynamic feature.

3   Dynamic Landscape Analysis

We analyze the dynamic landscape information of CMOPs using the Instance Space Analysis (ISA) methodology [12]. ISA is used for assessing the difficulty of a set of instances of a problem for a set of algorithms by visualizing the problem's instances in a 2D plane. Each instance is represented as a point, allowing for a visual comparison of their similarities and differences in terms of characteristics and algorithm performance. The ISA process involves the following:

1. collecting a meta-data set consisting of instances' features and performance measures of a group of algorithms on those instances;
2. selecting a set of unique features that are correlated and predictive of algorithm performance;
3. then, using a tailored dimensionality reduction method to project the meta-data into a 2D plane, called the *instance space*.

Our goal is to visualize the changes in a problem instance's features within the instance space and investigate their relationship with the algorithm's performance. We first explain how the 'original' instance space was created. Then, we present the proposed sampling method that utilizes the optimization algorithm trajectory. Finally, we provide details about the experimental setup.

3.1   The Instance Space of CMOPs by Using Offline Sampling

This paper builds upon the first ISA of CMOPs presented in [1]. Alsouly *et al.*, collected meta-data on 443 CMOPs and 15 CMOEAs. The meta-data consisted of 80 features that have been extracted by using two offline sampling techniques- random sampling to collect global features and random walks to sample the neighbourhood of solutions. Then, the features were processed using the Yeo-Johnson power transform method. Features that were not strongly correlated with algorithm performance were filtered out, leaving a subset of 14 uncorrelated (unique) features that could predict algorithm performance and show similarities and differences between instances. This subset of features was used to construct the 'original' instance space. Each instance was represented as a point in this space, and its location was defined by the projection matrix in Eq. 1, which we use here to demonstrate how instance features change as the algorithm progresses. The details of the features can be found in Table 1.

$$
\begin{bmatrix} z_1 \\ z_2 \end{bmatrix} = \begin{bmatrix} 0.2559 & 0.1348 \\ -0.2469 & -0.1649 \\ -0.0257 & -0.2703 \\ 0.2938 & -0.2278 \\ -0.2148 & -0.1338 \\ -0.1935 & -0.2210 \\ -0.1651 & 0.2998 \\ -0.2150 & 0.3137 \\ 0.3067 & 0.1382 \\ 0.0709 & 0.3047 \\ 0.2032 & -0.0515 \\ 0.1436 & 0.2869 \\ 0.1940 & 0.1154 \\ -0.0508 & -0.2466 \end{bmatrix}^{\mathsf{T}} \begin{bmatrix} corr\_cf \\ f\_mdl\_r2 \\ dist\_c\_corr \\ min\_cv \\ bhv\_avg\_rws \\ skew\_rnge \\ piz\_ob\_min \\ ps\_dist\_iqr\_mean \\ dist\_c\_dist\_x\_avg\_rws \\ cpo\_upo\_n \\ cv\_range\_coeff \\ corr\_obj \\ dist\_f\_dist\_x\_avg\_rws \\ cv\_mdl\_r2 \end{bmatrix}
\tag{1}
$$

3.2 Dynamic Sampling

A dynamic sampling method was used to sample the search space. The optimization algorithm was periodically halted every 10^3 function evaluation, and the most recent solutions were used as sample points. To ensure diversity, only unique solutions were retained from each sample set. To form the neighbourhoods, for each solution, we define its k closest individuals from the sample set as neighbours, where $k = 5$. Subsequently, the landscape analysis features were calculated. To facilitate consistent processing, the Yeo-Johnson power transform method was utilized with the same parameters that have been used in [1]. The feature collection and calculation were implemented using MATLAB [4], while preprocessing was carried out utilizing Scikit-learn [10].

3.3 Experimental Design

Benchmark Instances. Instead of using all 443 bi-objective instances from [1], which will be difficult to visualize, we have selected 18 instances randomly from across the space. This approach provides diverse characteristics. The selected benchmark instances are: C1-DTLZ1, C1-DTLZ3, CF7, DASCMOP3-1, DASCMOP6-1, DC2-DTLZ3, LIRCMOP1, LIRCMOP3, MW1, MW2, MW5, MW6, MW7 and MW9, where $n = 10$, DASCMOP1-11, DASCMOP6-12, LIRCMOP10 and LIRCMOP2 where $n = 5$. Figure 1 illustrates the 'original' 2D CMOP instance space from [1]. The selected instances are highlighted in bold.

Table 1. The subset of features used to characterize the landscape of CMOPs in this work. A full list of features and detailed description can be found in [1]

Feature	Description
corr_cf	Constraint violations and solutions' ranks correlation
f_mdl_r2	Adjusted coefficient of determination of a linear regression model for variables and unconstrained solutions' ranks
dist_c_corr	Violation-distance correlation
min_cv	Minimum of constraints violations
bhv_avg_rws	Average hypervolume-value of neighborhood's non-dominated solutions
skew_rnge	Range of objectives skewness
piz_ob_min	Minimum proportion of solutions in ideal zone per objectives
ps_dist_iqr_mean	Average difference between 75th and 25th percentiles of distances across PS
dist_c_dist_x_avg_rws	Ratio of average distance from neighbours in the constraints space to average distance from neighbours in the variable space
cpo_upo_n	Proportion of the *PF* to the unconstrained *PF* (*UPF*)
cv_range_coeff	Difference between maximum and minimum of the absolute value of the linear model coefficients
corr_obj	Correlation between objective values
dist_f_dist_x_avg_rws	Ratio of average distance from neighbours in the objective space to average distance from neighbours in the variable space
cv_mdl_r2	Adjusted coefficient of determination of a linear regression model for variables and violations

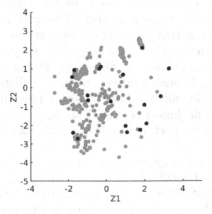

Fig. 1. The distribution of the instances in 2D space based on offline sampling and the projection matrix in Eq. 1 (adapted from [1]). The instances used in this work are highlighted in bold.

Algorithms. We have considered two algorithms, NSGAII [3] and CCMO [14], to collect the data during the run. The algorithms were selected based on their popularity and performance. NSGAII is a baseline CMOEA that uses the principle of constraint dominance to prioritize the constraints to direct the search toward a feasible area. However, due to the bias toward infeasible solutions, it may get trapped in a small part of the feasible region. Whilst, CCMO aims to balance objectives and constraints by using two populations, one to solve the original CMOP and another to solve a helper problem derived from the original one. We have used the default parameters in PlatEMO [13]. The population size was set to be 200 with all instances, while the total number of evaluations was set to 2×10^4. For each algorithm and each instance, 30 independent runs were conducted.

Performance Indicator. The most commonly used performance indicators when optimizing CMOPs are the hypervolume (HV), and the inverted generational distance (IGD^+), which evaluate the convergence and diversity of the approximated PF by using a reference point or set [2]. In this particular case of benchmark instances and algorithms, previous research has shown that there is no significant difference between the two metrics [1]. Consequently, we focus solely on the HV in this study. To normalize the approximated HV, we scale it to the range of [0,1] by dividing it by the true HV. A larger HV value indicates a better approximation of the true PF.

4   Results and Discussion

4.1   Feature Time Series Analysis: *piz_ob_min*

Our analysis starts with an illustration of the trajectory of one representative feature over time. Due to space limitations, we have showcased only *piz_ob_min*. This feature was selected because it is easily illustrated in a 2D space.

piz_ob_min is a metric that quantifies a characteristic in landscapes where fitness and violation functions are combined. It measures the minimum proportion of solutions in the ideal zone per objective, offering insights into the size of basins of attraction [6]. In such landscapes, the distribution of fitness and violation values can provide valuable information. Large concentrations of points in the lower left quadrant, known as the ideal zone, may indicate a larger basin of attraction, while few proportion of points in this zone may indicate a narrow basin.

Figure 2 illustrates an example of how *piz_ob_min* evolves during the search process, specifically for C1-DTLZ1 instance. The initial sample (first column), which comprises the first 10^3 solutions, reveals the presence of relatively large basins of attraction. However, as the optimization process progresses, the proportion of points in the ideal zone gradually diminishes, and the distribution of the fitness and the violation values shifts towards the upper right quadrant. This shift indicates that both the CCMO and NSGAII algorithms predominantly

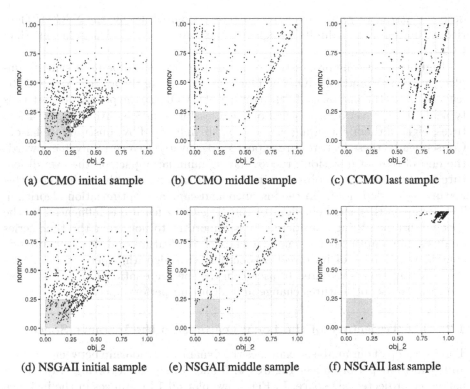

Fig. 2. Scatter plots to demonstrate the dynamics of *piz_ob_min* feature for C1-DTLZ1 instance. The normalized values of the fitness function against constraint violation for each sample point are projected. The highlighted square shows the concentrations of points in the ideal zone, which is defined by the minimum 25% of the lower left area per objectives. Each column represents a sample at a point in time. A decline in the proportion of points in the ideal zone occurs over time, along with a shift in the distribution of fitness values towards the upper right area.

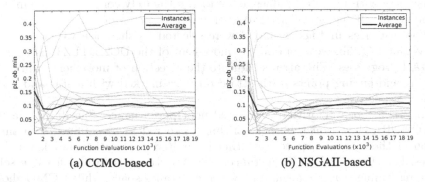

Fig. 3. The time series for *piz_ob_min* feature for 18 instances over time. The average value of the feature across the instances is represented by the black line, despite the presence of outliers.

detect a narrow or semi-narrow feasible area. Comparatively, the NSGAII algorithm (bottom row) exhibits a slightly larger proportion of ideal solutions than CCMO.

Figure 3 shows the progression of *piz_ob_min* over time for 18 instances sampled by NSGAII and CCMO. The black line represents the average value of the feature across the instances. In the context of CMOEAs, the primary objective is typically to discover a large set of optimal feasible solutions. Among the 14 features, eight of them, including *piz_ob_min*, are influenced by constraint violation. Consequently, these features are expected to converge to a certain value towards the end of the optimization process, having minimal impact on the overall feature vector. However, as observed in the *piz_ob_min* figures, the evolution of this feature varies depending on the instance landscape and optimization algorithm. This suggests that the behaviour of the *piz_ob_min* feature is influenced by the specific instance being considered. It is important to note that the time series analysis only captures the general trends of the feature and does not provide conclusive evidence of its direct association with algorithm performance. Further investigation is required to ascertain the specific relationship between how the local landscape features change and algorithm performance.

4.2   The Trajectory of the Local Features on the Instance Space

The next step of our analysis examines the dynamic relationship between features and algorithm performance. Figures 4, 5 and 6 map the evolving trajectories using the projected 2D space. In Fig. 4, we plot all 18 instances in the instance space and observe how the *HV* changes. We then 'zoom in' and provide a closer inspection of six instances in Figs. 5 and 6. Finally, we introduce two metrics that capture the relationship between algorithm performance and instance location.

Figure 4 displays the trajectory of instances in the instance space, accompanied by their corresponding *HV* values. Each marker represents an instance, and the colour density reflects the search stage or the *HV* value. Darker colours indicate the end of the search or high *HV*. Our analysis reveals that the majority of instances with high *HV* values, close to 1, ultimately converge to the middle of the upper left region, referred to as the *convergence zone*.

The instances in Figs. 5 and 6 were selected to showcase various scenarios. Within Fig. 5b, we zoom on the movement of the DC2_DTLZ3 instance as NSGAII progresses. The arrows indicate the direction of movement, while the starting and ending points of the trajectory are marked with a star and a red square, respectively.

In Fig. 5, both algorithms exhibited poor performance. Specifically, in the DASCMOP1_11 instance, both algorithms appeared to be confined to a small region of the instance space, resulting in limited exploration and lower *HV* values. For the DC2_DTLZ3 instance, NSGAII struggles to find feasible solutions, remaining stuck in a small part of the instance space, while CCMO shows slow progress towards the convergence zone. In the C1_DTLZ1 instance, both algorithms achieve less than 1% of the true *HV*. They encountered challenges in driving the feature vector towards the convergence zone, although NSGAII

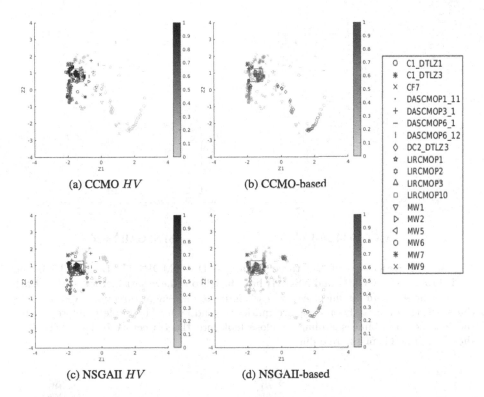

Fig. 4. The trajectory of 18 instances in the instance space. The algorithm data is used every 10^3 function evaluation for features and performance data. Each marker represents an instance. In the first column, the colour density corresponds to the normalized HV value, the darker the colour the better the HV. While in the second column, the colour density corresponds to the search stage, the colour becomes darker as the algorithm progresses. The convergence zone is marked with a red square. (Color figure online)

demonstrated some gradual movement towards the end of the search, leading to slightly improved performance. Moving to Fig. 6, both algorithms exhibit similar exploration trends and converge to a similar location. It is worth noting that the initial populations' approximated feature vectors resemble those estimated by random sampling. In the MW1 and MW5 instances, both algorithms achieve high performance, exceeding 94% of the true HV. As for the CF7 instance, NSGAII slightly outperforms CCMO, approximating around 64% of the true HV. The footprints further indicate that neither algorithm fully converges to the convergence zone.

In Fig. 7, we introduce the *distance metric*, measuring the Euclidean distance between an instance's current location and the estimated location of the convergence zone, with the centre located at $(-1.4, 1)$ and the boundaries at plus or minus 0.2. By calculating the correlation coefficient between the distance metric

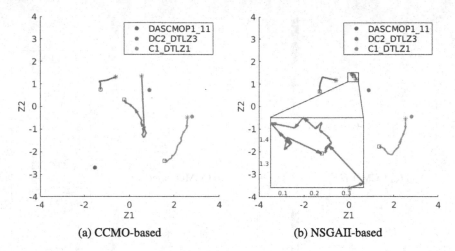

Fig. 5. The trajectory of the three instances: DASCMOP1_11, DC2_DTLZ3, and C1_DTLZ1 where CCMO and NSGAII have failed to achieve satisfactory results. Each colour represents an instance. Arrowheads indicate movement direction. A star marks the starting point, and a red square marks the endpoint. The circle point represents the instance by offline sampling. A close look into the trajectory of DC2_DTLZ3 is shown in (b). (Color figure online)

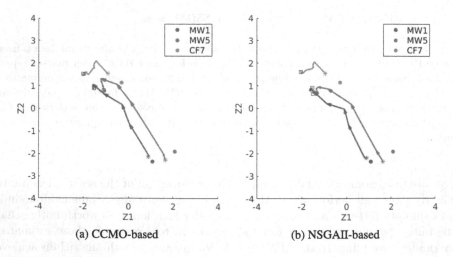

Fig. 6. The trajectory of the three instances: MW1, MW5, and CF7 where CCMO and NSGAII have achieved high performance.

and the HV, we gain valuable insights into the relationship between proximity to the convergence zone and algorithm performance. The correlation coefficients are -0.6472 for CCMO and -0.5981 for NSGAII, indicating that the closer an instance is to the convergence zone, the greater the chances of having better

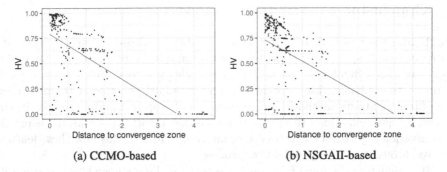

Fig. 7. Scatter plot showing the relationship between the algorithm's performance (HV) value and the Euclidean distance between the instance location and the convergence zone. The values are shown for each instance at each time step, with the distance to convergence represented by the x-axis and the HV value displayed on the y-axis. The blue line represents the linear regression line fitted to the data, indicating the overall trend. (Color figure online)

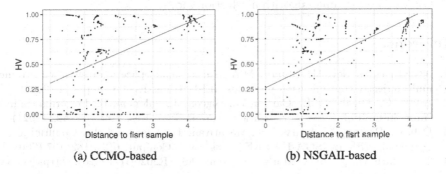

Fig. 8. Scatter plot showing the relationship between the algorithm's performance (HV) value and the Euclidean distance between the instance's current location and the location of the first instance. The values are shown for each instance at each time step, with the distance to the first sample represented by the x-axis and the HV value displayed on the y-axis. The blue line represents the linear regression line fitted to the data, indicating the overall trend. (Color figure online)

algorithm performance. Additionally, Fig. 8 illustrates the relationship between algorithm performance and the instance's displacement from its initial location. The correlation coefficients are 0.4791 for CCMO and 0.5890 for NSGAII, indicating that the further an instance moves from its initial location, the greater the chances of improved algorithm performance. Thus, dynamic landscape analysis features can be utilized to estimate algorithm performance.

5   Conclusion

Landscape analysis features refer to measurable topological characteristics of a problem search space. Early work has used these features as proxies for understanding the difficulty of problems. Later, the features came to be used to improve the optimization process [7]. The standard practice is to mine them via offline sampling methods, independent of the optimization process. In this paper, we have examined the measurement of local features of CMOPs using the optimization algorithm trajectory. Our analysis demonstrates how these features evolve throughout the optimization process.

By visualizing the local features trajectory on the instance space, our empirical analysis sheds light on the search behaviour of CCMO and NSGAII algorithms and their relationship to performance. Our findings imply that utilizing local features can serve as a valuable approach to approximate algorithm performance, especially when dealing with new instances lacking a known reference set. By understanding the dynamics of instance movement and their relation to the convergence zone, researchers and practitioners can design a landscape-aware CMOEA and a dynamic algorithm selection model.

References

1. Alsouly, H., Kirley, M., Muñoz, M.A.: An instance space analysis of constrained multi-objective optimization problems. IEEE Trans. Evol. Comput. 1 (2022)
2. Audet, C., Digabel, S.L., Cartier, D., Bigeon, J., Salomon, L.: Performance indicators in multiobjective optimization. Eur. J. Oper. Res. **292**, 397–422 (2021)
3. Deb, K., Pratap, A., Agarwal, S., Meyarivan, T.: A fast and elitist multiobjective genetic algorithm: NSGA-II. IEEE Trans. Evol. Comput. **6**(2), 182–197 (2002)
4. The MathWorks, Inc.: Matlab version: 9.9 (R2020B) (2020). https://www.mathworks.com
5. Janković, A., Doerr, C.: Adaptive landscape analysis. In: GECCO 2018 - Proceedings of the 2018 Genetic and Evolutionary Computation Conference, pp. 2032–2035. ACM, New York (2019)
6. Malan, K.M., Oberholzer, J.F., Engelbrecht, A.P.: Characterising constrained continuous optimisation problems. In: 2015 IEEE Congress on Evolutionary Computation - Proceedings, pp. 1351–1358 (2015)
7. Malan, K.M.: A survey of advances in landscape analysis for optimisation. Algorithms **14**(2), 40 (2021)
8. Mersmann, O., Bischl, B., Trautmann, H., Preuss, M., Weihs, C., Rudolph, G.: Exploratory landscape analysis. In: GECCO 2011 - Proceedings of the 2011 Genetic and Evolutionary Computation Conference, pp. 829–836 (2011)
9. Muñoz, M.A., Kirley, M., Halgamuge, S.K.: Landscape characterization of numerical optimization problems using biased scattered data. In: 2012 IEEE Congress on Evolutionary Computation - Proceedings, pp. 1–8 (2012)
10. Pedregosa, F., et al.: Scikit-learn: machine learning in Python. J. Mach. Learn. Res. **12**, 2825–2830 (2011)
11. Picard, C., Schiffmann, J.: Realistic constrained multiobjective optimization benchmark problems from design. IEEE Trans. Evol. Comput. **25**(2), 234–246 (2021)

12. Smith-Miles, K., Muñoz, M.A.: Instance space analysis for algorithm testing: methodology and software tools. ACM Comput. Surv. **55**, 1–31 (2022)
13. Tian, Y., Cheng, R., Zhang, X., Jin, Y.: PlatEMO: a MATLAB platform for evolutionary multi-objective optimization. IEEE Comput. Intell. Mag. **12**(4), 73–87 (2017)
14. Tian, Y., Zhang, T., Xiao, J., Zhang, X., Jin, Y.: A coevolutionary framework for constrained multiobjective optimization problems. IEEE Trans. Evol. Comput. **25**(1), 102–116 (2021)
15. Vodopija, A., Tušar, T., Filipič, B.: Characterization of constrained continuous multiobjective optimization problems: a feature space perspective. Info. Sci. **607**, 244–262 (2022)
16. Wang, M., Li, B., Zhang, G., Yao, X.: Population evolvability: dynamic fitness landscape analysis for population-based metaheuristic algorithms. IEEE Trans. Evol. Comput. **22**(4), 550–563 (2018)

Finding Maximum Weakly Stable Matchings for Hospitals/Residents with Ties Problem via Heuristic Search

Son Thanh Cao, Le Van Thanh, and Hoang Huu Viet[(✉)]

Faculty of Information Technology, Vinh University, Vinh City, Vietnam
{sonct,thanhlv,viethh}@vinhuni.edu.vn

Abstract. The Hospitals/Residents with Ties Problem is a many-one stable matching problem, in which residents need to be assigned to hospitals to meet their constraints. In this paper, we propose a simple heuristic algorithm but solve this problem efficiently. Our algorithm starts from an empty matching and gradually builds up a maximum stable matching of residents to hospitals. At each iteration, we propose a heuristic function to choose the best hospital for an active resident to form a resident-hospital pair for the matching. If the chosen hospital overcomes its offered capacity, we propose another heuristic function to remove the worst resident among residents assigned to the hospital in the matching. Our algorithm returns a stable matching if it finds no active resident. Experimental results show that our algorithm is efficient in execution time and solution quality for solving the problem.

Keywords: Gale-Shapley algorithm · Hospitals/Residents with Ties · Heuristic algorithm · Weakly stable matching

1 Introduction

The Hospitals/Residents problem, as defined by Gale and Shapley in 1962, was initially called the "College Admissions Problem" [3]. This classic problem addresses the issue of matching medical residents to hospitals in a way that is both stable and satisfactory for all parties involved. In the original formulation of the problem, there are a set of hospitals and a set of residents. Each hospital has a ranking of the residents based on their preferences, while each resident has a ranking of the hospitals. The goal is to find a *stable matching*, where no two hospitals and residents would prefer each other over their current assignments.

Since its introduction, the Gale-Shapley (GS) algorithm has inspired further research and variations on the Hospitals/Residents (HR) problem, including the consideration of ties in preference rankings and the exploration of different optimization objectives. These developments continue to shape the field and contribute to improving the allocation of medical residents to hospitals and other related matching problems. We can find HR applications in various contexts and systems worldwide. Notable examples include the National Resident Matching

T. Liu et al. (Eds.): AI 2023, LNAI 14471, pp. 442–454, 2024.
https://doi.org/10.1007/978-981-99-8388-9_36

Program (NRMP) in the United States, the Scottish Pre-registration House Officer Allocations (SPA) matching scheme, and the Canadian Resident Matching Service (CaRMS) in Canada [9].

The Hospitals/Residents problem with Ties (HRT) is an extension of the classic Hospitals/Residents problem that allows ties in the preferences of both residents and hospitals, in which participants can rank a subset of the other set with equal preference, indicating that they consider those options equally desirable [9]. Stability in HRT implies that there are no resident-hospital pairs who both prefer each other over their current assignments, considering the ties in their preferences. There are various criteria for stability, such as *weak-stability*, *strong-stability*, or *super-stability* [8,9], which determine the level of preference satisfaction and the absence of *blocking pairs*.

Solving the HRT problem poses computational challenges, as including ties in preferences increases the complexity of finding stable matchings. The problem of finding a weakly stable matching with the maximum number of residents assigned to hospitals, known as MAX-HRT, has been proven to be NP-hard [9]. Researchers have proposed different algorithms and approaches to tackle the HRT problem, including integer programming [1,12], local and adaptive search [4,13], approximation [11], and heuristic repair [2] algorithms. However, finding efficient solutions for large-sized instances of HRT remains an area of ongoing research.

This paper proposes a heuristic algorithm to deal with the MAX-HRT problem. Our algorithm starts from an empty matching and proceeds iteratively to achieve a maximum stable matching of residents and hospitals. At each iteration, we design a heuristic function for choosing a hospital to assign to an active resident such that the hospital has not only the minimum remaining capacity but also the minimum remaining preference list. If the chosen hospital is *over-subscribed*, we design another heuristic function for removing a resident from the hospital such that the removed resident has not only the maximum rank among residents assigned to the hospital but also the maximum remaining preference list. The algorithm repeats until it finds no active resident and returns a stable matching. Experimental results show that our algorithm is efficient in solving the large-sized MAX-HRT problem.

The remaining sections of this paper are organized as follows. Section 2 provides a brief background on HRT. Section 3 outlines the details of our algorithm. Section 4 shows the results obtained from our experiments. Finally, Sect. 5 presents the concluding remarks of our work.

2   Preliminaries

In this section, we remind a brief background on HRT taken from [9,12]. An instance I of HRT consists of two sets, a set $\mathcal{R} = \{r_1, r_2, \ldots, r_n\}$ of residents and a set $\mathcal{H} = \{h_1, h_2, \ldots, h_m\}$ of hospitals. Each resident $r_i \in \mathcal{R}$, $1 \leq i \leq n$, has a preference list that ranks a subset of hospitals in her/his preference with tie allowed. Similarly, each hospital $h_j \in \mathcal{H}$, $1 \leq j \leq m$, has a preference list that ranks a subset of residents in its preference with tie allowed. Each hospital h_j is assigned a *capacity* $c_j \in \mathbb{Z}^+$, indicating the maximum number of residents it can accommodate.

We denote the rank of $h_j \in \mathcal{H}$ (resp. $r_i \in \mathcal{R}$) in r_i's (resp. h_j's) preference list by $rank(r_i, h_j)$ (resp. $rank(h_j, r_i)$). Accordingly, if r_i (resp. h_j) ranks h_j (resp. r_i) in her/his (resp. its) preference list, then $1 \leq rank(r_i, h_j) \leq n$ (resp. $1 \leq rank(h_j, r_i) \leq n$), otherwise, $rank(r_i, h_j) = 0$ (resp. $rank(h_j, r_i) = 0$). If r_i (resp. h_j) strictly prefers h_j (resp. r_i) to h_k (resp. r_t), then we denote by $rank(r_i, h_j) < rank(r_i, h_k)$ (resp. $rank(h_j, r_i) < rank(h_j, r_t)$). If r_i (resp. h_j) prefers h_j (resp. r_i) and h_k (resp. r_t) equally, i.e. r_i (resp. h_j) ranks h_j (resp. r_i) and h_k (resp. r_t) with the same tie in her/his (resp. its) preference list, then we denote by $rank(r_i, h_j) = rank(r_i, h_k)$ (resp. $rank(h_j, r_i) = rank(h_j, r_t)$).

A pair $(r_i, h_j) \in \mathcal{R} \times \mathcal{H}$ is called an *acceptable* pair in I if $rank(r_i, h_j) \geq 1$ and $rank(h_j, r_i) \geq 1$, i.e. r_i and h_j rank each other in their preference lists. A set of *acceptable* pairs is denoted by $\mathcal{A} = \{(r_i, h_j) \in \mathcal{R} \times \mathcal{H} | rank(r_i, h_j) \geq 1$ and $rank(h_j, r_i) \geq 1\}$, i.e., both the resident r_i and the hospital h_j must rank each other in their preference lists.

An *assignment* M in I is defined as a subset of $\mathcal{A}$, i.e., $M = \{(r_i, h_j) \in \mathcal{A}\}$. If $(r_i, h_j) \in M$, we say that r_i is *assigned* to h_j and vice versa. For any hospital $h_j \in \mathcal{H}$, we denote $M(h_j)$ by the set of residents assigned to h_j (i.e., $M(h_j) = \{r_i | (r_i, h_j) \in M\}$) and $M(r_i)$ by the hospital h_j assigned to r_i (i.e., $M(r_i) = h_j$), respectively. We denote by $M(r_i) = \varnothing$ if r_i is unassigned in M. A hospital $h_j \in \mathcal{H}$ is referred to as *under-subscribed*, *full-subscribed*, or *over-subscribed* depending on the conditions $|M(h_j)| < c_j$, $|M(h_j)| = c_j$, or $|M(h_j)| > c_j$, respectively.

A *matching* M in I is an assignment such that $|M(r_i)| \leq 1$, $\forall r_i \in \mathcal{R}$, and $|M(h_j)| \leq c_j$, $\forall h_j \in \mathcal{H}$. These conditions ensure that each resident is assigned to at most one hospital, and no hospital exceeds its capacity. As we mentioned above, this paper only considers the problem of finding maximum weakly stable matchings in I. Hereafter, we use the term *matching* to refer to weak matching.

A pair $(r_i, h_j) \in \mathcal{R} \times \mathcal{H}$ is a *blocking pair* for a matching M if the following conditions hold: (*i*) $(r_i, h_j) \in \mathcal{A}$; (*ii*) $M(r_i) = \varnothing$ or $rank(r_i, h_j) < rank(r_i, M(r_i))$; and (*iii*) $|M(h_j)| < c_j$ or $rank(h_j, r_i) < rank(h_j, r_w)$, where r_w is the worst resident in $M(h_j)$. When a blocking pair exists, it implies that M is *unstable* because there are participants who would both prefer each other if given a chance, otherwise, M is called *stable*. The number of residents assigned to hospitals in M is denoted by $|M|$. If $|M| = n$, then M is called *perfect*, otherwise, M is called *non-perfect*.

Example 1. Let's consider the following example to provide greater clarity to mentioned notations. Given an instance of HRT consisting of a set $\mathcal{R} = \{r_1, r_2, r_3, r_4, r_5, r_6\}$ of six residents, and a set $\mathcal{H} = \{h_1, h_2, h_3\}$ of three hospitals. Additional details can be found in Table 1, which presents the preference lists of residents and hospitals, including any ties indicated within round brackets. In the hospitals' preference lists of Table 1, let's focus on the second row for the hospital h_2. The notation "$h_2 : r_2\ r_1\ r_6\ (r_4\ r_5)$" indicates that h_2 strictly prefers r_2 to r_1, r_1 to r_6, and r_6 to both r_4 and r_5. Also, in this particular case, h_2 considers r_4 and r_5

Table 1. An instance of HRT

Residents' preference list	Hospitals' preference list	Residents' rank list	Hospitals' rank list
r_1: h_1 h_2	h_1: r_1 r_2 r_3 r_6	r_1: 1 2 0	h_1: 1 2 3 0 0 4
r_2: h_1	h_2: r_2 r_1 r_6 (r_4 r_5)	r_2: 1 0 0	h_2: 2 1 0 4 4 3
r_3: h_1 h_3	h_3: r_5 (r_3 r_4)	r_3: 1 0 2	h_3: 0 0 2 2 1 0
r_4: h_2		r_4: 0 1 0	
r_5: h_2 (h_1 h_3)		r_5: 2 1 2	
r_6: h_1 h_2		r_6: 1 2 0	
Hospitals' capacities: $c_1 = c_2 = c_3 = 2$			

to be equally preferred. Accordingly, we have $rank(h_2, r_2) = 1$, $rank(h_2, r_1) = 2$, $rank(h_2, r_6) = 3$, and $rank(h_2, r_4) = rank(h_2, r_5) = 4$. The same notations are utilized for the residents' preference lists. The rank lists corresponding to both the residents' and hospitals' preference lists are presented on the right-hand side of Table 1. The matching $M = \{(r_1, h_2), (r_2, h_1), (r_3, h_3), (s_4, h_2), (r_5, h_3), (r_6, h_1)\}$ is unstable due to the presence of blocking pairs $\{(r_1, h_2), (r_3, h_3)\}$ for M. Particularly, we have $rank(r_1, h_1) < rank(r_1, h_2)$, indicating that (r_1, h_2) forms a blocking pair. A similar explanation holds for the pair (r_3, h_3). The matching $M = \{(r_1, h_1), (r_2, h_1), (r_3, h_3), (r_4, \varnothing), (r_5, h_2), (r_6, h_2)\}$ is stable because there are no blocking pairs for M. However, this matching is considered non-perfect due to its cardinality $|M| = 5$. Similarly, the matching $M = \{(r_1, h_1), (r_2, h_1), (r_3, h_3), (r_4, h_2), (r_5, h_3), (r_6, h_2)\}$ is stable since no blocking pair can be formed for M. In this case, M is a perfect matching since $|M| = 6$. ◁

3   Heuristic Algorithm for MAX-HRT

In this section, we propose a heuristic algorithm to deal with the MAX-HRT problem. We consider the resident-oriented GS algorithm for HR problem given in [7]. At the beginning, a matching M is initialized to be empty, meaning that every resident is unassigned to any hospital ranked by her/him. At each iteration, an unassigned resident $r_i \in \mathcal{R}$ with a non-empty preference list is provisionally assigned to the most preferred hospital h_j in her/his preference list to form a pair $(r_i, h_j) \in M$. If the hospital h_j is *over-subscribed*, then the worst resident r_w assigned to h_j in $M(h_j)$ becomes free and h_j deletes r_w in its preference list (i.e. $rank(h_j, r_w) = 0$). If the hospital h_j is full, then h_j deletes the successor residents of the worst resident r_w in its preference list to accelerate finding a stable matching. The algorithm terminates with a resident-optimal stable matching M.

It is evident that we can apply the resident-oriented GS algorithm to find a stable matching for an instance of HRT. However, the stable matchings of an instance of HRT may be of different sizes [7]. Therefore, we are to extend the resident-oriented GS algorithm to find maximum stable matchings of HRT instances. Since the preference lists of both residents and hospitals in HRT include ties, we need to solve two issues in iterations of the resident-oriented GS algorithm: (i) an unassigned resident $r_i \in \mathcal{R}$ with a non-empty preference list is provisionally assigned to which most preferred hospital if there exist at least two most preferred hospitals with the same ties in her/his preference list; and (ii) if the hospital h_j is *over-subscribed*, then which worst resident assigned to h_j in $M(h_j)$ should be removed if there exist at least two worst residents with the same ties in its preference list.

For the first issue, we propose a heuristic function for all the hospital h_j in the preference list of an unassigned resident r_i as follows:

$$f(h_j) = rank(r_i, h_j) + 0.5 \times (|M(h_j)|/(c_j + 1) + |rank(h_j, r_k)|/(n + 1)), \quad (1)$$

where $|rank(h_j, r_k)|$ is the number of residents $r_k \in \mathcal{R}$ ranked by h_j in its preference list. Then, a hospital h_j is chosen to assign to r_i as follows:

$$h_j = \underset{h_j}{\arg\min} \; f(h_j), \forall rank(r_i, h_j) > 0. \quad (2)$$

We can see $f(h_j)$ in Eq. (1) that $rank(r_i, h_j)$ is a positive integer and $0.5 \times (|M(h_j)|/(c_j + 1) + |rank(h_j, r_k)|/(n + 1)) < 1$. If a hospital h_j is chosen such that $f(h_j)$ is minimum, meaning that three following conditions are satisfied: (i) r_i prefers h_j most; (ii) h_j is being assigned to the least residents; and (iii) h_j ranks least residents in its preference list. These conditions ensure that the pair (r_i, h_j) is not only a stable pair in M, but also h_j with the least opportunities is prioritized to assign to r_j.

For the second issue, we propose another heuristic function for all the residents r_w being assigned to an *over-subscribed* hospital h_j as follows:

$$g(r_w) = rank(h_j, r_w) + |rank(r_w, h_t)|/(m + 1)), \forall r_w \in M(h_j), \quad (3)$$

where $|rank(r_w, h_t)|$ is the number of hospitals $h_t \in \mathcal{H}$ ranked by r_w in her/his preference list. Then, a resident r_w is chosen to remove from h_j as follows:

$$r_w = \underset{r_w}{\arg\max} \; g(r_w), \forall r_w \in M(h_j). \quad (4)$$

We can see $g(r_w)$ in Eq. (3) that $rank(h_j, r_w)$ is a positive integer and $|rank(r_w, h_t)|/(m + 1) < 1$. If a resident $r_w \in M(h_j)$ is chosen such that $g(r_w)$ is maximum, meaning that two following conditions are satisfied: (i) r_w is the worst resident assigned to h_j; and (ii) r_w ranks most hospitals in her/his preference list. In other words, r_w is not only the worst resident assigned to h_j, but also has the highest number of opportunities to choose a hospital from her/his preference list during the next iterations.

Based on two heuristic functions above, we propose a simple heuristic algorithm shown in Algorithm 1. Initially, we set $M = \varnothing$ and $active(r_i) = 1, \forall r_i \in \mathcal{R}$, meaning that all residents are unassigned to any hospitals and r_i is marked as active. At each iteration, our algorithm checks if some active resident $r_i \in \mathcal{R}$

Algorithm 1: Heuristic Algorithm for MAX-HRT

    Input   : An instance I of HRT
    Output : A stable matching M for HRT

1 $M := \varnothing$;
2 $active(r_i) := 1, \forall r_i \in \mathcal{R}$;
3 **while** $\exists r_i$ *such that* $active(r_i) > 0$ **do**
4   **if** (r_i's preference list is empty) **then**
5         $active(r_i) := 0$;
6   continue;
7   **end**
8     $f(h_j) = rank(r_i, h_j) + 0.5(|M(h_j)|/(c_j + 1) + |rank(h_j, r_k)|/(n + 1))$;
9     $h_j = \underset{h_j}{\operatorname{argmin}} \; f(h_j), \forall rank(r_i, h_j) > 0$;
10    $M := M \cup (r_i, h_j)$;
11    $active(r_i) := 0$;
12   **if** h_j *is over-subscribed* **then**
13       $g(r_w) := rank(h_j, r_w) + |rank(r_w, h_t)|/(m + 1)), \forall r_w \in M(h_j)$;
14       $r_w := \underset{r_w}{\operatorname{argmax}} \; g(r_w), \forall r_w \in M(h_j)$;
15       $M := M \setminus (r_w, h_j)$;
16       $rank(r_w, h_j) := 0$;
17       $rank(h_j, r_w) := 0$;
18       $active(r_w) := 1$;
19   **end**
20 **end**
21 **return** M;

---

has no remaining hospital in her/his preference list, i.e. $rank(r_i, h_t) = 0$ for $\forall h_t \in \mathcal{H}$, then r_i is marked as inactive permanently, and the algorithm runs the next iteration (lines 4–7). Otherwise, r_i remains unassigned. Next, the algorithm finds a hospital h_j such that $f(h_j)$ is minimum, assigns h_j to r_i to form a pair $(r_i, h_j) \in M$, and marks r_i as inactive (lines 8–11). If h_j is *over-subscribed*, the algorithm computes $g(r_w)$ for $\forall r_w \in M(h_j)$, finds a resident r_w such that $g(r_w)$ is maximum, and removes the pair (r_w, h_j) from M. If so, r_w deletes h_j from her/his preference list and vice versa, r_w is marked as active again (lines 12–19). The algorithm repeats until there exists no active resident and it returns a stable matching.

We consider the time complexity of the proposed algorithm. In the best case, if each resident proposes the first preferred hospital in her/his preference list and she/he is assigned to this hospital to form a stable matching, then our algorithm takes $O(n)$ time, where n is the number of residents. In the worst case, if each resident ranks m hospitals and proposes to all the hospitals in her/his preference list, then our algorithm takes $O(nm)$ time, which is a linear time complexity.

Example 2. This example shows how Algorithm 1 operates using the HRT instance given in Table 1. Algorithm 1 initializes with $M = \varnothing$, $active(r_i) = 1$ for

Table 2. The step-by-step of running Algorithm 1 for the instance given in Table 1

Iter.	r_i	h_j	Matching M $(M = M \cup \{r_i, h_j\})$	Over-subscribed	$M = M \setminus \{r_i, h_j\}$
1	r_1	h_1	$\{(r_1, h_1)\}$		
2	r_2	h_1	$\{(r_1, h_1), (r_2, h_1)\}$		
3	r_3	h_1	$\{(r_1, h_1), (r_2, h_1)\}$	h_1	$M = M \setminus \{r_3, h_1\}$
4	r_3	h_3	$\{(r_1, h_1), (r_2, h_1), (r_3, h_3)\}$		
5	r_4	h_2	$\{(r_1, h_1), (r_2, h_1), (r_3, h_3), (r_4, h_2)\}$		
6	r_5	h_2	$\{(r_1, h_1), (r_2, h_1), (r_3, h_3), (r_4, h_2), (r_5, h_2)\}$		
7	r_6	h_1	$\{(r_1, h_1), (r_2, h_1), (r_3, h_3), (r_4, h_2), (r_5, h_2)\}$	h_1	$M = M \setminus \{r_6, h_1\}$
8	r_6	h_2	$\{(r_1, h_1), (r_2, h_1), (r_3, h_3), (r_4, h_2), (r_6, h_2)\}$	h_2	$M = M \setminus \{r_5, h_2\}$
9	r_5	h_3	$\{(r_1, h_1), (r_2, h_1), (r_3, h_3), (r_4, h_2), (r_5, h_3), (r_6, h_2)\}$		

all $r_i \in \mathcal{R}, 1 \leq i \leq n$, and runs iterations shown in Table 2. At the ith iteration, the algorithm executes as follows:

(1) r_1 is assigned to h_1 since $f(h_1)$ is minimum. We have $M = \{(r_1, h_1)\}$ and r_1 is marked as inactive (i.e., $active(r_1) = 0$).

(2) r_2 is assigned to h_1 since $f(h_1)$ is minimum. We have $M = \{(r_1, h_1), (r_2, h_1)\}$ and r_2 is marked as inactive.

(3) r_3 is assigned to h_1 since $f(h_1)$ is minimum. We have $M = \{(r_1, h_1), (r_2, h_1), (r_3, h_1)\}$ and r_3 is marked as inactive. However, h_1 is over-subscribed and $g(r_3)$ is maximum, therefore the pair (r_3, h_1) is removed from M and r_3 is active again. So, we have $M = \{(r_1, h_1), (r_2, h_1)\}$.

(4) r_3 is assigned to h_3 since $f(h_3)$ is minimum. We have $M = \{(r_1, h_1), (r_2, h_1), (r_3, h_3)\}$ and r_3 is marked as inactive.

(5) r_4 is assigned to h_2 since $f(h_2)$ is minimum. We have $M = \{(r_1, h_1), (r_2, h_1), (r_3, h_3), (r_4, h_2)\}$ and r_4 is marked as inactive.

(6) r_5 is assigned to h_2 since $f(h_2)$ is minimum. We have $M = \{(r_1, h_1), (r_2, h_1), (r_3, h_3), (r_4, h_2), (r_5, h_2)\}$ and r_5 is marked as inactive.

(7) r_6 is assigned to h_1 since $f(h_1)$ is minimum. We have $M = \{(r_1, h_1), (r_2, h_1), (r_3, h_3), (r_4, h_2), (r_5, h_2), (r_6, h_1)\}$ and r_6 is marked as inactive. However, h_1 is over-subscribed and $g(r_6)$ is maximum, therefore the pair (r_6, h_1) is removed from M and r_6 is active again. So, we have $M = \{(r_1, h_1), (r_2, h_1), (r_3, h_3), (r_4, h_2), (r_5, h_2)\}$.

(8) r_6 is assigned to h_2 since $f(h_2)$ is minimum. We have $M = \{(r_1, h_1), (r_2, h_1), (r_3, h_3), (r_4, h_2), (r_5, h_2), (r_6, h_2)\}$ and r_6 is marked as inactive. However, h_2 is over-subscribed and $g(r_5)$ is maximum, therefore the pair (r_5, h_2) is removed from M and r_5 is active again. So, we have $M = \{(r_1, h_1), (r_2, h_1), (r_3, h_3), (r_4, h_2), (r_6, h_2)\}$.

(9) Finally, r_5 is assigned to h_3 since $f(h_3)$ is minimum and r_5 becomes inactive. After this step, all residents are inactive, the algorithm returns a stable matching $M = \{(r_1, h_1), (r_2, h_1), (r_3, h_3), (r_4, h_2), (r_5, h_3), (r_6, h_2)\}$. In this case, M is also a perfect matching.   ◁

4   Experiments

In this section, we present experiments to evaluate the performance of our heuristic algorithm, namely HA, for finding maximum stable matchings of HRT instances. We chose the heuristic repair (HR) algorithm [2] to compare with HA since HR outperformed LS [4,5] in terms of the execution time and solution quality as mentioned in [2].

Datasets: we adapted the SMTI generator given in [6] to generate random HRT instances $I(n, m, p_1, p_2, \{c_1, c_2, \cdots, c_m\})$, where n is the number of residents, m is the number of hospitals, p_1 is the probability of incompleteness in preference lists of residents and hospitals, and p_2 is the probability of ties in the preference lists of residents and hospitals, and c_j is the capacity of each hospital $h_j \in H, 1 \leq j \leq m$. This means that on average, each resident ranks about $m(1 - p_1)$ hospitals and each hospital ranks about $n(1 - p_1)$ residents in each HRT instance. Since stable matchings of HRT instances consist of acceptable pairs, we generated HRT instances in which the residents' and hospitals' preference lists of each instance have only acceptable pairs. We implemented HA and HR algorithms by Matlab 2019a on a computer with a Core i7-8550U CPU 1.8 GHz and 16 GB memory.

4.1   Experiment 1

In this experiment, we randomly generated 100 HRT instances for each combination of values (n, m, p_1, p_2), where $n = 500$, $m = 50$, $p_1 \in \{0.0, 0.1, \cdots, 0.9\}$, and $p_2 \in \{0.0, 0.1, \cdots, 1.0\}$. In each instance, we set $c_j = n/m$ for all $h_j \in H, 1 \leq j \leq m$, meaning that the total capacities of hospitals are equal to n. It is evident that this is a hard experiment for algorithms to find perfect matching in HRT instances since each resident has only a post to be assigned to each hospital in her/his preference list. We set the maximum number of iterations of HR to 500.

Figure 1(a) shows the percentage of perfect matchings found by HA and HR when p_1 varies from 0.6 to 0.9. It should be noted that when p_1 varies from 0.0 to 0.5, both HA and HR find 100% of perfect matchings and therefore, we do not depict in Fig. 1(a). We see that when p_2 varies from 0.0 to 1.0, the percentage of perfect matchings found by HA increases, while that found by HR decreases, meaning that HA found perfect matchings easier than HR when ties in the preference lists of residents and hospitals increase. When $p_1 = 0.6$, the percentage of perfect matchings found by HA is approximate to that found by HR. When $p_1 \in \{0.7, 0.8, 0.9\}$, the percentage of perfect matchings found by HA is much higher than that found by HR.

Figure 1(b) shows the average number of unassigned residents in stable matchings. When p_2 varies from 0.1 to 1.0, HA finds a much smaller number of unassigned residents than HR in stable matchings, meaning that HA finds much larger stable matchings than HR. It should be noted that when $p_2 = 0.0$, meaning that the residents' and hospitals' preference lists do not contain ties, both HA and HR find the same percentage of perfect matchings as well as the same number of unassigned residents since all stable matchings have the same size [14].

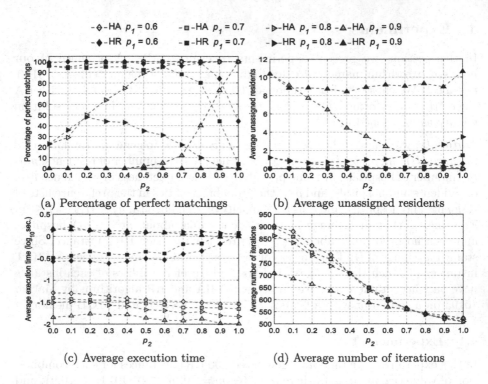

(a) Percentage of perfect matchings   (b) Average unassigned residents

(c) Average execution time   (d) Average number of iterations

Fig. 1. Comparing solution quality and execution time of HA and HR algorithms

Figure 1(c) shows the average execution time of HA and HR. When p_1 increases from 0.6 to 0.9, the average execution time of HA and HR slightly changed. When p_2 increases from 0.0 to 1.0, the average execution time of HA and HR slightly decreases. We can see that the average execution time of HA increases from about $10^{-1.99} = 0.01$ to $10^{-1.28} = 0.05$ s, while that of HR increases from about $10^{-0.62} = 0.24$ to $10^{0.22} = 1.66$ s for every value of p_1 and p_2. This means that HA runs from about 24 to 33 times faster than HR.

Figure 1(d) shows the average number of iterations used by HA. HA used about 900 down to 500 iterations when p_2 increases from 0.0 to 1.0. It should be noted that we do not show the average number of iterations found by HR since the iteration mechanism of HR is different from that of HA.

4.2   Experiment 2

In this experiment, we randomly generated 100 HRT instances for each combination of values (n, m, p_1, p_2) given in Experiment 1. In each instance, we set the total capacity of hospitals to n and randomly distributed n to the capacity c_j of each hospital h_j such that $1 \leq c_j \leq 20$. Besides, we set the maximum number of iterations of HR to 500.

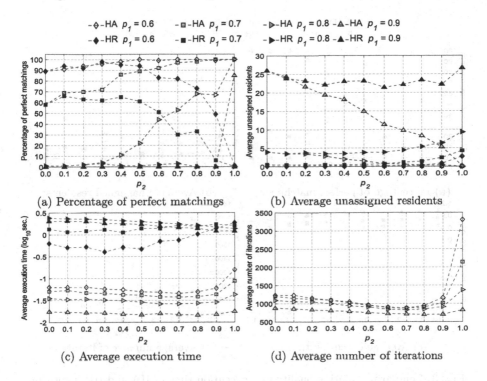

Fig. 2. Comparing solution quality and execution time of HA and HR algorithms

Figure 2(a) shows the percentage of perfect matchings found by HA and HR. Again, HA finds a much higher percentage of perfect matchings than HR. Moreover, both HA and HR find stable matchings more difficult when $c_j = n/m$ as shown in Experiment 1. Figure 2(b) shows the average number of unassigned residents in stable matchings. When p_2 varies from 0.1 to 1.0, HA finds a much smaller number of unassigned residents in stable matchings than HR, meaning that HA finds much larger stable matchings than HR.

Figure 2(c) shows the average execution time of HA and HR. When p_2 varies from 0.0 to 1.0, the average execution time of HA and HR slightly changed. When p_1 varies from 0.6 to 0.9, the average execution time of HA decreases from about $10^{-1.3} = 0.050$ down to $10^{-1.8} = 0.016$ s, while that of HR increases from about $10^{-0.2} = 0.631$ to $10^{0.4} = 2.512$ s. On average over p_1, the average execution time of HA and HR is with respective to $10^{-1.5} = 0.032$ and $10^0 = 1.0$ s, meaning that HA runs about 31 times faster than HR.

Figure 2(d) shows the average number of iterations used by HA. When p_2 varies from 0.0 to 0.9, HA runs from about 800 to 1200 iterations. When $p_2 = 1.0$, the average number of iterations used by HA increases significantly, but the average execution time of HA increases sightly. This indicates that at each iteration, HA used a small amount of time to find a resident-hospital pair for a stable matching.

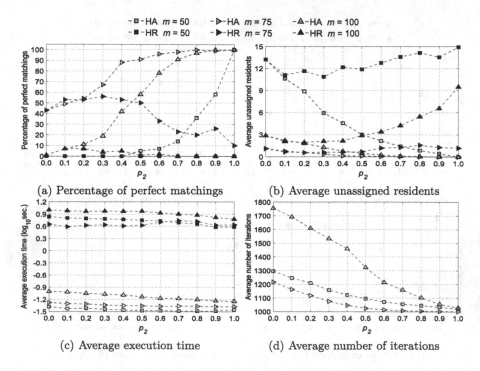

(a) Percentage of perfect matchings (b) Average unassigned residents

(c) Average execution time (d) Average number of iterations

Fig. 3. Comparing solution quality and execution time of HA and HR algorithms

4.3 Experiment 3

In this experiment, we varied the number of hospitals. Specifically, we randomly generated 100 HRT instances for each combination of parameters (n, m, p_1, p_2), where $n = 1000$, $m \in \{50, 75, 100\}$, $p_1 = 0.9$, and $p_2 \in \{0.0, 0.1, \cdots, 1.0\}$. This means that each resident ranks on average at 5, 7.5, and 10 hospitals when $m = 50, 75$, and 100, respectively, and each hospital ranks about 100 residents in the instances. In each instance, we set $c_j = n/m$ for all $h_j \in H, 1 \le j \le m$, meaning that each resident has one post assigned to each hospital. We set the maximum number of iterations of HR to 1000. Figure 3 shows the results of this experiment. Again, we see that our HA not only outperforms HR in matching quality shown by the percentage of perfect matchings and the number of unassigned residents, but also runs much faster than HR shown by the average execution time.

5 Conclusions

In this paper, we proposed a simple heuristic algorithm but it efficiently solved the MAX-HRT problem. Our algorithm starts from an empty matching to find a maximum stable matching of an HRT instance. At each iteration, the algorithm finds a hospital to assign to an active resident based on a heuristic function such that the hospital has not only the minimum remaining capacity but also the minimum preference list. If the chosen hospital overcomes its offered capacity, the algorithm finds a resident to remove her/him from the hospital based on another heuristic function such that the resident has not only the maximum rank among the residents being assigned to the hospital but also the maximum preference list. The algorithm repeats until it finds no active resident and returns a stable matching. Experiments showed that our algorithm is efficient in execution time and solution quality for large-sized HRT instances. In the future, we plan to extend this approach to find strongly- or super-stable matchings for HRT [9,10].

References

1. Biró, P., Manlove, D.F., McBride, I.: The hospitals/residents problem with couples: complexity and integer programming models. In: Proceeding of SEA 2014: 13th International Symposium on Experimental Algorithms, Copenhagen, Denmark, pp. 10–21 (2014)
2. Cao, S.T., Anh, L.Q., Viet, H.H.: A heuristic repair algorithm for the hospitals/residents problem with ties. In: Rutkowski, L., Scherer, R., Korytkowski, M., Pedrycz, W., Tadeusiewicz, R., Zurada, J.M. (eds.) ICAISC 2022. LNCS, vol. 13588, pp. 340–352. Springer, Cham (2023). https://doi.org/10.1007/978-3-031-23492-7_29
3. Gale, D., Shapley, L.S.: College admissions and the stability of marriage. Am. Math. Mon. **9**(1), 9–15 (1962)
4. Gelain, M., Pini, M.S., Rossi, F., Venable, K.B., Walsh, T.: Local search for stable marriage problems with ties and incomplete lists. In: Proceedings of 11th Pacific Rim International Conference on Artificial Intelligence, Daegu, Korea, pp. 64–75 (2010)
5. Gelain, M., Pini, M.S., Rossi, F., Venable, K.B., Walsh, T.: Local search approaches in stable matching problems. Algorithms **6**(4), 591–617 (2013)
6. Gent, I.P., Prosser, P.: An empirical study of the stable marriage problem with ties and incomplete lists. In: Proceedings of the 15th European Conference on Artificial Intelligence, Lyon, France, pp. 141–145 (2002)
7. Gusfield, D., Irving, R.W.: The Stable Marriage Problem: Structure and Algorithms. MIT Press Cambridge, New York (1989)
8. Irving, R.W.: Stable marriage and indifference. Discret. Appl. Math. **48**, 261–272 (1974)
9. Irving, R.W., Manlove, D.F., Scott, S.: The hospitals/residents problem with ties. In: Proceedings of the 7th Scandinavian Workshop on Algorithm Theory, Bergen, Norway, pp. 259–271 (2000)
10. Irving, R.W., Manlove, D.F., Scott, S.: Strong stability in the hospitals/residents problem. In: Alt, H., Habib, M. (eds.) STACS 2003. LNCS, vol. 2607, pp. 439–450. Springer, Heidelberg (2003). https://doi.org/10.1007/3-540-36494-3_39
11. Király, Z.: Linear time local approximation algorithm for maximum stable marriage. Algorithms **6**(1), 471–484 (2013)
12. Kwanashie, A., Manlove, D.F.: An integer programming approach to the hospitals/residents problem with ties. In: Huisman, D., Louwerse, I., Wagelmans, A.P.M. (eds.) Operations Research Proceedings 2013. ORP, pp. 263–269. Springer, Cham (2014). https://doi.org/10.1007/978-3-319-07001-8_36

13. Munera, D., Diaz, D., Abreu, S., Rossi, F., Saraswat, V., Codognet, P.: A local search algorithm for SMTI and its extension to HRT problems. In: Proceedings of the 3rd International Workshop on Matching Under Preferences, pp. 66–77. University of Glasgow (2015)
14. Irving, R.W., Manlove, D.F.: Finding large stable matchings. J. Exp. Algorithmics **14**(1), 1–2 (2009)

Approximating Solutions to the Knapsack Problem Using the Lagrangian Dual Framework

Mitchell Keegan[(✉)] and Mahdi Abolghasemi

The University of Queensland, Brisbane, Australia
m.keegan@uq.net.au, m.abolghasemi@uq.edu.au

Abstract. The Knapsack Problem is a classic problem in combinatorial optimisation. Solving these problems may be computationally expensive. Recent years have seen a growing interest in the use of deep learning methods to approximate the solutions to such problems. A core problem is how to enforce or encourage constraint satisfaction in predicted solutions. A promising approach for predicting solutions to constrained optimisation problems is the Lagrangian Dual Framework which builds on the method of Lagrangian Relaxation. In this paper we develop neural network models to approximate Knapsack Problem solutions using the Lagrangian Dual Framework while improving constraint satisfaction. We explore the problems of output interpretation and model selection within this context. Experimental results show strong constraint satisfaction with a minor reduction of optimality as compared to a baseline neural network which does not explicitly model the constraints.

Keywords: Knapsack Problem · Supervised Learning · Lagrangian Dual Framework · Neural Networks

1 Introduction

The Knapsack Problem (KP) is a classic problem in combinatorial optimisation (CO). The core statement is as follows; there are n items, each with weight w_i and value v_i. The goal is to select a set of items which maximise the total value, such that the total weights of the chosen items is not greater than some capacity W. It is formally defined as an integer programming (IP) problem as shown below:

$$
\begin{aligned}
x^* = \underset{x}{\operatorname{argmax}} \ & \sum_{i=1}^{n} x_i v_i \\
\text{s.t.} \ & \sum_{i=1}^{n} x_i w_i \leq W \\
& x_i \in \{0, 1\}, \ i = 1, ..., n
\end{aligned}
\tag{1}
$$

© The Author(s), under exclusive license to Springer Nature Singapore Pte Ltd. 2024
T. Liu et al. (Eds.): AI 2023, LNAI 14471, pp. 455–467, 2024.
https://doi.org/10.1007/978-981-99-8388-9_37

where $x_i = 1$ if the ith item is chosen or $x_i = 0$ if not. KP and its variants find applications when limited resources must be allocated efficiently. Examples include cutting stock problems, investment allocation, and transport and logistics. There exists effective exact and approximate algorithms to solve KP including dynamic programming, branch and bound, and a variety of metaheuristic algorithms. A detailed treatment of these algorithms for KP and its many variants can be found in [18]. A common drawback of these methods for KP and other CO problems is that the solving time may be too slow in applications where solutions are required under strict time constraints. A promising remedy is to use fast inference machine learning models to approximate solutions [2–4].

The idea of applying machine learning to CO is not a new one, although recent years have seen a resurgence in interest with the rise of deep learning. An early example in 1985 was the use of Hopfield Networks to approximate the objective of the Travelling Salesman Problem [13]. The last decade has seen explosive advances in machine learning, with progress in deep learning providing practical solutions to problems in the field of computer vision and natural language processing (among many others) which were once considered incredibly difficult or intractable. This has naturally led to renewed interest in the applications of machine learning to CO.

This paper focuses on using supervised deep learning to predict solutions to KP. For broad reviews of recent advancements in the applications of machine learning to CO we refer to [1,5], and [14]. A core challenge in machine learning for CO is data generation. Solving CO problems is often computationally expensive which makes generating target labels for large datasets difficult. We sidestep these challenges in this paper by focusing on KP for which we can generate solutions with reasonable speed, but this problem represents an interesting area of research [15]. For the purpose of approximating solutions to constrained optimisation problems, a key weakness of neural networks is the inability to enforce constraint satisfaction on the model outputs.

Several models have been developed to address this weakness. One approach proposes using an iterative training routine in which a learner step that trains a model as usual is interleaved with a master step that adjusts the target labels to something closer to the predicted solution while remaining feasible [6]. It makes no assumptions about the nature of the constraints and permits any supervised ML model in the learner step. Another approach is Deep Constraint Completion and Correction (DC3) [7]. DC3 uses differentiable processes, named completion and correction, to enforce the feasibility of solutions during training. It showed strong results with a high degree of constraint satisfaction on the AC optimal power flow problem (AC-OPF), which is a continuous optimisation problem. A third approach to integrating constraints is based on Lagrangian Relaxation.

Lagrangian Relaxation is a method of solving constrained optimisation problems by relaxing constraints into the objective function scaled by Lagrange Multipliers. [11] presents a generalisation of Lagrangian Relaxation and related concepts which allows it to be applied to arbitrary optimisation models. The Lagrangian Dual Framework [9,10] builds on this approach, using Lagrangian relaxation to encourage constraint satisfaction in neural network models during training. The LDF has been applied with encouraging results to the problems of AC-OPF [9], constrained predictors where constraints between samples are present [10], and the job shop scheduling problem [16].

In this paper, we apply the LDF to the Knapsack Problem. This is not the first attempt at predicting KP solutions using neural networks. For example GRU, CNN and feedforward neural networks have been designed for this purpose [17], but these models did not incorporate constraint satisfaction into the training procedure. Some research has been done to derive theoretical bounds on the depth and width of an RNN designed to mimic a dynamic programming algorithm for both exact and approximate solutions [12]. Directly incorporating constraints in the training process should theoretically improve performance since it explicitly models the constraints within the learning framework, instead of only appearing implicitly in the training data.

In this paper we develop three neural network models to approximate solutions to KP, one is a baseline fully connected network which is compared to another fully connected network that models the constraints using the LDF. A third model uses the LDF but also uses the baseline neural network as a pretrained model. We explore implementation details for applying the LDF to KP (and by extension IP problems), experimentally investigate the trade-off between the optimality of approximated solutions against constraint satisfaction, and discuss how to approach hyperparameter tuning and model selection in the context of this trade-off.

2 Lagrangian Dual Framework

We implement the LDF initially proposed in [10] to learn a model which approximates solutions to the Knapsack problem. This section restates the formulation of the LDF as relevant to KP. Consider a general constrained optimisation problem with inequality constraints:

$$\mathcal{O} = \operatorname*{argmin}_{y} \quad f(y)$$
$$\text{subject to} \quad g_i(y) \leq 0, \; i = 1, \ldots, m \tag{2}$$

The violation-based Lagrangian function is:

$$f_\lambda(y) = f(y) + \sum_{i=1}^{m} \lambda_i \max(0, g_i(y)) \tag{3}$$

where $\lambda_i \geq 0$ denote the Lagrange multipliers associated with the inequality constraints. Another approach is to consider the satisfiability-based Lagrangian function:

$$f_\lambda(y) = f(y) + \sum_{i=1}^{m} \lambda_i g_i(y) \tag{4}$$

Both of these can be generalised by considering the function $\nu(g_i)$ which returns either the constraint satisfiability or the degree of constraint violation as required.

$$f_\lambda(y) = f(y) + \sum_{i=1}^{m} \lambda_i \nu(g_i(y)) \tag{5}$$

In this paper $\nu(g_i)$ will always refer to the constraint violation degree $\max(0, g_i(y))$. The Lagrangian Relaxation is then:

$$LR_\lambda = \underset{y}{\mathrm{argmin}}\, f_\lambda(y) \tag{6}$$

for some set of Lagrangian multipliers $\lambda = \{\lambda_1, \ldots, \lambda_m\}$. The solution forms a lower bound on the original constrained problem, i.e. $f(LR_\lambda) \leq f(\mathcal{O})$.

To find the strongest Lagrangian relaxation of $\mathcal{O}$, we can find the best set of Lagrangian multipliers using the Lagrangian dual.

$$LD = \underset{\lambda \geq 0}{\mathrm{argmax}}\, f(LR_\lambda) \tag{7}$$

The LDF leverages the Lagrangian Relaxation to improve constraint satisfaction in neural network models [9]. Consider parametrising the original optimisation problem by the parameters of the objective function and constraints:

$$\mathcal{O}(d) = \underset{y}{\mathrm{argmin}} \quad f(y, d) \tag{8}$$
$$\text{subject to} \quad g_i(y, d) \leq 0, \ i = 1, \ldots, m$$

The goal is the learn some parametric model $\mathcal{M}_w$ with weights w such that $\mathcal{M}_w \approx \mathcal{O}$. In this paper $\mathcal{M}_w$ is always a feedforward neural network.

A set of data is denoted $D = \{(d_l, y_l = \mathcal{O}(d_l)\}$ for $l = 1, \ldots, n$. The learning problem is to solve:

$$w^* = \underset{w}{\mathrm{argmin}} \quad \sum_{l=1}^{n} \mathcal{L}(\mathcal{M}_w(d_l), y_l) \tag{9}$$
$$\text{subject to} \quad g_i(\mathcal{M}_w(d_l), d_l) \leq 0, \ i = 1, \ldots, m, \ l = 1, \ldots, n$$

for some loss function $\mathcal{L}$. In essence this states that $\mathcal{M}_w$ should have weights such that it minimises the loss over all samples, while being such that the output satisfies the constraints for all samples. To this end, relax the constraints into the loss function to form the Lagrangian loss function:

$$\mathcal{L}_\lambda(\hat{y}_l, y_l, d_l) = \mathcal{L}(\hat{y}_l, y_l) + \sum_{i=1}^{m} \lambda_i \nu(g_i(\hat{y}_l, d_l)) \tag{10}$$

where $\hat{y}_l = \mathcal{M}_w(d_l)$. For a given set of Lagrange multipliers λ, the Lagrangian relaxation is:

$$w^*(\lambda) = \underset{w}{\operatorname{argmin}} \sum_{l=1}^{n} \mathcal{L}_\lambda(\mathcal{M}_w(d_l), y_l, d_l) \tag{11}$$

The solution is an approximation $\mathcal{M}_{w^*(\lambda)}$ of $\mathcal{O}$. To find a stronger Lagrangian relaxation, the Lagrangian Dual is used to compute the optimal multipliers:

$$\lambda^* = \underset{\lambda}{\operatorname{argmax}} \ \underset{w}{\min} \sum_{l=1}^{n} \mathcal{L}_\lambda(\mathcal{M}_w(d_l), y_l, d_l) \tag{12}$$

The LDF implements subgradient optimisation to iteratively solve for w and λ. The training process is summarised in Algorithm 1. It takes in the following inputs: Training data D, number of training epochs n_{epochs}, Lagrangian step size s_i for $i = 1, ..., m$, and initial Lagrange multipliers λ_i^0 for $i = 1, ..., m$.

Algorithm 1. LDF Algorithm

Input: D, n_{epochs}, s_i, λ_i^0
   for $k = 0, 1, ..., n_{epochs}$ **do**
      for all $(y_l, d_l) \in D$ **do**
         $\hat{y} \leftarrow \mathcal{M}_w(d_l)$
         $w \leftarrow w - \alpha \nabla_w \mathcal{L}_{\lambda^k}(\hat{y}, y_l, d_l)$
      end for
      $\lambda_i^{k+1} \leftarrow \lambda_i^k + s_i \sum_{l=1}^{n} \nu_i(g_i(\hat{y}, d_l)), \quad i = 1, ..., m$
   end for

---

3 Experiment Setup

In this section, we describe the approach taken to approximate the solutions to the Knapsack Problem using neural networks within the Lagrangian Dual Framework. We provide the details of our methods for instance generation, introduce our approach for decoding and evaluating solutions predicted by our models, and discuss the network architectures and training procedures.

3.1 Instance Generation

Instances should be generated for training and testing which represent a wide variety of adequately difficult cases. This paper uses the size of the knapsack capacity relative to the total item weights as a proxy for instance difficulty. We generated 30,000 instances each with $n = 500$ items, from which we used 24,000 instances for training while 3000 were set aside for testing and validation sets. The item weights and values are uncorrelated and uniformly distributed on $[0, 1]$. A method proposed in [19] is used to generate the instance capacities, where the capacity of the jth instance is set to $W_j = \frac{j}{S+1} \sum_{i=1}^{n} w_i$, S being the total number of instances. This produces data with a full coverage of instance capacities and difficulties, from instances in which the optimal solution have few items to those which have almost all items. Target labels were generated for each instance using Gurobi 9.5.2 on a PC running Ubuntu Linux with an Intel i5-11400 processor.

3.2   Output Decoding and Evaluation

The output of the neural network is a vector in $\mathbb{R}^n$, where the ith element is a logit corresponding to the ith weight. Binary cross entropy is used for the label portion of the loss function $\mathcal{L}(\hat{y}_l, y_l)$.

Care must be taken in using the model outputs to evaluate the knapsack constraint. At training time the logit outputs must be mapped to binary decision variables. A naive approach is to apply a sigmoid at the output to scale it into the range $[0, 1]$ and then round to zero or one. The problem is that the derivative of the round function is zero everywhere, meaning that informative gradients will not flow back from the constraint evaluation. Instead, a surrogate gradient is substituted in during the backwards pass, similar to methods used to deal with uninformative gradients in spiking neural networks such as in [8]. The gradient of the sigmoid function centred at 0.5 is used as a surrogate with the form:

$$\frac{d\sigma}{dx} = \frac{ke^{-k(x-0.5)}}{(e^{-k(x-0.5)} + 1)^2} \tag{13}$$

The parameter k affects how tightly the function is distributed around 0.5 as shown in Fig. 1. The gradient will be higher when outputs are near 0.5 and tend towards zero near 0 and 1. This reflects the fact that near 0.5 small changes in the input can cause the output of the round function to jump between 0 and 1.

Rounding is also used at inference time to decode the model outputs. Another decoding scheme as proposed in [17], is a greedy algorithm which ensures the capacity constraint is satisfied. For general IP problems, it may not be obvious how to correct infeasible solutions, or it may be computationally expensive to do so. For this reason, although reasonable decoding schemes exist, this paper chooses to focus on a minimal decoding which simply rounds the outputs to investigate performance under minimal output post-processing. As in [17] the approximation ratio (AR) is used as a metric to evaluate the model. For a set of instances of size S it is defined as:

$$AR = \frac{1}{S} \sum_{j=1}^{S} max \left(\frac{f^*(x_j)}{f(x_j)}, \frac{f(x_j)}{f^*(x_j)} \right)$$

where $f^*(x_j)$ and $f(x_j)$ are the predicted and true optimal objective values for the jth instance. The approximation ratio is not well defined for instances where either $f^*(x_j)$ or $f(x_j)$ is equal to zero, but not both. Here we define the approximation ratio to be equal to two whenever this occurs. While arbitrary, the

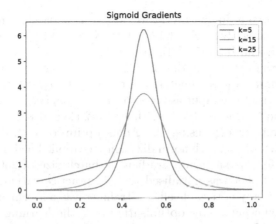

Fig. 1. Surrogate Gradient

idea is that a strong predictor should have an approximation ratio approaching one and this represents a small penalty for incorrect predictions. Regardless, instances where one of either $f^*(x_j) = 0$ or $f(x_j) = 0$ are rare and should not significantly influence results.

While the approximation ratio is important, for the purpose of model validation it only evaluates the optimality of predictions without considering constraint satisfaction. Since the Lagrangian multipliers update on each epoch, the loss $\mathcal{L}_\lambda(\hat{y}_l, y_l, d_l)$ may increase between epochs even as overall model performance is improving which makes it a flawed metric for comparing performance across epochs. To this end, we introduce the μ-loss, which replaces the Lagrange multiplier in the loss function with fixed μ values as shown below:

$$\mathcal{L}_\mu(\hat{y}_l, y_l, d_l) = \mathcal{L}(\hat{y}_l, y_l) + \mu\nu(g_i(\hat{y}_l, d_l)) \tag{14}$$

In choosing a model which minimises the μ-loss it should be possible to select for models which place higher relative priority on prediction optimality or constraint satisfaction as needed by varying μ. In practice model selection will depend on the specific application and the relative importance of optimality and constraint satisfaction.

3.3   Network Architecture and Training Procedure

We have trained three neural network models for approximating solutions to the Knapsack Problem: the first was a fully connected neural network (FC) used as a baseline model, the second model used the LDF to model the constraints, the third model used the LDF but also used the FC baseline as a pre-trained model. Equivalently the pre-trained LDF could be viewed as fixing the Lagrange multipliers at zero for some fixed amount of epochs or until the model converges before allowing them to be updated. To our knowledge there are currently no other suitable methods in the literature to use as a baseline model. All models were trained using PyTorch v1.12.1 with Numpy v1.23.5 on a PC running Ubuntu Linux with an Intel i5-11400 processor[1].

An input to the network is the parameters for a single instance. For a KP instance with n items, the parameters are the weights w_i, values v_i and capacity W. We then construct the input as a vector $[w_1, \ldots, w_n, v_1, \ldots, v_n, W] \in \mathbb{R}^{2n+1}$.

All models had two hidden layers with widths 2048 and 1024, ReLU activation functions, and were trained using the Adam optimiser with default β values. While training all models batch normalization was applied on both hidden layers, $k = 25$ was set for the surrogate gradient, a batch size of 256 was used, and the Lagrange multiplier was initialised to 1. In practice varying k in the range $[5, 35]$ did not have any noticeable effect. Training was performed for 500 epochs. We performed hyperparameter optimisation over the learning rate, maximum gradient norm, and Lagrangian step size. Values for hyperparameter optimisation are recorded in Table 1.

Table 1. Hyperparameter Optimisation Sets

Model	Learning Rate	Lagrangian Step	Grad Norm
FC	$\{10^{-4}, 10^{-3}\}$	N/A	$\{1, 10\}$
LDF	$\{10^{-5}, 10^{-4}, 10^{-3}\}$	$\{10^{-8}, 10^{-7}, 10^{-6}, 10^{-5}, 10^{-4}, 10^{-3}\}$	$\{0.5, 1, 10\}$
Pre-trained LDF	$\{10^{-4}, 10^{-3}\}$	$\{10^{-7}, 10^{-6}, 10^{-5}, 10^{-4}, 10^{-3}\}$	$\{1, 10\}$

We selected models based on performance on the validation set. The chosen FC model was the one which minimised the AR, both LDF models were chosen to minimise the 1-loss. All results are reported on the test set. The chosen FC model was trained with a learning rate of 10^{-3} and maximum gradient norm of 10. The chosen LDF model was trained with a learning rate of 10^{-4}, Lagrangian step size of 10^{-7}, and maximum gradient norm of 0.5. The chosen pre-trained LDF model was trained with a learning rate of 10^{-4}, Lagrangian step size of 10^{-4}, and maximum gradient norm of 10.

A common problem in training the LDF model was exploding gradients originating from the constraint evaluation. The default weight initialisation produces an output with approximately half of the knapsack weights active. As a result, predictions early in the training process can significantly violate the capacity constraint in low capacity instances. Gradient clipping alleviates this, but it's unclear if this may distort the training process and reduce the relative importance of constraint satisfaction in learned models. This issue is much less pronounced in the pre-trained model but is still present. This made the pre-trained LDF models easier to train and more robust to changes in hyperparameters since they did not have the same degree of instability as the LDF models.

---

[1] Code available at https://github.com/mitchellkeegan/knapsack-ldf.

4   Empirical Results

Table 2 reports results for the baseline FC model. All figures are reported as a percentage. For constraint violation, we report the percentage of instances in which the constraint was violated, and the average violation percentage. The average violation is calculated only over the instances in which the constraint is violated. It also filters out extreme outliers that can occur in very low capacity (those in which the knapsack capacity is extremely small compared to the sum of the item weights) instances by ignoring any instances in which the violation is more than six standard deviations from the mean violation. For the objective statistics, we report the percentage of instances in which the predicted objective was under and above the optimal objective, and the average undershoot and overshoot calculated only on the instances in which the predicted objective is under/over the optimal objective (labelled Avg-O and Avg-U). Lastly, the approximation ratio is reported. Performance is broken down into quintiles by instance capacity relative to total item weight denoted $\alpha = \frac{W_j}{\sum_{i=1}^{n} w_{ij}}$ where w_{ij} and W_j are the weight of the ith item and the knapsack capacity respectively in the jth instance.

Table 2. Baseline Neural Network Performance

α	Constraint Violation		Objective Statistics				
	% Violated	Mean Violation	% Under	% Over	Avg-O	Avg-U	AR
0–0.2	64.7	105.7	64.1	35.9	39.1	13.6	1.254
0.2–0.4	66.1	8.77	56.8	43.2	3.26	4.49	1.0417
0.4–0.6	95.42	9.9	14.92	85.1	3.17	1.36	1.029
0.6–0.8	99.84	10.4	8.6	91.4	1.97	0.55	1.0185
0.8–1	100	7.55	3.17	96.83	0.66	0.11	1.0064
All	84.93	21.14	30	70	5.89	7.76	1.0703

The FC model shows strong performance on the AR. Performance is relatively poor on low capacity instances. It would be expected that low capacity instances are inherently harder since the inclusion or exclusion of a single item could be the difference between a near optimal solution and a poor solution. This is reflected in the outsized mean violation on low capacity instances. This gives a distorted view of performance on these instances that is not present when the constraint violation is considered in absolute terms. Rates of constraint violation are extremely high across all quintiles, reaching 100% in the highest capacity quintile. This reflects that the constraints are not explicitly modelled during learning. Table 3 reports results on the LDF model.

Constraint satisfaction is significantly improved in the LDF model. Across all instances, the capacity constraint is violated only 2.13% of the time, a significant improvement on the 85% violation rate reported on the FC model. These

Table 3. LDF Neural Network Performance

α	Constraint Violation		Objective Statistics				
	% Violated	Mean Violation	% Under	% Over	Avg-O	Avg-U	AR
0–0.2	10.48	195	96.8	3.16	113.2	30.35	1.5643
0.2–0.4	0.16	2.93	100	0	N/A	12.73	1.1521
0.4–0.6	0	N/A	100	0	N/A	8.46	1.0931
0.6–0.8	0	N/A	100	0	N/A	5.88	1.0628
0.8–1	0	N/A	100	0	N/A	2.62	1.027
All	2.13	191	99.4	0.6	113.24	12	1.1811

constraint violations exclusively occur in lower capacity instances, moderate to high capacity instance recorded no constraint violations whatsoever. This comes at some cost to the approximation ratio, but the increase is relatively minor, on the order of 10%. Table 4 reports results on the pre-trained LDF model.

Table 4. Pre-trained LDF Neural Network Performance

α	Constraint Violation		Objective Statistics				
	% Violated	Mean Violation	% Under	% Over	Avg-O	Avg-U	AR
0–0.2	14	278	93.8	6.2	92.1	30.4	1.51
0.2–0.4	0.32	1.93	100	0	N/A	2.38	1.1894
0.4–0.6	0.17	1.38	100	0	N/A	8.78	1.09732
0.6–0.8	0	N/A	100	0	N/A	5.29	1.0562
0.8–1	0	N/A	100	0	N/A	2.37	1.0245
All	2.9	268	98.8	1.2	92.1	12.4	1.177

The performance of the pre-trained LDF model is very similar to the base LDF model, with no clear indication that either model is strictly better. As noted previously we found training to be significantly easier with the pre-trained model. In particular, it was found to be more robust with respect to hyperparameter changes, and alleviated the strong dependence on gradient clipping to deal with exploding gradients.

The computational time required for training and generating predictions is also of interest, particularly in comparison to traditional solvers. The total epochs and time taken for the three models to converge during training are listed in Table 5. For the pre-trained LDF model these values include the pre-training time. The LDF model takes significantly longer to converge fully. Prediction times for the neural network models, averaged over the full set of 30,000 instances, took 1 ms per instance. Gurobi on the same set of instances takes on average 5.7 ms per instance, noting that Gurobi is used here for convenience but other methods may be significantly faster at solving KP. Prediction time could likely be further reduced by using a GPU, with the caveat that in reality it may not be useful to solve more than one instance at a time. More generally for harder IP problems (or harder KP instances) we would expect predictions generated by neural networks to be significantly faster.

Table 5. Training Time for Neural Network Models

	FC	Pre-trained LDF	LDF
Epochs	16	46	250
Time (Minutes)	0.7	2.6	16

These results demonstrate an ability to trade-off optimality for constraint satisfaction by using the LDF. In LDF models poor performance is mostly concentrated in low capacity instances, moderate and high capacity instances achieve strong performance in terms of the AR without violating any constraints on the test set. It's likely that the unconstrained FC model achieves strong performance in terms of the approximation ratio by consistently violating the capacity constraint by a moderate amount. Enforcing constraint satisfaction reduces or removes these violations but does not directly help the model learn the weights in the optimal solution leading to a small increase in the AR. This gives rise to a trade-off between optimality and constraint satisfaction. It's not obvious whether this relationship would generalise to other IP problems, in fact experiments applying LDF to the job shop scheduling problems report impressive improvement in both optimality and constraint satisfaction [16].

5   Conclusions

In this paper, we investigated using neural networks to approximate solutions to the Knapsack Problem while committing to the constraints. It experimentally demonstrated that the Lagrangian Dual Framework could be used to strongly encourage satisfaction of the knapsack capacity constraint with a reasonably small reduction in the optimality of predicted solutions. It discussed issues in interpreting neural network outputs as solutions to integer programming problems and evaluating their performance, particularly in the context of model validation. It also demonstrated that using a pre-trained neural network as a base model may alleviate problems with exploding gradients found in the LDF model. While the scope of this work is limited to KP, it is hoped that some of the principles and techniques discussed will also be applicable to the approximation of solutions for other combinatorial optimisation problems using neural networks. An obvious research direction is applying LDF to more challenging combinatorial optimisation problems. In general solving combinatorial optimisation problem instances may itself be a difficult and computationally expensive task, which makes generating training sets challenging. In focusing on KP we sidestep these issues surrounding data generation, but this will be an interesting challenge in further research.

References

1. Abolghasemi, M.: The intersection of machine learning with forecasting and optimisation: theory and applications. In: Hamoudia, M., Makridakis, S., Spiliotis, E. (eds.) Forecasting with Artificial Intelligence. PAEIT, pp. 313–339. Palgrave Macmillan, Cham (2023). https://doi.org/10.1007/978-3-031-35879-1_12
2. Abolghasemi, M., Abbasi, B., Babaei, T., HosseiniFard, Z.: How to effectively use machine learning models to predict the solutions for optimization problems: lessons from loss function. arXiv preprint arXiv:2105.06618 (2021)
3. Abolghasemi, M., Abbasi, B., HosseiniFard, Z.: Machine learning for satisficing operational decision making: a case study in blood supply chain. Int. J. Forecast. (2023)
4. Abolghasemi, M., Esmaeilbeigi, R.: State-of-the-art predictive and prescriptive analytics for IEEE CIS 3rd technical challenge. arXiv preprint arXiv:2112.03595 (2021)
5. Bengio, Y., Lodi, A., Prouvost, A.: Machine learning for combinatorial optimization: a methodological tour d'Horizon. Eur. J. Oper. Res. **290**(2), 405–421 (2021)
6. Detassis, F., Lombardi, M., Milano, M.: Teaching the old dog new tricks: supervised learning with constraints. In: Proceedings of the AAAI Conference on Artificial Intelligence, vol. 35, no. 5, pp. 3742–3749 (2021)
7. Donti, P., Rolnick, D., Kolter, J.Z.: DC3: a learning method for optimization with hard constraints. In: International Conference on Learning Representations (2021)
8. Eshraghian, J.K., et al.: Training spiking neural networks using lessons from deep learning. arXiv preprint arXiv:2109.12894 (2021)
9. Fioretto, F., Mak, T.W., Van Hentenryck, P.: Predicting AC optimal power flows: combining deep learning and Lagrangian dual methods. In: Proceedings of the AAAI Conference on Artificial Intelligence, vol. 34, no. 1, pp. 630–637 (2020)
10. Fioretto, F., Van Hentenryck, P., Mak, T.W.K., Tran, C., Baldo, F., Lombardi, M.: Lagrangian duality for constrained deep learning. In: Dong, Y., Ifrim, G., Mladenić, D., Saunders, C., Van Hoecke, S. (eds.) ECML PKDD 2020. LNCS (LNAI), vol. 12461, pp. 118–135. Springer, Cham (2021). https://doi.org/10.1007/978-3-030-67670-4_8
11. Fontaine, D., LaurentMichel, Van Hentenryck, P.: Constraint-based Lagrangian relaxation. In: O'Sullivan, B. (ed.) CP 2014. LNCS, vol. 8656, pp. 324–339. Springer, Cham (2014). https://doi.org/10.1007/978-3-319-10428-7_25
12. Hertrich, C., Skutella, M.: Provably good solutions to the Knapsack problem via neural networks of bounded size. INFORMS J. Comput. (2023)
13. Hopfield, J., Tank, D.: "Neural" computation of decisions in optimization problems. Biol. Cybern. **52**(3), 141–152 (1985)
14. Kotary, J., Fioretto, F., van Hentenryck, P., Wilder, B.: End-to-end constrained optimization learning: a survey. In: IJCAI International Joint Conference on Artificial Intelligence, pp. 4475–4482 (2021)
15. Kotary, J., Fioretto, F., Hentenryck, P.V.: Learning hard optimization problems: a data generation perspective (2021)

16. Kotary, J., Fioretto, F., Hentenryck, P.V.: Fast approximations for job shop scheduling: a lagrangian dual deep learning method. In: Proceedings of the AAAI Conference on Artificial Intelligence, vol. 36, no. 7, pp. 7239–7246 (2022)
17. Nomer, H.A.A., Alnowibet, K.A., Elsayed, A., Mohamed, A.W.: Neural Knapsack: a neural network based solver for the Knapsack problem. IEEE Access **8**, 224200–224210 (2020)
18. Pisinger, D.D., Kellerer, H., Pferschy, U., Pisinger, D.: Knapsack Problems/Hans Kellerer, Ulrich Pferschy, David Pisinger. Springer, New York (2004)
19. Pisinger, D.: Core problems in Knapsack algorithms. Oper. Res. **47** (2002). https://doi.org/10.1287/opre.47.4.570

An Optimised Grid Search Based Framework for Robust Large-Scale Natural Soundscape Classification

Thomas Napier[✉][iD], Euijoon Ahn, Slade Allen-Ankins, and Ickjai Lee

James Cook University, Townsville, QLD 4811, Australia
{thomas.napier,euijoon.ahn,slade.allenankins,ickjai.lee}@jcu.edu.au

Abstract. Large-scale natural soundscapes are remarkably complex and offer invaluable insights into the biodiversity and health of ecosystems. Recent advances have shown promising results in automatically classifying the sounds captured using passive acoustic monitoring. However, the accuracy performance and lack of transferability across diverse environments remains a challenge. To rectify this, we propose a robust and flexible ecoacoustics sound classification grid search-based framework using optimised machine learning algorithms for the analysis of large-scale natural soundscapes. It consists of four steps: pre-processing including the application of spectral subtraction denoising to two distinct datasets extracted from the Australian Acoustic Observatory, feature extraction using Mel Frequency Cepstral Coefficients, feature reduction, and classification using a grid search approach for hyperparameter tuning across classifiers including Support Vector Machine, k-Nearest Neighbour, and Artificial Neural Networks. With 10-fold cross validation, our experimental results revealed that the best models obtained a classification accuracy of 96% and above in both datasets across the four major categories of sound (biophony, geophony, anthrophony, and silence). Furthermore, cross-dataset validation experiments using a pooled dataset highlight that our framework is rigorous and adaptable, despite the high variance in possible sounds at each site.

Keywords: Ecoacoustics · Signal Processing · Machine Learning · Optimised Grid Search

1 Introduction

Australia is one of the most biodiverse regions on Earth, yet many species are under threat [2]. Effective monitoring solutions and techniques have now become imperative for the tracking of at-risk species. Ecoacoustics serves as one such solution, which has gained recent attention for its potential in ecological conservation [1,8,16,17]. Leveraging modern advancements in low-cost sound recording and data storage solutions, remote sensor monitoring of natural soundscapes are now possible by way of large-scale Passive Acoustic Monitoring (PAM) [5]. This, in turn, allows ecoacoustics studies to now utilise the wide spatial and temporal soundscape coverage enabled by PAM [14].

In recognition of these conservation benefits, a new large-scale sensor network was established called the Australian Acoustic Observatory (A2O) [14]. The A2O seeks to capture sounds at an ecosystem level. To do so, over 360 listening stations are situated at 90 different sites across Australia to capture sounds from as many ecoregions as possible. Natural soundscapes are broadly composed of four sound groups including: biophony (the sounds produced by animals), geophony (natural non-biological sounds like wind or water), anthrophony (sounds induced by humans) and periods of silence. By segregating these recorded sounds into the four primary categories, researchers can gain a nuanced understanding of their relative balance which is useful for analysing the state of biodiversity and effects of human impact [6,7]. Despite this, much of the existing Machine Learning (ML) and Deep Learning (DL) ecoacoustics research relies on single-species, often non-ecological datasets [4,5,8,15]. While these are valuable, they don't provide the comprehensive insights needed for broader ecological conclusions. Analysing large-scale PAM datasets like those derived from the A2O, are uniquely challenging for several reasons. Firstly, they feature an exceptionally high variance in the types of sounds captured. Natural environments are profoundly complex and dynamic places, which change on a day-to-day basis. This is compounded by the sheer diversity of species calls, which often overlap each other in both time and frequency [13]. To address these intricacies, this research will explore the integration of select Feature Extraction (FE), Feature Reduction (FR), and denoising techniques. Our framework will need to be flexible and robust with these considerations in mind, while still maintaining high accuracy. We believe that using a grid search based approach for both the hyperparameter-sensitive FR techniques, and associated classifiers, will achieve these desired outcomes.

The contributions of this work can be summarised as follows:

- Proposal of a novel grid search based methodology that leverages ML algorithms to identify and categorise distinct sounds within diverse ecological environments;
- Utilisation of a human-inspired approach to feature representation, aimed at improving classification performance;
- Provision of a range of exploratory experiments using the real ecoacoustics data captured from two distinct ecosystems in Australia to evaluate the effectiveness of each classifier-based prediction model on unseen contextual test cases;
- Identification and suggestion of the most suitable algorithms and supervised learning techniques for classifying ecoacoustics sounds into the four broad categories in soundscape ecology.

2   Related Works

Natural soundscapes are inherently complex due to the vast diversity of species calls which can change on a day-to-day basis. For this reason, models must be generalisable with these complications considered in order to be genuinely useful in the real world. However, this is often not the case. Many existing approaches are trained and tested on single species datasets, or ones with low taxonomic variation, which does not accurately reflect the real-world [5]. Those which do use natural soundscape datasets such as one study in 2021 [16], have employed the use of summary statistics in the form of acoustic indices. By using overly summative features like acoustic indices, crucial details are lost which diminishes the richness of the recordings, resulting in a loss of overall accuracy down to 70% in some downstream classification tasks.

Another study using a natural soundscape dataset has shown some merit in classifying broad sound groupings. Here, the authors were able to achieve relatively accurate results across each of the major sound groups (biophony, geophony, and anthrophony), ranging from 88% to 95% accuracy using an Artificial Neural Network (ANN) classifier with Mel Frequency Cepstral Coefficients (MFCCs) [6]. However, the study was conducted using a dataset constructed from four recorders in a comparatively smaller study site compared to the A2O, situated on the border of France and Switzerland. Furthermore, sounds were collected episodically for 1 min, every 15 min, rather than continuously as with recordings derived from the large-scale initiatives like the A2O [14]. Thus, the flexibility of their approach is unknown as the authors did not experiment with data from alternative study sites.

Furthermore, several studies have used overly sanitised or non-representative datasets. For this reason, many approaches are capable of performing well on datasets containing specific species like frogs [3] and birds [8,15], however, they lack flexibility to different ecoregions, and are too specialised to assist in answering broader ecological questions. Accurate, adaptable ecoacoustics classification is pivotal for ecologists. Not only can it facilitate comprehensive biodiversity assessments, but it also plays a crucial role in early detection and prevention of species loss.

3   The Proposed Framework

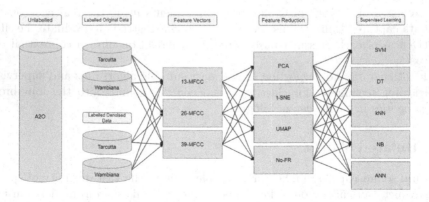

Fig. 1. A conceptual diagram for a robust grid search based framework for use with large-scale ecoacoustics data.

In this study, we propose a novel grid search based classification framework for use with large-scale natural soundscape audio signals based on MFCC feature vectors and ML techniques as shown in Fig. 1. Raw, unlabelled audio was collected directly from the A2O from two geographically different ecoregions to test for cross-site validity. The raw signals were segmented into 4.5-second-long non-overlapping sequences, resulting in a combined total of 8,841 samples, which were subsequently annotated by human experts. From this point we established two datasets: one was unaltered containing the original 8,841 labelled samples, and the second was a direct copy with spectral subtraction denoising applied to all samples. From this, we extracted the MFCC feature vectors as represented by the corresponding heatmap visualised in Fig. 2(b). To achieve this, we compute a Mel-Spectrogram with 128 Mel bands, as seen in Fig. 2(a). This was chosen due to its ability to closely mirror the human auditory system's frequency perception. Furthermore, transforming the spectrograms onto a decibel logarithmic scale has been shown to successfully capture the underlying signal properties. This has been validated across a range of ML tasks including bird song classification [11,19], as well as its usage in acoustic scene classification [6,12,20]. The extraction of 13 MFCC features provides a compact representation of the audio's spectral shape, allowing for high intra-class variability for class discrimination. In addition, we also include the first- and second-order derivatives of MFCCs in a 26- and 39-feature vector, respectively, as a way of capturing the audio's temporal dynamics. Finally, we apply a min-max normalisation scheme to avoid any single feature from disproportionately influencing the model due to its scale. Due to the high-dimensional nature of these feature vectors, we also used several FR techniques including Principle Component Analysis (PCA),

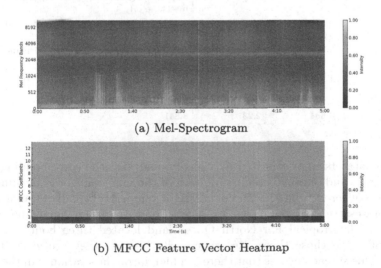

(a) Mel-Spectrogram

(b) MFCC Feature Vector Heatmap

Fig. 2. An example Mel-Spectrogram and associated 13-MFCC feature vector heatmap derived from the same 5 min of biophony audio from the Tarcutta A2O site.

t-distributed Stochastic Neighbour Embedding (t-SNE) and Uniform Manifold Approximation and Projection (UMAP) to measure their effect on downstream model performance and to counter potential cases of overfitting. For classifiers, we implemented a grid search approach for tuning the hyperparameters of several ML models including Artificial Neural Network (ANN) and k-Nearest Neighbour (k-NN), and used the default parameter settings for Support Vector Machine (SVM), Decision Tree (DT), and Naïve Bayes (NB). The performance of each classifier was evaluated using macro-based metrics for accuracy, precision, recall, and F1-score, specifically to account for the imbalanced nature of the dataset.

4   Datasets and Experiments

4.1   Datasets

Each A2O site consists of a group of four sensors, two in dry areas and two in wet areas, each recording continuously for 24 h per day [14]. In this study, we selected a two-week period during Australia's Autumn season for further analysis. For each day and sensor, a Long-Duration False Colour (LDFC) spectrogram was generated using a select combination of acoustic indices [18]. LDFC spectrograms provide a snapshot into the day's acoustic activity which can be visually scanned.

Table 1. Tarcutta and Wambiana dataset breakdown by class.

Sound Category	Tarcutta (# samples)	Wambiana (# samples)	Sounds Included
Biophony	2,379	685	Any sound generated by animals (birds, frogs, insects, etc.)
Geophony	2,014	573	Any sound from the earth (water, wind, fire, etc.)
Anthrophony	532	252	Any human-made sound (cars, airplanes, human speech, etc.)
Other/Silence	1,679	727	Mostly represents long periods of silence but can include sounds like "white noise" or "pink noise", electromagnetic interference, etc.
Total	6,604	2,237	8,841

Two datasets were collected by visually analysing the corresponding LDFC spectrogram and aurally listening for each of the main categories of sound. The first was collected from the Tarcutta site, a temperate woodland area located in south-western New South Wales. The second was from Wambiana, a small station in the tropical Far North Queensland located three hours outside of Townsville. We chose these sites specifically because they are geographically spread. This spread ensures that there is a high intra-class variance in the sounds captured, which is representative of the real-world. Furthermore, sounds vary greatly on a day-to-day basis, which we wanted to capture by taking samples from as many days as possible. As seen in Table 1, the final datasets were roughly equally distributed across three of the four major sound groupings (biophony, geophony and other/silence), with less anthrophony due to the seclusion of the sites. Importantly, no preprocessing was conducted prior to manual annotation to allow for the subsequent models to learn from real-world examples, with the noise and variability of environmental factors included.

4.2   Feature Reduction Experiments

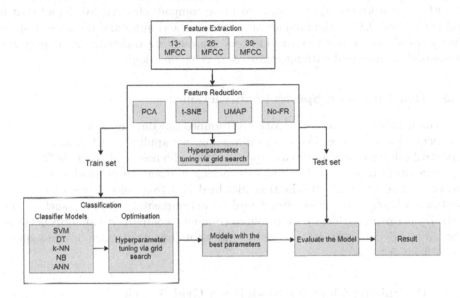

Fig. 3. The implementation workflow of the proposed framework.

Here we present the implementation workflow of the proposed framework as per Fig. 3. Firstly, to improve computational efficiency and mitigate the curse of dimensionality, we employed FR. Three techniques were selected for further analysis including PCA, t-SNE and UMAP, with No-FR as a baseline. A vital part of using PCA in practice is the ability to estimate how many components are needed to describe the data. Similarly, the results of t-SNE and UMAP can vary depending on the choice of hyperparameters, namely perplexity for t-SNE, and number of neighbours and minimum distance for UMAP. For PCA, we employed the cumulative explained variance ratio as a determinant for selecting the optimal number of components. By plotting this ratio against the number of components, we discerned that 7 components were sufficient to capture approximately 95% of the total variance in the data, across all 13, 26 and 39 feature vectors, striking a balance between data compression and information retention.

To find the optimal choice of hyperparameters for t-SNE and UMAP, we applied a grid search approach across both the Tarcutta and Wambiana datasets.

For t-SNE, optimal perplexity values typically fall between 5 and 50 [9]. Perplexity is a parameter designed to shift the attention of local versus global structures in the datasets. A low perplexity emphasises local structure, while a high value may reveal more global structure. For our datasets, we found that a perplexity value of 50 captured enough local distinctions while revealing the broader patterns. Similarly, UMAP relies on the tuning of the number of neighbours and minimum distance parameters to balance local versus global structures in the data embedding [10]. Here, a lower number of neighbours will focus on local structure, and the minimum distance controls how tightly UMAP packs points together, with lower values leading to more compact clusters. We found that a value of 6 and 0.1 for the number of neighbours and minimum distance, respectively, produced a result where clusters representing different sound patterns remained distinct and without excessive overlap or bridging.

4.3 Denoising with Spectral Subtraction

For the next step in our framework, we retained one dataset in its original state to serve as a reference. On a copied version, we applied spectral subtraction. Spectral subtraction is a common approach to audio noise reduction. It functions by generating a noise profile and subtracting it from the original signal. This process preserves vital information like bird and frog calls, while eliminating stationary background noise often found in environmental recordings such as rain [21]. By performing this, we were able to conduct a comparative analysis between the raw and denoised data, thereby understanding its impact on subsequent classification tasks.

4.4 Optimising Classifier Models via Grid Search

To effectively evaluate the classification performance on the uniquely challenging natural soundscape datasets, we employed a diverse set of supervised learning techniques including SVM, DT, k-NN, NB and ANN. To find this selection, we examined several techniques and chose based on their differences in learning principles and foundational algorithms. We wanted to showcase a range of strategies to investigate how they perform with the inherent complexities of the ecoacoustics datasets. SVM is well-known for handling high-dimensional data, making it an ideal candidate for this study. Given its ability to handle high-dimensional data and its efficacy in finding optimal hyperplanes for classification, it was a clear choice. For our purposes, we used the default parameters, as they offer a solid benchmark and are often optimised for a broad range of datasets.

We selected DT because they are interpretable, and their hierarchical structure allows for an intuitive understanding of decision processes. Using default parameters provides a baseline and avoids overfitting that might arise from excessively deep or complex trees. Similarly, NB was incorporated with default parameters to test how the model's underlying probabilistic assumptions perform with these datasets. With a range of algorithmic approaches selected, we identified the need for hyperparameter optimisation, where the model's sensitivity to them significantly influences the outcome. As such, we employed a grid search approach for both k-NN and ANN, as both require a thorough evaluation for performance optimisation. k-NN is particularly effective in situations where data might form natural clusters based on similarity. However, the choice of k is crucial. As such, we utilised a grid search approach to ascertain the optimal number of neighbours k to consider, ranging from ($k = 1, 3, 5, 7, ... 31$), ensuring our model was neither too generalised nor too specific. Similarly, for ANNs with their inherent flexibility, certain hyperparameters such as solver type and hidden layer sizes required tuning.

5 Results and Discussion

5.1 Classification Results

Experiment Setup. We applied the grid search method with 10-fold cross validation using several datasets: firstly, the unaltered datasets from the Tarcutta and Wambiana sites derived from the A2O containing 6,604 and 2,237 signal samples, respectively, as well as their denoised versions where spectral subtraction was applied. Additionally, we further constructed a combined dataset, pooling samples from both sites together. 10-fold cross validation ensured that each approach minimised the risk of overfitting and provided us with a more reliable assessment of their performance. Furthermore, to ascertain the model's capability for generalisation across different ecoregions, we designed two cross-dataset validation tests. Instead of using 10-fold cross validation, we used an 80%/20% train/test split, respectively. By training on one site, and testing it on another, we could evaluate how well the models adapted to new, unseen data, which is vital in large-scale ecoacoustics, where conditions can greatly vary between sites. We assess model performance using the following key evaluation metrics: accuracy (proportion of correct predictions to total predictions), precision (proportion of true positive predictions to total positive predictions), recall (proportion of true positive predictions to actual positives), and the F1-score (harmonic mean of precision and recall). After conducting several simulations, the classification accuracy performance of each combination of feature vector, tuned feature reduction approach, and optimised classifier for each dataset is showcased in Fig. 4.

Findings. For ANNs, we found the Adam optimiser with a hidden layer configuration of 10 neurons in two consecutive layers to be the best performer. Across our experiments, ANNs were consistent, positioning it around the midpoint in terms of performance out of all the classifiers as seen by the relatively even colouring in Fig. 4. Interestingly however, denoising exhibited mixed effects, improving accuracy by 2–3% in the Tarcutta dataset, but decreasing it by approximately the same for Wambiana. Conversely, NB performed the least effectively among the studied classifiers. Regardless of the MFCC vector dimensionality, it repeatedly underperformed. Despite this, some FR, especially UMAP, bolstered its

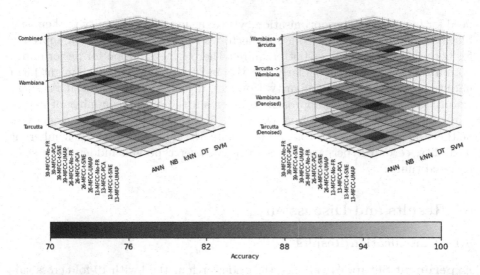

Fig. 4. Dual 3D stacked heatmaps showcasing the accuracy performance of each dataset on the Z-axis with respect to each classifier (ANN, NB, k-NN, DT, SVM) on the X-axis, MFCC feature vector (13, 26, 39) and FR combination (PCA, t-SNE, UMAP, No-FR) on the Y-axis.

capabilities, but it was still not able to elevate NB significantly enough. Similarly for DTs, although the 13-MFCC feature vector offered some increased performance when combined with t-SNE, it still lagged behind the leading classifiers. For the k-NN classifier algorithm we determined that the optimal k-value was $k = 5$, as it obtained the best classification performance across each dataset. With this, k-NN was able perform relatively well, particularly when paired with UMAP and t-SNE embeddings. However, SVM emerged as the strongest performer among the classifiers, achieving the highest accuracy in four out of the seven datasets used. Interestingly, the 13-dimensional MFCC vector, when combined with No-FR, maximised its class-separation capability. This is evidenced by 97.77%, 98.27%, 97.45% and 97.78% accuracy for the Tarcutta, Wambiana, their combination, and when training with Tarcutta and testing with Wambiana, respectively.

5.2 Discussion

In evaluating the multitude of algorithms employed in this study, SVM emerged as the standout performer in classifying the unique natural soundscape datasets, with k-NN in a close second. As seen in Fig. 5(a), while SVMs consistently achieved the highest accuracy, k-NN's effectiveness was more stable across the experiments as shown in Fig. 5(b). SVM's superior performance highlights its adaptability and efficiency in parsing the complexities inherent in such data. This goes against the findings of [6], who found ANNs with MFCCs to be as equally performant for their natural soundscape dataset. However, we have

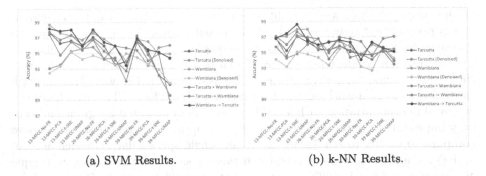

(a) SVM Results.   (b) k-NN Results.

Fig. 5. Comparative line charts highlighting the effectiveness of the two best performing classifiers of (a) SVM and (b) k-NN against each dataset with respect to each MFCC feature vector (13, 26, 39) and FR combination (PCA, t-SNE, UMAP, No-FR).

shown through our experimental design that our approach has more proven flexibility across a range of ecoregions, as opposed to a dataset derived from a single study site. As for the feature vectors, Table 2 shows that classification results were consistently the highest when utilising 13 MFCCs. This goes against our initial assumptions and indicates that these coefficients alone are capable of capturing the underlying audio features without the need for additional temporal detail. The only exception to this case is when signal detail is lost through other means, such as denoising, as signified by the best performing methods for the denoised datasets using higher-MFCC feature vectors.

Table 2. The best classification performance for each dataset using the optimal method.

Dataset	Method	Accuracy	Precision	Recall	F1-score
Tarcutta	13-MFCC+No-FR+SVM	97.77	96.82	95.90	96.31
Wambiana	13-MFCC+No-FR+SVM	98.27	98.38	97.70	97.98
Tarcutta (Denoised)	39-MFCC+UMAP+k-NN	97.03	97.18	97.08	97.08
Wambiana (Denoised)	26-MFCC+UMAP+k-NN	96.04	95.73	95.05	94.90
Tarcutta + Wambiana	13-MFCC+No-FR+SVM	97.45	96.40	95.85	96.09
Tarcutta → Wambiana	13-MFCC+No-FR+SVM	97.78	96.54	96.02	96.27
Wambiana → Tarcutta	13-MFCC+t-SNE+k-NN	98.66	98.63	98.22	98.42

With respect to FR, the efficacy of t-SNE and UMAP in comparison to PCA is conditional based on the type of dataset and classifier used, as evidenced by the varying results in Fig. 5. It is of interest to note, that while SVM had better outcomes with No-FR, other classifiers, like k-NN, consistently benefited from this step. Furthermore, the classifiers demonstrated consistent performance across both the Wambiana and Tarcutta datasets. Regardless of the wide variance in sounds, or the disparity in dataset sizes (6,604 compared to 2,238 samples), the models maintained consistency, further demonstrating the flexibility of the proposed framework. Additionally, our cross-dataset validation experiments revealed not only great robustness in our approach, but also affirmed its potential for generalising across diverse ecoregions. Our optimised models were purposefully not conditioned on a particular set of species, nor were the datasets they were trained on overly sanitised. Instead, our models were trained with the nuances of real-world natural soundscapes included. While this has been achieved in other studies, such as [16], we have demonstrated a significant accuracy improvement upon this. Despite this, our approach was still able to achieve comparable results to studies which do use single-species datasets such as [1,3] and [8]. Conversely, while spectral subtraction as a noise reduction technique appeared promising initially, in practice, it removed too much signal information. From our experiments using the denoised versions of the Tarcutta and Wambiana datasets, models generally saw higher accuracy performance without spectral subtraction. This reaffirms the inherent challenges in this approach, as it can be difficult to generate a single noise profile to cover the wide variance of sounds across two distinct ecosystems and may be more suitable for datasets with a narrower focus [21].

6 Conclusion

Australia has some of the richest biodiversity globally, and it is imperative to monitor its species using sound, particularly those at risk. Until now, there has been a lack of transferability in model design across multi-ecoregion and multi-species datasets. Our methodological design, incorporating a range of supervised classifiers, pivoted around these challenges through the use of two distinct natural soundscape datasets. From this, we were able to conclude that for the majority of cases, 13 MFCCs, with No-FR applied, with SVM as the classifier, is consistently superior for this task. Moreover, this indicates that large-scale ecoacoustics datasets, transformed under this proposed framework, may be linearly separable in high-dimensional space and implies that, given a similar dataset, SVM may provide reliable classifications. Since we focused on incorporating the nuances of real-world natural soundscapes across different ecoregions, we believe that our framework is generalisable. Cross-dataset validation reinforces this, as high accuracy was maintained despite the large sound variance at each site. With this, we have strong supporting evidence that shows our proposed framework improves upon pre-existing approaches, is accurate and robust, and may serve as an ideal base for future ecoacoustics classification tasks.

References

1. Bardeli, R., et al.: Detecting bird sounds in a complex acoustic environment and application to bioacoustic monitoring. Pattern Recogn. Lett. **31**(12), 1524–1534 (2010)
2. Cardinale, B.J., et al.: Biodiversity loss and its impact on humanity. Nature **486**, 59–67 (2012)

3. Colonna, J., et al.: Automatic classification of anuran sounds using convolutional neural networks. In: Proceedings of the Ninth International C* Conference on Computer Science & Software Engineering, C3S2E 2016, pp. 73–78. Association for Computing Machinery, New York (2016)
4. Eichinski, P., et al.: A convolutional neural network bird species recognizer built from little data by iteratively training, detecting, and labeling. Front. Ecol. Evol. **10** (2022)
5. Gibb, R., Browning, E., Glover-Kapfer, P., Jones, K.E.: Emerging opportunities and challenges for passive acoustics in ecological assessment and monitoring. Methods Ecol. Evol. **10**(2), 169–185 (2019)
6. Grinfeder, E., et al.: Soundscape dynamics of a cold protected forest: dominance of aircraft noise. Landscape Ecol. **37**(2), 567–582 (2022)
7. Krause, B.: Anatomy of the soundscape: evolving perspectives. J. Audio Eng. Soc. **56**(1/2), 73–80 (2008)
8. LeBien, J., et al.: A pipeline for identification of bird and frog species in tropical soundscape recordings using a convolutional neural network. Eco. Inform. **59**, 101113 (2020)
9. Van der Maaten, L., Hinton, G.: Visualizing data using t-SNE. J. Mach. Learn. Res. **9**(11) (2008)
10. McInnes, L., et al.: UMAP: uniform manifold approximation and projection. J. Open Source Softw. **3**(29), 861 (2018)
11. Mcloughlin, M.P., et al.: Automated bioacoustics: methods in ecology and conservation and their potential for animal welfare monitoring. J. R. Soc. Interface **16**(155), 20190225 (2019)
12. Mesaros, A., et al.: Detection and classification of acoustic scenes and events: outcome of the DCASE 2016 challenge. IEEE/ACM Trans. Audio Speech Lang. Process. **26**(2), 379–393 (2018)
13. Pijanowski, B.C., et al.: Soundscape ecology: the science of sound in the landscape. Bioscience **61**(3), 203–216 (2011)
14. Roe, P., et al.: The Australian acoustic observatory. Methods Ecol. Evol. **12**, 1802–1808 (2021)
15. Salamon, J., et al.: Towards the automatic classification of avian flight calls for bioacoustic monitoring. PLoS ONE **11**(11), 1–26 (2016)
16. Scarpelli, M.D.A., et al.: Multi-index ecoacoustics analysis for terrestrial soundscapes: a new semi-automated approach using time-series motif discovery and random forest classification. Front. Ecol. Evol. **9**, 738537 (2021)
17. Stowell, D., et al.: Automatic acoustic detection of birds through deep learning: the first bird audio detection challenge. Methods Ecol. Evol. **10**(3), 368–380 (2019)
18. Towsey, M., et al.: Long-duration, false-colour spectrograms for detecting species in large audio data-sets. J. Ecoacoustics **2**(1), 1–13 (2018)
19. Trawicki, M., Johnson, M., Osiejuk, T.: Automatic song-type classification and speaker identification of Norwegian ortolan bunting (Emberiza Hortulana) vocalizations. In: 2005 IEEE Workshop on ML for Signal Processing, pp. 277–282 (2005)
20. Wu, Z., Cao, Z.: Improved MFCC-based feature for robust speaker identification. Tsinghua Sci. Technol. **10**(2), 158–161 (2005)
21. Xie, J., Towsey, M., Zhang, J., Roe, P.: Adaptive frequency scaled wavelet packet decomposition for frog call classification. Eco. Inform. **32**, 134–144 (2016)

Medical AI

Interpretable 3D Multi-modal Residual Convolutional Neural Network for Mild Traumatic Brain Injury Diagnosis

Hanem Ellethy[1](✉) ⓘ, Viktor Vegh[2,3,4] ⓘ, and Shekhar S. Chandra[1] ⓘ

[1] School of Electrical Engineering and Computer Science, University of Queensland, Brisbane, QLD, Australia
h.elwaseif@uq.edu.au
[2] The Centre for Advanced Imaging, University of Queensland, Brisbane, QLD, Australia
[3] Australian Institute for Bioengineering and Nanotechnology, University of Queensland, Brisbane, QLD, Australia
[4] ARC Training Centre for Innovation in Biomedical Imaging Technology, Brisbane, QLD, Australia

Abstract. Mild Traumatic Brain Injury (mTBI) is a significant public health challenge due to its high prevalence and potential for long-term health effects. Despite Computed Tomography (CT) being the standard diagnostic tool for mTBI, it often yields normal results in mTBI patients despite symptomatic evidence. This fact underscores the complexity of accurate diagnosis. In this study, we introduce an interpretable 3D Multi-Modal Residual Convolutional Neural Network (MRCNN) for mTBI diagnostic model enhanced with Occlusion Sensitivity Maps (OSM). Our MRCNN model exhibits promising performance in mTBI diagnosis, demonstrating an average accuracy of 82.4%, sensitivity of 82.6%, and specificity of 81.6%, as validated by a five-fold cross-validation process. Notably, in comparison to the CT-based Residual Convolutional Neural Network (RCNN) model, the MRCNN shows an improvement of 4.4% in specificity and 9.0% in accuracy. We show that the OSM offers superior data-driven insights into CT images compared to the Grad-CAM approach. These results highlight the efficacy of the proposed multi-modal model in enhancing the diagnostic precision of mTBI.

Keywords: mTBI diagnosis · CNN · multi-modal · Occlusion sensitivity map · CT · Residual CNN · Deep learning

1 Introduction

Mild Traumatic Brain Injury (mTBI), representing the majority of annual global traumatic brain injuries (TBI), leads to serious and enduring consequence [1]. Despite its prevalence, the diagnosis of mTBI remains a substantial challenge [2]. Traditionally, mTBI diagnosis relies on subjective clinical evaluations, including symptoms and cognitive tests, which often lack consistency and accuracy [3]. Computed Tomography (CT), despite its limitations in detecting subtle or non-existent structural changes in mTBI cases [4], is the standard neuroimaging test due to its availability and speed in assessing intracranial lesions [5].

© The Author(s), under exclusive license to Springer Nature Singapore Pte Ltd. 2024
T. Liu et al. (Eds.): AI 2023, LNAI 14471, pp. 483–494, 2024.
https://doi.org/10.1007/978-981-99-8388-9_39

Machine learning, specifically deep learning (DL), has shown promise in medical imaging diagnosis and condition prediction [6, 7]. However, its application in diagnosing mTBI using CT data remains rare [8]. DL algorithms present a valuable opportunity to enhance mTBI diagnosis by extracting subtle, clinically meaningful information from complex CT data, which may not readily be apparent to human observers [9]. In our previous study [10], Artificial Neural Network (ANN) and Random Forest (RF) approaches demonstrated remarkable performance in diagnosing mTBI using clinical and image interpreted data. However, employing imaging data directly to build an automated mTBI diagnostic model promises to be more feasible, effective, and time saving. Radiologists screen and report mTBI-related findings from a CT scan, such as hemorrhage, hematomas, swelling, skull fracture, and any obvious tissue damage [11]. While conventional computer vision techniques such as the scale-invariant feature transform algorithm can perform feature detection/extraction [12], DL models have been shown to be more effective for extracting highly complex and task-specific features from images [13]. Therefore, an automated CT diagnosis system for mTBI could substantially improve diagnostic speed, decision-making, and resource efficiency, potentially reducing morbidity and mortality rates.

Several studies have applied DL techniques successfully, particularly using Convolutional Neural Networks (CNN), to construct TBI diagnostic models based on CT scans [14, 15]. These studies underline the potential of CNNs to segment and/or identify moderate and severe TBI hemorrhage lesions on CT images. However, the detection and interpretation of mTBI lesions on CT scans remain unclear and controversial. While many studies have used ML and DL for mTBI diagnosis using a range of data sources [16–18], these researches have not fully extended to the utilization of CT data. Thus, the potential of CT data for this purpose remains largely untapped.

Here, we introduce a novel Multi-Modal Residual Convolutional Neural Network (MRCNN) equipped with Occlusion Sensitivity Maps (OSM) for the diagnosis of mTBI, expanding the application of DL techniques in this field. Our work also provides an innovative comparison between OSM and Gradient-weighted Class Activation Mapping (Grad-CAM), offering new perspectives on their usage for data-driven insights into CT images.

The main contribution of this study can be summarized as follows:

– We developed a Multi-Modal Residual Convolutional Neural Network (MRCNN) enhanced with occlusion sensitivity. This represents a novel approach in the diagnosis of mTBI, effectively integrating CT imaging and clinical data to improve diagnostic precision.
– We introduce the application of OSM in the context of mTBI diagnosis, offering superior data-driven insights into CT images compared to the widely used Gradient-weighted Class Activation Mapping (Grad-CAM).
– We demonstrate the efficacy of integrating multiple modalities into the model to enhance diagnostic accuracy and precision. The effectiveness of this multi-modal approach is substantiated by the tangible improvements in specificity and accuracy compared to the proposed CT-based Residual Convolutional Neural Network (RCNN) model.

2   Methodology

We explored different DL architectures to extract and learn the spatial features of mTBI from 3D CT scans. Methods were selected based on their previously demonstrated ability to work with complex patterns within images. In the initial stage, clinical features were used to train the classifier employed in our previous study [10]. This was done to ensure that the clinical data used for the model's training aligned with evidence-based research pertinent to mTBI diagnosis. The selected features are the relevant clinical features that correspond with the feature ranking from our prior study of the pediatric emergency care applied research network (PECARN) dataset [10] for mTBI diagnosis. Following this, we applied DL models for mTBI diagnosis from CT imaging data.

2.1   Data Preparation

TRACK-TBI pilot study dataset [19] was employed in our research to develop deep learning-based diagnostic models for mTBI utilizing CT imaging data, and to assess the performance and integration capabilities of these models. TRACK-TBI pilot study is a prospective multi-center study that enrolled 600 patients older than seven years of age and having a positive TBI diagnosis. All patients had received a head CT scan while in the emergency department within 24 h of head injury. Data were collected in three level I trauma centers (University of California San Francisco, University Medical Centre Brackenridge, University of Pittsburgh Medical Centre) and one rehabilitation center (Mount Sinai Rehabilitation Centre) in the US between April 2010 and June 2012 [19]. We obtained access privileges to the data repository (i.e., FITBIR) from the Data Access and Quality (DAQ) committee. Our study using this data was approved by the human research ethics at the University of Queensland (no. 2020002583). The patients who had undergone non-contrasted CT scans within 48 h following closed head injury and were assigned Glasgow Coma Scale (GCS) $\geq$ 13 were selected for our study. A total of 296 de-identified records were sourced from the TRACK-TBI public dataset.

2.2   Dataset Pre-processing

We pre-processed CT images to ensure high-quality, consistent data were fed into our DL models. Initially, the dcm2nii tool [20] was used to convert the CT images from their original format to the Neuroimaging Informatics Technology Initiative (NIfTI) format, incorporating field-of-view alignment corrections to have spatial consistency across all images. Next, the appropriate brain window (window width = 80 HU and width level = 40 HU) [21] was applied to standardize CT images' dynamic range and to visually enhance potential mTBI lesions. Next, the Spline Interpolated Zoom (SIZ) resampling volume algorithm was used to create isotropic resolution images [22] resulting in 128 × 128 × 64 matrix size, and 1mm3 resolution images. This resizing and uniformization, aimed to reduce computational complexity and optimize GPU usage, utilized z-axis interpolation for uniform volume representation. This approach effectively captured data from multiple slices and ensured robust performance in line with GPU memory requirements [23]. Lastly, unnecessary background information in images was removed, e.g., the scanner bed, to focus the model solely on the brain. Figure 1 illustrates the pre-processing steps, and the Grad-CAM analysis.

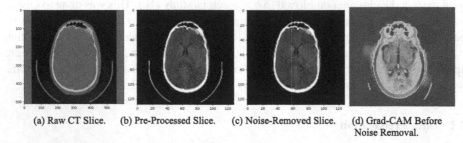

(a) Raw CT Slice.   (b) Pre-Processed Slice.   (c) Noise-Removed Slice.   (d) Grad-CAM Before Noise Removal.

Fig. 1. The image pre-processing steps applied to a CT slice and the impact of noise removal on the model's decision-making process. (a) An original, raw CT slice before any pre-processing steps are applied, (b) The same CT slice after it has undergone initial pre-processing steps, including standardizing the dynamic range for brain anatomical structures, (c) The CT slice has been further processed to remove high-intensity noise artifacts such as the scanner bed, and (d) illustrates how the model was significantly influenced by the high-intensity noise artifacts before their removal, as visualized by the Gradient-weighted Class Activation Mapping (Grad-CAM)

2.3 RCNN Model

In our initial attempt to construct an mTBI diagnostic DL model, we implemented a CNN model [22]. This was followed by trials with several other architectures, encompassing variations of 3D EfficientNet, ResNet, and DenseNet [24]. The unique challenge in developing mTBI diagnostic models arises from its inherent heterogeneity, with factors such as varied injury mechanisms and differing patient symptoms adding to the complexity. Consequently, the models tested did not yield satisfactory performance. Amidst these complexities, the model architecture depicted in Fig. 2 demonstrated superior performance. As such, we adopted a tuned Residual CNN as our primary CT diagnostic model for mTBI. The model involves six residual blocks, each comprised of three convolutional blocks and a 1 × 1 convolution layer. Each convolutional block integrates a 3d convolutional layer, an instance normalization layer, and a parametric rectified linear unit (PReLU) activation function. The output from these six residual blocks is then flattened and passed to three fully connected layers, with a sigmoid activation function in the output layer serving as the classifier output.

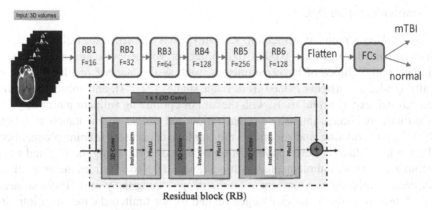

Fig. 2. Image-based mTBI diagnostic model architecture. FCs: fully connected layers

2.4 MRCNN

Multi-modal refers to the use of both clinical information in conjunction with CT scans. We chose a multi-modal approach that leverages both clinical data and 3D CT scans to enhance the diagnostic performance of our model. This clinical data encompasses patient demographics, medical history, and symptom profiles, as depicted in Fig. 3. We aimed to improve the classification accuracy by providing the model with comprehensive patient information by integrating the diversity of data sources into a DL framework. Our model, MRCNN, is based on the foundational architecture of RCNN. It includes an additional branch for clinical data and an infusion layer designed to merge the learned CT features with the clinical features. This combined data is then processed through an output layer equipped with a sigmoid activation function.

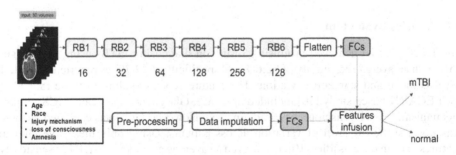

Fig. 3. MRCNN model architecture.

2.5 Implementation Details

Rectified Adaptive moment estimation (RAdam) [25] is adopted as the training optimizer with an initial learning rate of 0.0005, in conjunction with a cosine annealing scheduler [26]. The initial choice of hyperparameters, such as the number of filters used, was initially guided by insights gained from other studies [22, 24], and subsequently fine-tuned through experimental trials, with the aim of optimizing validation accuracy.

Furthermore, image intensity normalization and online augmentation techniques such as flipping, rotating, and zooming were employed during the training phase. Specifically, random rotations were applied within a range of $\pm 10°$ along the x, y, and z axes. Random zooming was implemented within a range of 0.9 to 1.2 times the original size. Also, vertical flipping was performed on the images along the y-axis. These strategies were adopted to address the challenges related to our limited dataset size, introduce additional variation, and to improve model generalizability [27].

We employed five-fold cross-validation as a strategy to evaluate our models' performance, individually assessing accuracy, sensitivity, and specificity using 95% confidence intervals (CI) as done previously [28]. This evaluation process helps in identifying the strengths and limitations of the model, thereby facilitating informed comparisons. Moreover, we enhanced visual understanding of the model's decision-making process using OSM [29], to provide an intuitive way for data interpretation for mTBI diagnosis compared to Grad-CAM [30]. The occlusion sensitivity map is a method to understand a model's decision-making process. It involves systematically occluding different parts of an input image and observing how these alterations affect the model's output. By doing so, it helps to identify the regions in the input image that the model considers critical for decision making. This approach adds depth to model evaluation by highlighting how image regions influence the decisions made by the model.

3 Results and Discussion

3.1 Metrics Evaluation

The RCNN model achieved promising results in diagnosing mTBI using CT images only with an average sensitivity of 82.6% and specificity of 72.7% as reported in Table 1 with its mean and standard deviation. The features corresponding to those ranked in the PECARN-based study [10] included age, race, Glasgow Coma Scale (GCS), injury mechanism, loss of consciousness, and amnesia. However, GCS was excluded as there was not a sufficient level of variation across patients. Using only these five clinical features to train the classifier [10], it achieved an average accuracy of 57.1%, a sensitivity of 69.1%, and a specificity of 50.0% for diagnosing mTBI using the clinical data only.

Notably, the MRCNN model exhibited an improvement in performance over RCNN with an average accuracy of 82.4% and specificity of 81.6%. With the integration of clinical data in the RCNN framework, accuracy and specificity increased by 4.4% and 9.0%, respectively. It does appear that the inclusion of clinical data offers a more comprehensive mTBI patient representation, extending beyond what can be seen in CT scans alone.

Table 1. Evaluation metrics of the five-fold cross-validation.

Metric	Clinical	Imaging	Multi-modal
	ANN [10]	RCNN	MRCNN
Accuracy ($\mu \pm \alpha$ in %)	57.1 ± 7.8	78.1 ± 6.6 ($p_{ANN} = 0.002$)	82.4 ± 1.7 ($p_{RCNN} = 0.024$)
Sensitivity ($\mu \pm \alpha$ in %)	69.1 ± 14.6	82.8 ± 5.1 ($p_{ANN} = 0.061$)	82.6 ± 5.1 ($p_{RCNN} = 0.294$)
Specificity ($\mu \pm \alpha$ in %)	50.0 ± 10.7	72.7 ± 8.7 ($p_{ANN} = 0.006$)	81.6 ± 5.4 ($p_{RCNN} = 0.022$)

ANN: Artificial Neural Network, μ: Mean, α: Standard Deviation. Statistically significant increase in the mean metric was achieved in all cases for $p < 0.05$, except for sensitivity between RCNN and MRCNN.

In addition to our primary metrics, we assessed the overall performance of our model using the Area Under the Curve (AUC) of the Receiver Operating Characteristic (ROC) curve. As illustrated in Fig. 4, the ROC curve graphically represents the trade-off between the true positive rate and false positive rate of the MRCNN model across varying discrimination thresholds. The obtained AUC score of 0.95 signifies an excellent ability of our model to differentiate between mTBI and normal CT scans.

Our results have profound implications as this may be one of the first attempts to employ DL in diagnosing mTBI using CT images. The diagnosis of mTBI is an exceptionally challenging task due to its subtle nature and the limited sensitivity of CT scans. However, the promising performance of the MRCNN model demonstrates the potential of DL to improve mTBI diagnosis using CT imaging and clinical data. It suggests that DL can potentially extract clinically relevant information from these scans that might otherwise be overlooked. Therefore, this work paves the way for further research on the application of DL models in leveraging CT images for mTBI diagnosis, offering new avenues for early detection and treatment [3].

3.2 Visual Assessment

A vital question that emerges with medical diagnostic models is whether the model truly identifies and relies on meaningful features, or if it is simply capitalizing on coincidental patterns. This underscores the importance of exploring visual explanations for the decision-making process involved in the model. For our RCNN model, we achieved this through the application of Grad-CAM and occlusion sensitivity, as depicted in Fig. 5.

While Grad-CAM is a well-established technique for exploring the decisions of CNNs within the scope of computer vision tasks, it does have certain limitations [31]. In our case, it primarily highlighted the skull because of its high intensity values, with less emphasis on the brain. This limitation is evident in the third column of Fig. 5, where the brain regions are under-represented compared to the skull and values with high intensities.

In contrast, OSM, generated by systematically blocking various parts of the input image, yields a more comprehensive, nuanced understanding. As the model output

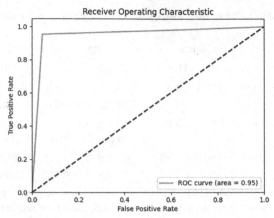

Fig. 4. The Area Under the Curve of the Receiver Operating Characteristic (AUC) curve of the MRCNN model.

changes due to the occlusion of certain regions, a region's importance in the decision-making process is scored [32]. The technique thus provides a more comprehensive visualization of scans, focusing on both the skull and the interior of the brain.

The comparative efficiency of occlusion sensitivity over Grad-CAM becomes clear in Fig. 5. While Grad-CAM primarily focuses on the skull area, occlusion sensitivity highlights the regions that substantially influence the decisions made by the model. The blue areas on the OSM represent the regions that, when occluded, have the most significant impact on the output, indicating their relevance to the model.

In the correctly classified mTBI CT scan, the occlusion sensitivity map highlights not only a part of the skull but also a significant portion of the brain. This indicates that the model correctly associates certain features within these regions with the diagnosis of mTBI. This contrasts with the Grad-CAM results, which primarily focus on the high-intensity areas of the skull.

For the correctly classified normal CT scan, there's a more significant focus on the skull and the brain areas adjacent to the skull in the occlusion sensitivity map. This suggests that the model correctly identifies these areas as important for a normal diagnosis, indicating a distinct difference from the features it associates with mTBI.

In the misclassified CT scan, attention is given to different areas, both in the skull and the brain, in the occlusion sensitivity map. It suggests that the model might be incorrectly focusing on these areas or missing capturing information on other important features, leading to misclassification. Understanding these less significant areas in terms of decision-making by the model could provide valuable insights for improving model performance.

Therefore, the OSM can reveal areas that substantially alter the output of the model when occluded, capturing vital details that could potentially be overlooked by Grad-CAM. These insights from OSM can greatly enhance the clinical utility of our model by significantly contributing to the understanding and diagnosis of mTBI.

Processed CT   Grad-CAM   Grad-CAM overlay   OSM   OSM overlay

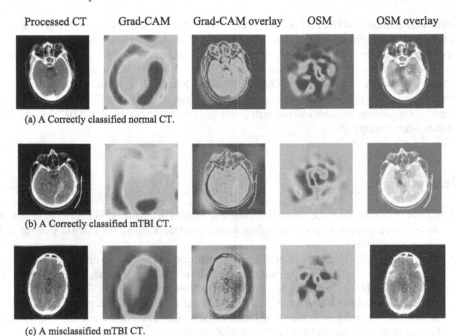

(a) A Correctly classified normal CT.

(b) A Correctly classified mTBI CT.

(c) A misclassified mTBI CT.

Fig. 5. The visual explanation of the model. From left to right, the columns respectively represent the processed CT scan, the Grad-CAM visualization, the overlay of the Grad-CAM and the processed CT scan, the occlusion sensitivity maps, and finally the overlay of the occlusion sensitivity maps with the processed CT scan.

4 Conclusions and Future Work

We introduced a CT-based interpretable Multi-modal Residual Convolutional Neural Network (MRCNN) model for mild traumatic brain injury (mTBI) diagnosis enhanced with Occlusion Sensitivity Maps (OSM). Remarkably, the MRCNN model improved the average specificity by an additional 4.4% and accuracy of 9.0%, indicating that the integration of clinical data with CT imaging offers a more thorough patient representation and improves the precision achieved by the MRCNN. Furthermore, the use of OSM for visual explanation adds significant value to model interpretability. This technique aids in understanding the regions of interest that the model utilizes for mTBI diagnosis, resulting in improved transparency and creating trust in the model output.

Despite our interesting and promising results for mTBI, the diagnostic model has several limitations that need to be addressed. It requires high-quality, accurately labelled CT scans for training. The performance of the model can be significantly impacted by the quality and diversity of the training data. Therefore, collecting and curating such datasets is challenging and requires significant effort. Future work should aim to address these limitations and further improve the model. Potential avenues could include the addition of other clinical data to create a more comprehensive diagnostic tool or exploring different neural network architectures achieving even better performance than presented here.

Acknowledgement. NITRC, NITRC-IR, and NITRC-CE have been funded in whole or in part with Federal funds from the Department of Health and Human Services, National Institute of Biomedical Imaging and Bioengineering, the National Institute of Neurological Disorders and Stroke, under the following NIH grants: 1R43NS074540, 2R44NS074540, and 1U24EB023398 and previously GSA Contract No. GS-00F-0034P, Order Number HHSN268200100090U. Moreover, we would like to acknowledge the principal investigators of the TRACK TBI Pilot research program, the sub-investigators and research teams that contributed to TRACK TBI Pilot, and the patients who participated.

References

1. Binder, S., Gerberding, J.L.: Report to congress on mild traumatic brain injury in the United States: steps to prevent a serious public health problem. Centers for Disease Control and Prevention (2003)
2. Diaz-Arrastia, R., et al.: Acute biomarkers of traumatic brain injury: relationship between plasma levels of ubiquitin C-terminal hydrolase-l1 and glial fibrillary acidic protein. J. Neurotrauma **31**, 19–25 (2014). https://doi.org/10.1089/neu.2013.3040
3. Korley, F.K., Kelen, G.D., Jones, C.M., Diaz-Arrastia, R.: Emergency department evaluation of traumatic brain injury in the United States, 2009–2010. J. Head Trauma Rehabil. **31**, 379 (2016). https://doi.org/10.1097/HTR.0000000000000187
4. Easter, J.S., Haukoos, J.S., Meehan, W.P., Novack, V., Edlow, J.A.: Will neuroimaging reveal a severe intracranial injury in this adult with minor head trauma?: The rational clinical examination systematic review. JAMA **314**(24), 2672–2681 (2015). https://doi.org/10.1001/jama.2015.16316
5. Schweitzer, A.D., Niogi, S.N., Whitlow, C.T., Tsiouris, A.J.: Traumatic brain injury: imaging patterns and complications. Radiographics **39**(6), 1571–1595 (2019). https://doi.org/10.1148/rg.2019190076
6. Litjens, G., et al.: A survey on deep learning in medical image analysis. Med. Image Anal. **42**, 60–88 (2017). https://doi.org/10.1016/j.media.2017.07.005
7. Ellethy, H., Chandra, S.S., Nasrallah, F.A.: Deep neural networks predict the need for CT in pediatric mild traumatic brain injury: a corroboration of the PECARN rule. J. Am. Coll. Radiol. **19**(6), 769–778 (2022). https://doi.org/10.1016/j.jacr.2022.02.024
8. Mohd Noor, N.S.E., Ibrahim, H.: Predicting Outcomes in patients with traumatic brain injury using machine learning models. In: Jamaludin, Z., Ali Mokhtar, M.N. (eds.) SympoSIMM 2019. LNME, pp. 12–20. Springer, Singapore (2020). https://doi.org/10.1007/978-981-13-9539-0_2
9. Yoon, B.C., et al.: Incorporating algorithmic uncertainty into a clinical machine deep learning algorithm for urgent head CTs. PLoS One **18**(3), e0281900 (2023)
10. Ellethy, H., Chandra, S.S., Nasrallah, F.A.: The detection of mild traumatic brain injury in paediatrics using artificial neural networks. Comput. Biol. Med. **135**, 104614 (2021). https://doi.org/10.1016/J.COMPBIOMED.2021.104614
11. Su, Y.R.S., Schuster, J.M., Smith, D.H., Stein, S.C.: Cost-effectiveness of biomarker screening for traumatic brain injury. J. Neurotrauma (2019). https://doi.org/10.1089/neu.2018.6020
12. Keshavamurthy, K.N., et al.: Machine learning algorithm for automatic detection of CT-identifiable hyperdense lesions associated with traumatic brain injury. In: Medical Imaging 2017: Computer-Aided Diagnosis (2017). https://doi.org/10.1117/12.2254227

13. Detone, D., Malisiewicz, T., Rabinovich, A.: SuperPoint: self-supervised interest point detection and description. In: IEEE Computer Society Conference on Computer Vision and Pattern Recognition Workshops (2018). https://doi.org/10.1109/CVPRW.2018.00060
14. Chilamkurthy, S., et al.: Development and validation of deep learning algorithms for detection of critical findings in head CT scans. arXiv Preprint arXiv:1803.05854 (2018)
15. Monteiro, M., et al.: Multiclass semantic segmentation and quantification of traumatic brain injury lesions on head CT using deep learning: an algorithm development and multicentre validation study. Lancet Digit. Heal. **2**, e314–e322 (2020). https://doi.org/10.1016/S2589-750 0(20)30085-6
16. Tamez-Peña, J., et al.: Post-concussive mTBI in student athletes: MRI features and machine learning. Front. Neurol. **12**, 2351 (2022)
17. Bostami, B., Espinoza, F.A., van der Horn, H.J., van der Naalt, J., Calhoun, V.D., Vergara, V.M.: Multi-site mild traumatic brain injury classification with machine learning and harmonization. In: 2022 44th Annual International Conference of the IEEE Engineering in Medicine & Biology Society (EMBC), pp. 537–540 (2022). https://doi.org/10.1109/EMB C48229.2022.9871869
18. Harrington, D.L., et al.: Detection of chronic blast-related mild traumatic brain injury with diffusion tensor imaging and support vector machines. Diagnostics **12**(4), 987 (2022)
19. Yue, J.K., et al.: Transforming research and clinical knowledge in traumatic brain injury pilot: Multicenter implementation of the common data elements for traumatic brain injury. J. Neurotrauma **30**, 1831–1844 (2013). https://doi.org/10.1089/neu.2013.2970
20. N I T R C. https://www.nitrc.org/. Accessed 25 Mar 2021
21. Lolli, V., Pezzullo, M., Delpierre, I., Sadeghi, N.: MDCT imaging of traumatic brain injury. Br. J. Radiol. **89**(1061), 20150849 (2016)
22. Zunair, H., Rahman, A., Mohammed, N., Cohen, J.P.: Uniformizing techniques to process CT scans with 3D CNNs for tuberculosis prediction. In: Rekik, I., Adeli, E., Park, S.H., Valdés Hernández, M.D.C. (eds.) PRIME 2020. LNCS, vol. 12329, pp. 156–168. Springer, Cham (2020). https://doi.org/10.1007/978-3-030-59354-4_15
23. Goenka, N., Tiwari, S.: AlzVNet: A volumetric convolutional neural network for multiclass classification of Alzheimer's disease through multiple neuroimaging computational approaches. Biomed. Signal Process. Control **74**, 103500 (2022). https://doi.org/10.1016/j. bspc.2022.103500
24. Cardoso, M.J., et al.: MONAI: an open-source framework for deep learning in healthcare. arXiv Preprint arXiv:2211.02701 (2022)
25. Liu, L., et al.: On the variance of the adaptive learning rate and beyond. arXiv Preprint arXiv: 1908.03265 (2019)
26. Loshchilov, I., Hutter, F.: SGDR: Stochastic gradient descent with warm restarts. arXiv Preprint arXiv:1608.03983 (2016)
27. Monteiro, M., et al.: TBI lesion segmentation in head CT: Impact of preprocessing and data augmentation. In: Crimi, A., Bakas, S. (eds.) BrainLes 2019. LNCS (LNAI and LNB), vol. 11992, pp. 13–22. Springer, Cham (2020)
28. Powers, D.M.W.: Evaluation: from precision, recall and F-measure to ROC, informedness, markedness & correlation. J. Mach. Learn. Technol. (2011)
29. Zeiler, M.D., Fergus, R.: Visualizing and understanding convolutional networks. In: Fleet, D., Pajdla, T., Schiele, B., Tuytelaars, T. (eds.) ECCV 2014. LNCS, vol. 8689, pp. 818–833. Springer, Cham (2014). https://doi.org/10.1007/978-3-319-10590-1_53
30. Selvaraju, R.R., Cogswell, M., Das, A., Vedantam, R., Parikh, D., Batra, D.: Grad-CAM: visual explanations from deep networks via gradient-based localization. In: Proceedings of the IEEE International Conference on Computer Vision, pp. 618–626 (2017)

31. Aminu, M., Ahmad, N.A., Mohd Noor, M.H.: Covid-19 detection via deep neural network and occlusion sensitivity maps. Alex. Eng. J. **60**(5), 4829–4855 (2021). https://doi.org/10.1016/j.aej.2021.03.052

32. Rieke, J., Eitel, F., Weygandt, M., Haynes, J.-D., Ritter, K.: Visualizing convolutional networks for MRI-based diagnosis of Alzheimer's disease. In: Stoyanov, D., et al. (eds.) MLCN DLF IMIMIC 2018. LNCS, vol. 11038, pp. 24–31. Springer, Cham (2018). https://doi.org/10.1007/978-3-030-02628-8_3

Comparative Assessment of Machine Learning Strategies for Electrocardiogram Denoising

Brenda Wang[1], Chirath Hettiarachchi[1](✉) [iD], Hanna Suominen[1,2] [iD],
and Elena Daskalaki[1] [iD]

[1] The Australian National University, Canberra, Australia
{chirath.hettiarachchi,hanna.suominen,eleni.daskalaki}@anu.edu.au
[2] University of Turku, Turku, Finland

Abstract. An electrocardiogram (ECG) is an important non-invasive predictor of cardiovascular disease (CVD) used to support early diagnosis and detection of various heart problems. Monitoring ECG continuously is expected to lower mortality from CVD, but achieving this aspiration is constrained by the high cost of medical-grade ECG. Although advancements in wearable devices have made ECG monitoring in everyday environments possible, the resulting recordings are affected by severe noise corruption, for which traditional signal processing techniques fall short. Therefore, in recent years, the focus has been on machine learning (ML) techniques for ECG denoising. Despite recent advances, many unanswered questions and unsolved challenges exist, and a comparative study is missing. To address this gap, we comparatively assessed state-of-the-art ML models, namely Denoising Autoencoder (DAE), Convolutional Neural Network (CNN), Long Short Term Memory (LSTM), and Generative Adversarial Network (GAN), for ECG denoising using the MIT-BIH Arrhythmia and ECG-ID datasets. Three noise types were considered, including baseline wander, electron motion, and motion artifacts. Performance was assessed explicitly by comparing denoised and clean signals, and implicitly through the balanced accuracy of downstream tasks (beat classification, person identification). Furthermore, we investigated the models' generalisation capabilities to unseen data (data transferability) and unseen noise types (noise transferability). Our findings suggested that in certain cases, explicit evaluation may be insufficient and implicit metrics need to be considered. Transfer learning improved data transferability, while all models could generalise to unseen noise types, albeit in different levels. Overall, CNN and GAN models achieved the best performance. Our results encourage the development of denoising and processing pipelines for healthcare applications based on wearable ECG.

Keywords: Biomedical Informatics · Electrocardiogram · Evaluation Study · Machine Learning · Signal Processing · Signal-to-Noise Ratio

T. Liu et al. (Eds.): AI 2023, LNAI 14471, pp. 495–506, 2024.
https://doi.org/10.1007/978-981-99-8388-9_40

1   Introduction

Cardiovascular diseases (CVD) are the leading cause of deaths, accounting for an estimated 17.92 million deaths worldwide in 2015 alone [16]. This statistic represents 31% of total global deaths [22], of which around 80% are estimated to be preventable [24]. Key to lowering mortality rates is early screening, detection, and continuous monitoring, reducing the severity of symptoms and increasing the success of treatment [4]. The electrocardiogram (ECG) is an important predictor of CVD [5], capturing the electrical changes of the heart. Analysis of this ECG waveform (Fig. 1) can be used to diagnose a wide variety of heart problems including, enlargement of the heart, arrhythmia, heart attacks, and heart inflammation. Although ECG is not the only way to monitor health, it is an extremely important one due to its convenience in contrast to more invasive approaches, such as measuring intra-arterial blood pressure. As a result of its powerful nature, ECG has been used in a variety of applications from supporting diagnosis [10] to monitoring [8, 19].

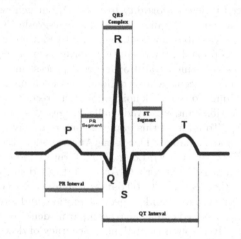

Fig. 1. An example of an ECG waveform and the labelled sections. In addition to illustrating its shape, we have depicted the important sections and intervals of the waveform by identifying and labeling them. Differences in length, height, or regularity of each section or interval can indicate a variety of heart conditions.

ECG signals are typically collected and analysed from medical-grade devices. Medical-grade ECG devices cost thousands of dollars and often require trained personnel to operate. Hence, they cannot be used for regular ECG monitoring during a person's daily life and activities. In recent years, wearable technology has entered the field of healthcare, with wearable ECG monitoring being an exemplary domain to supplement the limitations of traditional medical-grade ECG monitors [17]. Wearable devices are able to provide a smaller, cheaper, and more portable method of ECG monitoring that can be used in everyday environments. However, their signals suffer from much more severe noise corruption compared to the medical-grade counterparts, as well as from different types of noise.

Machine learning (ML) based approaches have been shown able to rectify the weaknesses of traditional signal processing techniques in all types of problems and applications, including signal denoising, and have demonstrated promising results for complex and severe noise types in a variety of different application settings [1]. They are also able to operate efficiently without expert intervention but require a significantly larger amount of data. Despite the recent advances, there is still a plethora of unanswered questions and unsolved challenges in ML-based ECG denoising and a comparative study of different ML techniques is missing. Moreover, studies tend to differ in their starting noise levels and scoring metrics, while little is known about the relation of those evaluation metrics to the performance of the end task using the denoised ECG (e.g., beat classification or person identification). Another gap in this area is the lack of research on how well models are able to cope with different sources of noise or generalisation to unseen datasets.

Consequently, the aim of this study was to comparatively assess the performance of state-of-the-art ML approaches for ECG denoising with respect to the explicit denoising task and the target application task, as well as with respect to their ability to generalise to unseen noise types and different datasets.

2   Related Work

Classical methods such as Finite Impulse Response (FIR) filter, Adaptive Notch Filter (ANF), and other filter-based approaches may not be suitable for noise removal from ECG coming from wearable devices. Similarly, wavelet transform, a popular method for signal denoising - including ECG denoising - cannot handle different noise types which contain a complex frequency content [1]. To this end, significant work has been performed on ECG denoising with the use of ML, the two most common strategies evolving around Denoising Auto Encoders (DAE) and Generative Adversarial Networks (GAN).

A common experimental methodology can be observed in the literature which is based on the use of the MIT-BIH Arrhythmia database [12] for the acquisition of clean ECG signals and the MIT-BIH NST database [13] for the acquisition of noise samples for three different noise types, namely baseline wander (BW), electron motion (EM), and motion artefacts (MA) which are used to artificially contaminate the clean signals. Studies explore the ability of their models to remove each type of noise and/or the mixture of all noise types. Xiong et al. [23] used a DAE combined with wavelets and achieved a Signal to Noise Ratio (SNR) of 22.74, 21.62, and 21.56 for BW, MA, and EM noise respectively with input SNR $= 0$ for all cases. However, they did not consider the mixture of all types of noise. Chiang et al. [3] proposed a convolutional DAE, which achieved SNR $= 15.49$ and Root Mean Squared Error (RMSE) of 0.063 for the mixture of the three noise types with input SNR $= -1$. Hou et al. [9] used a fully convolutional GAN and achieved a SNR $= 32$ and RMSE $= 0.0094$ for denoising a mixture of the three noise types. Singh et al [18] used a GAN framework on the MIT-BIH Arryhthmia dataset. Wang et al. [21] used a GAN and showed that it can remove mixed noise of three types with SNR = 30.5 and RMSE = 0.0137. The study additionally assessed the contribution of the GAN model on the task of classifying the ECG beats as normal or abnormal, based on their labelling in the MIT-BIH Arrhythmia dataset and showed that the classification accuracy of the denoised signal was close to that of the original clean signal. Other ML-based methods include the works by Wang et al. [20] to propose a deep factor analysis framework to denoise EM, BW, and MA noises, but not their mixture, and by Antczak et al. [1] to use deep recurrent neural networks on real and synthetic data contaminated with white noise.

From the state of the current research we recognise the efficient application of various ML methods for ECG denoising, however, it remains hard to assess which method performs best as a comprehensive comparative study is missing. At the same time, we observe that most studies assess the denoising quality based solely on the SNR and RMSE between clean and denoised signals, however, much less work has attended on the effect of denoising on the end task for which the denoising is taking place.

3   Methods

3.1   Datasets

We carefully selected two publicly available datasets and obtained necessary approvals from human research ethics committees. The datasets included clean ECG data collected for different target tasks as described below.

MIT-BIH Arrhythmia. The MIT-BIH Arrhythmia database [12] consists of 48 half-hour ECG recordings from 47 patients. These signals are digitised at 360 Hz and come with both peak and beat annotations. The target task associated with this dataset is beat classification, that is to classify the heartbeats into the five beat types. This database is the most frequently used database for ECG denoising, and a popular one for beat annotation.

ECG-ID. The ECG-ID database [11] consists of ECG records from 90 people, totalling 310 records. Each record is 20 s in length and is connected to one of the 90 individuals participating in the study. The signals are digitised at 500 Hz, and the target task associated is person identification, that is to match an ECG record to the corresponding person.

3.2   Noise

The data were subsequently contaminated with three types of simulated noise to generate noisy ECG signals using the MIT-BIH Noise Stress Test Database [13]. The MIT-BIH Noise Stress Test Database is composed of 3 half-hour recordings of noise in ambulatory ECG setting digitised at 360 Hz. These noise records include BW, MA, and EM noise types. In our study, the total noise n_k was generated as a weighted sum of the three noise types in equal proportions. The total noise was then mixed with the clean signal x_{clean} as shown in Eq. (1).

$$x_{noisy}(k) = x_{clean}(k) + n(k) \times \sqrt{\frac{\sum x_{clean}^2}{\sum x_{noise}^2 \times 10^{\frac{SNR}{10}}}}. \tag{1}$$

3.3   Algorithms

In this study, four different denoising models were developed and comparatively assessed; DAE, Convolutional Neural Network (CNN), Long Short Term Memory (LSTM) network, and GAN. All models have weights initialised randomly using the Xavier uniform initialisation [6]. The Adam optimiser was used with a learning rate of 0.0001. The loss function across all models except the GAN was the mean squared error loss function. The DAE model was based on [14] and was designed as a fully connected network with two dense layers (sizes 16 and 210) and ReLU activation functions. The CNN model developed in this study was based on [2]. The model is comprised of 4 convolutional layers (sizes 256, 64, 16, and 4), each followed by batch normalisation, ReLU activation, and pooling (size (2, 1), strides (4, 1)), and a final fully connected layer (size 256). The LSTM model was based on [7]. The model comprised of 3 LSTM layers (size 3) with *tanh* activation and a time distributed layer (size 1). The GAN was based on [21] and included a generator and a discriminator which were trained simultaneously using the loss function proposed in [21]. The discriminator model included three fully connected layers (150, 150, 1) with *tanh*, sigmoid, and linear activation, respectively. The generator model included four fully connected layers (250, 250, 250, 310) with *tanh*, tanh, sigmoid, and linear activation, respectively.

3.4   Evaluation Metrics

The SNR and RMSE between the original (clean) and the resulting (denoised) signal were used to assess the performance in the denoising task. For the cases where the target task of a dataset was considered (beat classification or person identification), the balanced accuracy (BA) was used to assess performance defined as the mean accuracy of all classes.

3.5   Experiment Design

We set the noise magnitude to a standard SNR of 0. For pre-processing, all experiments applied peak segmentation [15] with a window size of 310. This window size was chosen to reflect the size of a heartbeat, following [21]. Since the sample rate of the MIT-BIH Arrhythmia and the ECG-ID datasets are different, we chose the MIT-BIH Arrhythmia database's sample rate of 360 Hz as the base frequency and used interpolation to align the ECG-ID dataset. Three experiments were designed to explore the three main research questions of this study, as described below.

E1 - Implicit vs Explicit Evaluation. Although we can directly measure how close a denoised signal is to the clean signal (explicit evaluation) using various metrics such as SNR and RMSE, it is unclear as to how these metrics compare against the results from the target task for which the denoising is intended (implicit evaluation). Therefore for implicit evaluation, we assessed the models based on the BA of the beat classification and person identification tasks.

E2 - Dataset Transferability. With the amount of data required by ML and particularly neural network models, transferability is a pertinent concern. It assesses the capability of models to generalise to unseen or unknown data sources. This experiment aimed to determine whether denoising models can generalise to different datasets and to investigate whether transfer learning can be employed to improve denoising. To explore this, we set up three training scenarios in which the models were i) trained on the ECG-ID dataset only, ii) trained on the MIT-BIH Arrhythmia dataset only, and iii) trained first on the MIT-BIH Arrhythmia dataset and then on the ECG-ID dataset through transfer learning. In all scenarios the models were tested on the ECG-ID dataset based on their performance both on explicit (RMSE, SNR) and implicit (BA) denoising.

E3 - Noise Transferability. Similar to dataset transferability, this experiment explored how different models generalise to different types of noise. The aim of this experiment was to explore whether i) the models are learning the clean signals or learning to remove a particular type of noise, and ii) if the models can learn to remove noise with characteristics that they have not encountered before. To test this experiment, the three main sources (EM, BW, MA) of ECG noise were considered and transferability was tested for each noise type pair resulting in total nine experiments, where each noise type at a time was used in training and tested on all noise types (e.g., train on BW and test on BW, MA, and EM).

4   Results

4.1   Results of the Experiment *E1*: Implicit vs Explicit Evaluation

All four models experimented in the experiment *E1* were able to successfully produce a denoised signal with higher SNR and lower RMSE compared to the noisy one (Fig. 2). We also observed an apparent correlation between the RMSE and SNR metrics, where a reduction in RMSE is reflected by a corresponding increase in SNR. However, although the improvement of both these metrics suggests an increase in the value of the ECG signals, this did not always correlate with the BA of the tasks. Whilst the results from the BA agreed with the performance improvements of the GAN and CNN, the DAE and LSTM presented a different outcome. For the MIT-BIH Arrhythmia task, the DAE only resulted in a marginal BA improvement, whereas at the ECG-ID task, the DAE had the highest BA among all models. The LSTM underperformed with the noisy signal on both tasks, especially in the ECG-ID task.

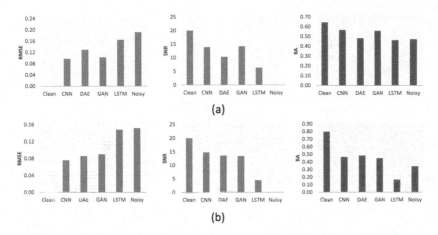

Fig. 2. The explicit (RMSE, SNR) and implicit (BA of the target task) evaluation of denoising results by the four experimented models for (a) MIT-BIH Arrhythmia and (b) ECG-ID datasets.

In order to shed more light into this disparity, we depicted an example of a signal in its clean, noisy, and denoised forms by the four models (Fig. 3). We could see that for the signal denoised by the DAE, the T-peak shifted from approximately 0.7 s to 0.6 s, while the Q wave also got malformed. This indicates that, despite a fairly successful reconstruction of the original signal from the noisy signal, the contamination of important features might counteract the improvements to other less integral features. On the other hand, although the signal denoised by the LSTM improved in RMSE and SNR compared to the noisy signal, it did not resolve the distorted shape and failed to represent well the original signal's morphology. This finding was observable in both examples from the Arrhythmia and ECG-ID datasets (Fig. 2 (a) and (b)). Although the DAE performed poorly on the MIT-BIH Arrhythmia dataset, it made a notable improvement to the BA on the ECG-ID dataset despite the distortions to the waveform. A reason for this difference could be the different ECG feature requirements for the two different tasks. Whilst arrhythmia detection might need non-malformed Q- and T-waves, they might not be required in identifying persons. As a result, slight deformations to the ECG waveform might not affect the resulting performance as much.

The results of both datasets implied that the RMSE and SNR tend to be good. Nonetheless, some caution should be taken when using these metrics since they do not take waveform distortion into consideration. For applications that are not as sensitive to waveform shapes, using the RMSE and SNR might be sufficient. However, for datasets where particular features, like the QRS complex, are crucial to the task performance, end-to-end evaluation, up to the accuracy of the target task, should be considered instead. In addition, neither the SNR nor the RMSE are tolerant of scaling; whilst shifting the scale of a signal is important when denoising, as in the case of removing baseline wander, it might not neces-

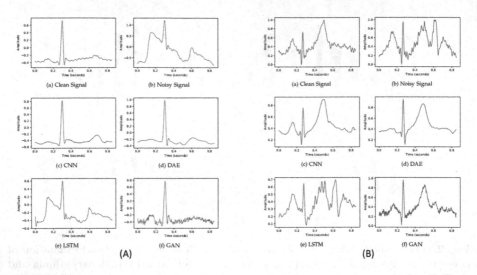

Fig. 3. Examples of a (a) clean, (b) noisy, and denoised (c-f) ECG waveform for (A) MIT BIH Arryhthmia and (B) ECG-ID datasets.

sarily improve the medical value of a signal. As a result, while implicit metrics like the RMSE and SNR are still good tools for evaluating task performance, an explicit evaluation metric should be more optimal.

4.2 Results of the Experiment *E2*: Data Transferability

Training on the larger MIT-BIH Arrhythmia dataset first and then on the ECG-ID (transfer learning) results in improved SNR and lower RMSE for all models compared to the other two training strategies included in the three data transferability experiments of *E2* (Fig. 4). Transfer learning was also beneficial for the BA in the person identification task when the CNN and GAN were used; however, for the DAE, it presented a performance decrease compared to the single ECG-ID training, while for the LSTM, the performance was comparable with training only on MIT-BIH.

This result could be associated with the complexity of those models. Due to the small size of the DAE model, few weights depended on training; hence, less training data was required. This means the DAE might not require the excess data from the MIT-BIH Arrhythmia dataset. In contrast, the LSTM had much more parameters, making the use of the MIT-BIH Arrhythmia dataset crucial for its effective training. This was reflected in the lower BA, lower SNR, and higher RMSE achieved when training the LSTM only on the ECG-ID dataset. Finally, the satisfactory performance achieved by the standalone MIT-BIH Arrhythmia training approach demonstrated that models presented a level of capability to generalise to new and unseen data types.

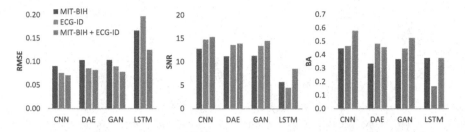

Fig. 4. Data transferability performance of the four models with respect to the RMSE, SNR, and BA in the three experiments (blue, grey, and orange). (Color figure online)

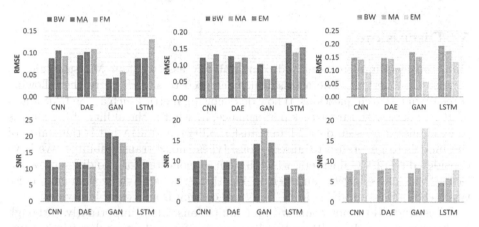

Fig. 5. Noise transferability performance of the models. The RMSE and SNR of the models are presented when tested on BW (blue), MA (grey), and EM (yellow). Each graph illustrates the results when the models were trained on the three noise types. (Color figure online)

4.3   Results of the Experiment *E3*: Noise Transferability

The four ML models were trained independently on the three different sources of noise (BW, EM, and MA) and tested on each of these noise types separately in the noise transferability experiment (*E3*, Fig. 5). The best testing outcome for each noise source occured when the models were trained on the same noise type, as expected. However, even when the training noise was different from the testing noise, the models performed in most cases at comparable levels. This indicated that they were able to generalise to unseen noise types. This was less prominent when testing with EM noise, in which case the models' performance when trained on other noise types was significantly poorer compared to training on EM noise. Moving on, BW noise seemed to be the easiest to remove regardless of which training noise was used (Fig. 6). It was clear in the two experimented cases that removal of BW noise presented the best performance.

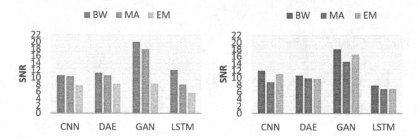

Fig. 6. Noise transferability performance (SNR) of the models when trained on MA (orange) and EM (blue). (Color figure online)

5   Discussion

Our study has implemented and comparatively assessed four ML strategies to denoise ECG signals. The assessment attempted to shed light to the following three main research questions: i) how do denoising metrics such as the RMSE and SNR correlate with target task performance, ii) what is the ability of the models to generalise to unseen data (data transferability), and iii) what is the ability of the models to generalise to unseen noise types (noise transferability). We have considered two ECG datasets, which have been collected for two different target tasks (beat classification and person identification) and three dominant noise types (BW, MA, and EM) in ECG signals coming from wearable devices.

This study does not come without limitations. Most importantly, although we have explored the SNR and RMSE metrics for evaluating denoising in two specific tasks, conducting a larger-scale experiment encompassing a wider range of ECG tasks would provide further insights into the implications of ECG signal distortion resulting from noise removal.

Our findings suggest that while the lower RMSE and higher SNR values usually correlate with better task performance, this is not always the case. This is because distortions caused by the denoising process can degrade certain features of the signal that may be necessary for the task. Hence, whilst these two metrics are still important indicators, target task performance should be considered for a safer and clearer assessment of the denoising strategy. As evidenced by our data transferability experiment, all models may be able to generalise but transfer learning could be the most beneficial approach for performance. The contribution of transfer learning seems to relate to the complexity of the models, with smaller and simpler models benefiting less from it and more complex models making substantial performance gains from the use of additional larger datasets. Another observed benefit of transfer learning is the reduction in training time; across the four models, we saved 18% in training time. The gain can be explained as an outcome of the more optimal starting weights that were transferred from the MIT-BIH trained model compared to the standard randomly initialised weights. In summary, regardless of each model's success in being able to adapt to previously unseen data in our experiments, this study showcases how transfer learning can be used to improve the denoising performance and reduce computational resources for datasets with limited annotations or models with long training times.

Finally, our results indicate that all models were able to successfully generalise to unseen noise types (but the CNN and GAN models respectively outperformed the other models in transferability to different datasets and types of noise, respectively). This means that the models were able to learn a generalised idea of a clean signal, rather than only learning to remove a singular source of noise. This is extremely important as there will always be noise characteristics that may not have been encountered previously.

6 Conclusions

This study provides valuable insights into the denoising process of ML models applied to ECG signals. It uncovers crucial information about evaluation metrics and models' generalisation capabilities, offering guidance for the development of more effective and well-adapted techniques for removing noise.

Acknowledgments. This research was funded in part by The Australian National University (ANU) and the Our Health in Our Hands initiative (OHIOH), a strategic initiative aiming to transform health care by developing new personalised health technologies and solutions in collaboration with patients, clinicians and healthcare providers.

References

1. Antczak, K.: Deep recurrent neural networks for ECG signal denoising. arXiv preprint arXiv:1807.11551 (2018)
2. Arsene, C.T., Hankins, R., Yin, H.: Deep learning models for denoising ECG signals. In: 2019 27th European Signal Processing Conference (EUSIPCO), pp. 1–5. IEEE (2019)
3. Chiang, H.T., Hsieh, Y.Y., Fu, S.W., Hung, K.H., Tsao, Y., Chien, S.Y.: Noise reduction in ECG signals using fully convolutional denoising autoencoders. IEEE Access **7**, 60806–60813 (2019)
4. Cohn, J.N., et al.: Screening for early detection of cardiovascular disease in asymptomatic individuals. Am. Heart J. **146**(4), 679–685 (2003)
5. De Bacquer, D., De Backer, G., Kornitzer, M., Blackburn, H.: Prognostic value of ECG findings for total, cardiovascular disease, and coronary heart disease death in men and women. Heart **80**(6), 570–577 (1998)
6. Glorot, X., Bengio, Y.: Understanding the difficulty of training deep feedforward neural networks. In: Proceedings of the Thirteenth International Conference on Artificial Intelligence and Statistics, pp. 249–256. JMLR Workshop and Conference Proceedings (2010)
7. Guan, J., Li, R., Li, R., Li, W., Wang, J., Xie, G.: Automated dynamic electrocardiogram noise reduction using multilayer LSTM network. In: Proceedings of the 15th EAI International Conference on Mobile and Ubiquitous Systems: Computing, Networking and Services, pp. 197–206 (2018)

8. Hadjem, M., Salem, O., Naït-Abdesselam, F.: An ECG monitoring system for prediction of cardiac anomalies using WBAN. In: 2014 IEEE 16th International Conference on e-Health Networking, Applications and Services (Healthcom), pp. 441–446. IEEE (2014)

9. Hou, Y., Liu, R., Shu, M., Chen, C.: An ECG denoising method based on adversarial denoising convolutional neural network. Biomed. Signal Process. Control **84**, 104964 (2023)

10. Jain, A., Tandri, H., Dalal, D., et al.: Diagnostic and prognostic utility of electrocardiography for left ventricular hypertrophy defined by magnetic resonance imaging in relationship to ethnicity: the multi-ethnic study of atherosclerosis (MESA). Am. Heart J. **159**(4), 652–658 (2010)

11. Lugovaya, T.S.: Biometric human identification based on electrocardiogram. Master's thesis, Faculty of Computing Technologies and Informatics, Electrotechnical University 'LETI', Saint-Petersburg, Russian Federation (2005)

12. Moody, G.B., Mark, R.G.: The impact of the MIT-BIH arrhythmia database. IEEE Eng. Med. Biol. Mag. **20**(3), 45–50 (2001)

13. Moody, G.B., Muldrow, W., Mark, R.G.: A noise stress test for arrhythmia detectors. Comput. Cardiol. **11**(3), 381–384 (1984)

14. Nurmaini, S., Darmawahyuni, A., et al.: Deep learning-based stacked denoising and autoencoder for ECG heartbeat classification. Electronics **9**(1), 135 (2020)

15. Pan, J., Tompkins, W.J.: A real-time QRS detection algorithm. IEEE Trans. Biomed. Eng. **3**, 230–236 (1985)

16. Roth, G.A., Johnson, C., et al.: Global, regional, and national burden of cardiovascular diseases for 10 causes, 1990 to 2015. J. Am. Coll. Cardiol. **70**(1), 1–25 (2017)

17. Samol, A., Bischof, K., et al.: Single-lead ECG recordings including Einthoven and Wilson leads by a smartwatch: a new era of patient directed early ECG differential diagnosis of cardiac diseases? Sensors **19**(20), 4377 (2019)

18. Singh, P., Pradhan, G.: A new ECG denoising framework using generative adversarial network. IEEE/ACM Trans. Comput. Biol. Bioinf. **18**(2), 759–764 (2020)

19. Stern, S., Tzivoni, D., Stern, Z.: Diagnostic accuracy of ambulatory ECG monitoring in ischemic heart disease. Circulation **52**(6), 1045–1049 (1975)

20. Wang, G., Yang, L., et al.: ECG signal denoising based on deep factor analysis. Biomed. Signal Process. Control **57**, 101824 (2020)

21. Wang, J., et al.: Adversarial de-noising of electrocardiogram. Neurocomputing **349**, 212–224 (2019)

22. WHO: Cardiovascular diseases (cvds) (2017). https://www.who.int/news-room/fact-sheets/detail/cardiovascular-diseases-(cvds)

23. Xiong, P., Wang, H., et al.: ECG signal enhancement based on improved denoising auto-encoder. Eng. Appl. Artif. Intell. **52**, 194–202 (2016)

24. Yang, Q., Cogswell, M.E., Flanders, W.D., et al.: Trends in cardiovascular health metrics and associations with all-cause and CVD mortality among us adults. JAMA **307**(12), 1273–1283 (2012)

COVID-19 Fake News Detection Using Cross-Domain Classification Techniques

Arnav Sharma[1], Subhanjali Sharma[1], Utkarsh Bhardwaj[1], Sajib Mistry[2(✉)], Novarun Deb[1], and Aneesh Krishna[2]

[1] Indian Institute of Information Technology, Vadodara, Gandhinagar, India
{201951030,201952236,201952238,novarun_deb}@iiitvadodara.ac.in
[2] Curtin University, Perth, Australia
{sajib.mistry,A.Krishna}@curtin.edu.au

Abstract. The recent pandemic has witnessed a parallel infodemic happening on social media platforms, leading to fear and anxiety within the population. Traditional machine learning (ML) frameworks for fake news detection are limited by the availability of data for training the model. By the time sufficient labeled datasets are available, the existing infodemic may itself come to an end. We propose a COVID-19 fake news detection framework using cross-domain classification techniques to achieve high levels of accuracy while reducing the waiting time for large training datasets to become available. We investigate the effectiveness of three approaches: Domain Adaptive Training, Transfer Learning, and Knowledge Distillation that reuse ML models from past infodemics to improve the accuracy in detecting COVID-19 fake news. Experiments with real-world datasets depict that Transfer Learning performs better than Domain Adaptive Training and Knowledge Distillation techniques.

1 Introduction

In the last couple of years during the COVID-19 pandemic, the world has witnessed a deluge of misleading information and news articles that have been circulated on social media platforms resulting in spreading widespread panic, fear, and anxiety among the masses. These fake news articles have ranged from the origin of the novel coronavirus to its mutations, potential threats, efficacy of vaccines and their after-effects, as well as other diseases related to the black fungus or yellow fungus. The need for AI/ML-based solutions was promptly identified by the community to fight against this infodemic. Fake news classification for social media platforms has been explored over the last 6–7 years. The problem with conventional fake news detection models is that they require large quantities of labeled datasets for better training of the classification models. High levels of accuracy, precision, and recall are achieved when a sufficient amount of datasets is available. For instance, if we assimilate all the Covid-19-related fake news articles that have been circulated over the last couple of years and use them for training a fake news classifier, then the model is expected to perform very

© The Author(s), under exclusive license to Springer Nature Singapore Pte Ltd. 2024
T. Liu et al. (Eds.): AI 2023, LNAI 14471, pp. 507–519, 2024.
https://doi.org/10.1007/978-981-99-8388-9_41

efficiently. Unfortunately, by the time we have sufficient data to create an efficient model, we see that the Covid-19 infodemic has already waned significantly. The major research challenge being addressed is as follows - *"Is it possible to obtain an accurate fake news detection model that reuses the knowledge obtained from previous infodemics and combines it with the relatively small datasets that are available at the very initial phases of the current infodemic?"* This would greatly reduce the waiting time for collecting sufficient amounts of training data and building the model from scratch. Conventional infrastructures for text classification have gotten more sophisticated over time and have shown exceptional results for benchmark datasets. However, for datasets having disparities in terms of their domain origin and size, these infrastructures need to be paired with various procedures which would lessen the gap between source data and target data in order to improve a model's inter domain classification ability.

Cross domain text classification is classification of target data belonging to one domain using a model trained on source data belonging to a different domain. In our case, we attempted to use a general fake news dataset (the source data) to classify COVID-19 news (the target data). For our curated source dataset, ISOT [1] and a dataset from Kaggle [2] were used. COVID-19 data for fake-news classification is sparse so we had to curate it from scratch. COVID-19-rumour-dataset [3] and Co-AID dataset [4] were used for this purpose. The main contributions of this paper are as follows:

1. A cross domain classification using the general fake news and COVID-19 fake news datasets.
2. A fine-tuning approach tailored for COVID-19 fake news classification given the progression of COVID-19 infodemic over time.
3. Using the standard BERT-BASE-UNCASED, a pre-trained BERT model named CT-BERT [22], a Transfer Learning approach named Gradient Reversal [18] and a Knowledge Distillation [20] infrastructure for cross domain text classification.
4. Comparing the performance of the above approaches to identify most suitable approach for cross-domain classification task.

The rest of this paper is structured as follows. Section 2 discusses the recent works in the literature that try to address this issue. In Sect. 3, we elaborate the cross-domain classification techniques in detail. This is followed by our experiments and results in Sect. 4. Section 5 concludes the paper.

2   Related Work

Fake news detection is a Natural Language Processing task that involves determining the legitimacy of a news article. This task necessitates a classifier capable of understanding a news article semantically. Several Natural Language Processing techniques can be implemented for this task to come to fruition. To enable the model to understand the article's language, preprocessing of the article's text is performed. This involves converting the article's tokens into features using techniques such as Term Frequency-Inverse Document Frequency

(TF-IDF) vectorization. Another approach is rendering the article's text using a learned representation, which is also known as word embedding. Recent advances in word-embedding include GloVe [5] and Word2Vec [6] which have a vocabulary of millions of words and have been trained on huge datasets.

Machine Learning models, such as Support Vector Machines [7] (SVMs) and Naive Bayes Classifiers [8] (NBCs), have been employed for Fake News detection. These models generally outperform other techniques, such as Logistic Regression [9] (LR) and Random Forest Classifiers [10] (RFCs). Deep Learning models, like RNNs and CNNs, can also serve the same purpose. One of the architectures of RNNs, named LSTM [11] (Long Short-Term Memory), has shown promising results for fake-news datasets. An architecture [12] comprising distinct Deep Learning models has also been utilized. Transformers like BERT [13] have also been compared with standard Machine and Deep Learning models for the classification of U.S. General Presidential Election-2016 news.

Talking about cross domain classification techniques, a survey [14] collated various domain adaptation techniques for several downstream tasks. The text classification task was performed using several state-of-the art procedures and some of them like Domain Adversarial Neural Networks and Adaptive Pre-Training served as a starting point for us to look in the right direction for classifying COVID-19 fake news. Another approach which aimed at reducing the gap between source data and target data for a model was Knowledge Distillation. A Student-Teacher [15] model performed very well on benchmark datasets in terms of Text Classification which urged us to implement Knowledge Distillation.

3 Methodologies

This section is mainly focused on the methodologies we explored and approaches we propose for COVID-19 fake news detection.

3.1 Problem Definition

Given the source domain data (general news) is denoted as $D_s\{(X_i, Y_i)|1 \leq i \leq N\}$ and the target domain data (COVID-19 news) is denoted as $D_t\{(X_j, Y_j)|1 \leq j \leq M\}$, the problem is to find Y_j using X_j after training a classifier on $D_s\{(X_i, Y_i)\}$. Here N is the size of the source dataset and M is the size of the target dataset.

3.2 Cross Domain Text Classification Using BERT

To detect whether a piece of information is authentic or not, contextual learning is very important. For a language model, capturing the semantic nature of the language to perform a downstream task is always the primary goal. This semantic nature is represented in the form of an entity called embeddings. These embeddings are actually sequences of probability distributions of tokens that are learned after feeding textual examples. This learning in existing architectures is usually done in a unidirectional/forward fashion, i.e., left-to-right or right-to-left. Mathematically for a piece of text that contains N tokens, uni-directional embeddings for a position k would look like [16]:

$$p(t_1, t_2, ..., t_N) = \prod_{k=1}^{N} p(t_k | t_1, t_2, ..., t_{k-1}) \tag{1}$$

Recent advances in the language modeling avenue have resulted in architectures capable of capturing context from both directions of a token. BERT (Bidirectional Encoder Representations from Transformers) involves two tasks, namely, pre-training and fine-tuning. It is pre-training that enables BERT to formulate embeddings for a particular token. The two main objectives of pre-training are MLM (masked language model) and NSP (next sentence prediction), which aid in understanding inter-word and inter-sentence relationships.

1. MLM (Masked Language Modeling): This process has 3 outcomes with fixed probabilities. In the first outcome, a random token of a sentence is replaced with a (MASK) token. In the second outcome, a random token of a sentence is replaced with another token that does not belong to that sentence. In the third outcome, no tokens are changed. The model is trained on these three outcomes by predicting the (MASK) token based on the rest of the tokens present. This helps the model understand the overall contextual relationships between tokens in a sentence.

2. NSP (Next Sentence Prediction): This process is a classification task that involves only two labels. In the first outcome, the next sentence of a particular sentence is not changed, and in the second outcome, the next sentence is replaced with a random sentence from the corpus. The model is trained on these outcomes to determine if a sentence succeeds a particular sentence. With the help of these pre-training tasks, BERT is able to gain an understanding of a language's semantics. For a model to distinguish real news from fake news, it needs to understand how sentences are formulated in each class, how words are interlinked with each other, and what impact these relationships could have on determining the authenticity of the sentence. BERT becomes capable of understanding context in a sentence after pre-training and becomes ideal for our task of fake news detection.

Pre-training on domain-specific corpora may improve a model's ability to perform downstream tasks on domain-specific target data. We implemented an approach where we fine-tuned a pretrained BERT model named COVID-TWITTER-BERT [22] using the general fake news data and the COVID-19 fake news data paired with the proposed fine-tuning approach. COVID-TWITTER-BERT or CT-BERT, is a model based on BERT-LARGE (trained mainly on raw text data from Wikipedia, 3.5B words, and a free book corpus, 0.8B words) that has been trained on 22.5M tweets related to the coronavirus, with a total vocabulary of 0.6B words.

3.3   Transfer Learning

Another avenue that we explored for cross-domain text classification was Transfer Learning. For instance, certain researchers studied adults and gathered relevant information to diagnose heart-related problems in infants. When two datasets from similar distributions for the same task of classification are available, the only challenge is to overcome the difference between the domains and train a classifier on source data and somehow transfer the knowledge of this classifier to work on target data. This classifier is called a domain-adaptive classifier [17] or transfer learner. As the name suggests, this domain-adaptive classifier, in a way, adapts itself to changing domains to perform classification. Domain Adaptation can be achieved by many processes, one of them being Adversarial Alignment [18]. This process focuses on two aspects, which are:

1. Discriminativeness, i.e., ability to distinguish two different entities.
2. Domain Invariance (ability of the model to classify data regardless of the data's domain)

For the first aspect two classifiers namely label classifier and domain classifier are included in the model's architecture. For the second aspect, the label classifier's loss is minimised and its adversary, the domain classifier's loss is maximised during optimisation. What the adversary does is that it tries to increase domain confusion in the model which helps in improving the model's classification for an unfamiliar domain. To define this problem [19], $f = G_f(x; \theta_f)$ is considered a vector of distribution of parameters, $G_y(f)$ is a mapping between the label classifier and vector of parameters parameters, $G_d(f)$ is a mapping between the domain classifier and vector of parameters, $S(f) = \{G_f(x; \theta_f) | x \sim S(x)\}$ is the distribution of source data and $T(f) = \{G_f(x; \theta_f) | x \sim T(x)\}$ is the distribution of target data. The models responsibility is to lessen the gap between S(f) and T(f) for it to be independent of the domain of the data. This dissimilarity gap can actually be represented by the domain classifier's loss (a quantification of how accurately a model could predict the correct domain). To put it mathematically:

$$E(\theta_f, \theta_y, \theta_d) = \sum_{\substack{i=1...N \\ d_i=0}} L_y^i(\theta_f, \theta_y) - \lambda \sum_{\substack{i=1...N \\ d_i=0}} L_d^i(\theta_f, \theta_d) \tag{2}$$

$E(\theta_f, \theta_y, \theta_d)$ needs to be minimised by optimisation of the model. θ_f, θ_y and θ_d are parameters corresponding to the overall classifier, label classifier and the domain classifier respectively.

Optimisation [19] for θ_f, θ_y, θ_d would look like this:

$$\theta_f \leftarrow \theta_f - \mu \left(\frac{\partial L_y^i}{\partial \theta_f} - \lambda \frac{\partial L_d^i}{\partial \theta_f} \right) \tag{3}$$

$$\theta_y \leftarrow \theta_y - \mu \frac{\partial L_y^i}{\partial \theta_y} \tag{4}$$

$$\theta_d \leftarrow \theta_d - \mu \frac{\partial L_d^i}{\partial \theta_d} \tag{5}$$

This technique is also called Gradient Reversal, since the domain classifier's loss is being made negative. The main reason why the loss of the domain classifier needs to be maximized during optimization is to make the model's ability independent of domain variance. The maximized loss will reduce the overall loss of the domain during optimization, which improves its ability to classify accurately.

3.4   Knowledge Distillation

A similar approach that allows the concept of cross-domain learning is Knowledge Distillation. This approach is used to save computation power, where a model's knowledge trained on a large dataset is inherited by a model for which training data is very limited. Knowledge Distillation involves a Teacher model and a Student model. This dynamic can be better understood by comparing the Student model to an actual student learning how to perform a task from the Teacher model's already existing vast knowledge. For cross-domain classification, online distillation [20], a procedure that involves the Student learning from the Teacher simultaneously as the Teacher is being trained, is ideal, as opposed to offline distillation, which involves the use of an existing large Teacher model (a pre-trained classifier trained on a large corpus for fake news detection in our case) that wasn't available. Loss [21] in general would look like:

$$L_{ResD}(zt, zs) = \zeta_R(zt, zs) \tag{6}$$

where $\zeta_R(.)$ indicates the divergence loss of logits (quantization of how much logits differ from each other), and zt and zs are logits of teacher and student, respectively. Logits are a representation of the output of a particular model. The Distillation loss and the Student loss is back-propagated for the Student model to learn from the Teacher model.

3.5   Fine-Tuning

The mercurial nature of COVID-19 information is a problem because the context of COVID-19 information keeps changing over time with new events occurring on a daily basis. To address this problem, we devised an approach to train the model iteratively to enhance its ability to detect future COVID-19 infodemics.

The data was split into n number of batches $[x_1, x_2, x_3, ..., x_n]$ which was a representation of COVID-19 data collected over a period of n time frames. These batches were then used to re-train our model for the number of time frames as the number of epochs. In other words, the model was first trained on the batch $[x_1]$ followed by $[x_1, x_2]$, $[x_1, x_2, x_3]$, ... until $[x_1, x_2, x_3, x_4, ..., x_n]$ was the data fed to our model. By this approach, we desired our model to learn the trend COVID-19 infodemic follows to make sure the model doesn't become obsolete.

4   Experiments and Results

We conducted some experiments to assess the performance of the proposed fine-tuning method and different models. These experiments were performed to answer the following pertinent questions:

Q1. What should the time quantum be to obtain the best performance for the proposed fine-tuning approach for COVID-19 fake news detection?
Q2. Which cross domain technique is actually effective for domain adaptation?

The details of the datasets we used for our experiments are shown in Tables 1 and 2.

Table 1. Number of labels for each dataset

Datasets	Real News	Fake News
General News	31797	31085
Covid-19 News	12187	12168

Table 2. Number of labels in Multi-Domain Amazon Reviews dataset

Categories	Positive Reviews	Negative Reviews
books	1000	1000
dvd	1000	1000
electronics	1000	1000
kitchen	1000	1000

4.1   Implementation

For the General Fake news dataset, the ISOT dataset and the Kaggle dataset were simply combined. For the COVID-19 fake news dataset, both datasets had tweet IDs that had to be used to obtain the actual tweets. Twint, a tool for scraping Twitter, was used to collect all the data. The available data also included pre-existing news sentences. These datasets were labeled.

The training dataset for COVID-19 fake news detection was the General Fake news dataset. For the proposed fine-tuning approach, the COVID-19 fake news data was split into 2 parts (50% each). The first part was used for fine-tuning the trained model following the proposed procedure. The other half was used to test all the models. For the Multi-Domain Amazon Reviews Sentiment Dataset, available processed data was used. The format of the document was:

$$feature : \ <count> ....feature : \ <count> \ \#label\# : \ <label> \qquad (7)$$

where feature is referred to as a token's value, count refers to the number of times the token occurred in the review and label refers to the sentiment(Positive or Negative) of the review. The features were extracted after opening the available file as an XML file. Training dataset was curated using the reviews of books, dvd and electronics categories. Testing dataset was curated using the kitchen category. For comparing how well our models performed, the accuracy metric, the precision metric and the recall metric were used. For calculating each metric, the number of True Positives, True Negatives, False Positives and False Negatives should be known. The Positive in the COVID-19 fake news detection is the 1 or the Fake label and the Negative is the 0 or the Real label. The Positive in the Amazon Reviews Sentiment Analysis is the 1 or the Negative label and the Negative is the 0 or the Positive label.

We provide the link[1] to our GitHub repository for the reproduction of results presented in the following sections.

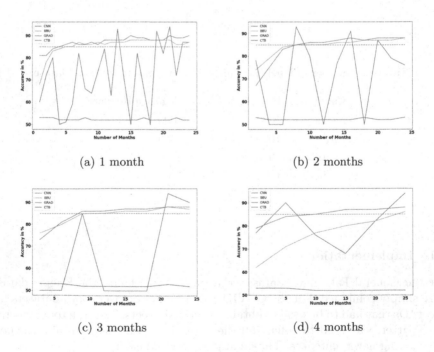

(a) 1 month   (b) 2 months

(c) 3 months   (d) 4 months

Fig. 1. Accuracy of data accumulated over different time quanta

[1] https://github.com/ArnavS1102/Fake-News-Detection.

4.2 Baselines

We used some baseline models to compare the performance of our fine-tuning approach and cross domain text classification techniques. For the task of Q1, a Convolutional Neural Network(CNN) paired with GLoVe word embeddings was used as a baseline along with BERT_BASE_UNCASED (BBU), BBU paired with a Transfer Learning approach named Gradient Reversal (GRAD) and a pre-trained BERT architecture (CTB). For the task of Q2, architecture of Knowledge Distillation (KD) was used to compare the extent to which the BBU and GRAD models become domain adaptive.

4.3 Time Quantum for Fine-Tuning (Q1)

The proposed fine-tuning approach was used for using data slices of 1 month, 2 months, 3 months and 4 months. Lineplots were obtained to observe how accuracy of models fine-tuned for each quantum progresses over multiple iterations (refer Figs. 1a–1d).

Among the models used, we observed subpar performance by CTB for all the time quanta. CNN, on the other hand, exhibited variability in performance. This variance in its performance can be explained by the fact that Convolutional Networks need a large amount of data to be trained, and passing only slices of data wasn't enough to ensure improvement. For all the quanta, its accuracy wasn't consistent. BBU and GRAD showed consistent improvement in accuracy over the months for all the quanta.

The final accuracy of the models after the fine-tuning process were observed as follows (refer Table 3):

Table 3. Accuracy of different models using data slices of different time quantum

Time Quantum	CNN	BBU	GRAD	CTB
1 month	0.87	0.87	0.9	0.52
2 months	0.76	0.88	0.88	0.53
3 months	0.9	0.87	0.88	0.52
4 months	0.94	0.86	0.88	0.52

CNN showed a lot of variance in its accuracy. Hence, we decided to calculate the variance for each model for each quantum to understand which model actually ensured improvement, rather than occasional pitfalls in performance. (refer Table 4). If the variance metric was considered to determine a model's consistency it would be easy to say that 1 month quantum is better suited for our proposed fine-tuning approach. CNN has the lowest variance for the 1 month quantum and the most accurate model, GRAD also has the lowest variance for 1 month quantum.

Table 4. Variance of each model for different time quanta

Time Quantum	CNN	BBU	GRAD	CTB
1 month	0.0204	0.0018	0.0006	0.00001
2 months	0.028	0.0016	0.0041	0.00002
3 months	0.042	0.0016	0.0026	0.00002
4 months	0.009	0.007	0.001	0.00002

Another approach to determine the ideal time quantum would be how quickly a model reaches a certain threshold for high accuracy. If 85% were considered this threshold, the ideal quantum of time would be the one which helps models achieve 85% of accuracy in the smallest period of time. To determine this, lineplots were obtained for BBU and GRAD models using different time quanta. CNN and CTB were not used because of CNN's high variance in performance which would have caused a lot of confusion in making a conclusion and the end result would have been ambiguous. On the other hand, the accuracy of CTB saturates between 50–55% (as seen in Figs. 1a–1d).

The lineplots in Fig. 2 show that with the 1 month time quantum, each model was able to reach higher accuracy in a shorter span of time. Both BBU and GRAD are able to achieve 85–90% accuracy within 6 months with the proposed fine-tuning approach and a time quantum of 1 month compared to other time quanta of data slices.

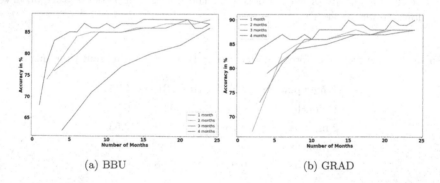

(a) BBU   (b) GRAD

Fig. 2. Accuracy of BBU and GRAD for each time quantum

4.4 Comparison of Effectiveness of Cross Domain Techniques (Q2)

To determine the effectiveness of the proposed fine-tuning approach of retraining the model over batches of data for a particular quantum of time and cross domain text classification techniques, we compared the accuracy of BBU, GRAD, CTB with fine-tuning vis-a-vis using Knowledge Distillation. The General and

Table 5. Metrics for COVID-19 dataset

MODEL	ACCURACY	PRECISION	RECALL
BBU	0.85	0.80	0.94
CTB	0.52	0.51	0.62
GRAD	0.9	0.84	0.91
KD	0.88	0.84	0.93

COVID-19 Fake News dataset were used for training and fine-tuning respectively. The metrics obtained for the different models have been shown in Table 5. After performing this experiment and obtaining the relevant metrics for comparison, it was apparent that GRAD, the model based on the Transfer Learning approach of Gradient Reversal, outperformed all the other models in the case of COVID-19 fake news classification. When comparing the accuracy of the BBU model and GRAD it was clear that GRAD was able to induce domain confusion. KD, the baseline model for this task did not outperform GRAD in terms of accuracy, indicating GRAD is a better approach for domain adaptation in case of COVID-19 fake news detection. For checking our models' ability to correctly classify data from different domains, BBU, GRAD and KD were also trained on the Multi-Domain Amazon Reviews Sentiment Dataset. CTB wasn't trained on this dataset because it has been pre-trained on COVID-19 corpus and using it for a sentiment dataset of utilities was irrational. The metrics for Multi-domain Sentiment Amazon Reviews dataset can be seen in Table 6. The BBU, GRAD and KD which weren't COVID-19 specific models unlike CTB which had been pre-trained on COVID-19 related corpus were used for cross domain text classification for the Multi-Domain Amazon Reviews Sentiment Dataset. Relevant metrics were found for each model. GRAD was better than the baseline model BBU in terms of accuracy. KD showed better characteristics than all the other models in all regards indicating it's a better approach for domain adaptation.

Table 6. Metrics for Multi-Domain Amazon Reviews Dataset

MODEL	ACCURACY	PRECISION	RECALL
BBU	0.61	0.61	0.61
GRAD	0.63	0.66	0.55
KD	0.81	0.89	0.70

5   Conclusion

In this paper, we focused on the problem of detecting fake COVID-19 information using intricate procedures and sophisticated models. We proposed a fine-tuning approach specifically for COVID-19 information because of its volatile nature. We studied how each model utilised the general fake news data and the limited COVID-19 data to train itself for text classification. We also studied state-of-the-art approaches for inducing domain adaptation in our models to bridge the disparities between the general fake news domain and the COVID-19 fake news domain. We performed experiments for COVID-19 fake news detection and Domain Adaptation in general. From the experiments that we conducted, for COVID-19 fake news detection Gradient Reversal(Transfer Learning) is observed to be the most accurate approach with a time quantum of 1 month for fine-tuning. Apart from this we also tested these models on a multi-domain Amazon Reviews dataset to determine their domain adaptation abilities. From this experiment, Knowledge Distillation was observed to be the most accurate approach. In the future, we aim to delve deeper into these approaches and try to achieve faster convergence of accuracy estimates.

Acknowledgement. This work is partly made possible by Regional Collaborations Programme COVID-19 Digital Grant from the Australian Academy of Science. The statements made herein are solely the responsibility of the authors.

References

1. Fake News Detection Datasets - University of Victoria. https://www.uvic.ca/ecs/ece/isot/datasets/fake-news/index.php
2. Fake News. https://kaggle.com/competitions/fake-news
3. Wang, S.: File structure (2022). https://github.com/MickeysClubhouse/COVID-19-rumor-dataset. Accessed 24 May 2022
4. Cui, L., Lee, D.: CoAID: COVID-19 healthcare misinformation dataset. arXiv (2020). http://arxiv.org/abs/2006.00885. Accessed 24 May 2022
5. Pennington, J., Socher, R., Manning, C.: GloVe: global vectors for word representation. In: Proceedings of the 2014 Conference on Empirical Methods in Natural Language Processing (EMNLP), Doha, Qatar, pp. 1532–1543 (2014). https://doi.org/10.3115/v1/D14-1162
6. Mikolov, T., Chen, K., Corrado, G., Dean, J.: Efficient estimation of word representations in vector space. arXiv (2013). http://arxiv.org/abs/1301.3781. Accessed 30 May 2022
7. Conroy, N.J., Rubin, V.L., Chen, Y.: Automatic deception detection: methods for finding fake news. Proc. Assoc. Inf. Sci. Technol. **52**(1), 1–4 (2015)
8. Khurana, U., Intelligentie, B.O.K.: The linguistic features of fake news headlines and statements (2017)
9. Bhattacharjee, S.D., Talukder, A., Balantrapu, B.V.: Active learning based news veracity detection with feature weighting and deep-shallow fusion. In: 2017 IEEE International Conference on Big Data (Big Data), pp. 556–565. IEEE (2017)
10. Hassan, N., Arslan, F., Li, C., Tremayne, M.: Toward automated fact-checking: detecting check-worthy factual claims by ClaimBuster. In: SIGKDD, pp. 1803–1812 (2017)
11. Rashkin, H., Choi, E., Jang, J.Y., Volkova, S., Choi, Y.: Truth of varying shades: analyzing language in fake news and political fact-checking. In: Proceedings of the 2017 Conference on Empirical Methods in Natural Language Processing, pp. 2931–2937 (2017)

12. Wang, W.Y.: "Liar, liar pants on fire": a new benchmark dataset for fake news detection. arXiv preprint arXiv:1705.00648 (2017)
13. Kaliyar, R.K., Goswami, A., Narang, P.: FakeBERT: fake news detection in social media with a BERT-based deep learning approach. Multimed. Tools Appl. **80**(8), 11765–11788 (2021). https://doi.org/10.1007/s11042-020-10183-2
14. Ramponi, A., Plank, B.: Neural unsupervised domain adaptation in NLP–a survey. arXiv (2020)
15. Zhang, B., Zhang, X., Liu, Y., Cheng, L., Li, Z.: Matching distributions between model and data: cross-domain knowledge distillation for unsupervised domain adaptation. In: ICONIP, pp. 5423–5433 (2021)
16. Peters, M.E., et al.: Deep contextualized word representations. In: HLT, pp. 2227–2237 (2018)
17. Kouw, W.M., Loog, M.: An introduction to domain adaptation and transfer learning. arXiv (2019). http://arxiv.org/abs/1812.11806. Accessed 24 May 2022
18. Ganin, Y., et al.: Domain-adversarial training of neural networks. arXiv (2016). http://arxiv.org/abs/1505.07818. Accessed 25 May 2022
19. Ganin, Y., Lempitsky, V.: Unsupervised domain adaptation by backpropagation. arXiv (2015). http://arxiv.org/abs/1409.7495. Accessed 25 May 2022
20. Gou, J., Yu, B., Maybank, S.J., Tao, D.: Knowledge distillation: a survey. Int. J. Comput. Vision **129**(6), 1789–1819 (2021). https://doi.org/10.1007/s11263-021-01453-z
21. Müller, M., Salathé, M., Kummervold, P.E.: COVID-twitter-BERT: a natural language processing model to analyse COVID-19 content on twitter. arXiv (2020)
22. Multi-Domain Sentiment Dataset. https://www.cs.jhu.edu/~mdredze/datasets/sentiment/

Context-Based Masking for Spontaneous Venous Pulsations Detection

Hongwei Sheng[1,2] , Xin Yu[1(✉)] , Xue Li[1] , and Mojtaba Golzan[2]

[1] The University of Queensland, Brisbane, QLD 4067, Australia
Xin.yu@uq.edu.au
[2] University of Technology Sydney, Sydney, NSW 2007, Australia

Abstract. Spontaneous retinal venous pulsations (SVP) serve as vital dynamic biomarkers, representing rhythmic changes of the central retinal vein observed at the optic disc region (ODR) within an eye. SVPs serve as vital dynamic biomarkers, representing rhythmic changes of the central retinal vein observed at the optic disc region (ODR) within an eye. In light of their crucial clinical role, automatic detection of SVPs from fundus videos has become an area of burgeoning research. However, the inherent eye movements and the variability in retinal video quality present significant challenges to direct SVP detection via existing deep learning models. In response, we devise a spatio-temporal context-based masking approach (STC Masking), exploiting the spatiotemporal characteristics of SVPs to enhance their detection in retinal videos. We first apply a spatio-temporal mask to clip the video into an ODR-focused video tube. Diverging from conventional masking with gray or black blocks, we then employ a context masking method which using the original pixel values from video frames as the mask fill-in. The context mask map temporally transforms the dynamic video tubes into static tubes, thus changing the pulsation status of SVPs. Correspondingly, we adjust the SVP video labels based on the changing extent of masked regions to avoid ambiguity in data labelling. This innovative strategy provides more vivid videos which are similar to unmasked videos pixel-wise but having contrast semantics in SVP presenting regions. This enables network to capture the most discriminating regions through spatio-temporal variations, allowing explicit detection on SVP existence in the video. Our experiments illustrate the efficacy of our STC masking strategy, outperforming baseline methods. This work, thereby, underscores the potential of grid context-based masking for more accurate SVP detection in retinal video analysis.

Keywords: Medical image analysis · Spontaneous venous pulsations · video Recognition · Data augmentation · Grid masking

1 Introduction

Spontaneous retinal Venous Pulsations (SVP) [11] are rhythmic changes in the central retinal vein and can be observed in the optic disc region (ODR) of the

© The Author(s), under exclusive license to Springer Nature Singapore Pte Ltd. 2024
T. Liu et al. (Eds.): AI 2023, LNAI 14471, pp. 520–532, 2024.
https://doi.org/10.1007/978-981-99-8388-9_42

retina as shown in Fig. 1. It is a physiological phenomenon that present in most healthy individuals and is caused by variations in ocular blood flow and pressure. The absence of SVP can be associated with rapid progression in glaucoma or increased intracranial pressure [6,20,27,36]. Performing regular checks of the status of SVP can greatly aid in the early detection of related diseases and avoid subsequent complications [10]. Recently, smartphone-based fundus cameras are becoming popular for visualizing the retina and assessing ocular conditions [10,12,13,25,28,37]. The low cost and accessibility of smartphone-based fundus cameras allow clinicians to perform retinal imaging in various environments. However, images captured by hand-held devices are frequently impacted by various types of noise, including shaking hands, changes in illumination within the environment, and noise generated by the device itself [2,10,18,36]. These not only hinder human perception but also prevent automatic SVP diagnoses, necessitating the development of an efficient SVP detection method.

Video recognition has been extensively investigated over the past two decades and various methods have been proposed [16,30,40,41]. Recently, many deep learning based methods have been developed [3,7,24,29,33,35,39] to classify real-world videos, and also achieve promising performance in different video tasks. However, to the best of our knowledge, no prior methods for SVP detection have been identified. Furthermore, existing video recognition networks do not take the clinical-specific domain information into account, and thus fail to provide a comprehensive and robust solution for processing real clinical datasets.

We observe two challenges that prevent the existing methods to be applied to automatic SVP detection: (i) The quality of fundus videos captured by smartphones can be greatly compromised by practical factors, such as caused by inevitable eye movements, and unintended shakes. Moreover, video jittering effects will further impair the spatio-temporal information analysis for deep neural networks. (ii) Unlike generic actions, the size of SVP is relatively small in both spatial and temporal dimensions. Additionally, SVP appears in only a few vessel segments within the majority of blood vessels, indicating a complex multi-vessel scenario. The presence of SVP is thus frequently overlooked by conventional network architectures. As a result, current video processing methods are struggling to localize SVP spatially and temporally.

To address these challenges, we propose our spatio-temporal context-based masking method (STC Masking) to facilitate SVP recognition networks. Our method first spatially centralizes the ODR by masking out ODR surrounding regions. Temporally, our STC masking then pinpoints the SVP segments by masking superfluous frames for a single SVP. Simultaneously, we use existing video semantics as masks to skillfully conceal new context information. This technique refines the network's sensitivity to spontaneous venous pulsations (SVPs), enabling it to more precisely distinguish between various vascular scenarios. In a nutshell, we implement spatio-temporal and context-based masking to enhance the performance of SVP recognition networks.

In our SVP recognition training pipeline, we initially perform a pre-processing to clean the clinical data and retrieve the video clips with continuously ODR-

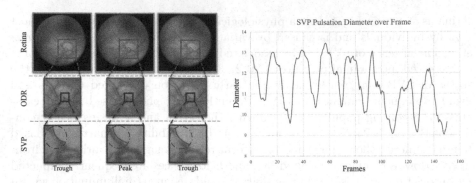

Spontaneous retinal Venous Pulsations (SVP) Sequence

Fig. 1. Left: A visualization of SVP in ODR region. An SVP is occurring in the Optic Disc Region (ODR) in the retina. Please note the changes in the thickness of the blood vessels near our blue guide lines. A complete cycle of SVP entails the distinct presentation of a "peak" and a "trough" in the variation of vessel diameter. Right: A polygraph reflecting the diameter changing over frames at SVP position. We employ ImageJ [26] to measure the diameters. In each cycle, we categorize the highest value as "peak" and the lowest value as "trough". (Color figure online)

visible frames. In the spatio-temporal masking step, we spatially stabilizes ODR in the center of the footage across the video sequence via a template matching algorithm and mask out the ODR-surrounding areas, then randomly sample sequences of 30 continuous frames to guarantee the complete display of one SVP cycle. In the context masking step, we apply random sampling on a grid video frame and utilize different sampled grids as our grid mask maps in our mask training network. Our proposed strategies significantly improve the input semantics and enforce our model to explicitly identify spatial regions and temporal duration of SVP. Therefore, this significantly facilitates our recognition network to achieve better recognition performance. Extensive experiments on our newly curated masking strategy demonstrate the effectiveness of our network in recognizing SVP on such practical and challenging data.

2   Related Work

Some researchers and clinicians [37] have commenced using smartphone-based fundus photography to observe retinal conditions, allowing frequent SVP assessment. However, the quality of retinal fundus images captured by smartphones can be affected greatly by real-world issues, such as unpredictable environments, and uncooperative patients. Manual analysis of these noisy smartphone data is time-consuming and inaccurate. More importantly, the processing highly relies on the skill of the medical professional, thus restricting the usage of smartphone-based fundus videos in practice. Though some automatic methods [22,23] have been proposed, they still require manual pre-screening of informative high-quality data prior to processing. Despite the rapid development in computer assist medical techniques [21], automatic and efficient monitoring of SVPs still remains an unaddressed challenge.

Exploring advances in deep learning, video recognition and detection neural networks are well-developed and have the ability to handle real-life video tasks. Action recognition networks, including I3D [3], X3D [7] and ResNet3D [29], extend advantages of the 2D network design to the 3D network, which makes the network perform well in video tasks. TSM [19] exceeds on architecture by introducing a shift operation in the 3D convolutional layers. In addition, a recently proposed action recognition network called VTN [24] utilizes transformer architecture [31]. VTN is able to model long-range temporal dependencies in videos, which is a challenging task for traditional convolutional neural networks.

All of these action recognition networks have exhibited robust performance across diverse benchmarks. We can harness these networks for the specialized task of SVP detection. While the existence of domain gaps is an unavoidable challenge, alternative methods can be deployed to enhance the network's performance [9]. This enhancement makes it feasible to train on heavily noisy medical data, utilizing existing networks to extract valuable insights.

Data Augmentation is a method to expand the training dataset by creating new samples through minor transformations of existing ones, thereby enhancing data diversity. It is widely used in the field of deep learning, especially in computer vision and natural language processing [1,5,38]. In the growing field of semantic understanding, various masking techniques have been developed to bolster the comprehension of network semantics [4,17,42]. Video learning foundation models based on masking have also been introduced, providing ample evidence of the effectiveness of the masking mechanism [8,32,34]. Inspired by these pioneering efforts, we design our STC masking method specifically tailored for SVP detection.

3 Method

3.1 Dataset

We carried out our experiments using the RVD dataset, which is the first and sole benchmarking dataset for fundus videos [14]. This video dataset contains 635 fundus videos from 4 clinics with different handheld devices. For each video, the dataset provides the label of SVP status (i.e., SVP-present or SVP-absent). Additionally, it provides the time period of presenting SVPs and also contains "Peak and Trough" segmentation annotation which actually indicates the spatial positions of SVPs. These annotations could help us retrieve the correct SVP label after modifying the video semantic via our masking. The videos offered by this dataset are of 1800×1800 resolution. As mentioned, our concerned biomarker SVP is only presenting in the ODR, while the diameter of ODRs in this dataset is around 450 pixels and the diameter of vessels in ODR is around 25 pixels. This indicates the videos have considerable spatial redundancy. Moreover, as a clinical dataset, it contains many challenges from real clinical settings, such as

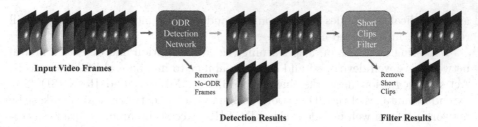

Input Video Frames   Remove No-ODR Frames   Remove Short Clips

Detection Results   **Filter Results**

Fig. 2. The pipeline of our pre-processing. We employ a detection algorithm to detect the ODR in the fundus videos and remove frames without ODR detected. After obtaining the clips with only ODR-visible frames, we filter out those with fewer than 30 remaining frames to avoid bias caused by an incomplete SVP cycle.

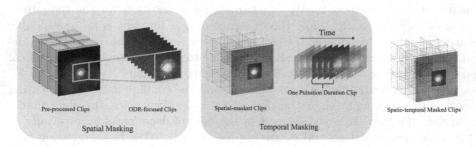

Fig. 3. An example of our spatial-temporal masking. The spatial masking part retrieves a video tube with ODR by masking out the surrounding regions outside ODR. The temporal masking part reduces temporal redundancy by preserving only a segment with sufficient duration to display one complete cycle of SVP.

eyelid blinking and highlight spots. These challenges increase the difficulty for neural network to detect SVP efficiently.

3.2 Pre-processing

In data pre-processing, we first clip the videos utilizing the temporal annotation (*i.e.*, the time period during which SVPs are present), thereby creating two distinct sets of clips: SVP-present video clips and SVP-absent video clips respectively. The clipped videos retain the same background contextual information as the original video, thus making the network insensitive to noisy background redundancy. To avoid interference from clinical noise, we then remove the video frames without ODR presenting. As shown in Fig. 2, we perform an ODR detection to temporally localize ODR regions across the video sequence. Note that the video frames in which ODR cannot be localized will be removed and the remaining video clips are required to consist of more than 30 consecutive frames (*i.e.*, 1 s) to ensure a complete cycle of SVP. After this processing, we significantly improve the quality of each video clip and facilitate SVP observation for both humans and machines.

3.3   Spatio-Temporal Masking

To further remove redundancy and emphasis the dynamic characteristics of SVP, we employ a spatio-temporal masking for each pre-processed video.

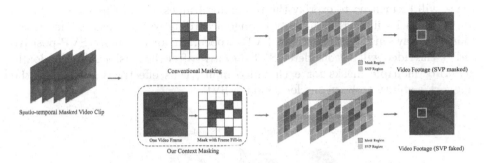

Fig. 4. Top: Conventional mask with gray fill-in. Bottom: Our context-based mask using image patches from video frames as fill-in. Note that SVP-presenting positions are masked in both cases, thus making both labels change from SVP-present to SVP-absent. Conventional masking inevitably shifts the network's attention to redundant regions without veins and pulsations. Compared with it, our context-based strategy provides simulated retina regions to retain the network's attention in potential SVP regions, allowing network to better grasp the relationship between feature changes in SVP regions and SVP label variations. (Color figure online)

Spatial Masking. We first stabilize the ODR in a fixed position in each frame via template matching algorithm and then mask out all the surrounding area around ODR.

Temporal Masking. Our masking strategy randomly chooses a continuous sequence of 30 frames and masks out the rest frames to accentuate the dynamic characteristics of SVP.

Via the spatio-temporal masking, we can obtain ODR-focused video tubes with enough frames for a full SVP cycle. As shown in Fig. 3, these videos are significantly lightweight and more capable to highlight spatio-temporal features of SVP. This masking strategy effectively mitigates the challenge posed by SVP's small size in both spatial and temporal dimensions, preventing it from being easily overlooked.

3.4   Context Masking

We then devise a novel grid-masking strategy to make the neural network focus on the spatial location where the SVP occurs. Thanks to the stabilized videos, positions of vessels are fixed over through the video frames. For any given grid in video footage, we can substitute the grid from the previous frame for the grid in the current frame without introducing vessel discontinuity or distortion. This capability allows us to manipulate temporal information without introducing spatial anomalies, enabling us to edit the status of the dynamic biomarker SVP. Specifically, we divide an image into multiple grids and then randomly select some grids to keep the image content within them as a context mask map as in Fig. 4. A context mask can be applied to all frames in a video clip. The generated mask will be utilized to overlay the remaining frames within this video segment and the label will correspondingly change according to the mask extent. If the mask overlay all locations where SVPs appear in an originally SVP-positive video, the video will be considered SVP-negative after the masking. We randomly generate multiple masks for each video input, thus effectively enhancing the network's ability to learn the locations of SVPs.

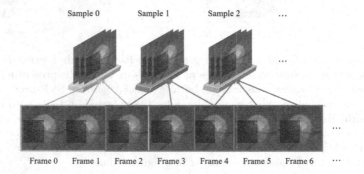

Fig. 5. Example of our SVP-emphasized sampling. We employ this sampling approach to accentuate the dynamic features of SVP. Employing a straightforward sliding window technique, we amalgamate consecutive frames into a single sample before feeding it into the network.

3.5   Network Training

Data Augmentation. We first perform pre-processing on the input video data and then apply our STC masking method onto the pre-processed data. To further enrich the training set and enhance its diversity, we also employ several data augmentation techniques after masking, such as rotating and flipping, etc. To mitigate the risk of erroneous labels, operations that can directly change the semantic integrity of SVP regions, such as random cropping and drastic translation, have been omitted. Additionally, we adopt color augmentations to overcome the inner domain gaps caused by different fundus recording devices in this dataset.

SVP-Emphasized Sampling. Considering the pulse-like dynamic characteristics of SVP as shown in Fig. 1, the pulsations at their peak values might be confined to just a few brief frames. Conventional dense sampling methods may mix a small number of "trough" value frames with a large number of "peak" value frames, thereby causing the most prominent features to be obscured. To endow the network with the capability of capturing the dynamic nature of SVPs, we need to amplify the prominence of these "trough" temporal patterns. Thus we employ a sliding window technique to emphasize the motion of SVP. As illustrated in Fig. 5, a sliding window is applied to compile frame segments from a video, and each frame segment is treated as an individual sample in the neural network training. While this approach may increase temporal redundancy, it also allows the features of the "trough" values to be dispersed across the time dimension. As a result, it leads to a more balanced distribution of samples between "peak" and "trough" values.

Loss Function. To accelerate convergence and enhance the ability to differentiate between the two classes, we then adopt binary cross-entropy (BCE) as the loss function for the network. BCE loss is a widely used loss function for binary classification tasks. Compared with conventional cross-entropy loss, it measures the dissimilarity between the predicted output and the actual label in a binary classification scenario with faster training and convergence of the model. Binary cross-entropy loss calculates the logarithmic loss between the predicted probability and the actual label. The loss is expressed as:

$$Loss_{BCE} = -(y\log(p) + (1-y)log(1-p)), \tag{1}$$

where y is the actual label (*i.e.*, 0 or 1), p is the predicted probability of the positive class, and log is the natural logarithm.

4   Experiments

In order to validate the effectiveness of the proposed context-based masking strategy, we evaluated it with the use of several well-adopted action recognition models, including I3d [3], X3D [7], TSN [33], and VTN [24]. These models are integrated into the training pipeline with our masking strategy. We split the dataset into training, validation, and testing sets in a 7:1:2 ratio.

4.1   Evaluation Metrics

We employ the ROC curve (Receiver Operating Characteristic) and PR curve (Precision-Recall), which are widely used in binary classification tasks. The ROC curve provides a graphical representation of the trade-off between true positive rate and false positive rate for different classification thresholds. It measures how well the classifier is performing by plotting the true positive rate against the false positive rate. Similarly, the PR curve measures how well the classifier is identifying true positive cases by plotting precision against recall. The area under the curve (AUC) is used to compare the performance of different classifiers. In general, a larger AUC indicates better performance of the classifier.

4.2   Implementation Details

We divide the footage into a 10×10 grid and apply the context-based masking stage with the mask ratio of 0.3. Furthermore, we adopt data augmentation methods including resizing and rotating. Throughout the experiments, we employed dual NVIDIA RTX 4090 with Adam [15] for 200 epochs of training.

4.3   Quantitative Results

Baselines. We first train our dataset on several popular video classification networks plainly as baselines, I3d, X3D, TSN, and VTN respectively. As shown in Fig. 6, we can clearly see that these models perform very poorly. The AUC results suggest the performance of neural networks is close to pure random classification.

Main Experiment. The main experiments are performed with our masking strategies. Since the absence of SVP may have serious consequences in medical practice, we prioritize the recall rate of the SVP-negative class and tolerate some potential misclassification of SVP-positive. As shown in Fig. 6, our SVP-DET pipeline outperforms the baselines by 50%.

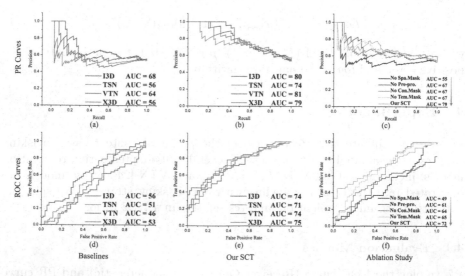

Fig. 6. Experiment results on the test set. **(a)**, **(d)** PR curves & ROC curves of baselines. **(b)**, **(e)** PR curves & ROC curves after applying our SCT masking. **(c)**, **(f)** PR curves & ROC curves when removing specific modules.

4.4   Ablation Study

Our network consists of three main parts to improve SVP detection, namely the data pre-processing part, the spatio-temporal masking part, and the context-based masking part. To fully demonstrate the effectiveness of these three modules, we conduct three sets of ablation experiments separately. The backbone of ablation experiments is I3D. For equality, our experiments ensure that the networks have the same receptive field of blood vessels.

Impacts of Pre-processing. The pre-processing in our method first split the videos according to the temporal annotation and remove the frames without visible ODR. By obtaining new SVP-present video clips and SVP-absent video clips with corresponding labels, it helps reduce the ambiguity in SVP existence while training. We then train the model with original videos with clinical noises.

Impacts of Spatio-Temporal Masking. Spatio-temporal masking includes spatial masking and temporal masking. The spatial masking is to center the ODR and remove the surrounding redundancy, while the temporal masking helps removing the redundant frames to avoid distraction. If spatio-temporal masking is omitted, the context masking will be directly applied on the original videos at 1800×1800 resolution. We present the results for the removal of spatial masking and temporal masking, respectively.

Impacts of Context-Based Masking. Context masking is designed to modify the semantic in the video tube. By dropping out this masking, we would only train the model on the video tubes from spatio-temporal masking.

As shown in Fig. 6, the absence of each module leads to a reduction in the performance of the pipeline. This indirectly attests to the effectiveness of each of our modules. Removing the pre-processing has the greatest affection on overall performance. The absence of pre-processing results in increased false positives due to greater ambiguity in training.

5   Conclusion

In conclusion, this work presents a novel pre-masking approach to retinal video data augmentation. Our proposed context-based masking ingeniously leverages the video's intrinsic semantics as a mask, modifying the dynamic semantics of the video to create similar but differently labeled data, thereby enhancing the network's ability to discern distinct vascular scenarios. The success of our approach also demonstrates the potential of context-based masking techniques in the realm of deep learning, paving the way for innovative solutions to complex detection tasks, particularly in the challenging field of medical video analysis.

Acknowledgements. This research is funded in part by ARC-Discovery grant (DP220100800 to XY) and ARC-DECRA grant (DE230100477 to XY). We thank all anonymous reviewers and ACs for their constructive suggestions.

References

1. Alomar, K., Aysel, H.I., Cai, X.: Data augmentation in classification and segmentation: a survey and new strategies. J. Imaging **9**, 46 (2023)
2. Beede, E., et al.: A human-centered evaluation of a deep learning system deployed in clinics for the detection of diabetic retinopathy. In: Proceedings of the 2020 CHI Conference on Human Factors in Computing Systems, pp. 1–12 (2020)
3. Carreira, J., Zisserman, A.: Quo vadis, action recognition? A new model and the kinetics dataset. In: Proceedings of the IEEE Conference on Computer Vision and Pattern Recognition, pp. 6299–6308 (2017)
4. Chen, C., Hammernik, K., Ouyang, C., Qin, C., Bai, W., Rueckert, D.: Cooperative training and latent space data augmentation for robust medical image segmentation. In: de Bruijne, M., et al. (eds.) MICCAI 2021, Part III. LNCS, vol. 12903, pp. 149–159. Springer, Cham (2021). https://doi.org/10.1007/978-3-030-87199-4_14
5. DeVries, T., Taylor, G.W.: Improved regularization of convolutional neural networks with cutout. arXiv preprint arXiv:1708.04552 (2017)
6. D'Antona, L., et al.: Association of intracranial pressure and spontaneous retinal venous pulsation. JAMA Neurol. **76**(12), 1502–1505 (2019)
7. Feichtenhofer, C.: X3D: expanding architectures for efficient video recognition. In: Proceedings of the IEEE/CVF Conference on Computer Vision and Pattern Recognition, pp. 203–213 (2020)
8. Feichtenhofer, C., Fan, H., Li, Y., He, K.: Masked autoencoders as spatiotemporal learners. In: NeurIPS (2022)
9. Guan, H., Liu, M.: Domain adaptation for medical image analysis: a survey. IEEE Trans. Biomed. Eng. **69**(3), 1173–1185 (2022)
10. Hamann, T., Wiest, M., Mislevics, A., Bondarenko, A., Zweifel, S.: At the pulse of time: machine vision in retinal videos. In: Staartjes, V.E., Regli, L., Serra, C. (eds.) Machine Learning in Clinical Neuroscience. ANS, vol. 134, pp. 303–311. Springer, Cham (2022). https://doi.org/10.1007/978-3-030-85292-4_34
11. Hedges Jr., T.R., Baron, E.M., Hedges III, T.R., Sinclair, S.H.: The retinal venous pulse: its relation to optic disc characteristics and choroidal pulse. Ophthalmology **101**(3), 542–547 (1994)
12. Hogarty, D.T., Hogarty, J.P., Hewitt, A.W.: Smartphone use in ophthalmology: what is their place in clinical practice? Surv. Ophthalmol. **65**(2), 250–262 (2020)
13. Iqbal, U.: Smartphone fundus photography: a narrative review. Int. J. Retina Vitreous **7**(1), 44 (2021)
14. Khan, M., et al.: RVD: a handheld device-based fundus video dataset for retinal vessel segmentation. arXiv preprint arXiv:2307.06577 (2023)
15. Kingma, D.P., Ba, J.: Adam: a method for stochastic optimization. arXiv preprint arXiv:1412.6980 (2014)
16. Kong, Y., Fu, Y.: Human action recognition and prediction: a survey. Int. J. Comput. Vision **130**(5), 1366–1401 (2022)
17. Kumar Singh, K., Jae Lee, Y.: Hide-and-seek: forcing a network to be meticulous for weakly-supervised object and action localization. In: Proceedings of the IEEE International Conference on Computer Vision, pp. 3524–3533 (2017)
18. Laurent, C., Hong, S.C., Cheyne, K.R., Ogbuehi, K.C.: The detection of spontaneous venous pulsation with smartphone video ophthalmoscopy. Clin. Ophthalmol. (Auckland NZ) **14**, 331 (2020)
19. Lin, J., Gan, C., Han, S.: TSM: temporal shift module for efficient video understanding. In: Proceedings of the IEEE/CVF International Conference on Computer Vision, pp. 7083–7093 (2019)

20. Liu, J., Yu, X.: Few-shot weighted style matching for glaucoma detection. In: Fang, L., Chen, Y., Zhai, G., Wang, J., Wang, R., Dong, W. (eds.) CICAI 2021. LNCS, vol. 13069, pp. 289–300. Springer, Cham (2021). https://doi.org/10.1007/978-3-030-93046-2_25

21. McHugh, J.A., D'Antona, L., Toma, A.K., Bremner, F.D.: Spontaneous venous pulsations detected with infrared videography. J. Neuroophthalmol. 40(2), 174–177 (2020)

22. Monjur, M., Hoque, I.T., Hashem, T., Rakib, M.A., Kim, J.E., Ahamed, S.I.: Smartphone based fundus camera for the diagnosis of retinal diseases. Smart Health 19, 100177 (2021)

23. Mueller, S., Karpova, S., Wintergerst, M.W.M., Murali, K., Shanmugam, M.P., Finger, R.P., Schultz, T.: Automated detection of diabetic retinopathy from smartphone fundus videos. In: Fu, H., Garvin, M.K., MacGillivray, T., Xu, Y., Zheng, Y. (eds.) OMIA 2020. LNCS, vol. 12069, pp. 83–92. Springer, Cham (2020). https://doi.org/10.1007/978-3-030-63419-3_9

24. Neimark, D., Bar, O., Zohar, M., Asselmann, D.: Video transformer network. arXiv preprint arXiv:2102.00719 (2021)

25. Pujari, A., et al.: Clinical role of smartphone fundus imaging in diabetic retinopathy and other neuro-retinal diseases. Curr. Eye Res. 46(11), 1605–1613 (2021)

26. Schneider, C.A., Rasband, W.S., Eliceiri, K.W.: NIH image to ImageJ: 25 years of image analysis. Nat. Methods 9(7), 671–675 (2012)

27. Seo, J.H., Kim, T.W., Weinreb, R.N., Kim, Y.A., Kim, M.: Relationship of intraocular pressure and frequency of spontaneous retinal venous pulsation in primary open-angle glaucoma. Ophthalmology 119(11), 2254–2260 (2012)

28. Sheng, H., et al.: Autonomous stabilization of retinal videos for streamlining assessment of spontaneous venous pulsations. arXiv preprint arXiv:2305.06043 (2023)

29. Simonyan, K., Zisserman, A.: Very deep convolutional networks for large-scale image recognition. arXiv preprint arXiv:1409.1556 (2014)

30. Sun, Z., Ke, Q., Rahmani, H., Bennamoun, M., Wang, G., Liu, J.: Human action recognition from various data modalities: a review. IEEE Trans. Pattern Anal. Mach. Intell. 45(3), 3200–3225 (2023)

31. Vaswani, A., et al.: Attention is all you need. In: NeurIPS, vol. 30 (2017)

32. Wang, L., et al.: VideoMAE V2: scaling video masked autoencoders with dual masking. In: Proceedings of the IEEE/CVF Conference on Computer Vision and Pattern Recognition, pp. 14549–14560 (2023)

33. Wang, L., et al.: Temporal segment networks: towards good practices for deep action recognition. In: Leibe, B., Matas, J., Sebe, N., Welling, M. (eds.) ECCV 2016. LNCS, vol. 9912, pp. 20–36. Springer, Cham (2016). https://doi.org/10.1007/978-3-319-46484-8_2

34. Wang, R., et al.: BEVT: BERT pretraining of video transformers. In: IEEE/CVF Conference on Computer Vision and Pattern Recognition, CVPR 2022, pp. 14713–14723 (2022)

35. Wei, Y., et al.: MPP-net: multi-perspective perception network for dense video captioning. Neurocomputing 552, 126523 (2023)

36. Wintergerst, M.W., Jansen, L.G., Holz, F.G., Finger, R.P.: Smartphone-based fundus imaging-where are we now? Asia-Pac. J. Ophthalmol. 9(4), 308–314 (2020)

37. Wintergerst, M.W., et al.: Diabetic retinopathy screening using smartphone-based fundus imaging in India. Ophthalmology 127(11), 1529–1538 (2020)

38. Yang, S., Xiao, W., Zhang, M., Guo, S., Zhao, J., Shen, F.: Image data augmentation for deep learning: a survey. arXiv preprint arXiv:2204.08610 (2022)

39. Yao, Y., Wang, T., Du, H., Zheng, L., Gedeon, T.: Spotting visual keywords from temporal sliding windows. In: 2019 International Conference on Multimodal Interaction, pp. 536–539 (2019)
40. Zhang, H.B., et al.: A comprehensive survey of vision-based human action recognition methods. Sensors **19**(5), 1005 (2019)
41. Zhang, H., Zhu, L., Wang, X., Yang, Y.: Divide and retain: a dual-phase modeling for long-tailed visual recognition. IEEE Trans. Neural Netw. Learn. Syst. (2023)
42. Zhong, Z., Zheng, L., Kang, G., Li, S., Yang, Y.: Random erasing data augmentation. In: Proceedings of the AAAI Conference on Artificial Intelligence, pp. 13001–13008 (2020)

Beyond Model Accuracy: Identifying Hidden Underlying Issues in Chest X-ray Classification

Richard Wainwright[✉], Danny Wang, Harrison Layton, and Alina Bialkowski

School of Electrical Engineering and Computer Science, The University
of Queensland, St Lucia, QLD 4072, Australia
{r.wainwright,danny.wang,harrison.layton}@uq.net.au,
alina.bialkowski@uq.edu.au

Abstract. As deep learning model performance continues to advance in detecting and classifying disease, it is important to show that these models are trustworthy and reliable for use by medical professionals. However, reporting of model accuracy alone can conceal underlying issues with the utilised data and model training, which could lead to serious consequences in practical applications. In this paper, we investigate machine learning models for chest X-ray disease classification using the COVID-19 Radiography Database (CRD), COVIDx and ChestX-ray14 (CXR-14) datasets. Existing literature has identified issues with these datasets, including spurious correlations and incorrect ground truth labels in the data. Through the utilisation of model attention visualisation, uncertainty measures, and low-dimensionality data representations, we underscore a suite of techniques capable of detecting such issues. Our procedure offers a means to visualise data quality issues and the uncertainty of model predictions that extend beyond numerical values, helping to improve understanding of true performance and trust (or distrust) in machine learning models.

Keywords: Interpretability · Chest X-rays · Deep learning · Data quality

1 Introduction

In recent years, the performance of deep learning models has shown continuous improvement in disease detection and classification. Some models even report accuracy comparable to human experts [16]. However, accuracy alone doesn't tell the whole story and might obscure underlying issues present in the data or modelling. These issues can significantly impact model performance in practical scenarios. In addition, trained models might appear accurate but could be relying on spurious features present in the data, rather than learning how the disease manifests in medical images. In such cases, it becomes challenging for medical professionals to trust the model's predictions, even if the model is highly

R. Wainwright and D. Wang—Equal contribution.

accurate. Moreover, this can lead to highly accurate models failing to generalise, resulting in misdiagnoses when applied to new data. Such issues could remain completely hidden without employing model inspection and interpretability techniques.

In this paper, we present a novel procedure comprising a suite of tools designed to detect and visually address underlying issues within datasets and trained models, which can remain hidden when relying solely on reported model accuracy. This procedure enriches the comprehension of model performance and accuracy, fostering increased confidence in machine learning model output and knowing when they can be reliably applied to assist medical diagnosis. Furthermore, this approach enhances the interpretability of machine learning models by providing insights into their performance across datasets, transcending beyond accuracy. It provides a human-understandable tool which supports the decisions made by the model, and offers explanations for where it may go wrong. We demonstrate the utility of the procedure in Chest X-Ray classification. We also believe its utility extends to a diverse field of medical and other imaging contexts.

2   Related Work

Convolutional Neural Networks (CNNs) have found wide applications in the domain of medical imaging including organ segmentation and diagnosis of disease [24]. Rajpurkar et al. [16] developed the CheXNet model, a CNN trained on the ChestX-ray 14 dataset [23] that diagnosed a range of diseases from chest X-rays (CXRs). Accuracy metrics measured on an unseen test set showed performance exceeding that of working radiologists. However, based on interpretability analysis [25] it was shown that CheXNet relied on spurious information, such as the presence of radiologist's tags in an X-ray, to make its classification.

Interpretability techniques have been developed to help understand what neural networks are using for their classification. Saliency or attention mapping methods are a family of interpretability techniques which aim to highlight regions of an input that are important in determining a model's output. These include Grad-CAM [18], Integrated Gradients [20] and SmoothGrad [19]. The SmoothGrad technique [19] has been shown to reduce results in Integrated Gradient's saliency or attention maps associated with irrelevant noisy regions while being more defined than Grad-CAM. Other methods include Layer-Wise Relevance Propagation (LRP) [2] which uses back propagation to produce pixel-wise decomposition to illustrate the classification output of a machine learning model, and SHapley Additive exPlanations (SHAP) [10] which uses Shapley equations from game theory to produce a map labelling each pixel on a scale of how much it contributed to the output classification, either positively or negatively.

While saliency or attention maps are useful for quick visual inspection, it has been shown that some saliency mapping methods are invariant to the learned weights of the network and are biased towards highlighting object edges within an image [1]. This can give the false impression that a model is using the correct visual features to make its classification. Alone, these methods are insufficient, and additional techniques are needed to verify the generalization performance of a model.

Calculating the numerical uncertainty associated with model predictions increases the confidence users have in a model's outputs. Gal et al. [7] showed that a deep neural network trained with dropout layers before every weighted layer is equivalent to a Bayesian Neural Network (BNN) and provides uncertainty quantification through a technique called MC-Dropout. Ghoshal et al. [8] developed a framework MC-DropWeights to approximate Bayesian inference and quantify uncertainty by applying DropWeights to the fully connected layers of a network. This framework was applied to a model that diagnosed the presence of COVID-19 in chest X-rays, demonstrating that the calculated uncertainty was strongly correlated with the accuracy of a prediction.

Dimensionality reduction techniques allow humans to gain some understanding of the structure and distribution of high-dimensional datasets. T-distributed Stochastic Neighbor Embedding (t-SNE) [11] and Uniform Manifold Approximation and Projection (UMAP) [12] both map higher dimensional data into two or three dimensions in such a way that similar datapoints are embedded near each other in the low-dimensional map, while UMAP provides better preservation of global structure [12]. Colouring the data points according to class labels or other metadata allows for quick visual inspection of a dataset.

3   Method

In this section, we detail the three components and our proposed procedure for identifying underlying issues in machine learning models for medical imaging. In brief, the procedure for analysing model performance beyond traditional accuracy measurements is as follows:

1. Perform dimensionality reduction on the raw pixel values of the dataset
2. Train the desired model on the data
3. Adapt the trained model into a Bayesian Neural Network and use this model to make predictions, applying MC-Dropout to assess model uncertainty
4. Perform dimensionality reduction on features extracted from the trained model
5. Produce attention maps on test data.

Leveraging dimensionality reduction techniques, we are able to visually observe the data distribution, encompassing both the raw pixel data and feature representations. These representations can unveil biases and correlations within the dataset. Moreover, attention maps help to illustrate regions of the input image used for classification, while uncertainty estimation offers a quantification of the model's confidence in its predictions. Uncertainty measures act as an audit for both the datasets and the model itself. Figure 1 demonstrates our three-pronged approach.

Furthermore, as a means of assessing the effectiveness of our procedure in identifying underlying issues within medical imaging, particularly Chest X-ray

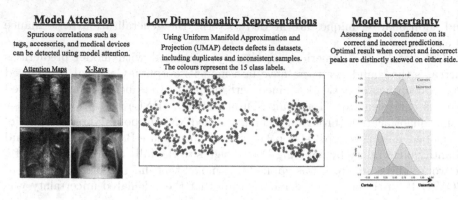

Fig. 1. Method summary

classification, we selected 3 datasets with known problems related to data collection or labeling. Oakden-Rayner [14] identified that the ChestX-ray14 (CXR-14) dataset contains a high proportion of incorrectly labelled images. Additionally, we selected two COVID-19 datasets (COVIDx and CRD) to provide deeper insights into publicly available chest X-ray datasets. COVIDx has been recognised for containing duplicated data and source-related issues stemming from its amalgamated nature [17].

3.1   Datasets

To evaluate our proposed procedure, we investigate three chest x-ray datasets as detailed in Table 1.

Table 1. Chest X-ray datasets evaluated in this work

Dataset	#Images	#Patients	Labels
Chest-Xray-14 [23]	112,120	30,805	15 classes
COVIDx [13,22]	13,975	13,870	3 classes
COVID-19 Radiography Database (CRD) [5,15]	10,507	Unknown	3 classes

3.2   InceptionV3 Training on ChestX-ray14

To provide a model that gives uncertainty estimates, and ensure full control, we trained several neural networks using TensorFlow 2.8 on an A100 GPU. Moreover, an InceptionV3 [21] model was fine-tuned to fit the ChestX-ray14 dataset. As the number of samples in the dataset is quite large, minimal data augmentation (i.e. no flips) was applied in order to keep the inputs accurate to the nature of chest X-rays. In addition, the dataset was split by patient with 70% of the set used for training, 10% for validation and 20% for testing.

The model's performance was found to be on par with InceptionV3 on the ChestX-ray14 dataset documented in previous research [9]. Furthermore, when evaluated using ROC curves, the Area Under the Curve (AUC) score for the test dataset was 0.716. The performance varied across the different classes, with AUC scores ranging from 0.56 for pneumonia to 0.84 for edema.

3.3   Transfer Learning on COVID-19 Datasets

A pre-trained DenseNet model [4] was modified to incorporate dropout and was fine-tuned using a combination of CRD and COVIDx, utilising the same parameters as the ChestX-ray14 training. Additionally, to account for the imbalanced distribution of the datasets, class weights were introduced during training, inversely proportional to their frequency. As a result of these adjustments, the final model achieved an accuracy of 86.58% on COVIDx and 87.24% on the CRD dataset.

3.4   Low-Dimensional Representations

To assist in interpreting the data and models, Uniform Manifold Approximation and Projection (UMAP) [12] was utilised to create low-dimensional representations of the pixel data and the features learnt by the model. Each datapoint is projected to a location on a two dimensional map, where similar datapoints get mapped close to each other while preserving global structure. The pixel distribution was computed from the vectorised raw pixel values of the images, while the feature distribution was produced from the feature activations of the last convolutional layer of the trained models. A user interface was created using the Bokeh 2.4.2 library [3] as in Fig. 4, providing the user with a tool to view the images corresponding to each point, as well as buttons to change the colour of the points based on metadata such as class labels and data source (available pre-training), and predicted labels and uncertainty measures (after model training), enabling fast visual inspection of the data. This method was also carried out using t-SNE [11], providing similar results on the datasets tested, without a strong case to prefer either dimensionality reduction technique.

3.5   Model Uncertainty

Model uncertainty was measured by adapting each of our trained models into a Bayesian Neural Network. This was achieved by including Monte Carlo dropout in the models' prediction process as done in [8], providing the epistemic uncertainty for each image prediction. Each prediction was carried out 50 times with dropout ratio of 0.2, giving each prediction an uncertainty score. Scores were then normalised using the mean and standard deviation. Consequently, this provided a quantitative analysis of model performance by providing a score for each prediction. Distributions of scores can then be compared between correct and incorrect model predictions.

3.6   Model Attention

To investigate how models utilised X-rays from each dataset for making predictions, we employed the SmoothGrad Integrated Gradients technique [19] on the CNN models. This approach was chosen due to its capability to generate distinct, smoothed attention maps that retain pixel-level details. The output of the technique produces a rough localisation map, which shows the regions in the image which are important to the model prediction. Attention maps were produced for each X-ray in the test datasets. These maps visually depict the impact of a slight change on model predictions for each pixel, effectively illustrating the model's learned significant features for accurate classification.

4   Results and Discussion

4.1   Low-Dimensional Representations - Pixel Distribution

The evaluation can begin prior to the training and analysis of the models, in which UMAP plots were created to investigate the distribution of image pixel values of the datasets. From this, unusual clustering and anomalies can be quickly identified. For example, several abnormal images (different image proportions and resolutions) were identified in the pixel distribution plot created with random samples of the ChestX-ray14 dataset based on the image labels as in Fig. 2.

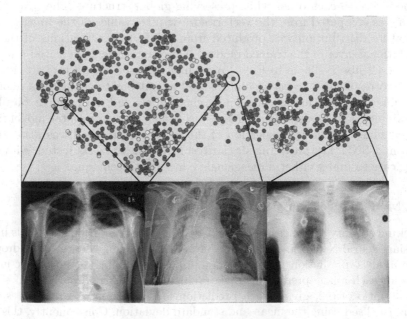

Fig. 2. UMAP Pixel distribution of ChestX-ray14 images with selected X-ray images shown below. The colours represent the 15 class labels. Abnormalities including poor image quality and spurious features (accessories and electronic devices) were identified.

With access to a distribution of the raw pixel data, image inspection becomes possible by hovering over data points. This is particularly important when the dataset size is over the limit for pure human inspection. Images situated near the edges or isolated from the rest often indicate abnormalities within the dataset. From this, flaws and spurious features within the dataset can be quickly detected prior to training and analysis. Additionally, this method can shed light on data collection flaws, such as clusters suggesting a correlation between data source and class. This process can also provide insights into potential data augmentation strategies for addressing data quality issues. For instance, Fig. 2 depicts a regular distribution for the dataset, in which there are no visible clustering of images based on pixel values and class labels (i.e. no evident biases in data collection). Nonetheless, the discernible cluster of images based on only pixel values in the figure draws attention to the diverse quality of data in this dataset.

4.2 Model Uncertainty

After acquiring a trained model, we can analyse the epistemic uncertainty of the predictions, both across the entirety of the dataset and for each individual class contained within it. This approach yields a quantifiable indicator of the model's prediction confidence. While the outcomes are expressed numerically, presenting them visually as shown in Fig. 3 enables simpler inspection and comprehension, even for those without specialised expertise.

According to work by Ghoshal et al. [8], a model's uncertainty should be higher for incorrect predictions and lower for correct samples. This pattern may not hold with insufficient training data or errors in ground truth labels as shown in [6]. On the ChestX-ray14 dataset, this pattern did not hold true. Our findings revealed that the uncertainty distribution produced by the InceptionV3 model on the test data remained similar for both accurate and incorrect predictions. Similarly, we anticipated that the model would exhibit more certainty when predicting diseases with higher accuracy (such as cardiomegaly) than for those with lower performance (like pneumonia). Nonetheless, this was also not the case, as the uncertainty distribution demonstrated little variation across all classes. Thus, the observations indicate the model's lack of robustness, casting doubt on the accuracy of even correct predictions. These observations revealed through our proposed procedure, suggest issues with the dataset ground truth labels which has been identified in literature [14].

For the COVID-19 data, we observed similar results to the ChestX-ray 14 analysis. Figure 3(a) and (b) illustrate significant overlaps between correct and incorrect predictions for all classes and Covid-19 specifically, suggesting a poor model due to equal uncertainty levels for accurate and inaccurate predictions. However, a positive pattern emerges in Fig. 3(d), where distinct peaks appear for correct and incorrect results in pneumonia classification. This skew indicates that the model is capable of predicting pneumonia, exhibiting confidence in its correct predictions while being uncertain for incorrect ones. Nevertheless, it's crucial to ensure that the correct features are being utilised. This can be achieved with the final component of our procedure, which identifies what features are utilised by the model.

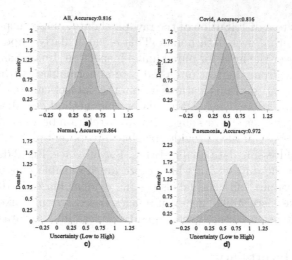

Fig. 3. COVID-19 model uncertainty distributions of correct (green) and incorrect (red) predictions. Greater distinction between distributions indicates better performance. (Color figure online)

4.3   Low-Dimensional Representations - Feature Distribution

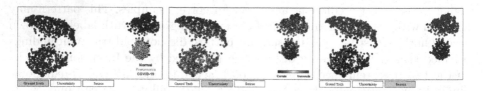

Fig. 4. Feature distribution of CRD dataset: coloured by (1) Ground truth labels, (2) Uncertainty score, (3) Data source, in the interactive Bokeh interface. Reveals two distinct distributions for the "normal" class. (Color figure online)

In conjunction with the quantified uncertainty results, we can delve into the learned feature distributions. Illustrated in Fig. 4, we present the feature distribution for the COVID model with the CRD dataset. Notably, we were able to identify four distinct clusters corresponding to the three diseases, as indicated by the colour code in the legend. The clear separation of classes into these clusters suggests accurate model performance. However, the bifurcation of the 'Normal' class into two clusters signifies a systematic difference in the data. Furthermore, the clusters formed based on data source may indicate issues in how the data was sourced [17].

The uncertainty scores for each image have been plotted in Fig. 4(b). The gradual decrease in darkness of colours corresponds the decreasing certainty in the model's predictions. Additionally, while there are two clusters of normal data, a visible pattern emerges as both exhibiting a descending colour gradient of uncertainty from right to left or top to bottom. Moreover, the plot highlights that the COVID cluster primarily consists of brighter colors (indicating higher uncertainty), while the pneumonia cluster is predominantly composed of darker colors (indicating higher certainty). This alignment reinforces the conclusions drawn from the earlier uncertainty analysis.

4.4 Model Attention

In order to gain deeper insight into the uncertainty and clustering patterns, we can visualise the model's attention to understand *what* the important features are. Specifically, we expect a model to pay attention to the chest and lungs of the images. While we did observe this behaviour in some images (see Fig. 5(a)), our more prevalent finding was the model's tendency to emphasise medical equipment (Fig. 5(b)), X-ray labels (Fig. 5(c)), image boarders, and negative space (Fig. 5(d)). This indicates that this model is not effectively learning the distinctive characteristics of various diseases as they present in X-ray images. Instead, it relies on spurious features within the X-rays to make its predictions. These findings remained consistent across all tested models and datasets that we examined. Incorporating this with the uncertainty measures and feature distributions, it may provide a visual explanation for the uncertainty in the model's predictions, facilitating an understanding of correlations with samples sharing similar features.

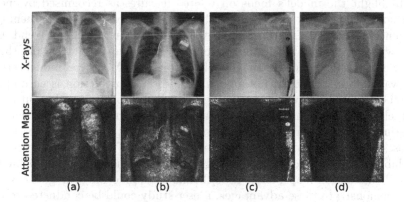

Fig. 5. X-rays and corresponding attention maps using SmoothGrad [19]

By paying attention to medical equipment and labels within the images, the model could learn to associate these with certain classes more strongly. Subsequently, the model utilises this learned information rather than evidence of the disease. Furthermore, the edges of the images offer additional information to the model. For example, X-rays from different hospitals could exhibit varying configurations, showing more or less of the overall patient anatomy. If the model can detect this pattern and correlate it with the fact that a particular disease tends to come from a specific source, it might prioritise this correlation over the actual evidence of the diseases when making predictions.

5   Conclusion

Utilising low-dimensionality representations of data, uncertainty measures and analysis of model attention, we demonstrated a procedure which is capable of detecting defects within datasets such as spurious correlations and poor quality samples in the data. Through this approach, we presented a novel way to enable visual explanation of model performance, going beyond mere accuracy reporting. We demonstrated the effectiveness of these techniques in detecting issues within chest X-ray datasets that would have eluded detection solely through accuracy measurements. This suite of techniques offers valuable insights into data quality and model performance when conducting research and model training. In addition to assessing the data and models, it also suggests where improvements can be made, including identifying areas necessitating more data or data augmentation.

Beyond its utility for the machine learning and AI community, our methodology can serve as a supportive mechanism for utilising machine learning models. The presented approach enables the visualisation of potential anomalies in data and the level of confidence in the model's predictions, making it comprehensible even to individuals without extensive knowledge in the field. Attention maps that highlight the model's focus on related features as recognised by medical professionals signal the model's desirable utilisation of data, strengthening the confidence of medical experts. Similarly, if the model is showing a trend in certainty where accurate predictions are marked by higher certainty levels, this can further enhance confidence in the model's predictions. Moreover, model certainty offers valuable information; notably, when a disease prediction is accompanied by high certainty, it serves as a warning to medical professional to scrutinise the result closely. Additionally, a low-dimensional pixel representation that lacks clustering by data source or class suggests a dataset's suitability for training purposes. Finally, a low-dimensional representation of features that clusters by class label rather than data source, together with plausible features being highlighted by attention maps, show that the model is indeed learning as desired.

To demonstrate these advantages, a user study could be conducted to quantify the benefits of the approach as future work. For instance, this may involve comparing speed of manual inspection of images to identify defects in the dataset, duplicated and spurious images, compared to employing the method presented here. In addition, exploring alternative attention mapping techniques could prove insightful. For example, incorporating SHAP values within the SmoothGrad Integrated Gradients technique may help to reveal specific attention areas that increase or decrease classification likelihood. Moreover, further advancements towards automating and adapting this procedure for diverse machine learning models and datasets would be highly valuable.

Acknowledgments. We would like to acknowledge the UQ AI Collaboratory for their support, resources and funding which made this study possible.

References

1. Adebayo, J., Gilmer, J., Muelly, M., Goodfellow, I., Hardt, M., Kim, B.: Sanity checks for saliency maps. In: Advances in Neural Information Processing Systems (NeurIPS), vol. 31 (2018). https://doi.org/10.48550/arXiv.1810.03292
2. Bach, S., Binder, A., Montavon, G., Klauschen, F., Müller, K.R., Samek, W.: On pixel-wise explanations for non-linear classifier decisions by layer-wise relevance propagation. PLoS ONE **10**(7), 1–46 (2015). https://doi.org/10.1371/journal.pone.0130140
3. Bokeh: Bokeh documentation (2022). https://docs.bokeh.org/en/2.4.2/index.html#bokeh-documentation. Accessed 16 Feb 2022
4. Chou, B.: Chexnet-keras (2020). https://github.com/brucechou1983/CheXNet-Keras. Accessed 16 Feb 2022
5. Chowdhury, M.E.H., et al.: Can AI help in screening viral and COVID-19 pneumonia? IEEE Access **8**, 132665–132676 (2020). https://doi.org/10.1109/ACCESS.2020.3010287
6. Cusack, H., Bialkowski, A.: The effect of training data quantity on Monte Carlo dropout uncertainty quantification in deep learning. In: International Joint Conference on Neural Networks (IJCNN), pp. 1–8 (2023). https://doi.org/10.1109/IJCNN54540.2023.10191327
7. Gal, Y., Ghahramani, Z.: Dropout as a Bayesian approximation: representing model uncertainty in deep learning (2016). https://doi.org/10.48550/arXiv.1506.02142
8. Ghoshal, B., Tucker, A., Sanghera, B., Lup Wong, W.: Estimating uncertainty in deep learning for reporting confidence to clinicians in medical image segmentation and diseases detection. Computat. Intell. **37**(2), 701–734 (2021). https://onlinelibrary.wiley.com/doi/abs/10.1111/coin.12411
9. Ho, T.K.K., Gwak, J.: Utilizing knowledge distillation in deep learning for classification of chest x-ray abnormalities. IEEE Access **8**, 160749–160761 (2020). https://doi.org/10.1109/ACCESS.2020.3020802
10. Lundberg, S.M., Lee, S.I.: A unified approach to interpreting model predictions. In: Guyon, I., Luxburg, U.V., Bengio, S., Wallach, H., Fergus, R., Vishwanathan, S., Garnett, R. (eds.) Advances in Neural Information Processing Systems 30, pp. 4765–4774. Curran Associates, Inc. (2017). https://papers.nips.cc/paper/7062-a-unified-approach-to-interpreting-model-predictions.pdf
11. van der Maaten, L., Hinton, G.: Visualizing data using t-SNE. J. Mach. Learn. Res. **9**(86), 2579–2605 (2008). https://jmlr.org/papers/v9/vandermaaten08a.html
12. McInnes, L., Healy, J., Melville, J.: Umap: uniform manifold approximation and projection for dimension reduction (2018). https://doi.org/10.48550/arXiv.1802.03426
13. Nisar, Z.: Detecting and visualising the infectious regions of covid-19 in x-ray images and CT scans using different pretrained-networks in tensorflow 2.x. (2020). https://github.com/zeeshannisar/COVID-19. Accessed 12 Feb 2022
14. Oakden-Rayner, L.: Exploring the chestxray14 dataset: problems (2017). https://laurenoakdenrayner.com/2017/12/18/the-chestxray14-dataset-problems/

15. Rahman, T., et al.: Exploring the effect of image enhancement techniques on covid-19 detection using chest x-ray images. Comput. Biol. Med. **132**, 104319 (2021). https://www.sciencedirect.com/science/article/pii/S001048252100113X
16. Rajpurkar, P., et al.: CheXNet: radiologist-level pneumonia detection on chest x-rays with deep learning (2017). https://doi.org/10.48550/arXiv.1711.05225
17. Roberts, M., et al.: AIX-COVNET: common pitfalls and recommendations for using machine learning to detect and prognosticate for COVID-19 using chest radiographs and CT scans. Nat. Mach. Intell. **3**(3), 199–217 (2021). https://doi.org/10.1038/s42256-021-00307-0
18. Selvaraju, R.R., Cogswell, M., Das, A., Vedantam, R., Parikh, D., Batra, D.: Grad-CAM: visual explanations from deep networks via gradient-based localization. Int. J. Comput. Vis. **128**(2), 336–359 (2019). https://doi.org/10.1007/s11263-019-01228-7
19. Smilkov, D., Thorat, N., Kim, B., Viégas, F., Wattenberg, M.: Smoothgrad: removing noise by adding noise. arXiv preprint arXiv:1706.03825 (2017). https://doi.org/10.48550/arXiv.1706.03825
20. Sundararajan, M., Taly, A., Yan, Q.: Axiomatic attribution for deep networks. In: International Conference on Machine Learning, pp. 3319–3328. PMLR (2017). https://doi.org/10.48550/arXiv.1703.01365
21. Szegedy, C., Vanhoucke, V., Ioffe, S., Shlens, J., Wojna, Z.: Rethinking the inception architecture for computer vision (2015). https://doi.org/10.48550/arXiv.1512.00567
22. Wang, L., Lin, Z.Q., Wong, A.: COVID-Net: a tailored deep convolutional neural network design for detection of COVID-19 cases from chest x-ray images. Nat. Sci. Rep. **10**(1) (2020). https://doi.org/10.1038/s41598-020-76550-z
23. Wang, X., Peng, Y., Lu, L., Lu, Z., Bagheri, M., Summers, R.M.: Chestx-ray8: hospital-scale chest x-ray database and benchmarks on weakly-supervised classification and localization of common thorax diseases. In: 2017 IEEE Conference on Computer Vision and Pattern Recognition (CVPR) (2017). https://doi.org/10.1109/CVPR.2017.369
24. Yamashita, R., Nishio, M., Do, R.K.G., Togashi, K.: Convolutional neural networks: an overview and application in radiology. Insights Imaging **9**(4), 611–629 (2018). https://insightsimaging.springeropen.com/track/pdf/10.1007/s13244-018-0639-9
25. Zech, J.R., Badgeley, M.A., Liu, M., Costa, A.B., Titano, J.J., Oermann, E.K.: Variable generalization performance of a deep learning model to detect pneumonia in chest radiographs: a cross-sectional study. PLOS Med. **15**(11), e1002683 (2018). https://www.ncbi.nlm.nih.gov/pmc/articles/PMC6219764

Enhance Reading Comprehension from EEG-Based Brain-Computer Interface

Xinping Liu[1] and Zehong Cao[2(✉)]

[1] School of ICT, University of Tasmania, Hobart, Australia
[2] STEM, University of South Australia, Adelaide, Australia
Jimmy.Cao@unisa.edu.au

Abstract. Electroencephalography (EEG)-based brain-computer interfaces (BCIs) have emerged as a valuable technology for decoding human cognitive processes, including reading attention and cognitive loads. While previous studies have explored eye fixations during word recognition, the intricacies of brain dynamics involved in sentence comprehension in the temporal or spectral domains still need to be discovered. Addressing this gap is crucial for enhancing learning processes; thus, in this study, we propose the first acquisition and recognition of event-related potentials and spectral perturbations using channel and independent component analysis, based on sentence-level simultaneous EEG and eye-tracking recorded from human natural reading tasks. Our results showed peaks of brain activation evoked at around 162 ms (approaching 200 ms) after the stimulus (starting to read each sentence) in the occipital area, indicating the onset timing of human retrieving lexical and semantic visual information processing. Approximately 200 ms occipital area presented increased alpha power and decreased beta and gamma power, relative to the baseline. Our results implied that most semantic-perception responses occurred around 200 ms in alpha, beta and gamma bands to facilitate human reading representation. The implications of our study underscore the significance of EEG-based BCI applications in reading tasks, serving as a potential catalyst for improving cognitive attention and comprehension in end-user reading and learning experiences. By retrieving the intricate cognitive mechanisms underlying sentence comprehension, we pave the way for developing brain-computer learning strategies to optimise reading instruction and support a diverse range of users.

Keywords: Brain Dynamics · EEG · BCI · Sentence Reading

1 Introduction

The deployment of Electroencephalography (EEG)-based brain-computer interfaces (BCIs) to capture human brain activities [4] plays a pivotal role in recognising and understanding human reading representation and comprehension [3,8].

This technology has demonstrated its efficacy in enhancing reading attention and managing cognitive loads across various multimedia applications [10]. Integrating EEG with eye-tracking techniques for reading tasks could open up new avenues for investigating the intricate human brain dynamics of reading comprehension.

Recognising brain responses, such as those evoked by reading stimuli, requires quantifying the timeline of event-related potentials (ERPs) and event-related spectral perturbations (ERSPs) of EEG data. They allow monitoring of brain dynamics across the time and frequency spectrum. However, the previous research primarily focused on word-level reading tasks [6] in a short period, which is significantly different from human cognitive comprehension in the reading design of sentence stimulus.

To address this gap, our work aims to provide a deeper understanding of how the brain processes and represents sentences, shedding light on the cognitive mechanisms involved in higher-level linguistic comprehension. Thus, we developed an EEG-based BCI to discover brain dynamics based on natural reading tasks with sentence stimulus segments [7]. We could retrieve both lexical and semantic information through the channel and independent component analysis of EEG and eye-tracking data from 18 participants. This research not only contributes to our comprehension of natural sentence reading processes, but also demonstrates the potential for extracting detailed lexical and semantic information from EEG signals. These insights hold promise for enhancing our understanding of human cognition and opening avenues for developing innovative EEG-based BCI with reading applications across various domains.

2   Related Work

BCIs have emerged as a transformative technology, augmenting the interaction capabilities of the human brain with various environmental stimuli, including multimedia content such as video, images, audio, and text. Extracting representative EEG patterns to discern human cognitive states has paved the way for closed-loop systems that empower individuals to engage in parallel or independent tasks concurrently, including reading tasks. The integration of EEG-based BCIs can significantly enhance operational efficiency in decision-making while processing multimedia information sources [12].

ERPs have played a pivotal role in elucidating neural responses to specific stimuli or events, such as reading tasks. Notably, the N200 component manifests as a negative peak approximately 200 ms after the onset of a stimulus, exemplified by the commencement of the first word in a sentence. An early seminal study [2] discovered neural dynamics associated with cognitive processing during short observation tasks, and recent studies have explored the applicability of fixation-related potentials in text-based emotion analysis. For instance, [11] demonstrated the potential of these potentials to extract naturalistic reading skills. However, their focus remained confined to word-based reading tasks, leaving a gap in comprehending more complex sentence-level cognitive dynamics.

ERSPs have garnered attention for their capacity to reveal spectral variations in EEG signals during cognitive tasks. Zhang et al. [13] ventured into ERSP analyses, particularly within the emotion and working memory tasks in reading contexts. This work was more specialised, targeting specific, small scope of stimulus, rather than encompassing the broader context of natural reading. An additional exploration by Sharmistha et al. [9] explored into the retrieval of lexical and semantic information within linguistic content based on word embedding. Nevertheless, this work is inadvertently to preserve essential reading patterns critical for adequate comprehension. This suggested that valuable insights into the cognitive intricacies of more complex reading tasks, such as sentence reading tasks, are insufficient, so it is necessary to be more in understanding the holistic brain dynamics when humans naturally require reading texts in sentence format.

3   Method

3.1   Experiment Paradigm

The study involved the participation of 18 healthy individuals who were native English speakers originating from countries such as Canada, USA, UK, and Australia, as detailed in [7]. The experiment asked for natural reading in sentences, and the participants were presented with sentences individually on a monitor. The text appeared consistently at the same position, characterised by a light grey background with black text using a 20-point Arial font, resulting in a letter height of 0.8 mm. Each line displayed a total of 80 letters or 13 words, and the participants had to read each sentence at their own pace, fostering a natural reading paradigm.

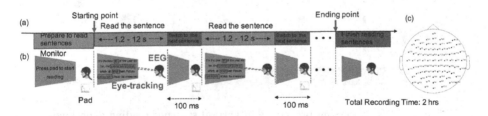

Fig. 1. Natural sentence reading experiment. (a) The overview of natural reading tasks. (b) Each sentence traces the onset, offset and switch set (switch to the following sentence) by a control pad. Of note, the switching period of each sentence lasts 100 ms, and the reading period lasts between 1.2 to 12 s. (c) Location map of multi-channel EEG.

During the task, a set of 349 sentences were sequentially displayed at the identical position on the monitor for each participant. The data collection apparatus encompassed both a 128-channel EEG system and an eye-tracking device.

This comprehensive setup enabled the simultaneous collection of brain cortex signals and eye motion data, while participants engaged in the natural sentence reading task. The synergy of EEG and eye-tracking data was pivotal in capturing the interplay between brain dynamics and eye movements during the reading process, as visually depicted in Fig. 1.

In the context of this study, each EEG trial corresponded to the act of reading an individual sentence. Consequently, the dataset comprised a total of 6,282 trials, calculated by multiplying the number of participants (18) by the number of sentences (349) within the natural reading paradigm. This extensive dataset facilitated a robust exploration of the neural and cognitive processes underpinning sentence comprehension, offering a nuanced understanding of how the brain responds during natural reading tasks.

3.2   EEG Processing and Analysis

Figure 2 presents an overview of EEG processing with channel and component analysis, and ERP/ERSP images.

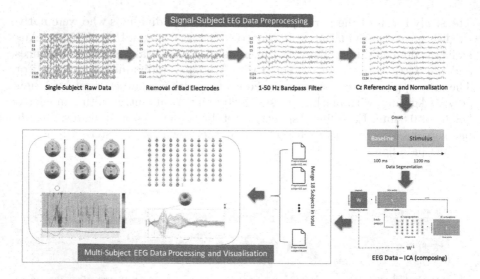

Fig. 2. EEG processing procedure in natural sentence reading experiment

EEG Processing Pipeline. The raw EEG data underwent several critical steps to enhance the quality of EEG signals and extract meaningful neural activity patterns. The processing was conducted using the EEGLAB toolbox [1], following a series of systematic procedures.

Initially, the raw EEG data were loaded into EEGLAB, where a careful selection process resulted in the retention of 105 EEG channels after eliminating poor-quality electrodes. Subsequently, potential artifacts arising from electromyogram and electrooculogram were removed, and additional artifacts were mitigated by applying a finite impulse response filter ranging from 1 Hz to 50 Hz. This filter configuration effectively removed unwanted noise while retaining the neural signals of interest.

To establish a consistent reference point for the EEG data, all electrodes were re-referenced using the vertex electrode (Cz). Z-score normalisation was then performed to ensure uniform scaling across all electrodes, facilitating reliable inter-electrode comparisons. EEG trials were segmented based on two critical time-locking events: the onset and offset of each sentence reading instance. For event extraction, a temporal window was defined from 100 ms, prior to the commencement of sentence reading (acting as the baseline) to 1200 ms after the reading onset. This time window was carefully chosen to encompass the entire span of relevant brain responses, while remaining within the confines of the minimum sentence offset time of 1223 ms.

To accurately capture the underlying brain activation patterns during sentence reading, a decomposition step was introduced to the pre-processed EEG data. Independent Channel Analysis (ICA) was employed to identify regional cerebral cortex information, focusing on individual EEG channels. Simultaneously, ICA was leveraged to delve into deeper source information represented by the components. This dual analysis approach aimed to understand the neural dynamics occurring during sentence comprehension thoroughly. To account for inter-participant variability and ensure the robustness of the findings, the pre-processed EEG data from all 18 participants were combined. A statistical group analysis was subsequently conducted to validate the presence of consistent reading patterns across all participants. This collaborative approach reinforced the reliability and generalisability of the observed brain activation patterns during sentence reading, enhancing the validity of our findings.

EEG Analysis. To provide an intuitive visualisation of the EEG dynamics, event latency scalp maps were generated alongside ERP and ERSP images. These visual representations aimed to vividly portray the temporal and spectral neural responses associated with sentence comprehension.

The quest for identifying the most impactful components drove an in-depth investigation into the EEG data. A crucial step is calculating and selecting components, which exhibited a pronounced influence on the power spectrum and spectral perturbations. This selection process was guided by the goal of isolating the components that contributed significantly to the observed neural dynamics.

To establish a clear criterion for component selection, a significance threshold was set at 0.01, corresponding to a p value of less than 0.01. Components that met this significance level were identified and retained for further analysis. As part of the pre-processing procedure, the baseline was removed to enhance the clarity and focus of the ERP and ERSP images. By adopting this rigorous selection, we distil the most pertinent components from the EEG data, ultimately enhancing the precision and interpretability of the subsequent analyses. This methodological strategy ensured that the extracted EEG dynamics were visually impactful and statistically robust, contributing to a comprehensive understanding of the cognitive processes underpinning natural sentence reading tasks.

4   Results

Our results showed the comprehensive analysis of EEG dynamics for natural sentence reading tasks across 18 participants, including event latency potentials plots and ERP/ERSP channel and component images.

4.1   Event Latency Potentials

As depicted in Fig. 3, the visual representation of our findings reveals a prominent peak occurring at approximately 162 ms following the onset of natural sentence reading. This observation holds significant implications, as it underscores substantial EEG changes across all EEG channels, which continue to evolve and stabilise around 200 ms. This temporal pattern aligns seamlessly with the recognised N200 pattern, a well-documented neural phenomenon [5,11]. Of note, we observed N200 pattern is characterised by a distinct reduction in power spectral activity within the occipital region. This intriguing discovery towards an early and coherent cortical response corresponds to initiating both lexical and semantic language processing. Remarkably, it signifies that human cognitive processes are poised to retrieve lexical and semantic information almost simultaneously within the central occipital region (Oz) merely 200 ms after the onset of sentence reading.

This finding elucidates the rapidity with which our cognitive faculties engage with and assimilate the linguistic content of sentences. Also, it supports the notion that the central occipital region plays a pivotal role in orchestrating this early-stage lexical and semantic retrieval process. Collectively, it contributes to a more profound understanding of the intricate neural mechanisms underpinning language processing and comprehension, enriching the discourse within cognitive neuroscience.

Furthermore, a meticulous selection process was employed to identify critical components within the event latency potentials. As illustrated in Fig. 4, a total of seven components emerged as particularly influential. Among these, the component located in the occipital region garnered the highest rank (rank 1). This specific component exhibited the most pronounced contribution to the cognitive dynamics associated with natural sentence reading tasks.

The prominence of the occipital component in this selection underscores the key role in the cognitive processes underpinning semantic information retrieval. Notably, this component's robust power spectral response within the occipital region suggests a concentrated neural activity pattern during the sentiment information retrieval process. This heightened activity is juxtaposed against the neural responses observed across other brain regions.

By singling out the occipital component as the most impactful, our findings accentuate the significance of the occipital region in processing sentiment information during natural sentence reading. This discovery offers valuable insights

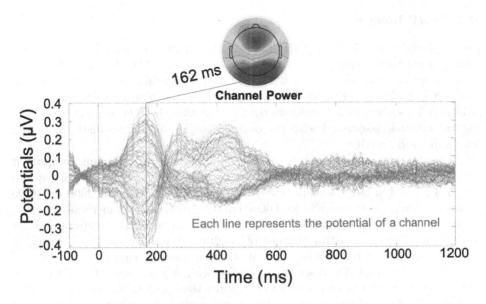

Fig. 3. Event latency potentials with the scalp map in all EEG channels across 18 participants for natural sentence reading task.

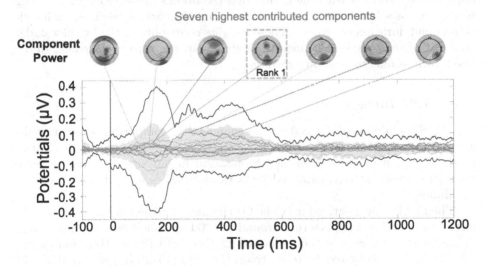

Fig. 4. Significant components from event latency potentials with the scalp map across 18 participants for natural sentence reading task.

into the specialised neural pathways involved in the interpretation of emotional and semantic content within sentences. As a result, our study could contribute to the ongoing exploration of neural networks engaged in higher-order cognitive functions, further enriching our understanding of the intricate processes that underlie language comprehension.

4.2   ERP Images

As displayed in Fig. 5-A, the ERP image in the occipital region offers a visualisation of the neural response patterns captured by channel Oz. During a relatively rapid reaction time, specifically between 1 and 2000 trials, the ERP image indicates a notable reduction in power, depicted by the blue colouration, around 200 ms. This observation suggests that within this short timeframe, the brain neural network associated with the occipital region exhibits a distinct pattern of diminished activity.

An alternate picture emerges when examining the corresponding component within the occipital region, portrayed in Fig. 5-B. The ERP image portrays an opposing trend in reaction times categorised as medium (between 2000 and 4000 trials) and long (between 4000 and 6000 trials). A conspicuous increase in power, indicated by the red colouration, is evident around the same temporal marker of 200 ms. This contrasting pattern of heightened activity implies a divergent neural response, in relation to different stimulus reaction times.

Our discovery of this discrepancy in the brain's processing of reading stimuli across varying reaction times underscores the complex nature of cognitive dynamics. The findings indicate that the occipital region's responsiveness is modulated by the temporal context within which stimuli are presented. This observation resonates with the notion that the brain's intricate interplay of cognitive processes is influenced by factors beyond the content itself, including the timing and duration of stimuli. These insights contribute to the broader understanding of the temporal nuances that shape our cognitive engagement during tasks such as natural sentence reading.

4.3   ERSP Images

The ERSP analysis, performed at the channel and component levels, revealed a striking congruence in the observed trends, particularly around the 200 ms mark, as shown in Fig. 6. This synchronisation underscores the consistency of neural response patterns captured by both approaches and adds credibility to the findings.

The ERSP image derived from the Oz-channel (depicted in Fig. 6-A) unveiled distinct spectral power patterns. Around the 200 ms timeframe, the image highlighted elevated power in the alpha (10–12 Hz), beta (25–30 Hz), and gamma (40–50 Hz) frequency bands. In contrast, the beta (13–25 Hz) and gamma (30–40 Hz) frequency bands exhibited decreased power relative to the baseline (the interval between sentence reading onset and offset). This spectral distribution suggests a neural state optimised for cognitive engagement and processing during the interval around 200 ms.

Similarly, the ERSP image based on the occipital component (shown in Fig. 6-B) mirrored the observed spectral patterns. Around the 200 ms period, the image showcased heightened power in the alpha (10–12 Hz) and gamma (40–50 Hz) frequency bands. The beta (13–25 Hz) and gamma (30–40 Hz) frequency bands

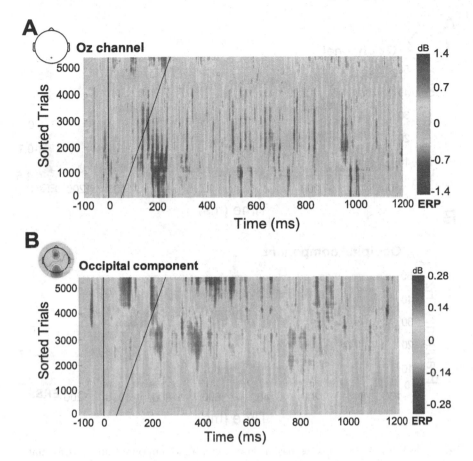

Fig. 5. Occipital ERP images based on channel and component analysis for natural reading tasks.

displayed reduced power relative to the baseline, reinforcing the consistent neural response pattern during the interval around 200 ms.

Furthermore, the ERSP images at 200 ms within the occipital region extended beyond the immediate latency, encompassing longer durations ranging from 300 ms to 1200 ms (the offset point of sentence reading). Particularly notable within this timeframe were spectral patterns dominated by the high gamma (40–50 Hz) frequency band, as well as the low beta (13–25 Hz) and gamma (30–40 Hz) frequency bands. These spectral brain signatures, relative to the baseline, align with the neural processes involved in retrieving lexical and semantic understanding during sentence reading.

The parallelism observed between the channel and component ERSP analyses underscores the consistency of the findings and bolsters the interpretation of the cognitive mechanisms underpinning natural sentence comprehension. This align-

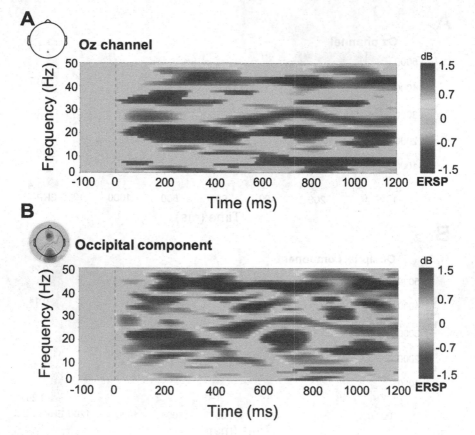

Fig. 6. Occipital ERSP images based on channel and component analysis for natural reading tasks.

ment underscores the multifaceted nature of cognitive processing, as evidenced by the convergence of distinct analytical perspectives.

5   Conclusion

Our study introduces an EEG-based BCI application that integrates high-density EEG and eye-tracking data within the context of natural sentence reading tasks. The discovery of channel and component analysis allowed us to unravel the intricate spatio-temporal dynamics of the human brain's processing of sentence reading content. Our investigation unveiled a pivotal temporal milestone in the cognitive process of lexical and semantic information retrieval. Notably, this critical phase occurs around 200 ms after the onset of sentence reading. This temporal marker is characterised by distinct neural signatures, marked by elevated high gamma and decreased low beta and low gamma power. Tapping into brain signatures of human language understanding lays the foundation for enhancing semantic representations of sentence reading and developing sophisticated cognitive models. In essence, our work could contribute to the broader landscape of combining artificial intelligence and neuroscience to detect or predict complex information processing in linguistic content. The implications of our findings can extend to various multimedia applications, particularly in enhancing users' reading attention and managing cognitive loads.

References

1. Delorme, A., Makeig, S.: EEGLAB: an open source toolbox for analysis of single-trial EEG dynamics including independent component analysis. J. Neurosci. Methods **134**(1), 9–21 (2004). https://doi.org/10.1016/j.jneumeth.2003.10.009
2. Frank, S.L., Otten, L.J., Galli, G., Vigliocco, G.: The ERP response to the amount of information conveyed by words in sentences. Brain Lang. **140**, 1–11 (2015)
3. Giurgiu, I., Schumann, A.: Additive explanations for anomalies detected from multivariate temporal data. In: Proceedings of the 28th ACM International Conference on Information and Knowledge Management, pp. 2245–2248 (2019)
4. Gu, X., et al.: EEG-based brain-computer interfaces (BCIS): a survey of recent studies on signal sensing technologies and computational intelligence approaches and their applications. IEEE/ACM Trans. Comput. Biol. Bioinf. **18**(5), 1645–1666 (2021)
5. Hauk, O., Coutout, C., Holden, A., Chen, Y.: The time-course of single-word reading: evidence from fast behavioral and brain responses. Neuroimage **60**(2), 1462–1477 (2012). https://doi.org/10.1016/j.neuroimage.2012.01.061
6. Hollenstein, N., Rotsztejn, J., Troendle, M., Pedroni, A., Zhang, C., Langer, N.: ZuCo, a simultaneous EEG and eye-tracking resource for natural sentence reading. Sci. Data **5**, 1–13 (2018)
7. Hollenstein, N., Troendle, M., Zhang, C., Langer, N.: ZuCo 2.0: a dataset of physiological recordings during natural reading and annotation. In: Proceedings of the 12th Language Resources and Evaluation Conference, pp. 138–146. European Language Resources Association, Marseille (2020). https://www.aclweb.org/anthology/2020.lrec-1.18
8. Hollenstein, N., Zhang, C.: Entity recognition at first sight improving NER with eye movement information. CoRR abs/1902.10068 (2019). arxiv.org/abs/1902.10068
9. Jat, S., Tang, H., Talukdar, P.P., Mitchell, T.M.: Relating simple sentence representations in deep neural networks and the brain. CoRR abs/1906.11861 (2019). arxiv.org/abs/1906.11861
10. Li, C., et al.: Effective emotion recognition by learning discriminative graph topologies in EEG brain networks. IEEE Trans. Neural Netw. Learn. Syst. (2023)
11. Pfeiffer, C., Hollenstein, N., Zhang, C., Langer, N.: Neural dynamics of sentiment processing during naturalistic sentence reading. Neuroimage **218**, 116934 (2020). https://doi.org/10.1016/j.neuroimage.2020.116934, https://www.sciencedirect.com/science/article/pii/S1053811920304201
12. Si, Y., et al.: Predicting individual decision-making responses based on single-trial EEG. Neuroimage **206**, 116333 (2020)
13. Zhang, Y., Zhang, G., Liu, B.: Investigation of the influence of emotions on working memory capacity using ERP and ERSP. Neuroscience **357**, 338–348 (2017)

Author Index

T. Liu et al. (Eds.): AI 2023, LNAI 14471, pp. 557–560, 2024.
https://doi.org/10.1007/978-981-99-8388-9